Mechanics and Physics of Structured Media

Mechanics and Physics of Structured Media

Asymptotic and Integral Equations Methods of Leonid Filshtinsky

Edited by

Igor Andrianov
RWTH Aachen University
Aachen, Germany

Simon Gluzman
Toronto, ON, Canada

Vladimir Mityushev
Research Group Materialica+
Faculty of Computer Science and Telecommunications
Cracow University of Technology
Kraków, Poland

Academic Press is an imprint of Elsevier
125 London Wall, London EC2Y 5AS, United Kingdom
525 B Street, Suite 1650, San Diego, CA 92101, United States
50 Hampshire Street, 5th Floor, Cambridge, MA 02139, United States
The Boulevard, Langford Lane, Kidlington, Oxford OX5 1GB, United Kingdom

Copyright © 2022 Elsevier Inc. All rights reserved.

No part of this publication may be reproduced or transmitted in any form or by any means, electronic or mechanical, including photocopying, recording, or any information storage and retrieval system, without permission in writing from the publisher. Details on how to seek permission, further information about the Publisher's permissions policies and our arrangements with organizations such as the Copyright Clearance Center and the Copyright Licensing Agency, can be found at our website: www.elsevier.com/permissions.

This book and the individual contributions contained in it are protected under copyright by the Publisher (other than as may be noted herein).

Notices

Knowledge and best practice in this field are constantly changing. As new research and experience broaden our understanding, changes in research methods, professional practices, or medical treatment may become necessary.

Practitioners and researchers must always rely on their own experience and knowledge in evaluating and using any information, methods, compounds, or experiments described herein. In using such information or methods they should be mindful of their own safety and the safety of others, including parties for whom they have a professional responsibility.

To the fullest extent of the law, neither the Publisher nor the authors, contributors, or editors, assume any liability for any injury and/or damage to persons or property as a matter of products liability, negligence or otherwise, or from any use or operation of any methods, products, instructions, or ideas contained in the material herein.

Library of Congress Cataloging-in-Publication Data
A catalog record for this book is available from the Library of Congress

British Library Cataloguing-in-Publication Data
A catalogue record for this book is available from the British Library

ISBN: 978-0-323-90543-5

For information on all Academic Press publications
visit our website at https://www.elsevier.com/books-and-journals

Publisher: Matthew Deans
Acquisitions Editor: Dennis McGonagle
Editorial Project Manager: Mariana L. Kuhl
Production Project Manager: Prasanna Kalyanaraman
Designer: Mark Rogers

Typeset by VTeX

Leonid Anshelovich Filshtinsky

Contents

List of contributors	xvii
Acknowledgements	xxi

1. L.A. Filshtinsky's contribution to Applied Mathematics and Mechanics of Solids
Vladimir Mityushev, Igor Andrianov, and Simon Gluzman

1.1 Introduction	1
1.1.1 Personality and career	1
1.1.2 Lessons of collaboration (V. Mityushev)	3
1.1.3 Filshtinsky's contribution to the theory of integral equations	5
1.2 Double periodic array of circular inclusions. Founders	7
1.2.1 Preliminaries	7
1.2.2 Contribution by Eisenstein	10
1.2.3 Contribution by Rayleigh	12
1.2.4 Contribution by Natanzon	14
1.2.5 Contribution by Filshtinsky	18
1.3 Synthesis. Retrospective view from the year 2021	22
1.4 Filshtinsky's contribution to the theory of magneto-electro-elasticity	27
1.5 Filshtinsky's contribution to the homogenization theory	30
1.6 Filshtinsky's contribution to the theory of shells	32
1.7 Decent and creative endeavor	33
Acknowledgment	34
References	34

2. Cracks in two-dimensional magneto-electro-elastic medium
Dmytro Nosov, Leonid Filshtinsky, and Vladimir Mityushev

2.1 Introduction	41
2.2 Boundary-value problems for an unbounded domain	42
2.3 Integral equations for an unbounded domain	45
2.4 Asymptotic solution at the ends of cracks	50

viii Contents

2.5 Stress intensity factors	52
A crack in MME plane	55
2.6 Numerical example	56
2.7 Conclusion	60
References	60

3. Two-dimensional equations of magneto-electro-elasticity

Vladimir Mityushev, Dmytro Nosov, and Ryszard Wojnar

3.1 Introduction	63
3.2 2D equations of magneto-electro-elasticity	65
3.2.1 Linear equations of magneto-electro-elasticity and potentials	66
3.2.2 Complex representation of field values	67
3.3 Boundary value problem	70
3.4 Dielectrics	72
3.5 Circular hole	77
Numerical example	80
3.6 MEE equations and homogenization	82
3.7 Homogenization of 2D composites by decomposition of coupled fields	89
3.7.1 Straley-Milgrom decomposition	90
3.7.2 Rylko decomposition	91
3.7.3 Example	94
3.8 Conclusion	95
References	96

4. Hashin-Shtrikman assemblage of inhomogeneous spheres

Andrej Cherkaev and Vladimir Mityushev

4.1 Introduction	99
4.2 The classic Hashin-Shtrikman assemblage	100
4.3 HSA-type structure	102
4.4 Conclusion	107
Acknowledgments	107
References	107

5. Inverse conductivity problem for spherical particles

Vladimir Mityushev, Natalia Rylko, Zhanat Zhunussova, and Yeskendyr Ashimov

5.1 Introduction	109
5.2 Modified Dirichlet problem	110
5.2.1 Reduction to functional equations	110

Contents **ix**

5.2.2 Explicit asymptotic formulas	114
5.3 Inverse boundary value problem	116
5.4 Discussion and conclusion	119
Acknowledgments	120
References	120

6. Compatibility conditions: number of independent equations and boundary conditions

Igor Andrianov and Heiko Topol

6.1 Introduction	123
6.2 Governing relations and Southwell's paradox	126
6.3 System of ninth order	128
6.4 Counterexamples proposed by Pobedrya and Georgievskii	129
6.5 Various formulations of the linear theory of elasticity problems in stresses	130
6.6 Other approximations	131
6.7 Generalization	132
6.8 Concluding remarks	133
Conflict of interest	134
Acknowledgments	134
References	134

7. Critical index for conductivity, elasticity, superconductivity. Results and methods

Simon Gluzman

7.1 Introduction	141
7.2 Critical index in 2D percolation. Root approximants	144
7.2.1 Minimal difference condition according to original	145
7.2.2 Iterated roots. Conditions imposed on thresholds	145
7.2.3 Conditions imposed on the critical index	146
7.2.4 Conditions imposed on amplitudes	147
7.2.5 Minimal derivative (sensitivity) condition	148
7.3 3D Conductivity and elasticity	149
7.3.1 3D elasticity, or high-frequency viscosity	150
7.4 Compressibility factor of hard-disks fluids	151
7.5 Sedimentation coefficient of rigid spheres	152
7.6 Susceptibility of 2D Ising model	153
7.7 Susceptibility of three-dimensional Ising model. Root approximants of higher orders	155
7.7.1 Comment on unbiased estimates. Iterated roots	155
7.8 3D Superconductivity critical index of random composite	157
7.9 Effective conductivity of graphene-type composites	159
7.10 Expansion factor of three-dimensional polymer chain	165
7.11 Concluding remarks	167

x Contents

Appendix 7.A Failure of the $DLog$ Padé method   168
Appendix 7.B Polynomials for the effective conductivity of
      graphene-type composites with vacancies   170
References   172

8. Double periodic bianalytic functions
Piotr Drygaś

8.1 Introduction   177
8.2 Weierstrass and Natanzon-Filshtinsky functions   178
8.3 Properties of the generalized Natanzon-Filshtinsky functions   180
8.4 The function $\wp_{1,2}$ 182
8.5 Relation between the generalized Natanzon-Filshtinsky and
    Eisenstein functions   185
8.6 Double periodic bianalytic functions via the Eisenstein series   186
8.7 Conclusion   188
References   189

9. The slowdown of group velocity in periodic waveguides
Yuri A. Godin and Boris Vainberg

9.1 Introduction   191
9.2 Acoustic waves   192
    9.2.1 Equal impedances   194
    9.2.2 Small scatterers   194
    9.2.3 Highly mismatched impedances   196
9.3 Electromagnetic waves   197
9.4 Elastic waves   198
9.5 Discussion   199
Acknowledgments   199
References   199

10. Some aspects of wave propagation in a fluid-loaded membrane
Julius Kaplunov, Ludmila Prikazchikova, and Sheeru Shamsi

10.1 Introduction   201
10.2 Statement of the problem   202
10.3 Dispersion relation   203
10.4 Moving load problem   205
10.5 Subsonic regime   207
10.6 Supersonic regime   209
10.7 Concluding remarks   210
Acknowledgment   211
References   211

11. Parametric vibrations of axially compressed functionally graded sandwich plates with a complex plan form

Lidiya Kurpa and Tetyana Shmatko

11.1 Introduction	213
11.2 Mathematical problem	215
11.3 Method of solution	219
11.4 Numerical results	221
11.5 Conclusions	229
Conflict of interest	230
References	231

12. Application of volume integral equations for numerical calculation of local fields and effective properties of elastic composites

Sergei Kanaun and Anatoly Markov

12.1 Introduction	233
12.2 Integral equations for elastic fields in heterogeneous media	235
12.2.1 Heterogeneous inclusions in a homogeneous host medium	235
12.2.2 Cracks in homogeneous elastic media	236
12.2.3 Medium with cracks and inclusions	237
12.3 The effective field method	238
12.3.1 The effective external field acting on a representative volume element	238
12.3.2 The effective compliance tensor of heterogeneous media	241
12.4 Numerical solution of the integral equations for the RVE	241
12.5 Numerical examples and optimal choice of the RVE	246
12.5.1 Periodic system of penny-shaped cracks of the same orientation	246
12.5.2 Periodic system of rigid spherical inclusions	248
12.6 Conclusions	250
References	251

13. A slipping zone model for a conducting interface crack in a piezoelectric bimaterial

Volodymyr Loboda, Alla Sheveleva, and Oleksandr Mykhail

13.1 Introduction	253
13.2 Formulation of the problem	254
13.3 An interface crack with slipping zones at the crack tips	258
13.4 Slipping zone length	262

xii Contents

13.5 The crack faces free from electrodes 263
13.6 Numerical results and discussion 265
13.7 Conclusion 267
References 268

14. Dependence of effective properties upon regular perturbations

Matteo Dalla Riva, Paolo Luzzini, Paolo Musolino, and Roman Pukhtaievych

14.1 Introduction 271
14.2 The geometric setting 272
14.3 The average longitudinal flow along a periodic array of cylinders 275
14.4 The effective conductivity of a two-phase periodic composite with ideal contact condition 278
14.5 The effective conductivity of a two-phase periodic composite with nonideal contact condition 280
14.6 Proof of Theorem 14.5.2 284
    14.6.1 Preliminaries 284
    14.6.2 An integral equation formulation of problem (14.7) 286
    14.6.3 Analyticity of the solution of the integral equation 291
    14.6.4 Analyticity of the effective conductivity 296
14.7 Conclusions 297
Acknowledgments 298
References 298

15. Riemann-Hilbert problems with coefficients in compact Lie groups

Gia Giorgadze and Giorgi Khimshiashvili

15.1 Introduction 303
15.2 Recollections on classical Riemann-Hilbert problems 304
15.3 Generalized Riemann-Hilbert transmission problem 307
15.4 Lie groups and principal bundles 311
15.5 Riemann-Hilbert monodromy problem for a compact Lie group 316
References 325

16. When risks and uncertainties collide: quantum mechanical formulation of mathematical finance for arbitrage markets

Simone Farinelli and Hideyuki Takada

16.1 Introduction 327
16.2 Geometric arbitrage theory background 329

16.2.1	The classical market model	329
16.2.2	Geometric reformulation of the market model: primitives	331
16.2.3	Geometric reformulation of the market model: portfolios	332
16.2.4	Arbitrage theory in a differential geometric framework	333

16.3 Asset and market portfolio dynamics as a constrained Lagrangian system — 338

16.4 Asset and market portfolio dynamics as solution of the Schrödinger equation: the quantization of the deterministic constrained Hamiltonian system — 341

16.5 The (numerical) solution of the Schrödinger equation via Feynman integrals — 348

16.5.1	From the stochastic Euler-Lagrangian equations to Schrödinger's equation: Nelson's method	348
16.5.2	Solution to Schrödinger's equation via Feynman's path integral	350
16.5.3	Application to geometric arbitrage theory	350

16.6 Conclusion — 351

Appendix 16.A Generalized derivatives of stochastic processes — 352

References — 354

17. Thermodynamics and stability of metallic nano-ensembles

Michael Vigdorowitsch

17.1 Introduction — 357

17.1.1	Nano-substance: inception	357
17.1.2	Nano-substance: thermodynamics basics	358
17.1.3	Nano-substance: kinetics basics	360

17.2 Vacancy-related reduction of the metallic nano-ensemble's TPs — 362

17.2.1	Solution in quadrature of the problem of vacancy-related reduction of TPs	362
17.2.2	Particle distributions on their radii	363
17.2.3	Derivation of equations for TPs reduction	370
17.2.4	Reduction of TPs: results	375

17.3 Increase of the metallic nano-ensemble's TPs due to surface tension — 378

17.3.1	Solution in quadrature of the problem of the TP increase due to surface tension	379
17.3.2	Derivation of equations for TPs increase	381
17.3.3	Increase of TPs: results	387

17.4 Balance of the vacancy-related and surface-tension effects — 390

17.5 Conclusions — 391

References — 392

xiv Contents

18. Comparative analysis of local stresses in unidirectional and cross-reinforced composites
Alexander G. Kolpakov and Sergei I. Rakin

18.1 Introduction	395
18.2 Homogenization method as applied to composite reinforced with systems of fibers	397
18.3 Numerical analysis of the microscopic stress-strain state of the composite material	400
18.3.1 Macroscopic strain ε_{11} (tension-compression along the Ox-axis)	400
18.3.2 Macroscopic strain ε_{33} (tension-compression along the Oz-axis)	402
18.3.3 Macroscopic deformations ε_{22} (tension-compression along the Oy-axis)	403
18.3.4 Macroscopic deformations ε_{13} (shift in the Oxz-plane)	404
18.3.5 Macroscopic strain ε_{12} (shift in the Oxy-plane)	406
18.3.6 Macroscopic strain ε_{23} (shift in the Oyz-plane)	407
18.4 The "anisotropic layers" approach	408
18.4.1 Axial overall elastic moduli A_{1111} and A_{3333}	408
18.4.2 Axial overall elastic modulus A_{2222}	409
18.4.3 Shift elastic moduli A_{1212} and A_{2323}	409
18.4.4 Shift elastic modulus A_{1313}	409
18.4.5 The local stresses	409
18.5 The "multicomponent" approach by Panasenko	409
18.6 Solution to the periodicity cell problem for laminated composite	410
18.7 The homogenized strength criterion of composite laminae	411
18.8 Conclusions	413
References	414

19. Statistical theory of structures with extended defects
Vyacheslav Yukalov and Elizaveta Yukalova

19.1 Introduction	417
19.2 Spatial separation of phases	418
19.3 Statistical operator of mixture	419
19.4 Quasiequilibrium snapshot picture	421
19.5 Averaging over phase configurations	423
19.6 Geometric phase probabilities	425
19.7 Classical heterophase systems	428
19.8 Quasiaverages in classical statistics	432
19.9 Surface free energy	434
19.10 Crystal with regions of disorder	435
19.11 System existence and stability	439
19.12 Conclusion	441
References	442

20. Effective conductivity of 2D composites and circle packing approximations

Roman Czapla and Wojciech Nawalaniec

20.1 Introduction	445
20.2 General polydispersed structure of disks	446
20.3 Approximation of hexagonal array of disks	449
20.4 Checkerboard	451
20.5 Regular array of triangles	454
20.6 Discussion and conclusions	454
References	457

21. Asymptotic homogenization approach applied to Cosserat heterogeneous media

*Victor Yanes, Federico J. Sabina, Yoanh Espinosa-Almeyda,
José A. Otero, and Reinaldo Rodríguez-Ramos*

21.1 Introduction	459
21.2 Basic equations for micropolar media. Statement of the problem	461
21.2.1 Two-scale asymptotic expansions	463
21.3 Example. Effective properties of heterogeneous periodic Cosserat laminate media	473
21.4 Numerical results	480
21.4.1 Cosserat laminated composite with cubic constituents	480
21.5 Conclusions	486
Acknowledgments	489
References	489

A. Finite clusters in composites

V. Mityushev

Index	495

List of contributors

Igor Andrianov, Institute of General Mechanics, RWTH Aachen University, Aachen, Germany

Yeskendyr Ashimov, Institute of Mathematics and Mathematical Modeling, Almaty, Kazakhstan
al-Farabi Kazakh National University, Almaty, Kazakhstan

Andrej Cherkaev, Department of Mathematics, University of Utah, Salt Lake City, UT, United States

Roman Czapla, Institute of Computer Science, Pedagogical University of Cracow, Kraków, Poland

Matteo Dalla Riva, Department of Mathematics, The University of Tulsa, Tulsa, OK, United States

Piotr Drygaś, University of Rzeszow, Rzeszow, Poland

Yoanh Espinosa-Almeyda, Applied Mathematics and Systems Research Institute, National Autonomous University of Mexico (UNAM), Mexico City, Mexico

Simone Farinelli, Core Dynamics GmbH, Zurich, Switzerland

Leonid Filshtinsky, Sumy State University, Sumy, Ukraine

Gia Giorgadze, Faculty of Exact and Natural Sciences, Tbilisi State University, Tbilisi, Georgia

Simon Gluzman, Research Group Materialica+, ON, Toronto, Canada

Yuri A. Godin, Department of Mathematics and Statistics, University of North Carolina at Charlotte, Charlotte, NC, United States

Sergei Kanaun, Tecnologico de Monterrey, School of Engineering and Science, Monterrey, Mexico

Julius Kaplunov, School of Computing and Mathematics, Keele University, Keele, Staffordshire, United Kingdom

Giorgi Khimshiashvili, Institute of Fundamental and Interdisciplinary Mathematical Research, Ilia State University, Tbilisi, Georgia

Alexander G. Kolpakov, SysAn, Novosibirsk, Russia

xviii List of contributors

Lidiya Kurpa, Department of Applied Mathematics, National Technical University "Kharkiv Polytechnic Institute", Kharkiv, Ukraine

Volodymyr Loboda, Department of Theoretical and Computational Mechanics, Oles Honchar Dnipro National University, Dnipro, Ukraine

Paolo Luzzini, EPFL, SB Institute of Mathematics, Station 8, Lausanne, Switzerland

Anatoly Markov, Tecnologico de Monterrey, School of Engineering and Science, Monterrey, Mexico

Vladimir Mityushev, Research Group Materialica+, Faculty of Computer Science and Telecommunications, Cracow University of Technology, Kraków, Poland

Paolo Musolino, Dipartimento di Scienze Molecolari e Nanosistemi, Università Ca' Foscari Venezia, Venezia Mestre, Italy

Oleksandr Mykhail, Department of Theoretical and Computational Mechanics, Oles Honchar Dnipro National University, Dnipro, Ukraine

Wojciech Nawalaniec, Institute of Computer Science, Pedagogical University of Cracow, Kraków, Poland

Dmytro Nosov, Constantine the Philosopher University, Nitra, Slovakia Pedagogical University, Kraków, Poland

José A. Otero, Tecnologico de Monterrey, School of Engineering and Sciences, Mexico State, Nuevo León, Mexico

Ludmila Prikazchikova, School of Computing and Mathematics, Keele University, Keele, Staffordshire, United Kingdom

Roman Pukhtaievych, Department of Complex Analysis and Potential Theory, Institute of Mathematics of the National Academy of Sciences of Ukraine, Kyiv, Ukraine

Sergei I. Rakin, SysAn, Novosibirsk, Russia Siberian Transport University, Novosibirsk, Russia

Reinaldo Rodríguez-Ramos, Faculty of Mathematics and Computer Science, University of Havana, Havana, Cuba

Natalia Rylko, Research Group Materialica+, Faculty of Computer Science and Telecommunications, Cracow University of Technology, Kraków, Poland

Federico J. Sabina, Applied Mathematics and Systems Research Institute, National Autonomous University of Mexico (UNAM), Mexico City, Mexico

Sheeru Shamsi, School of Computing and Mathematics, Keele University, Keele, Staffordshire, United Kingdom

Alla Sheveleva, Department of Computational Mathematics and Mathematical Cybernetics, Oles Honchar Dnipro National University, Dnipro, Ukraine

Tetyana Shmatko, Department of Higher Mathematics, National Technical University "Kharkiv Polytechnic Institute", Kharkiv, Ukraine

List of contributors   **xix**

Hideyuki Takada, Department of Information Science, Narashino Campus, Toho University, Funabashi-Shi, Chiba, Japan

Heiko Topol, Institute of General Mechanics, RWTH Aachen University, Aachen, Germany

Boris Vainberg, Department of Mathematics and Statistics, University of North Carolina at Charlotte, Charlotte, NC, United States

Michael Vigdorowitsch, Angara GmbH, Düsseldorf, Germany
All-Russian Scientific Research Institute for the Use of Machinery and Oil Products in Agriculture, Tambov, Russia
Tambov State Technical University, Tambov, Russia

Ryszard Wojnar, Institute of Fundamental Technological Research PAS, Warszawa, Poland

Victor Yanes, Faculty of Physics, University of Havana, Havana, Cuba

Vyacheslav Yukalov, Bogolubov Laboratory of Theoretical Physics, Joint Institute for Nuclear Research, Dubna, Russia
Instituto de Fisica de São Carlos, Universidade de São Paulo, São Carlos, São Paulo, Brazil

Elizaveta Yukalova, Laboratory of Information Technologies, Joint Institute for Nuclear Research, Dubna, Russia

Zhanat Zhunussova, Institute of Mathematics and Mathematical Modeling, Almaty, Kazakhstan
al-Farabi Kazakh National University, Almaty, Kazakhstan

Acknowledgements

The editors express their deep gratitude to Professor, Academician of the Russian Academy of Sciences A.T. Fomenko for kind permission to use his painting "The remarkable numbers pi and e, II" for the design of the book cover.

Chapter 1

L.A. Filshtinsky's contribution to Applied Mathematics and Mechanics of Solids

Vladimir Mityushev[a], Igor Andrianov[b], and Simon Gluzman[c]

[a]*Research Group Materialica+, Faculty of Computer Science and Telecommunications, Cracow University of Technology, Kraków, Poland,* [b]*Institute of General Mechanics, RWTH Aachen University, Aachen, Germany,* [c]*Research Group Materialica+, ON, Toronto, Canada*

1.1 Introduction

1.1.1 Personality and career

Leonid Anshelovich Filshtinsky (Fil'shtinskii, Фильштинский) was born on October 19, 1930 in Kharkiv and passed away on May 19, 2019 in Sumy. We can trace the origin of his name to the small town Felsztyn, now in Ukraine.

Filshtinsky was a prominent applied mathematician who worked on the problems of the theory of elasticity and mathematical physics. Many of his works are devoted to studies of the elastic plane fields in composites and fractured media.

Filshtinsky received his engineering diploma at the Kharkiv Aviation Institute in 1956. Filshtinsky studied in Kharkiv, one of the prominent scientific and industrial centers of Ukraine. After a few years spent in Siberian cities of Irkutsk and Novosibirsk, he returned to the Ukrainian town Sumy. Here he lived and worked till the end.

Then in 1956–1959 he worked at the Aviation Plant in Irkutsk. Education on all levels was free in Soviet Union, but the caveat was that one had to work at a state-prescribed institution for a few years. Thus, after receiving his Master degree, Filshtinsky headed to Irkutsk. (See Fig. 1.1.)

It was a time when mathematics was considered as Queen of Sciences (Carl Friedrich Gauss), and to hold the crown for all engineering disciplines. Emmanuel Kant philosophy was ostracized in Soviet Union, but the catch phrase "In any special doctrine of nature there can be only as much proper science as there is mathematics therein" was approved and much quoted.

Specialized technical universities and aviation institutes had a focus on advanced mathematical education. Having worked as an aircraft engineer Filshtin-

Mechanics and Physics of Structured Media. https://doi.org/10.1016/B978-0-32-390543-5.00006-2
Copyright © 2022 Elsevier Inc. All rights reserved.

2 Mechanics and Physics of Structured Media

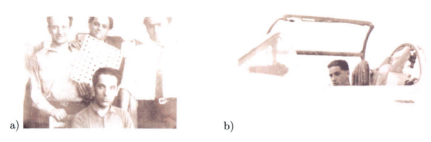

FIGURE 1.1 a) Filshtinsky with colleagues on aircraft engineering; b) Filshtinsky in the cockpit.

sky had experienced firsthand the technological processes related to mathematical modeling and computations. The title of his first book "Perforated Plates and Shells" [60] is rather closely motivated by the problems arising in the course of engineering an aircraft, though its content was mainly mathematical, see vivid illustration in Fig. 1.1a. He arrived to the notion of homogenization as the way to evaluate the macroscopic, averaged response of the perforated plate.

His scientific work began in 1961 at the famous Khristianovich Institute of Theoretical and Applied Mechanics (Novosibirsk). In 1964 he defended his PhD thesis under the supervision of E.I. Grigolyuk, and in 1971 presented his habilitation thesis (D.Sc. thesis) before becoming a full professor in 1974. Grigolyuk and Filshtinsky coauthored many books; according to Russian alphabet, Grigolyuk always was the first author.

In the years 1959–1974 Filshtinsky worked at Chaplygin Siberian Scientific Research Institute of Aviation at Novosibirsk.

In 1969–1974 he worked in Novosibirsk Electrotechnical Institute (now Novosibirsk State Technical University (NGTU),) as a lecturer and professor at the Department of Aircraft Strength. Consider that Novosibirsk is located in Siberia, the Russian analog to the Wild West. His lectures were quite popular according to the reminiscences of his students and coworkers.[1]

From 1974 Filshtinsky worked at the Sumy branch campus of the Kharkiv Polytechnical Institute, which became the Sumy Institute of Physics and Technology in 1990. In 1993 it was reorganized into Sumy State University. In 1991 Ukraine became independent. Filshtinsky stayed in Ukraine, working at Sumy at the same university till his last days.

Filshtinsky organized and headed departments of Applied Mathematics. Under his supervision 30 PhD students and three D.Sc. students defended their theses. It is perhaps not the final number, because his last PhD students who did not complete their studies in Filshtinsky lifetime, still continue working on their thesis until today. (See Figs. 1.2–1.5.)

---

[1] See the Book: Science. Technology. Innovation, Part 9. Novosibirsk, NGTU, 2019.

FIGURE 1.2 a) Filshtinsky (right) with colleagues Dr. S. Shapovalov (sitting) and Prof. V. Maksimenko (center); b) Filshtinsky (left) and Prof. D.I. Bardzokas from the National Technical University of Athens.

FIGURE 1.3 Filshtinsky (bottom row, center) with his wife T.L. Misina to the left, his brother Vadim A. Filshtinsky and brother's son Stas, and colleagues from the department of Applied Mathematics and Mechanics in Sumy in 2005. Among them in the first row from the left are D. Kushnir and Yu. Shramko. In the second row from the left second is L. Bratsihina, third is E.I. Ogloblina, sixth is T. Sushko, eighth is A. Bondar, ninth is T. Kirichok, next is Yu. Kovalev.

1.1.2 Lessons of collaboration (V. Mityushev)

My (Mityushev) interest in this problem dates back to 1992 and is due to the influence of my respected Polish colleagues, Barbara Gambin, Joachim Telega, Ryszard Wojnar. Ross McPhedran (Sydney University) introduced me to the

4 Mechanics and Physics of Structured Media

FIGURE 1.4 Filshtinsky with students in Sumy.

FIGURE 1.5 a) Filshtinsky at the conference in Alushta in 2004; b) Filshtinsky with Mityushev in Rzeszów (Poland) during the invited lecture by Filshtinsky at the Rzeszów University on June 6, 2012.

method of Rayleigh in 1999 and to the extensive research of his group on the double periodic array of cylinders. Before I met Filshtinsky in person in 2010 at the conference in Dnepropetrovsk organized by Igor Andrianov and Vladislav Danishevskyy (Fig. 1.6), I thought that his main contributions were related only to the method of integral equations [60,66,67]. Filshtinsky had introduced me to his earlier work (1964), as well as to Natanzon's paper (1935) devoted to extensions of the Rayleigh result (1892) to elastic problems. Up to our meeting in 2010, Filshtinsky was not familiar with Rayleigh's paper and McPhedran's results.

It became clear to me how to merge the efforts of Eisenstein, Rayleigh, Natanzon, Filshtinsky, McPhedran, and others and to derive analytical formu-

FIGURE 1.6 The conference in Dnepropetrovsk in 2010: Filshtinsky with the brown briefcase, the tallest participant; Andrianov is the third left in the top row. Mityushev is between Filshtinsky and Andrianov. Danishevskyy is on the right side of the middle row.

las for the random 2D elastic composites and viscous suspensions. The plan was recently realized and summarized in the books [22,55]. Of course, it was technically difficult to get access to the sources of original works. Nowadays, the Internet helps a lot, but the most part of Filshtinsky life had passed in the pre-Internet era. Fortunately, I do have now an unlimited access to the sources, and it is possible to find the roots of Filshtinsky work and establish his predecessors...

Sections 1.2–1.4 are general, historical reviews where attention is paid to the results rigorously prioritized, together with precise citations or dates of publications. Many authors, colleagues, and pupils around Filshtinsky are only named. Their results have published only in Russian, yet they are surely worth mentioning.

1.1.3 Filshtinsky's contribution to the theory of integral equations

The main results by Filshtinsky concern integral equations in a class of double periodic functions summarized in the books [60,66,67] and outlined in the publication [7], [49] in English. Filshtinsky simply was convinced of the high efficiency of integral equations in the boundary value problems of mathematical physics. He thought that the proper integral representations were able to capture singularities and, in combination with asymptotic methods, would allow to match physical fields at different scales.

6 Mechanics and Physics of Structured Media

As an example, he brought up the computation of the stress intensity factors. The section devoted to Filshtinsky's contribution to integral equations might be added to the present chapter. However, it could be viewed as a rerun of the comprehensive historical notes by Filshtinsky himself [66], which contained a review of many significant results published, in particular, in 1950–1977. We can only add that Filshtinsky was the leading mind in this area. There are several variants of integral equations in the plane elasticity. In the beginning of the computer era a specialist in numerical methods fought for every bite and for each millisecond of computational time by optimization of algorithms. Filshtinsky followed these lines, by constructing a set of equations and developed the corresponding robust algorithms. Perhaps, now his plots, tables, and schemes illustrating the elastic fields look somewhat archaic. This situation can be easily remedied by means of the modern computational tools. The most important is that his methods work, and are still fast and precise.

We would like to raise the question about the gap between obtaining the result and presenting it properly, which is reduced largely to the matter of belief. In the scope of publications on computational mechanics, one can with ease find a paper devoted to a "novel numerical method", which solves "any problem". Sometimes, such papers are supported by a meager example with the typical comments "you see, it works". If the example is not overly impressive, you do or do not believe the author. But we always believe Filshtinsky, and trust the effectiveness of his method of integral equations. In this note, we would also like to raise the question of quality of publications on computational mechanics. It seems that every paper with a claim of a "novel numerical method" has to contain a supplement with codes or the precise information about computational resources, time, and memory involved in computation, etc. It is rather impossible to work out any universal standards of how to check a theoretical method and its implementation, but even pure standardization of the way to present the results of computations may be very useful.

It is proper to recall the pioneers, who considered and solved the problems relevant to our discussion. That is why we have to recall L.I. Sedov, who in 1950 [131] applied the integrals with a double periodic kernel to fluid dynamics. The paper by L.I. Chibrikova published in 1956 [18], initiated the study of the Riemann-Hilbert problems on flat torus, i.e. in a class of doubly periodic functions, and its extension to Riemann surfaces. This theory was completed by E.I. Zverovich in 1971 [141]. W.T. Koiter [72,73] was the first to reduce a plane periodic elastic problem to a Fredholm integral equation. The books [60,66,67] contain a review of hundreds of significant results published later, in particular, in 1964–1977.

1.2 Double periodic array of circular inclusions. Founders

1.2.1 Preliminaries

The present chapter is devoted to the contributions of Filshtinsky and other outstanding scientists to the two-dimensional problems of elasticity and related mathematical problems for doubly periodic complex potentials and polyharmonic functions. His works remain almost unknown outside of the former Soviet Union, since his major publications were written in Russian, and often published in local, hardly accessible journals of the former Soviet Union and Ukraine. His most influential books [60,66,67] exist only in Russian. The few papers published in English after 2005 give the misleading impression that they treat formerly existing problems of applied mathematics that have been solved by someone else. It was typical of Filshtinsky to rarely mention himself as the founding father. He rarely cited his works published before the 1980s because of the better computational quality of the results he obtained with his coauthors after 1990. Reading his papers it is difficult to identify an original contribution, therefore a comprehensive review of Filshtinsky's works has been long overdue.

In the present section, we outline the problem of two-dimensional stationary conductivity and elasticity for a doubly periodic array of circular inclusions. We review the seminal works from the historical perspective and relate celebrated achievements of the famous scientists Eisenstein, Weierstrass, Rayleigh to the less known results by Natanzon (Natanson) and Filshtinsky (Fil'shtinskii). Filshtinsky contributed to the solution of considered problem et par with the Greats. First, we have to remember that this problem is not so easy, since it depends on specifics of physical formulation. To follow the historical line is also not so straightforward, since all of the above scientists did not refer to each other and worked independently, with the notable exception of Filshtinsky's works based on Natanzon's paper.

First, we discuss the stationary conductivity problem governed by the 2D Laplace equation $\nabla^2 u = 0$. Further discussion will concern elastic composites governed by biharmonic functions for which $\nabla^4 u = 0$. The final notes are devoted to polyharmonic functions satisfying the equation $\nabla^{2p} u = 0$ for a natural number p. We consider doubly periodic harmonic and analytic functions for conductivity problems, biharmonic and bianalytic functions related to the doubly periodic elastic tensors.

The first attempt at the problem solution is always accomplished from the standpoint of physics. Consider a double periodic array of unidirectional infinite cylinders of conductivity σ_1 embedded in a host material of conductivity σ. Problems for fibrous composites refer to the plane conductivity. The latter serves as a 2D approximation of the 3D fibrous composites. An array of infinitely long, aligned cylinders in a matrix is a physically relevant model applicable to a perforated plate when the cylinders are hollow. More precisely, the deformation of a matrix with hollow, penny-shaped platelets under a constant load along the generatrix of the cylinders is described by the generalized plane stress state of

8 Mechanics and Physics of Structured Media

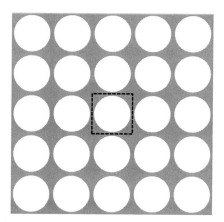

FIGURE 1.7 Square array of inclusions with $\omega_1 = 1$ and $\omega_2 = i$. The periodicity cell is shown by dashed line.

a perforated plate. This yields a boundary value problem on the interface for the limit values of the local fields described below. The square array of circular inclusions is displayed in Fig. 1.7.

The coordinates (x_1, x_2) will be used as well as the complex numbers $z = x_1 + ix_2$. Let ω_1 and ω_2 denote the fundamental translation vectors expressed by means of complex numbers. Let $\text{Im} \frac{\omega_2}{\omega_1} > 0$ for definiteness. The points $m_1\omega_1 + m_2\omega_2$ ($m_1, m_2 \in \mathbb{Z}$) generate a doubly periodic lattice $\mathcal{Q}$, where $\mathbb{Z}$ stands for the set of integer numbers. The zeroth cell is the parallelogram

$$Q_{00} = \left\{ z = t_1\omega_1 + t_2\omega_2 \in \mathbb{C} : -\frac{1}{2} < t_1, t_2 < \frac{1}{2} \right\},$$

and its area is normalized to unity, i.e.,

$$|Q_{00}| = \text{Im}(\overline{\omega_1}\omega_2) = 1. \tag{1.1}$$

Introduce the cells $Q_{(m_1, m_2)} = Q_{00} + m_1\omega_1 + m_2\omega_2$. Let the disk $|z| < r$ be occupied by a material of conductivity σ_1. Let the domain G, the complement of $|z| \leq r$ to the cell Q_{00}, be occupied by a material of conductivity σ. Let the potentials $u(z)$ and $u_1(z)$ satisfy the Laplace equation in G and in $|z| < r$, respectively, and be continuously differentiable in the closures of the considered domains.

The perfect contact between the components is expressed by the boundary conditions

$$u = u_1, \quad \sigma \frac{\partial u}{\partial \mathbf{n}} = \sigma_1 \frac{\partial u_1}{\partial \mathbf{n}}, \quad \text{on } |z| = r. \tag{1.2}$$

Let the external field be applied in the x_1-direction and normalized in such a way that

$$u(z + \omega_1) - u(z) = |\omega_1|, \quad u(z + \omega_2) - u(z) = 0. \tag{1.3}$$

The limit cases of holes ($\sigma_1 = 0$) and perfectly conducting disks ($\sigma_1 = \infty$, $u_1 = constant$) yield the classic Neumann and Dirichlet boundary conditions. Thus, our problem can be stated as the problem of local fields in a doubly periodic medium.

The next natural question, on what are the effective properties of fibrous composites, leads to the problem of homogenization. The problem is reduced to the determination of averaged fields over the representative parallelogram called RVE (representative volume element) in the theory of composites.

The classical Maxwell formula (MF) (Maxwell, 1873) [97] is one of the most often applied formulas of the theory of composites. The MF for arbitrarily nonoverlapping disks can be written as follows

$$\sigma_e \approx \sigma \, \frac{1 + \varrho f}{1 - \varrho f}, \tag{1.4}$$

where f stands for the concentration of disks, σ_e the effective conductivity and

$$\varrho = \frac{\sigma_1 - \sigma}{\sigma_1 + \sigma}, \tag{1.5}$$

is the contrast parameter. The MF (1.4) is also referred to as the Clausius-Mossotti approximation, Maxwell Garnett [98], Maxwell-Odelevskii, Lorenz-Lorentz, Landauer [80], and Wiener-Wagner formula, see [81,86,103] and others. It combines simplicity and satisfactory accuracy for dilute and low contrast parameter composites. The MF (1.4) is valid up to $O(f^5)$ for the square array and up to $O(f^7)$ for the hexagonal array [55]. Its numerous modifications for general composites based on various semiempirical extensions have been arising all the time. Recently, it became clear [114] that all attempts to improve (1.4) and the corresponding formulas for 3D [116] and elasticity [115] by effective medium approximations, differential scheme, Mori–Tanaka approach, etc., lead either to wrong formulas or to the same approximation (1.4), valid up to $O(\varrho^2 f^2)$ [113] formally written with redundant high-order tails. The approximate formula (1.4) was deduced by Maxwell's self-consistent approach when interactions among particles are neglected. Even when one takes into account all interactions between neighboring inclusions in a finite cluster, the results are wrong. We will return to the discussion of Maxwell's approach at the end of this section.

It is worth noting that formulas for neutral inclusions, for instance for Hashin-Shtrikman assemblages and their extensions [17], [16], are exact. The approximate formula (1.4) is valid for noninteracting inclusions, hence, actually it refers to approximately neutral inclusions. Formulas derived from solution to

10 Mechanics and Physics of Structured Media

the problems in a periodic statement can give high order formulas in f for the local fields and the effective constants. This is in agreement with the theory of homogenization when the boundary effects are excluded.

Let us now look at this physical problem from the formal point of view of the theory of functions on Riemann surfaces. First, we note that a 2D boundary value problem for harmonic functions is reduced to the problem for an analytic function [111], [55]; for a biharmonic function to a pair of analytic functions [118]; for a polyharmonic function it can be reduced to a system of analytic functions [53]. Second, mind that any problem in a class of doubly periodic functions can be considered as a problem on the flat torus obtained from the fundamental parallelogram by welding of the opposite sides [1], [69]. Therefore, the conductivity problem for doubly periodic functions can be stated as a boundary value problem on a Riemann surface.

1.2.2 Contribution by Eisenstein

Now, we are prepared to recite the story, beginning with the paper [25] published by Gotthold Eisenstein in 1847. The modern theory of the Eisenstein functions and analysis of his approach can be found in [138]. It should be noted that the special elliptic sine function was introduced in 1799 by Carl Friedrich Gauss. Carl Gustav Jacob Jacobi introduced the elliptic functions in 1827 as the inverse functions to the elliptic integral.

We now follow the Eisenstein approach based on the double periodicity on the complex plane. Consider the lattice sums

$$S_m = \sum_{m_1, m_2}' \frac{1}{(m_1 \omega_1 + i m_2 \omega_2)^m}, \quad m = 1, 2, 3, \dots, \tag{1.6}$$

where m_1, m_2 run over $\mathbb{Z}$. The prime means that the term $m_1 = m_2 = 0$ is skipped in the summation. The double series (1.6) is conditionally convergent for $m = 1$ and $m = 2$, hence, its value depends on the order of summation. Eisenstein proposed the iterated summation method

$$\sum_{m_1, m_2}^{e} := \lim_{M_2 \to +\infty} \lim_{M_1 \to +\infty} \sum_{m_2 = -M_2}^{M_2} \sum_{m_1 = -M_1}^{M_1}. \tag{1.7}$$

The symbol $\sum^e$ stands for the Eisenstein summation. It could be checked that $S_m = 0$ for odd m.

Using the summation (1.7), Eisenstein introduced the following functions (which are called now Eisenstein functions)

$$E_m(z) = \sum_{m_1, m_2 \in \mathbb{Z}}^{e} \frac{1}{(z + m_1 \omega_1 + m_2 \omega_2)^m}. \tag{1.8}$$

Eisenstein calculated the jumps of $E_1(z)$ along the periodicity cell

$$E_1(z+\omega_1) - E_1(z) = 0, \quad E_1(z+\omega_2) - E_1(z) = -\frac{2\pi i}{\omega_1} \quad (1.9)$$

and established the formula for the derivatives

$$E'_m(z) = -m E_{m+1}(z). \quad (1.10)$$

It was proved that the function $E_m(z)$ is doubly periodic for $m > 1$ and has a pole of order m at $z = 0$. The Laurent series of the even order Eisenstein function has the form

$$E_{2m}(z) = \frac{1}{z^{2m}} + \sum_{k=1}^{\infty} \frac{(2m+2k-3)!}{(2m-1)!(2k-2)!} S_{2(m+k-1)} z^{2(k-1)}. \quad (1.11)$$

Analogous formulas were established for the functions of odd order. For instance,

$$E_1(z) = \frac{1}{z} - \sum_{k=1}^{\infty} S_{2k} z^{2k-1}. \quad (1.12)$$

Eisenstein was the first to establish formulas concerning doubly periodic meromorphic (analytic) functions, see [138] for more formulas. His brilliant insights could be justified by introduction of uniform convergence. It is necessary for differentiating his series term by term. Though the idea of uniform convergence was in the air in the first half of the XIX century, the complete understanding was achieved by Karl Weierstrass between 1861 and 1886 [137]. The concept was not widespread at the Eisenstein lifetime, as he passed away in the year 1852, at the age of 29.

The absence of uniform convergence argument in Eisenstein's study was, perhaps, the reason why Weierstrass ignored his work. But the notion could have been introduced quite easily... Maybe, it looks "easy" for those of us who have studied mathematical and functional analysis. Mathematicians of the XIX century used uniform convergence in a Banach space similar to Monsieur Jourdain from Molière's comedy. Mind that the respected gentleman was quite surprised that all his life he spoke in "high" prose.

The Eisenstein functions are related to the Weierstrass ζ- and $\wp$-functions by the following identities [138]

$$E_1(z) = \zeta(z) - S_2 z, \quad E_2(z) = \wp(z) + S_2. \quad (1.13)$$

Weierstrass defined the $\wp$-function by means of the absolutely convergent series

$$\wp(z) = \frac{1}{z^2} + {\sum_{P}}' \left[\frac{1}{(z-P)^2} - \frac{1}{P^2} \right], \quad (1.14)$$

12 Mechanics and Physics of Structured Media

where the notations $P = m_1\omega_1 + m_2\omega_2$ and $\sum'_{m_1,m_2\in\mathbb{Z}} = \sum'_P$ are used for shortness. The prime means that the term $P = 0$ is skipped. One can see that the conditionally convergent lattice sum S_2 is dropped out from the definition (1.14).

1.2.3 Contribution by Rayleigh

Lord Rayleigh [126] stated the problem expressed by Eqs. (1.2)–(1.3), and developed the method of its solution for a rectangular array when $\omega_1 \in \mathbb{R}$ and $\omega_2 \in i\mathbb{R}$ in 1892. His method is based on the representations of the unknown potentials by the Laurent and Taylor series

$$u(z) = A_0 + \mathrm{Re}\sum_{l=1}^{\infty}\left[A_l z^l + B_l z^{-l}\right], \quad z \in G, \quad u_1(z) = \mathrm{Re}\sum_{l=0}^{\infty} C_l z^l, \quad |z| < r.$$

$$(1.15)$$

Rayleigh used the trigonometric form of the series (1.15) and the symmetry of rectangular lattice when the unknown coefficients are real. Using the boundary conditions (1.2) Rayleigh established linear relations between A_l, B_l, C_l and reduced the problem to the infinite system of linear algebraic equations.

Let us introduce the complex flux

$$\psi_1(z) = \frac{\partial u_1}{\partial x_1} - i\frac{\partial u_1}{\partial x_2}, \quad |z| < r. \qquad (1.16)$$

Consider representing the function $\psi_1(z)$ analytic in the disk $|z| < r$ by its Taylor series $\psi_1(z) = \sum_{l=0}^{\infty}\alpha_l z^l$. Let us write Rayleigh's system in the form [127], [55, formula (4.4.73)], valid for an arbitrary cell

$$\alpha_l = \rho\sum_{m=0}^{\infty}(-1)^m\frac{(l+m+1)!}{l!(m+1)!}S_{l+m+2}r^{2(m+1)}\overline{\alpha_m}+\delta_{l0}, \quad l = 0, 1, \ldots, \quad (1.17)$$

where the bar stands for the complex conjugation and δ_{l0} denotes the delta Kronecker symbol.

The system (1.17) is equivalent to Rayleigh's system usually written for the coefficients $B_l = \frac{\sigma-\sigma_1}{2\sigma l}r^{2l}\overline{\alpha_{l-1}}$, sometimes separately for the real and imaginary parts of B_l. McPhedran with coworkers had developed the method of Rayleigh and extended it to some other fundamental physical problems [99,100,123,124]. The infinite system (1.17) was truncated [126] to get approximate formulas for the coefficient $\alpha_1 = \psi_1(0)$, needed to calculate the effective conductivity.

Rayleigh's system is presented in the form (1.17) in order to demonstrate its equivalence to the functional equation [104], [55, (4.4.70)]

$$\psi_1(z) = \varrho\sum_P^{'e}\frac{r^2}{(z-P)^2}\overline{\psi_1\left(\frac{r^2}{z-P}\right)} + 1, \quad |z| \leq r, \qquad (1.18)$$

where the Eisenstein summation $\sum_P'^e$ with the skipped term $P = 0$ is used. It was proved in [104], that Eq. (1.18) is an equation with a compact operator in a Banach space which can be solved by the method of successive approximations convergent for any $|\varrho| < 1$. This justifies the application of Rayleigh's system. The exact formula for the local field and the effective conductivity of an array of circular holes was established by Mityushev in 1998 and written in the general form in [55,109] for inclusions. For illustration, we write the asymptotic formula

$$\psi_1(z) = 1 + \varrho r^2 \sum_P'^e \frac{1}{(z - P)^2} + O(\varrho^2 r^4), \quad |z| \le r. \tag{1.19}$$

The series $\sum_P^e (z - P)^{-2}$ is the Eisenstein function $E_2(z)$.

Rayleigh (1892) did not refer to Eisenstein (1847) and quoted only the Weierstrass theory of elliptic functions (1856). Nevertheless, Rayleigh used the Eisenstein summation and proved the fascinating formula $S_2 = \pi$ for the square array not known to Eisenstein.

Rayleigh replaced the lattice sum S_2 by the double integral

$$S_2 = \int \int_{\mathbb{R}^2} \frac{dx_1 dx_2}{(x_1 + ix_2)^2}, \tag{1.20}$$

and extended the Eisenstein method by performing integration

$$S_2 = 2 \int_v^\infty \left[\int_{-v}^v \frac{dx_2}{(x_1 + ix_2)^2} \right] dx_1 = \pi. \tag{1.21}$$

Conversion of the double integral (1.20) into the iterated integral (1.21) corresponds to the Eisenstein summation of the series. The direct application of the Eisenstein summation (1.7) to the lattice sum S_2 from (1.6) for the square array requires the following more advanced transformations (write m instead of m_1 and n instead of m_2)

$$S_2 = \sum_{m \ne 0} \frac{1}{m^2} + \sum_{\substack{m,n \\ n \ne 0}}^e \frac{1}{(m + in)^2} = \sum_{m \ne 0} \frac{1}{m^2} - \sum_{n \ne 0} \frac{1}{n^2} \tag{1.22}$$

$$+ \lim_{N \to +\infty} \sum_{\substack{n = -N \\ n \ne 0}}^N \sum_{m=1}^\infty \left(\frac{1}{(m + in)^2} - \frac{1}{(m - in)^2} \right) =$$

$$\lim_{N \to +\infty} \sum_{\substack{n = -N \\ n \ne 0}}^N 2 \sum_{m=1}^\infty \frac{m^2 - n^2}{(m^2 + n^2)^2} = 2 \sum_{n=1}^\infty \left(\frac{1}{n^2} - \frac{\pi^2}{\mathrm{sh}^2 \pi n} \right) = \pi.$$

The known formula $\sum_{n=1}^\infty \frac{1}{\mathrm{sh}^2 \pi n} = \frac{1}{6} - \frac{1}{2\pi}$ is used to find the final equality.

Rayleigh did not explain the application of the Eisenstein summation. It was accomplished in [100] and [106].

14 Mechanics and Physics of Structured Media

1.2.4 Contribution by Natanzon

The paper by V.Ya. Natanzon (Wl.Ya. Natanson) [119] is devoted to the plane elasticity problem for a regular array of circular holes. His outstanding paper extends Rayleigh's result to biharmonic problems. It is accessible only in Russian. In this section, we outline the main ideas of [119]. In order to systematically study the doubly periodic complex potentials and their applications to plane elasticity, we recommend the book [66] by Grigolyuk and Filshtinsky, also accessible only in Russian. Natanzon was rather unaware of the Eisenstein and Rayleigh publications. He dwelt on Goursat's short chapter and Hurwitz's book [69] translated into Russian in 1933, to work with the Weierstrass elliptic functions. The first edition of Muskhelishvili's book [118] was in 1933. The situation for explosive development of complex analysis to doubly periodic problems on the Riemann surface, was created in the Soviet Union in 1933. In 1933–1935 M.V. Keldysh and M.A. Lavrentiev published their first joint papers, summarized in the great book [83] by Lavrentiev and Shabat, existing in Russian, German, and Spanish. We know of only a single Natanzon's paper. The bang caused by the emergence of Filshtinsky was postponed to 1961. It was caused to a large extent by Akhiezer's book [1] first published in 1948.

The works of N.I. Akhiezer deserve special attention. He had only fragmentary information about works published abroad concerning the entire functions and approximation theory. One can say the same regarding his works on the elliptic functions. Akhiezer independently developed the theory of elliptic functions and discovered new formulas, see the example in [22, Appendix A.2]. Akhiezer constructed the Schwarz-Christoffel mapping for doubly connected polygons in the paper published in a local journal in 1929 in Ukrainian, see this result in English translation [1]. It is worth noting that boundary value problems for doubly connected domains are conformally equivalent to the problems on the plane torus, see complete solution to the problem of two perfectly conducting disks [130] with references to U. Dini (1913), M. Lagally (1929), C. Ferrari (1930) which paralleled Akhiezer's study. The most general form of the Schwarz-Christoffel integral for an arbitrary multiply-connected polygon was derived in [13,110]. Although the books [1] and [69] belong to the treasures of the constructive theory of elliptic functions and may be completed by modern trends of mathematics outlined in [52], one can still find Akhiezer's formulas being rediscovered in publications after 2000.

So, we come back to the beginning of XX century. The complex potentials method was first proposed by G.V. Kolosov in his Doctor thesis [74] published in 1909 and finally published with corrections in 1910. The method lead to the formulas expressing the 2D strain and stress tensors through the two analytic functions $\Phi(z)$ and $\Psi(z)$

$$\sigma_{11} + \sigma_{22} = 4\mathrm{Re}\ \Phi(z), \quad \sigma_{22} - \sigma_{11} + 2i\sigma_{12} = 2\left[z\overline{\Phi'(z)} + \overline{\Psi(z)}\right], \quad (1.23)$$

where σ_{ij} denotes the components of the stress tensor for isotropic elastic media. Analogous formulas hold for the strain tensor.

It is worth noting that S.A. Chaplygin in 1900 derived formulas for the plane deformations similar to Kolosov's formulas. These formulas were discovered in Chaplygin's drafts and published 50 years later in 1950 [15]. Chaplygin also discussed the main tenets of Kolosov talks given in 1909–1910, and in his book. Kolosov presented his results sloppily and not clearly, with some mistakes. One can only imagine that Chaplygin assumed that Kolosov was able to correct the mistakes and improve presentation, and was his and Steklov staunch supporter. N.I. Muskhelishvili prepared the Magisterial thesis equivalent to the modern PhD under Kolosov's supervision in 1915. Muskhelishvili performed a titanic job. He developed the theory of boundary value problems for analytic functions and singular integral equations. Muskhelishvili also developed an approach to solve the plane elastic problems based on the representations (1.23). In 1932–1933, Muskhelishvili delivered a course of lectures for advanced researchers and PhD students at Seismological and Physical-Mathematical Institutes in Leningrad. The results on the plane elasticity were brought up into systematic form in the first edition of [118] in 1933.

Consider now the double periodic domain consisting of the cells Q_{m_1,m_2} ($m_1, m_2 \in \mathbb{Z}$). Natanzon did not consider the general lattice, but only a rhombus having set $\omega_1 = \overline{\omega_2}$. Such simplifications allowed him to reduce the amount of calculations, performed for the general case later by Filshtinsky. The given external stress tensor yields the double periodic stress tensor in the plane, i.e., the same tensor in each cell Q_{m_1,m_2}. Natanzon extended (1.23) to the considered doubly periodic case having applied the following arguments. According to him it follows from the first formula (1.23) that $\Phi(z)$ is doubly periodic

$$\Phi(z + \omega_1) = \Phi(z), \quad \Phi(z + \omega_2) = \Phi(z). \tag{1.24}$$

The periodicity of the second combination (1.23) yields the jump conditions per cell for the second function

$$\Psi(z + \omega_1) = \Psi(z) - \overline{\omega_1}\Phi'(z), \quad \Psi(z + \omega_2) = \Psi(z) - \overline{\omega_2}\Phi'(z). \tag{1.25}$$

Filshtinsky explained later [66, p. 11] that the conditions (1.24)–(1.25) imply the quasiperiodicity of displacement, i.e., the displacement has constant jumps per periodicity cell. Perhaps, Natanzon was the first who understood that the external stress tensor has to be given as the averaged stress tensor per cell instead of stresses at infinity. This observation leads to the proper constructive homogenization with the given averaged stress tensor. Later on, Filshtinsky properly calculated the averaged strain tensor.

The function $\Phi(z)$ is expanded into the Laurent type series for doubly periodic functions

$$\Phi(z) = A + \sum_{k=0}^{\infty} \frac{A_k}{k!} \wp^{(k)}(z) \tag{1.26}$$

16 Mechanics and Physics of Structured Media

with undetermined coefficients A, A_k. The higher order derivatives of the Weierstrass function are expressed through $\wp(z)$ and $\wp'(z)$ by algebraic expressions [1].

The coefficients A_{2l+1} vanish due to symmetry of the problem. To obtain the similar representation for $\Psi(z)$ we follow Natanzon. Using (1.14) we find

$$\overline{z}\Phi'(z) = -2A_0\left[\frac{\overline{z}}{z^3} + {\sum_P}'\frac{\overline{z}}{(z-P)^3}\right] \tag{1.27}$$

$$-2\sum_{k=1}^{\infty}(k+1)A_{2k}\left[\frac{\overline{z}}{z^{2k+3}} + {\sum_P}'\frac{\overline{z}}{(z-P)^{2k+3}}\right].$$

Let us introduce an auxiliary double periodic function analytic in G

$$\Phi_1(z) = 2A_0{\sum_P}'\left[\frac{\overline{P}}{(z-P)^3} + \frac{\overline{P}}{P^3}\right] + 2\sum_{k=1}^{\infty}(k+1)A_{2k}{\sum_P}'\frac{\overline{P}}{(z-P)^{2k+3}}, \tag{1.28}$$

and introduce the function similar to the Weierstrass function (1.14)

$$\wp_1'(z) = -2{\sum_P}'\left[\frac{\overline{P}}{(z-P)^3} + \frac{\overline{P}}{P^3}\right]. \tag{1.29}$$

More precisely, we introduce the derivative of the function $\wp_1(z)$, introduced later by Filshtinsky. This function is meromorphic in $\mathbb{C}$ and has the poles at $z = P$ except $z = 0$. All the sums considered by Natanzon converge absolutely and almost uniformly in $\mathbb{C}\backslash\mathcal{Q}$. Hence, the term by term differentiation is justified. This implies that the representation (1.28) can be written in the form

$$\Phi_1(z) = -\sum_{k=0}^{\infty}\frac{A_{2k}}{(2k+1)!}\wp_1^{(2k+1)}(z). \tag{1.30}$$

It can be checked that the function $\Psi(z) - \Phi_1(z)$ is doubly periodic. Hence, $\Psi(z)$ can be represented in the form

$$\Psi(z) = \Phi_1(z) + B + \sum_{k=0}^{\infty}\frac{B_{2k}}{(2k+1)!}\wp^{(2k)}(z) \tag{1.31}$$

with undetermined coefficients B, B_{2k}.

Using (1.27) and (1.30) we obtain

$$\overline{z}\Phi'(z) + \Phi_1(z) = -2A_0\left\{\frac{\overline{z}}{z^3} + {\sum_P}'\left[\frac{\overline{z}-\overline{P}}{(z-P)^3} - \frac{\overline{P}}{P^3}\right]\right\} \tag{1.32}$$

$$-2\sum_{k=1}^{\infty}(k+1)A_{2k}\left[\frac{\overline{z}}{z^{2k+3}} + {\sum_P}'\frac{\overline{z}-\overline{P}}{(z-P)^{2k+3}}\right].$$

The latter function depends on z and $\bar{z}$. Hence, it is in general not analytic, but double periodic. Therefore, the function $\bar{z}\Phi'(z) + \Phi_1(z) + \Psi(z) - \Phi_1(z)$ is double periodic. This yields the conditions (1.25).

Thus, the functions $\Phi(z)$ and $\Psi(z)$ are represented by the series (1.26) and (1.30)–(1.31), respectively. The function $\Phi(z)$ is double periodic and analytic in G. The function $\Psi(z)$ is not analytic and satisfies the jump conditions (1.25).

The condition of vanishing stresses on the boundary circle $|z| = r$, can be written, for instance, in the Kolosov's form [118]

$$\Phi(z) + \overline{\Phi(z)} - e^{2i\theta}[\bar{z}\Phi'(z) + \Psi(z)] = 0, \quad |z| = r, \tag{1.33}$$

where $z = re^{i\theta}$.

Substitute the series (1.26) and (1.30)–(1.31) into (1.33), replace $\bar{z}$ by $\frac{r^2}{z}$ and $e^{2i\theta}$ by $\frac{z^2}{r^2}$. Selecting the coefficients with the same powers of z we arrive at an infinite system of linear algebraic equations on the coefficients A, A_{2k}, B, B_{2k}. Three real equations for the averaged stress tensor over the cell should be added. The obtained infinite system is solved by truncation. Besides the Rayleigh lattice sums (1.6), the coefficients of the system depend on the sums

$$S_m^{(j)} = {\sum_{m_1,m_2}}' \frac{(m_1\omega_1 + im_2\omega_2)^j}{(m_1\omega_1 + im_2\omega_2)^m} \equiv {\sum_P}' \frac{\bar{P}^j}{P^m}, \quad j, m \in \mathbb{Z}. \tag{1.34}$$

The lattice sums (1.34) with $j + 2 \leq m$ ($m = 3, 4, \ldots$) enter Natanzon's system. Just outlined result from [119] is the next crucial step made after [126]. It extends the method of Rayleigh to the elastic plane problem.

Maybe, if the paper [119] was prepared in our computerized time, the computation of slowly convergent sums (1.34) would not constitute a problem. But in order to construct a numerical example in the year 1935, Natanzon had to invent the method of their fast computation, alike to Weierstrass' method summarized in [55, Section B.3.3]. His brilliant idea is presented below.

Let $\zeta(z)$ denote the Weierstrass ζ-function for which

$$\zeta(z + \omega_j) - \zeta(z) = \delta_j \quad (j = 1, 2), \tag{1.35}$$

where the constants $\delta_j = 2\zeta\left(\frac{\omega_j}{2}\right)$ satisfy the Legendre identity

$$\delta_1\omega_2 - \delta_2\omega_1 = 2\pi i. \tag{1.36}$$

Having used the series (1.14) and (1.29) Natanzon checked the identities

$$\wp'_1(z + \omega_1) = \wp'_1(z) - \overline{\omega_1}\wp'(z), \quad \wp'_1(z + \omega_2) = \wp'_1(z) - \overline{\omega_2}\wp'(z). \tag{1.37}$$

Consider the function

$$f(z) = \alpha\wp'_1(z) + \beta z\wp'(z) - \zeta(z)\wp'(z), \tag{1.38}$$

18 Mechanics and Physics of Structured Media

where α and β are constants. Calculate the jumps per cell

$$
\begin{aligned}
f(z+\omega_1) - f(z) &= (\alpha\overline{\omega_1} + \beta\omega_1 - \delta_1)\wp'(z), \\
f(z+\omega_2) - f(z) &= (\alpha\overline{\omega_2} + \beta\omega_2 - \delta_2)\wp'(z).
\end{aligned} \tag{1.39}
$$

The function $f(z)$ is doubly periodic if α and β satisfy the uniquely resolved system of equations

$$
\alpha\overline{\omega_1} + \beta\omega_1 - \delta_1 = 0, \quad \alpha\overline{\omega_2} + \beta\omega_2 - \delta_2 = 0. \tag{1.40}
$$

One can see that in this case the function $f(z)$ is an even elliptic function having one pole at zero, where its principal part is equal to $\frac{2}{z^4} - \frac{2\beta}{z^2}$. The Liouville theorem on torus implies that

$$
f(z) = \frac{1}{3}\wp''(z) - 2\beta\wp(z) + c, \tag{1.41}
$$

where c is a constant. Substitution of (1.38) into (1.41) yields an expression for $\wp_1'(z)$ in terms of the Weierstrass elliptic functions. The constant c was also found.

1.2.5 Contribution by Filshtinsky

The first influential papers by Filshtinsky were published in 1961 [26] and in 1964–1967 [27–30,57], and summarized afterwards in his books [60,66]. Filshtinsky developed and extended Natanzon's method to an arbitrary double periodic cell. The paper [27] is devoted to the steady heat conduction problem for a double periodic array of circular holes, i.e., to the problem considered by Rayleigh for the square array.

Filshtinsky extended the Natanzon representations of the complex potentials (1.26), (1.31) to an arbitrary cell

$$
\varphi(z) = Az - \sum_{k=0}^{\infty} A_k \frac{\zeta^{(k)}(z)}{k!}, \tag{1.42}
$$

$$
\psi(z) = Bz - \sum_{k=0}^{\infty} B_k \frac{\zeta^{(k)}}{k!} - \sum_{k=0}^{\infty} A_k \frac{\wp_1^{(k)}(z)}{k!},
$$

where A, A_k, B, B_k are undetermined complex constants. The derivative of the complex potentials (1.42) coincides with Natanzon's complex potentials $\Phi(z) = \varphi'(z)$ and $\Psi(z) = \psi'(z)$.

First, Filshtinsky (1964) considered the conductivity and elasticity boundary value problems for perforated domains. Later, Filshtinsky extended his result [27] to an arbitrary double periodic array of inclusions in the books [60,66].

The representation (1.42) of the potential $\varphi(z)$ used for the conductivity problem assumes its quasiperiodicity and reduces the set of undetermined constants A_l, B_l, C_l in the series (1.15) used by Rayleigh [126].

Filshtinsky considered various plane elastic problems including the antiplane statement with circular inclusions, equivalent to the heat conduction problem, as a particular example. The effective constants were calculated analytically and numerically. One can consider Filshtinsky's paper [27] and the further developments [60,66] as an extension of Rayleigh's result to an arbitrary double periodic array of circular conductive and elastic inclusions. Mind that Rayleigh also considered the cubic array of spheres.

Filshtinsky did not refer neither to Eisenstein nor to Rayleigh before the year 2014, and was not aware that beginning from the year 1978, Ross McPhedran et al. further developed the method of Rayleigh. Vice versa, McPhedran et al., did not cite Natanzon and Filshtinsky. One can think that such phenomenon of mutual noncitation of the truly relevant works is the consequence of the now almost abandoned Iron Curtain. It was sometimes partial, sometimes total, hindering the development of Science for the bulk of XX century. We strongly follow history, i.e., follow the time line passing through the Eisenstein-Rayleigh-Natanzon-Filshtinsky. McPhedran's group branches out of the Rayleigh vertex, independently and successfully complementing the time line.

If we add a mystery voile rather than the curtain existing between Moscow and the rest of the Soviet Union, we may understand problems of local scientists. The geography of the Soviet Union was endowed by the topology in which Moscow was connected to the rest of the world, and other lands like Belarus and Ukraine were situated not to the west of Moscow but rather somewhere to the east in the Siberian forests.

There are publications where the representation (1.42) of complex potentials for double periodic problems is ascribed mistakenly to B.E. Pobedrya & V.A. Mol'kov (1984), or to Balagurov & Kashin (2001), and Yu. Godin (2012, 2013) and to others, who just reproduced some part of Filshtinsky works [27,60,66]. One can find an extended discussion with corresponding references in [6].

We continue with the functions introduced by Filshtinsky

$$\wp_i(z) = \sum_P{}' \left[\frac{\overline{P}^i}{(z-P)^2} - \sum_{s=0}^i (s+1) \frac{\overline{P}^i z^s}{P^{s+2}} \right], \quad i = 0, 1, 2, \ldots. \quad (1.43)$$

The series (1.43) converges absolutely and almost uniformly in $\mathbb{C}\backslash\mathcal{Q}$, since the absolute values of the terms are asymptotically equivalent to $|P|^{-3}$. The function $\wp_0(z)$ is the Weierstrass function (1.14); $\wp_1'(z)$ is Natanzon's function (1.29). Filshtinsky established that the derivatives of (1.43) for $k \geq i+1$ are represented by the series

$$\wp_i^{(k)}(z) = (-1)^k (k+1)! \sum_P{}' \frac{\overline{P}^i}{(z-P)^{k+2}}, \quad i = 0, 1, 2, \ldots \quad (1.44)$$

20 Mechanics and Physics of Structured Media

and satisfy the jump conditions per cell

$$\wp_i^{(k)}(z+\omega_j) - \wp_i^{(k)}(z) = \sum_{s=1}^{i} C_i^s \overline{\omega_j}^s \wp_{i-s}^{(k)}(z), \qquad (1.45)$$

where $C_i^s = \frac{i!}{s!(i-s)!}$ denotes the binomial coefficient, and $j=1,2,\,k \geq i+1$, $i=1,2,\ldots$. The proof of (1.45) is based on (1.44) with subsequent substitution of $z+\omega_j$ into (1.44) and summation of the obtained expressions. The relation (1.45) for $k=i$ is proved separately.

The definition of (1.44) implies the symmetry conditions

$$\wp_i^{(k)}(-z) = (-1)^{i+k}\wp_i^{(k)}(z). \qquad (1.46)$$

After substitution $z = \frac{1}{2}\omega_j$, we arrive at the expressions

$$2\wp_i^{(2k+i+1)}\left(\frac{\omega_j}{2}\right) = \sum_{s=1}^{i}(-1)^{s+1} C_i^s \overline{\omega_j}^s \wp_{i-s}^{(2k+i+1)}\left(\frac{\omega_j}{2}\right), \qquad (1.47)$$

where $j=1,2,\,k=0,1,\ldots,\,i=1,2,\ldots$.

In particular,

$$2\wp_1^{(2k)}\left(\frac{\omega_j}{2}\right) = \overline{\omega_j}\wp^{(2k)}\left(\frac{\omega_j}{2}\right), \quad k=1,2\ldots\ (j=1,2). \qquad (1.48)$$

One can see that the values $\wp_1^{(2k)}\left(\frac{\omega_j}{2}\right)$ are found in terms of the derivatives of Weierstrass function $\wp(z)$ at the points $z = \frac{1}{2}\omega_j$. It turns out that the values $\wp^{(2k)}\left(\frac{\omega_j}{2}\right)$ can be expressed by algebraic relations through the fundamental values $e_1 = \wp\left(\frac{\omega_1}{2}\right)$, $e_2 = \wp\left(\frac{\omega_1+\omega_2}{2}\right)$, $e_3 = \wp\left(\frac{\omega_2}{2}\right)$ [1,69]. It follows from (1.47) and $\wp'\left(\frac{\omega_1}{2}\right) = \wp'\left(\frac{\omega_1+\omega_2}{2}\right) = \wp'\left(\frac{\omega_2}{2}\right) = 0$, that such representation holds for any $\wp_i^{(2k+i+1)}\left(\frac{\omega_j}{2}\right)$.

The identity analogous to the Legendre identity (1.36) was derived for $\wp_1(z)$. The Laurent series for the functions (1.47) were written exactly in terms of the lattice sums (1.6) and (1.34). One can see that the outlined scheme reminds the classic Weierstrass approach to describe elliptic functions through algebraic relations including elementary functions. Natanzon and Filshtinsky's way is to describe the polyanalytic functions through algebraic relations including the Weierstrass functions.

A function is called polyharmonic in a 2D domain G if it satisfies in G the differential equation

$$\nabla^{2n} u = 0, \qquad (1.49)$$

where ∇^2 is Laplace's operator. Double periodic polyharmonic functions were introduced by Filshtinsky [70] in 1971, intertwined like in a double spiral with the classic elliptic functions and Kolosov-Muskhelishvili formulas on the zeroth loop. The local representation of a complex-valued polyharmonic function in terms of n pairs of analytic functions is well known

$$u_n(z, \bar{z}) = \sum_{k=0}^{n-1} [\bar{z}^k \varphi_k(z) + z^k \psi_k(\bar{z})], \tag{1.50}$$

where $\varphi_k(z)$ is analytic, $\psi_k(\bar{z})$ analytic in $\bar{z}$ in the domain G.

The double periodicity means that

$$u_n(z + \omega_j, \overline{z + \omega_j}) = u_n(z, \bar{z}), \quad j = 1, 2. \tag{1.51}$$

Filshtinsky introduced the double periodic polyharmonic function

$$f_n(z, \bar{z}) = \sum_{k=0}^{n-1} (-1)^k C_{n-1}^k \bar{z}^{n-k-1} \wp_k^{(n-1)}(z) \tag{1.52}$$

and derived the following general representation for the double periodic polyharmonic functions

$$u_n(z, \bar{z}) = \sum_{m=0}^{\infty} \left[\frac{f_n^{(m)}(z, \bar{z})}{(m+n-1)!} A_{m+n-1} + \frac{f_n^{(m)}(\bar{z}, z)}{(m+n-1)!} B_{m+n-1} \right], \tag{1.53}$$

where A_k and B_k are arbitrary constants. The functions represented by (1.50), after using of (1.52), can be presented as the series

$$\varphi_k(z) = (-1)^{n-k-1} C_{n-1}^k \sum_{s=n-1}^{\infty} \frac{A_s}{s!} \wp_s^{(n-k-1)}(z), \tag{1.54}$$

$$\psi_k(\bar{z}) = (-1)^{n-k-1} C_{n-1}^k \sum_{s=n-1}^{\infty} \frac{A_s}{s!} \wp_s^{(n-k-1)}(\bar{z}).$$

The obtained series representations are used in the plane boundary value problem of elasticity. The Riemann boundary value problem ($\mathbb{C}$-linear problem) for doubly periodic polyanalytic functions was solved in [24].

The problem of double periodic coated inclusions deserves a separate discussion. In 1965–1971 Van Fo Fy [135] extended the Natanzon-Filshtinsky approach [27,28,119] to the double periodic arrays of coated inclusions by application of the series (1.42). The problem of antiplane deformation (torsion) which is equivalent to conductivity problem, and the plane elastic problem were all reduced to an infinite system of linear algebraic equations. N.A. Nicorovici, R.C. McPhedran, and G.W. Milton [122] in 1993 extended the Rayleigh method to the square arrays of coated inclusions and reduced the problem to the same infinite system of linear algebraic equations. The eigenvalues for this system were

22 Mechanics and Physics of Structured Media

carefully investigated in [122]. The group of authors produced the effective conductivity in the form suitable for the problem with the complex conductivities of components.

The paper [122] refers to the theoretical foundations of metamaterials. In the last decade the number of publications in the field had passed through the stage of explosive growth, see for instance [101,102,125,139].

The asymptotic analytical formulas for the effective longitudinal shear moduli of the square and hexagonal array were derived by the method of boundary shape perturbation in 2008 [5].

1.3 Synthesis. Retrospective view from the year 2021

We discuss here how Eisenstein's approach complements Natanzon's and Filshtinsky works and how to pass from finite number of inclusions to infinite. We suggest a unified view of the results due to Eisenstein (1847), Rayleigh (1892), Natanzon (1935), and Filshtinsky (1961, 1964).

Consider the problem of the perfect contact similar to (1.2), for a finite number of mutually disjoint disks $|z - a_k| < r$ $(k = 1, 2, \ldots, n)$ on the complex plane. Here, a_k is the complex coordinate of center and r denotes radius. Let ϱ stand for the contrast parameter (1.5). The exact solution to the considered problem for any $|\varrho| < 1$ was found in the form of absolutely and uniformly convergent Poincaré type series [55,104,105]. Let the external flux parallel to the x_1-axis be given at infinity. The complex flux in the matrix, the complement of disks $|z - a_k| \leq r$ $(k = 1, 2, \ldots, n)$ to the extended complex plane, can be approximated by the formula

$$\varphi'(z) = 1 + \sum_{k=1}^{n} \frac{\varrho r^2}{(z - a_k)^2} + \sum_{k=1}^{n} \sum_{k \neq m} \frac{\varrho^2 r^4}{(z - a_k)^2 (a_k - a_m)^2} + O(|\varrho|^3 r^6).$$

$$(1.55)$$

The complex flux in the kth inclusion can be estimated by the expression

$$\varphi_k'(z) = \frac{\sigma_1 + \sigma}{2\sigma} \left[1 + \varrho r^2 \sum_{m \neq k}^{n} \frac{1}{(z - a_m)^2} \right] + O(\varrho^2 r^4), \quad |z - a_k| < r. \quad (1.56)$$

Consider now an infinite regular array of cylinders when the centers $a_k = \omega_1 m_1 + \omega_2 m_2$ are enumerated by the double index $(m_1, m_2) \in \mathbb{Z}^2$. The limit sum in (1.55) is not correctly defined as n tends to infinity, since the obtained series is absolutely divergent. In order to be consistent with Rayleigh we have to apply the Eisenstein summation (1.7). Then, the local complex flux around the inclusions is approximately calculated by means of the Eisenstein function

$$\varphi'(z) = 1 + \varrho r^2 E_2(z) + \varrho^2 r^4 E_2(z) \overline{S_2} + O(|\varrho|^3 r^6). \quad (1.57)$$

Filshtinsky's contribution to Applied Mathematics & Mechanics **Chapter | 1 23**

In the inclusion the complex flux is equal to $\varphi'_k(z) = \frac{\sigma_1+\sigma}{2\sigma}\psi_1(z)$, where the same function as in (1.19) is used. We find after the summation that

$$\psi_1(z) = 1 + \varrho r^2 \left(E_2(z) - \frac{1}{z^2} \right) + O(\varrho^2 r^4), \quad |z| < r. \qquad (1.58)$$

The components of the effective conductivity tensor can be calculated by the Rayleigh formula in the complex form

$$\frac{1}{\sigma}(\sigma_e^{(11)} - i\sigma_e^{(12)}) = 1 + 2\varrho\pi r^2 \psi_1(0). \qquad (1.59)$$

The general formula (1.59) was derived in [55, formula (3.2.44)]. Substituting $z = 0$ into (1.58) and using (1.11) for $E_2(z)$, we get $\psi_1(0) = 1 + \varrho r^2 S_2 + O(\varrho^2 r^4)$ and

$$\frac{1}{\sigma}(\sigma_e^{(11)} - i\sigma_e^{(12)}) = 1 + 2\varrho f + 2\varrho^2 f^2 \frac{S_2}{\pi} + O(\varrho^2 f^3), \qquad (1.60)$$

where $f = \pi r^2$ is the concentration of inclusions.

Consider the square array of cylinders. Using Rayleigh's formula (1.21) we obtain from (1.60) the approximation (1.4) for the macroscopically isotropic composite

$$\frac{\sigma_e}{\sigma} = 1 + 2\varrho f + 2\varrho^2 f^2 + O(\varrho^2 f^3) = \frac{1 + \varrho f}{1 - \varrho f} + O(\varrho^2 f^3). \qquad (1.61)$$

Rayleigh calculated the next terms in (1.61) by using the truncated system (1.17). Besides the Eisenstein lattice series (1.6) Rayleigh could have possibly arrived at the Eisenstein functions (1.8), if only he had tried to determine the local fields. Indeed, Filshtinsky determined the local fields and derived an analytical formula similar to (1.57) for the elastic stress tensor.

As an example, we consider the square array of holes and the given averaged stress tensor $\langle\sigma_{11}\rangle = \langle\sigma_{22}\rangle = 1$, $\langle\sigma_{12}\rangle = 0$. Filshtinsky derived the complex potentials written here in the truncated form

$$\Phi(z) \approx 1 + \pi r^2 \beta_2, \quad \Psi(z) \approx 1 + \pi r^2 \beta_2 \wp(z), \qquad (1.62)$$

where the constant β_2 was calculated numerically from the truncated system of linear algebraic equations, representing an analog of Rayleigh's system for conductivity.

Filshtinsky in addition found the next terms of the approximation (1.62) as a linear combination of $\wp^{(k)}(z)$ and $\wp_1^{(k)}(z)$ with the coefficients determined numerically. The effective elastic tensor was calculated by the numerical averaging of the constructed series. The radius r was considered in a symbolic form and the elastic constants were taken into account numerically.

24 Mechanics and Physics of Structured Media

In order to better understand Filshtinsky's method we write (1.57), formally following his method

$$\varphi'(z) \approx 1 + \pi r^2 \beta_2 + \pi r^2 \beta_2 \wp(z). \tag{1.63}$$

After numerical computations the constant β_2 should be equal to the numerically given contrast parameter ϱ. Now, it is clear that actually Filshtinsky (1964–1970) computed numerically the conditionally convergent series S_2 and $S_3^{(1)}$ from (1.34). The high-order lattice sums (1.6) were calculated by the Weierstrass algebraic formulas [1] and (1.34) by development of Natanzon's approach. An alternative numerical method for computing the sum (1.34) was developed by A.B. Movchan, N.A. Nicorovici, and R.C. McPhedran [117] in 1997.

The numerical experiments led to the final, exact formula $S_3^{(1)} = \frac{\pi}{2}$ for the hexagonal array proved in [140], and to some other fascinating formulas [14] in the theory of lattice sums [11]. See also Chapter [23] in the present book. The above historical arguments enable us to extend the set of functions (1.43)–(1.44) by application of the Eisenstein summation. Authors of the book [22] coined them *Eisenstein-Natanzon-Filshtinsky functions*. This term covers double periodic polyharmonic (polyanalytic) functions.

The name of Rayleigh is omitted here, since he used only the lattice sums in the Eisenstein summation. On the other hand, the term *Eisenstein-Rayleigh-Natanzon-Filshtinsky lattice sums* could be fully justified. The term *lattice sums* is used in literature and seems to be appropriate.

Formally, one can omit the Eisenstein summation of S_2 and $S_3^{(1)}$ in this theory, and calculate properties only from the Weierstrass functions. For instance, one can define the Rayleigh sum as $S_2 = \frac{\delta_1}{\omega_1}$ [107]. But such a technique becomes too complicated when attempted for random composites. Moreover, it can lead to methodologically wrong extensions in the spirit of self-consistent theory.

One can find in the literature a lot of descriptions with illustrations when a finite cluster $\mathcal{C}_n$ consisting of n inclusions is plucked out from periodic or infinite random structure $\mathcal{C}$. Next, the local fields are computed and averaged over the cluster. The averaged constants $c_e^{(n)}$ are calculated and presented as the effective constants. It is tacitly assumed that for sufficiently large n the obtained constants $c_e^{(n)}$ approximate the effective constants c_e of $\mathcal{C}$. However, it is not so. Though the local fields in the interior $\mathcal{C}_n$ and $\mathcal{C}$ may be similar, for any fixed n the constants $c_e^{(n)}$ are the effective constants for clusters $\mathcal{C}_n$ diluted in the host, nothing more. The limit $\mathcal{C}_n \to \mathcal{C}$, as $n \to \infty$, is just an imaginative attribute which in general does not yield $c_e^{(n)} \not\to c_e$. Moreover, $\lim_{n\to\infty} c_e^{(n)}$ does not exist, since an additional charge (in the terminology of electric fields) is induced near the boundary of the cluster changing with the growth of the cluster.

In the case of a square array, such a "methodology" means that one calculates the conditionally divergent Rayleigh sum S_2 by some "self-consistent method"

Filshtinsky's contribution to Applied Mathematics & Mechanics Chapter | 1 **25**

and call this a "novel model".[2] Rayleigh did not explain clearly why he applied the Eisenstein summation. He did not compare his formula with MF (1.4). More precisely, he invented his method, applied it to the square array and saw the same low order term as in MF. Rayleigh did not develop MF as well, just as Maxwell and others recalled after the formula (1.5). At the end of this section, we describe the proper formalism on the example of square arrays. We show that S_2 may be equal to π or equal to 0, depending on the summation method. In the elasticity theory both types of summation should be performed.

We now proceed with the outline of the extension to random composites and, by default, to the square array which can be treated as a probabilistic event in a class of random composites. First, we note that the normalization of the area cell $|Q_{00}| = 1$ is important, since $S_2 = \pi$ only for square and hexagonal arrays and $S_3^{(1)} = \frac{\pi}{2}$ only for hexagonal arrays. The constants α and β from (1.40) can be written in the simple form

$$\alpha = \pi, \quad \beta = S_2 + 2\pi \arg \omega_1. \tag{1.64}$$

The other condition $\omega_1 > 0$, obtained by a rotation, simplifies β. Of course, S_2 changes with the rotation by the multiplier $2 \arg \omega_1$.

Consider now a periodic (representative) cell Q_{00} with many nonoverlapping circular inclusions $|z - a_k| < r$ $(k = 1, 2, \dots, N)$, where N is an arbitrary finite number, and the concentration of inclusions $f = N\pi r^2$. Instead of (1.60) we have [108]

$$\frac{1}{\sigma}(\sigma_e^{(11)} - i\sigma_e^{(12)}) = 1 + 2\varrho f + 2\varrho^2 f^2 \frac{e_2}{\pi} + O(\varrho^2 f^3), \tag{1.65}$$

where

$$e_2 := \frac{1}{N^2} \sum_{k=1}^{N} \sum_{m=1}^{N} E_2(a_k - a_m). \tag{1.66}$$

Here, the Eisenstein function is redefined at zero in such a way that $E_2(a_k - a_m) := S_2$ for $a_m = a_k$. One can see that $e_2 = S_2$ for $N = 1$. Moreover, it follows from (1.65) that the homogenized material is macroscopically isotropic for small f, if $e_2 = \pi$. The same statement holds for the structural sum $e_3^{(1)}$ used in elasticity [22].

This computationally simple criterion of isotropy and its extensions [112] by far supersedes recommended two-point correlation functions in the task of quantifying anisotropy. The study of macroscopic anisotropy by means of e_2 and its application to some technological processes can be found in [78,79] including 3D extensions [129].

---

[2] It is a modern fashion to solve a rigorously stated mathematical problem in the theory of random composites by a doubtful method and call this *solution* a "novel *model*". Nevertheless, some results seem to be reasonable and could be useful in carefully selected special cases.

26 Mechanics and Physics of Structured Media

We now proceed with an outline of Maxwell's self-consistent approach. For shortness consider the square array of disks. Consider a finite cluster of a sufficiently large radius R_0, composed from the square cells $\cup_{(m_1,m_2)\in Z}\ Q_{m_1,m_2}$, where $Z = \{(m_1, m_2) \in \mathbb{Z}^2 : m_1^2 + m_2^2 < R_0^2\}$. Let the disks $|z - a_k| < r$ be linearly numerated by the numbers $k = 1, 2, \ldots, n$, where n is the number of elements of Z. The local concentration of inclusions in the cluster is equal to the concentration of the homogenized medium $f = \pi r^2$.

The important characteristic of the cluster is its dipole moment $\mathcal{M}^{(n)}$ [117], defined as the coefficient of $\varphi'(z)$ on $r^2 z^{-2}$. It follows from (1.55) that

$$\mathcal{M}^{(n)} = \varrho \left(1 + \varrho r^2 e_2^{(n)}\right) + O(r^4), \tag{1.67}$$

where

$$e_2^{(n)} = \sum_{k=1}^{n} \sum_{m \neq k} \frac{1}{(a_k - a_m)^2}. \tag{1.68}$$

Maxwell's homogenization insists that the dipole moment of the cluster is equal to the dipole moment of the homogenized square array

$$f\mathcal{M}^{(n)} = \frac{\sigma_e^{(n)} - \sigma}{\sigma_e^{(n)} + \sigma}. \tag{1.69}$$

Here, $\sigma_e^{(n)}$ is the averaged effective conductivity of the cluster approximated by the sufficiently large disk $|z| < R_0$ of conductivity $\sigma_e^{(n)}$. Substitute then (1.67) into (1.69) and find that

$$\frac{\sigma_e^{(n)}}{\sigma} = \frac{1 + \varrho f \left(1 + \varrho r^2 e_2^{(n)}\right)}{1 - \varrho f \left(1 + \varrho r^2 e_2^{(n)}\right)} + O(f^3). \tag{1.70}$$

The sum $e_2^{(n)}$ tends to S_2, as $n \to \infty$, and (1.70) yields

$$\frac{\sigma_e^{(n)}}{\sigma} = \frac{1 + \varrho f \left(1 + \varrho f \frac{S_2}{\pi}\right)}{1 - \varrho f \left(1 + \varrho f \frac{S_2}{\pi}\right)} + O(f^3) = 1 + 2\varrho f + 2\varrho^2 f^2 \left(1 + \frac{S_2}{\pi}\right) + O(f^3). \tag{1.71}$$

Therefore, we are bound to accept such method of summation for which $S_2 = 0$, in order to be in a complete agreement with (1.61).

The term $O(f^2)$ includes the conditionally convergent sum e_2. The next terms in the effective conductivity of random composites depend on the multiple structural sums computed in [54].

1.4 Filshtinsky's contribution to the theory of magneto-electro-elasticity

Section 1.4 is concerned with 2D equations of magneto-electro-elasticity. This is the last problem extensively investigated by Filshtinsky. He published a few papers on piezoelectricity with his PhD students L.V. Shramko and D.M. Nosov [8,48,51] and prepared the tutorial in Russian [50], published as a preprint. The separate chapters "Cracks in two-dimensional magneto-electro-elastic composites" and "Two-dimensional equations of magneto-electro-elasticity" of the present book are based on [50] and other Filshtinsky's works.

Interacting physical fields are called coupled. Typical coupled fields are magnetic, electric and elastic discussed in [71]. Other examples of coupled fields encompass thermoelasticity [121], magnetohydrodynamics, magnetogasdynamics, electromechanics [134], piezo-optic and elasto-optic effects, pyroelectricity, etc. [82].

The main static equations of magneto-electro-elastic (MEE) fields follow from the Maxwell's and elasticity equations [7]. The electric field $\mathbf{E}$ and the displacement caused by application of the electric field $\mathbf{D}$, satisfy the following equations

$$\nabla \times \mathbf{E} = \mathbf{0}, \quad \nabla \cdot \mathbf{D} = 0. \tag{1.72}$$

The magnetic induction $\mathbf{B}$ and displacement $\mathbf{H}$ satisfy the same equations

$$\nabla \times \mathbf{H} = \mathbf{0}, \quad \nabla \cdot \mathbf{B} = 0. \tag{1.73}$$

The stress equilibrium equations for the stress tensor $\boldsymbol{\sigma}$ in the absence of body forces have the form

$$\nabla \cdot \boldsymbol{\sigma} = \mathbf{0}. \tag{1.74}$$

For small deformations the strain tensor $\boldsymbol{\epsilon}$ is expressed by means of the displacement $\mathbf{u}$ as follows

$$\boldsymbol{\epsilon} = \frac{1}{2}(\nabla \mathbf{u} + \nabla \mathbf{u}^T). \tag{1.75}$$

The linear constitutive equations for the coupled fields are written in the following form:

$$\begin{pmatrix} \boldsymbol{\epsilon} \\ \mathbf{E} \\ \mathbf{H} \end{pmatrix} = \begin{pmatrix} \mathcal{S} & \mathbf{g} & \mathbf{p} \\ -\mathbf{g}^T & \beta & \nu \\ -\mathbf{p}^T & \nu^T & \chi \end{pmatrix} \begin{pmatrix} \boldsymbol{\sigma} \\ \mathbf{D} \\ \mathbf{B} \end{pmatrix}. \tag{1.76}$$

Here, $\mathcal{S}$ denotes the compliance tensor which relates the strain and stress tensors in the case of pure elastic deformation. This leads to Hooke's law

$$\boldsymbol{\epsilon} = \mathcal{S}\boldsymbol{\sigma}. \tag{1.77}$$

28 Mechanics and Physics of Structured Media

The tensors $\boldsymbol{\beta}$, $\boldsymbol{v}$, and $\boldsymbol{\chi}$ describe dielectric, magnetic, and electromagnetic material properties; $\mathbf{g}$ and $\mathbf{p}$ correspond to the piezoelectric and piezomagnetic material properties.

The formally written Eq. (1.76) requires explanations concerning the different tensors rank. The strain and stress tensors $\boldsymbol{\epsilon}$ and $\boldsymbol{\sigma}$ are of rank 2; the tensor S is of rank 4. One can consider the tensor S as a linear transformation between the components of rank-2 tensors $\boldsymbol{\epsilon}$ and $\boldsymbol{\sigma}$. This observation leads to the vector-matrix representation of Hooke's law (1.77). The symmetric rank-2 tensors having 6 independent components are written as six-dimensional vectors by their linear ordering, e.g., $(\sigma_{11}, \sigma_{12}, \ldots, \sigma_{33}) = (\sigma_1, \sigma_2, \ldots, \sigma_6)$. In the same way the tensor S can be written as a matrix of dimension 6×6. There are different ways to perform such an operation, see for details [136].

For the coupled rank-2 tensors $\boldsymbol{\beta}$, $\boldsymbol{v}$, and $\boldsymbol{\chi}$ matrix representations are available. They relate the electromagnetic fields expressed by vectors. Therefore, the electromagnetic interactions are directly related by the standard vector-matrix relations, as expressed in the corresponding part of Eq. (1.76). The tensors $\mathbf{g}$ and $\mathbf{p}$ relate rank-2 elastic tensors $\boldsymbol{\epsilon}$, $\boldsymbol{\sigma}$ and rank-1 electric and magnetic tensors (vectors). Hence, the tensors $\mathbf{g}$ and $\mathbf{p}$ have the rank 3 and can be written as matrices of dimension 3×6.

The constitutive equations (1.76) can be considered as a linear approximation of the interactions between the MEE fields in anisotropic media. The representations outlined above need profound justifications on the level of tensor analysis and frequently lead to unusual properties of the coupled constituent tensors.

We now leave the general theory and descend to the 2D pure elasticity problem for anisotropic materials in the stationary case. It is based on the seminal paper by S.G. Lekhnitskii [84] published in 1939 in Russian and his book [85]. S.G. Lekhnitskii extended the complex potentials method [118] applied before only to isotropic elastic media. It is worth noting that an elegant interpretation of Lekhnitskii's approach was proposed by A.N. Stroh in 1958 and summarized later by T.C.T. Ting [133]. The stresses in the domain G are expressed in terms of two analytic functions by Kolosov-Muskhelishvili formulas (1.23). Analogous Lekhnitskii's representation for anisotropic media has the form

$$\sigma_{11} = 2\text{Re}\,[\mu_1^2 \Phi_1(z_1) + \mu_2^2 \Phi_2(z_2)], \quad \sigma_{22} = 2\text{Re}\,[\Phi_1(z_1) + \Phi_2(z_2)],$$
$$\sigma_{12} = 2\text{Re}\,[\mu_1 \Phi_1(z_1) + \mu_2 \Phi_2(z_2)], \tag{1.78}$$

where μ_1 and μ_2 are complex roots with positive imaginary parts of Lekhnitskii's polynomial equation of fourth order $\ell_4(\mu) = 0$ and $z_k = x_1 + \mu_k x_2$ ($k = 1, 2$). Analogous formulas hold for the strain tensor. The real coefficients of equation $\ell_4(\mu) = 0$ can be exactly written in terms of the components of the compliance tensor S [85]. The representations (1.78) hold for different roots μ_1 and μ_2. If $\mu_1 = \mu_2$, we arrive at the Kolosov-Muskhelishvili formulas (1.23).

Filshtinsky's contribution to Applied Mathematics & Mechanics Chapter | 1 **29**

Therefore, the isotropic elastic materials refer to a degenerate class of all materials. The degenerate classes are not limited to isotropic materials [133].

The anti-plane deformation of anisotropic materials can be considered separately by introduction of the third complex potential $\Phi_3(z_3)$ where $z_3 = x_1 + \mu_3 x_2$, where μ_3 satisfies the quadratic equation [85]. The same formalism holds for the 2D anisotropic conductivity [66, Chapter 5].

The first step in the extension of Lekhnitskii's formalism to the 2D piezoelectric materials was made by A.S. Kosmodamianskii and V.N. Lozhkin in 1975 [75], where the coupled piezoelectric fields were represented in terms of three analytic functions similar to (1.78). B.A. Kudryavtsev et al., in 1975 investigated a crack on the boundary of a two-phase piezoelectrics [77]. These papers deliver broad studies of the plane problems with holes, cracks and inclusions in piezoelectrics. First works were devoted to particular boundary value problems such as problems for an elliptical hole, for a point source in half-plane, etc. Beginning with the year 1979, Filshtinsky became one of the key figures in the field, see [9] and [47]. He extended his own method of integral equations to general problems of plane piezoelectricity. Due to his contribution, the theory of static two-dimensional piezoelectrics became a part of the classic theory of boundary value problems for analytic functions, see fundamental book [67] and the short survey [49].

Regretfully, hundreds of valuable papers and books of this Gold Rush era of Soviet mechanics were published in local journals. The insufficient paper supply in the Soviet Union after 1985 led to reduction of the volumes mainly of local journals. Only a brief summary was published with the complete manuscript collected in Moscow and called the deposited publication (something like ArXiv without Internet). It was possible to acquire a copy having waited a couple of months. The situation smoothly changed with the advent of a new century due to the total reduction of scientific work, emigration of scientists and possibility to publish abroad.

Filshtinsky took advantage of the latter option, by publishing [43] and [8]. The book [7] summarizes the above and many others results concerning the crack problems for 2D piezoelectric media. Besides the static problem, the major part of the book [7] is devoted to dynamic problems. The book is rich in interesting examples demonstrating the complicated physics of piezoelectricity. Though magnetism was also discussed in connection with elasticity, the mathematical theory of coupled electro-magneto-elasticity was not developed. In a separate book [50] Filshtinsky extended the static problem considered in [7], to the full electro-magneto-elastic equations. This book was considered by him as the general plan to develop his method of integral equations to the coupled problems. Partially, this plan begins to realize by Dmytro Nosov, PhD student of Filshtinsky and Mityushev, whose thesis is based on Chapters "Cracks in two-dimensional magneto-electro-elastic composites" and "Two-dimensional equations of magneto-electro-elasticity" of the present book, and [128]. Dmytro Nosov improved Filshtinsky's codes and extended them to electro-magneto-

30 Mechanics and Physics of Structured Media

elasticity. Some corrections of the numerical schemes were also performed by him.

1.5 Filshtinsky's contribution to the homogenization theory

Let's comment on one of the most impressive aspects of Filshtinsky's activities. He began his career as an aircraft engineer (Natanzon also worked for the aircraft industry). When studying aircraft construction, one has to deal with periodically inhomogeneous plates and shells – ribbed, corrugated, perforated, multisupported, riveted, etc. Filshtinsky, having a good mathematical background, a prerequisite for Soviet technical universities (the level which current reformers consider excessive), quickly understood that the complex analysis, and especially the Weierstrass double-periodic functions, is adequate to the problems at hand.

The results obtained by Filshtinsky using these approaches are described above in this article. However, they were of little use for practical purposes. Before the computer revolution and wide use of finite element methods, the engineer relied on simple calculation formulas (they are still important to the engineers of today, especially in optimal design). At that time, the standard method for obtaining such formulas was to reduce the original problem for periodically heterogeneous plates and shells to orthotropic or anisotropic structures ("structurally-orthotropic theory" in Russian terminology or "smeared stiffener theory" in Western terminology).

The main problem here is to obtain analytically the so-called effective parameters. Anisotropic structure with such parameters is equivalent to the original heterogeneous structure in certain sense, e.g., has the same elastic energy or maximum displacements. Such an approximation was mathematically justified within the framework of homogenization theory [96]. It is important that the schemes of structural orthotropy used by engineers (see, e.g., a lot of approximations made in the same spirit in [132]) turned out to be the first approximations to some asymptotic processes. Filshtinsky was well aware that engineers would not be able to use the complicated solutions he received. Therefore, he paid considerable attention to the approximate determination of the effective parameters, which would be described by simple formulas.

He used the following technique. Suppose we have a "formally exact solution" [6], which reduces to an infinite system of linear algebraic equations. Let the regularity of such a system be proved, i.e. the possibility of its approximate reduction with controlled accuracy to a finite system of equations [6]. Then we can estimate how many equations should be left in the system in order to obtain a solution with a given accuracy, e.g. 5%, for a practically important range of parameters. Filshtinsky used this approach to obtain the effective parameters.

Consider an excellent book [60], which can serve as an example of a monograph that is significant both from a theoretical and an engineering point of view. Along with purely mathematical results, which have not lost their importance in

Filshtinsky's contribution to Applied Mathematics & Mechanics Chapter | 1 **31**

our time, each section of this book contains a paragraph devoted to the analytical expressions of the effective parameters, as well as a summary of numerical results. More than 200 pages of the 556-page book are tables and graphs, important for engineers of that time (the book was published in 1970, and the results presented in it were obtained much earlier).

Of course, there is no need now for such a representation of numerical data, but the analytical expressions of the effective parameters have not lost their value. In modern homogenization theory, the main mathematical tools are multiple scales asymptotic approaches or Bakhvalov's ansatz [96]. Filshtinsky's approach is close to the Fourier homogenization method [12]. Thus, Filshtinsky made a certain contribution to the homogenization theory. As he noted, these results were not rigorous from a mathematical point of view. However, let us recall the words of the famous mathematician Littlewood about the work of not less famous and even a little mysterious Ramanujan: "He had no strict logical justification for his operations. He was not interested in rigor, which for that matter is not of first-rate importance in analysis beyond the undergraduate stage, and can be supplied, given a real idea, by any competent professional", see p. 88 of [87].

Based on the above, we would like to emphasize the importance of practical problems for the development of mathematics. From this point of view, for academics in the field of mechanics the close links with industry, especially with such high-tech industries as aircraft, rocketry, shipbuilding, could be fruitful. Looking at the current dominance of formally correct, but useless from a theoretical and practical point of view articles, one cannot but recall the warning of J. Neumann: "When a branch of mathematics deviates far enough from its empirical source and keeps being inspired only indirectly by ideas which come from reality, it is subjected to a very serious danger. It becomes more and more alike an aimless exercise in aesthetics and art for art.... On this stage the only remedy, in my opinion, is to return to the source and to inject empirical ideas more or less directly" [120].

On the other hand, "Everything should be made as simple as possible, but no simpler" (saying attributed to Einstein). These words of caution can be applied to a naive empirical approach discussed in Section 1.3, where it is explained that the extension of a relatively simple Maxwell's approach is far from being naive and should be based on the deep insights due to Eisenstein, Rayleigh, Natanzon and Filshtinsky. Fortunately, theories of modern composite, functionally-graded, active materials, as well as nanomaterials and metamaterials, create a lot of new interesting mathematical problems [2].

The development of homogenization theory made it possible to evaluate accuracy of the results obtained by Filshtinsky for the effective characteristics of perforated plates [3,4,76]. For small and medium-sized holes, the effective parameters obtained by homogenization, practically coincide with those obtained by Filshtinsky. For large hole sizes, the homogenization approach and singular perturbation asymptotics allows us to obtain more accurate solutions. In addition, homogenization theory allows to calculate the rapidly oscillating corrector

32 Mechanics and Physics of Structured Media

needed to correctly determine the total stress-strain state of the perforated plates and shells. One of us (Andrianov) had a pleasure to discuss these issues directly with Filshtinsky after the presentation at his seminar, and complete understanding was reached which warranted a subsequent celebration, so much liked by Filshtinsky!

1.6 Filshtinsky's contribution to the theory of shells

During his work in the aircraft industry Filshtinsky had engineered not only the perforated plates and shells [31,59,62], but also many other constructions with periodically repeating elements. Among them there were plates and shells with cracks and cuts [36,39,88], regular piece-inhomogeneous structures [19,64,65], ribbed plates, shells, and disks [20,21,45,46,90,91], composite structures [68], plates on the multipoint supports [56,58]. By employing the doubly periodic functions, in particular the Weierstrass elliptic functions, he found integral representations of the solutions and reduced the original problem to a singular integral or integro-differential equation, or a system of Fredholm integral equations of the second kind.

While solving boundary value problems from the theory of shells it is necessary to have a set of special solutions of the corresponding equations. The set can constitute a fundamental solution or a complete system of particular solutions adapted for the considered domain. Integral representations of the solutions in the theory of shallow shells can be obtained provided that the coefficients of these equations are analytic functions of coordinates. The kernels included in the integral representations are given by the equations of the Volterra type in a two-dimensional domain. In [32,35], a general solution of the equations of the applied theory of shallow shells was obtained, and the kernels included in the integral representations were written explicitly.

Using representations similar to the Kolosov-Muskhelishvili formulas made it possible to derive the series solutions of the boundary value problems for shallow shells, for simply and multiply connected domains [33]. On the basis of the integral representations of the solutions to the equations of the theory of shallow shells, Filshtinsky found the systems of regular solutions which are complete with respect to an arbitrary simply connected domain on the surface of the shell [63], as well as the fundamental solutions of the shallow shell theory equations [34,61].

Several works [40,44,89,94] are devoted to the study of the response of anisotropic shells to static and dynamic actions concentrated at the points or along the lines. Fundamental periodical solutions of both static and dynamic equations of the theory of anisotropic shallow shells were constructed.

Filshtinsky then formulated the theory of shallow piezoceramic shells polarized along the generatrices. Corresponding equations are used to construct the Green's matrix and to investigate the response of an infinite and a finite shell to concentrated perturbations. The asymptotic solution is written down for the cases of mechanical forces and moments as well as for the electrical field potential generated in the neighborhood of the points where the load is applied [41].

Much attention in the works of Filshtinsky was paid to such important issues in aircraft construction as the contact problems of the theory of elasticity [93], as well as to the problems of transmitting the load from elastic ribs to anisotropic plates and shells. Important for application case of a rigid connection of the ribs and skin [10,37,38,42,92,95], with or without an adhesive layer were investigated. The functionals that ensure the existence of the solutions to the equations lead to the systems of Fredholm integral equations of the second kind, and the conjugation condition lead to a singular integro-differential equation. The solution was represented in the form of a series. Their convergence was thoroughly investigated. The problem of reinforcement with equal strength stringers was considered as well [92].

The problem of the rib coming out on the boundary of the plate or shell was studied in detail. In particular, the interaction of an anisotropic semiinfinite plate with an elastic rib located at an angle to the boundary was considered. In this case, the order of emerging singularity of contact forces at the end of the rib depends on the nature of the anisotropy and the angle of inclination of the rib. The problem was reduced to a singular integro-differential equation, which was solved numerically [95]. In all instances just discussed, the solution to the problem was given in the sufficiently accurate numerical form. Reduction of various problems to the integral equations or to infinite systems of linear algebraic equations was a distinct trend in applied mathematics of the 20th century. Various advanced studies often ended with such reduction, and it was taken for granted that the equations could somehow be solved by someone else. Many mathematicians even believed that the integral equation was in fact the solution to the problem.

Contrary to the trend, Filshtinsky developed robust numerical algorithms for solving integral equations and applied asymptotic methods to calculate stress intensity factors and other singularities brought up by the solutions. Multiple numerical examples could be found in Filshtinsky's articles. They are well thought off and systematized to the high degree of making it possible for engineers to understand completely the process behind their practical needs.

1.7 Decent and creative endeavor

Life of Filshtinsky never was a sugar tale. But the Professor, or Chef, spent most of his time in his own world of science. How to convey the image of Filshtinsky? We feel that Filshtinsky should be remembered as described by Boris Pasternak:
"In scarf, with hand before my eyes,
I'll shout outdoors and ask the kids:
Oh tell me, dear ones, if you please,
Just what millennium this is?

Who beat a pathway to my door,
That hole all blocked with snow,
While I with Byron had a smoke
And drank with Edgar Poe?" ∎

Acknowledgment

The authors are deeply indebted to Dr. Heiko Topol for his useful comments in the course of the paper preparation. We are grateful to Dr. Elena Ogloblina and Dr. Tatiana Sushko, longtime collaborators of Filshtinsky, for all their help with photos and incredible support. Mityushev seizes the occasion to thank Peter Yuditskii (Petro Yudytskiy) for the nice workshop at Linz, Johannes Kepler University, May 12–13, 2014.

References

[1] N.I. Akhiezer, Elements of the Theory of Elliptic Functions, American Mathematical Society, Providence, R.I., 1990 (first Russian edition 1948).

[2] I.V. Andrianov, J. Awrejcewicz, V.V. Danishevskyy, Linear and Nonlinear Waves in Microstructured Solids: Homogenization and Asymptotic Approaches, Taylor & Francis, 2021.

[3] I.V. Andrianov, V.V. Danishevs'kyy, A.L. Kalamkarov, Asymptotic analysis of perforated plates and membranes. Part 1: static problems for small holes, International Journal of Solids and Structures 49 (2012) 298–310.

[4] I.V. Andrianov, V.V. Danishevs'kyy, A.L. Kalamkarov, Asymptotic analysis of perforated plates and membranes. Part 2: static and dynamic problems for large holes, International Journal of Solids and Structures 49 (2012) 311–317.

[5] I.V. Andrianov, V.V. Danishevs'kyy, A.L. Kalamkarov, Micromechanical analysis of fiber-reinforced composites on account of influence of fiber coatings, Composites. Part B, Engineering 39 (2008) 874–881.

[6] I. Andrianov, V. Mityushev, Exact and "exact" formulae in the theory of composites, in: P. Drygaś, S. Rogosin (Eds.), Modern Problems in Applied Analysis, Birkhäuzer, Springer International Publishing AG, Cham, 2018, pp. 15–34.

[7] D.I. Bardzokas, M.L. Filshtinsky, L.A. Filshtinsky, Mathematical Methods in Electro-Magneto-Elasticity, Springer, Berlin, 2007.

[8] D.I. Bardzokas, L.A. Filshtinskii, L.V. Shramko, Homogeneous solutions of the electroelasticity equations for piezoceramic layers in $\mathbb{R}^3$, Acta Mechanica 209 (2010) 27–41.

[9] L.V. Belokopytova, L.A. Fil'shtinskii, Two-dimensional boundary value problem of electroelasticity for a piezoelectric medium with cuts, Journal of Applied Mathematics and Mechanics 43 (1979) 147–153.

[10] L.V. Belokopytova, O.A. Ivanenko, L.A. Fil'shtinskii, Load transfer from elastic ribs to a semi-infinite piezoceramic plate, Izvestiâ Akademii Nauk Armânskoj SSR. Mehanika 34 (5) (1981) 41–51 (in Russian).

[11] M. Borwein, M.L. Glasser, R.C. McPhedran, J.G. Wan, I.J. Zucker, Lattice Sums Then and Now, Encyclopedia of Mathematics and its Applications, vol. 150, Cambridge University Press, Cambridge, 2013.

[12] C. Conca, F. Lund, Fourier homogenization method and the propagation of acoustic waves through a periodic vortex array, SIAM Journal on Applied Mathematics 59 (5) (1999) 1573–1581.

[13] R. Czapla, V. Mityushev, N. Rylko, Conformal mapping of circular multiply connected domains onto slit domains, Electronic Transactions on Numerical Analysis 39 (2012) 286–297.

[14] P.Y. Chen, M.J.A. Smith, R.C. McPhedran, Evaluation and regularization of phase-modulated Eisenstein series and application to double Schlomilch-type sums, Journal of Mathematics and Physics 59 (2018) 072902.

Filshtinsky's contribution to Applied Mathematics & Mechanics Chapter | 1 **35**

[15] S.A. Chaplygin, Deformation in two dimensions, in: L.S. Leibenzon (Ed.), Collected Works, v.3, Gostehizdat, Moscow-Leningrad, 1950, pp. 306–316.

[16] A. Cherkaev, Variational Methods for Structural Optimization, Springer Verlag, New York, 2012.

[17] A. Cherkaev, Optimal three-material wheel assemblage of conducting and elastic composites, International Journal of Engineering Science 59 (2012) 27–39.

[18] L.I. Chibrikova, On the Riemann boundary value problem for automorphic functions, Uchenye Zapiski Kazanskogo Universiteta 116 (1956) 59–109.

[19] V.N. Dolgikh, L.A. Filshtinskii, A model of a regular piecewise-homogeneous medium, Mechanics of Solids 11 (2) (1976) 158–164.

[20] V.N. Dolgikh, L.A. Filshtinskii, Theory of linearly reinforced composite material with anisotropic structural components, Mechanics of Solids 13 (6) (1978) 53–63.

[21] V.N. Dolgikh, L.A. Filshtinskii, Model of an anisotropic medium reinforced by thin tapes, Soviet Applied Mechanics 15 (4) (1979) 292–296.

[22] P. Drygaś, S. Gluzman, V. Mityushev, W. Nawalaniec, Applied Analysis of Composite Media. Analytical and Computational Results for Materials Scientists and Engineers, Elsevier, Oxford, 2020.

[23] P. Drygaś, Double periodic bianalytic functions, in the present book.

[24] P. Duan, Y. Wang, Y. Wang, Riemann boundary value problem for doubly periodic polyanalytic functions, Complex Variables and Elliptic Equations 65 (2020) 1882–1901.

[25] G. Eisenstein, Beiträge zur Theorie der elliptischen Functionen, Crelles Journal 35 (1847) 153–247.

[26] L.A. Filshtinskii, L.M. Kurshin, Determination of the averaged elastic modulus of isotropic plane weaken by a double periodic array of equal circular holes, Izvestiya Akademii Nauk SSSR. Department of Technical Sciences. Mechanical Engineering 6 (1961) 10–14.

[27] L.A. Filshtinskii, Problems of thermal conduction and thermoelasticity for a plane weaken by a double periodic array of equal circular holes, in: Thermal Stresses in Elements of Constructions, Naukova Dumka, Institute of Mechanics, Academy of Sciences USSR, Kiev, 1964, pp. 103–112.

[28] L.A. Fil'shtinskii, Stresses and displacements in an elastic plane weaken by a double periodic array of equal circular holes, Journal of Applied Mathematics and Mechanics 28 (1964) 1530–1543.

[29] L.A. Filshtinskii, E.I. Grigolyuk, L.M. Kurshin, On a method to solve double periodic elastic problems of the elasticity theory, Prikladnaya Mehanika 1 (1965) 22–31.

[30] L.A. Filshtinsky, Stresses in regular double-periodic lattices, Mechanics of Solids 2 (1) (1967) 112–114.

[31] L.A. Filshtinskii, Doubly periodic problem in the theory of a circular closed cylindrical shell, Soviet Physics. Doklady 12 (1967) 739–741.

[32] L.A. Filshtinskii, Complete systems of particular solutions in shallow-shell theory, Journal of Applied Mathematics and Mechanics 33 (4) (1969) 666–675.

[33] L.A. Filshtinskii, Boundary value problems of the theory of shallow shells, Journal of Applied Mathematics and Mechanics 34 (2) (1970) 244–255.

[34] L.A. Filshtinskii, General solution of the equations of the technical theory of shallow shells, Mechanics of Deformable Body and Design of Structures, Novosibirsk 96 (1970) 166–168 (in Russian).

[35] L.A. Filshtinskii, Integral representations of solutions in shallow shell theory, Soviet Physics. Doklady 15 (1970/1971) 188–189.

[36] L.A. Filshtinskii, Elastic equilibrium of a plane anisotropic medium weakened by arbitrary curvilinear cracks. Limiting passage to anisotropic medium, Mechanics of Solids 11 (5) (1976) 91–97.

[37] L.A. Filshtinskii, Singularities of the stress field in an elastic anisotropic half-plane with a stiffener terminating at the surface, Soviet Applied Mechanics 17 (10) (1981) 944–947.

36 Mechanics and Physics of Structured Media

[38] L.A. Filshtinskii, On load transfer from an elastic rib to a semi-infinite anisotropic shell, Mechanics of Solids 17 (1) (1982) 150–158.

[39] L.A. Filshtinskii, L.B. Belokopytova, O.A. Ivanenko, Conjugated electric and mechanical fields in piezoelastic bodies with cuts or inclusions, Dinamika i Prochnost Mashin. Kharkov, Osnova 34 (1981) 16–21 (in Russian).

[40] L.A. Filshtinskii, A.T. Gumennii, Dynamic response of an anisotropic shell to concentrated harmonic loads, Issledovaniâ po Teorii Plastin i Obolochek. Kazan, Kazan State Univ. 19 (1985) 145–158 (in Russian).

[41] L.A. Filshtinskii, L.A. Khizhniak, Reaction of a piezoceramic shell to concentrated effects, Journal of Applied Mathematics and Mechanics 47 (3) (1983) 403–407.

[42] L.A. Filshtinskii, L.A. Khizhniak, Load transfer from elastic ribs to the finite piezoceramic shell, Izvestiâ Akademii Nauk Armânskoj SSR. Mehanika 40 (2) (1987) 43–48 (in Russian).

[43] L.A. Filshtinskii, Yu.D. Kovalyov, Concentration of mechanical stresses near a hole in a piezoceramic layer, Mechanics of Composite Materials 38 (2002) 121–124.

[44] L.A. Fil'shtinskii, V.A. Lyubchak, Elastic behavior of semi-infinite anisotropic plates and shells under the action of concentrated loads, Dinamika i Prochnost Mashin. Kharkov, Osnova 33 (1981) 11–15 (in Russian).

[45] L.A. Filshtinskii, N.I. Volkov, V.A. Lyubchak, On the solution of some contact problems in the theory of plates and anisotropic shells, Issledovaniâ po Teorii Plastin i Obolochek. Kazan, Kazan State Univ. 16 (1981) 90–99 (in Russian).

[46] L.A. Filshtinskii, N.I. Volkov, On the stressed state of a rotating ribbed disk of complex configuration in the presence of holes and cracks, Mechanics of Solids 17 (5) (1982) 124–129.

[47] L.A. Filshtinskii, L.V. Volkova, O.A. Ivanenko, V.A. Lyubchak, Some static and dynamic boundary value problem of elasticity and electro-elasticity for media with cuts and elastic inclusions, in: Proceedings of All-Soviet Union Conference on the Theory of Elasticity, Erevan, 1979, pp. 56–58.

[48] L.A. Fil'shtyns'kyi, Yu.V. Shramko, D.S. Kovalenko, Averaging of the magnetic properties of fibrous ferromagnetic composites, Materials Science 46 (2011) 808–818.

[49] L.A. Filshtinsky, V. Mityushev, Mathematical models of elastic and piezoelectric fields in two-dimensional composites, in: Panos M. Pardalos, Themistocles M. Rassias (Eds.), Mathematics Without Boundaries, Springer, New York, 2014, pp. 217–262.

[50] L.A. Filshtinskii (Filshtinsky), Tutorial to the implementation of master's theses on the topic: "Boundary value problems of fracture mechanics of new composite materials", Sumy University Publ., 2015 (in Russian).

[51] L.A. Filshtinskii (Filshtinsky), D.M. Nosov, H.A. Eremenko, Boundary value problem of the mechanics of distraction of new magnetoelectroelastic materials weakened by cracks, Bulletin Kherson National Technical University (ISSN 2078-4481) 3 (58) (2016) 438–443.

[52] E. Freitag, Complex Analysis 2: Riemann Surfaces, Several Complex Variables, Abelian Functions, Higher Modular Functions, Springer Science & Business Media, 2011.

[53] F.D. Gakhov, Boundary Value Problems, Pergamon Press, Oxford, 1966; Third edition, Moscow, Nauka, 1977.

[54] S. Gluzman, V. Mityushev, W. Nawalaniec, G. Sokal, Random composite: stirred or shaken?, Archives of Mechanics 68 (2016) 229–241.

[55] S. Gluzman, V. Mityushev, W. Nawalaniec, Computational Analysis of Structured Media, Elsevier, London, 2018.

[56] E.I. Grigolyuk, L.A. Filshtinskii, Transverse bending of an isotropic plane resting on a doubly periodic system of point supports, Doklady Akademii Nauk SSSR 157 (6) (1964) 1316–1318 (in Russian).

[57] E.I. Grigolyuk, L.A. Filshtinskii, Elastic equilibrium of an isotropic plane with a doubly periodic system of inclusions, Soviet Applied Mechanics 2 (1966) 1–5.

[58] E.I. Grigolyuk, L.A. Filshtinskii, The elastic equilibrium of an isotropic plane supported at a doubly periodic set of points, under the action of an arbitrary doubly periodic transverse load, Soviet Physics. Doklady 10 (1966) 1230–1231.

Filshtinsky's contribution to Applied Mathematics & Mechanics Chapter | 1 **37**

[59] E.I. Grigolyuk, L.A. Filshtinskii, Perforated plates and shells and related problems. Results overview, in: Itogi Nauki. Mekhanika. Uprugost i Plastichnost, Moscow, VINITI, 1965, pp. 7–163, 1967 (in Russian).

[60] E.I. Grigolyuk, L.A. Filshtinskii, Perforated Plates and Shells, Nauka, Moscow, 1970 (in Russian).

[61] E.I. Grigolyuk, L.A. Filshtinskii, General solutions of the equations of the theory of shallow shells in displacements, Mechanics of Solids 5 (2) (1970) 75–82.

[62] E.I. Grigolyuk, L.A. Filshtinskii, V.E. Kats, On the construction of periodic solutions in the problem of stress concentration in a circular cylindrical shell with holes, Issledovaniâ po Teorii Plastin i Obolochek. Kazan, Kazan State Univ. 6–7 (1970) 65–76 (in Russian).

[63] E.I. Grigolyuk, L.A. Filshtinskii, A complete system of solutions in the theory of shallow shells, Soviet Physics. Doklady 15 (1970/1971) 77–78.

[64] E.I. Grigolyuk, M.G. Gryngauz, V.N. Dolgikh, L.A. Filshtinskii, On the bending of elastic plates with a regular structure, Mechanics of Solids 17 (3) (1982) 124–130.

[65] E.I. Grigolyuk, M.G. Gryngauz, L.A. Filshtinskii, Solution of two-dimensional elasticity-theory problems for domains with a piecewise-smooth boundary, Soviet Physics. Doklady 29 (1984) 342–344.

[66] E.I. Grigolyuk, L.A. Filshtinskii, Periodical Piece–Homogeneous Elastic Structures, Nauka, Moscow, 1992 (in Russian).

[67] E.I. Grigolyuk, L.A. Filshtinskii, Regular Piece-Homogeneous Structures with defects, Fiziko-Matematicheskaja Literatura, Moscow, 1994 (in Russian).

[68] M.G. Gringauz, L.A. Filshtinskii, Theory of an elastic linearly reinforced composite, Journal of Applied Mathematics and Mechanics 39 (3) (1975) 510–519.

[69] A. Hurwitz, R. Courant, Vorlesungen über Allgemeine Funktionentheorie und Elliptische Funktionen, Springer-Verlag, Berlin, 1964.

[70] V.E. Kats, L.A. Filshtinsky, On a method of construction of double periodic polyharmonic functions, Prikladnaya Mehanika 7 (1971) 83–88.

[71] T. Ikeda, Fundamentals of Piezoelectricity, Oxford Science Publications, Oxford, 1996.

[72] W.T. Koiter, Some general theorems on doubly-periodic and quasi-periodic functions, Koninklijke Nederlandsche Akademie van Wetenschappen, Proceedings Series A 62 (1959) 120–128.

[73] W.T. Koiter, Stress distribution in an infinite elastic sheet with a doubly-periodic set of equal holes, in: R.E. Langer (Ed.), Boundary Problems in Differential Equations, The University of Wisconsin Press, Madison, Wisconsin, 1960, pp. 191–213.

[74] G.N. Kolosov, On Application of the Theory of Complex Functions to Plane Problem of the Mathematical Theory of Elasticity, Mattisen, Yur'ev, 1909.

[75] A.S. Kosmodamianskii, V.N. Lozhkin, General two-dimensional stressed state of thin piezoelectric slabs, Soviet Applied Mechanics 11 (5) (1975) 495–501.

[76] A.L. Kalamkarov, I.V. Andrianov, D. Weichert, Asymptotic analysis of perforated shallow shells, International Journal of Engineering Science 53 (2012) 1–18.

[77] B.A. Kudryavcev, V.Z. Parton, V.I. Rakitin, Fracture mechanics of piezoelectric materials. Rectilinear tunnel crack on the boundary with a conductor, Prikladnaya Matematika i Mehanika 39 (1975) 149–159, Journal of Applied Mathematics and Mechanics 39 (1975) 136–146.

[78] P. Kurtyka, N. Rylko, Structure analysis of the modified cast metal matrix composites by use of the RVE theory, Archives of Metallurgy and Materials 58 (2013) 357–360.

[79] P. Kurtyka, N. Rylko, Quantitative analysis of the particles distributions in reinforced composites, Composite Structures 182 (2017) 412–419.

[80] R. Landauer, The electrical resistance of binary metallic mixture, Journal of Applied Physics 23 (7) (1952) 779–784.

[81] R. Landauer, Electrical conductivity in inhomogeneous media, in: J.C. Garland, D.B. Tanner (Eds.), Electrical, Transport and Optical Properties of Inhomogeneous Media, American Institute of Physics, New York, 1978, pp. 2–43.

38 Mechanics and Physics of Structured Media

[82] M. Laso, N. Jimeno, Representation Surfaces for Physical Properties of Materials: A Visual Approach to Understanding Anisotropic Materials, Springer International Publishing, 2020.

[83] M.A. Lavrent'ev, B.V. Shabat, Methoden der komplexen Funktionentheorie, Deutsch. Verlag Wissenschaft, 1967.

[84] S.G. Lekhnitskii, Some cases of elastic equilibrium of homogeneous cylinder with arbitrary anisotropy, Prikladnaya Matematika i Mehanika 2 (1939) 345–367.

[85] S.G. Lekhnitskii, Theory of Elasticity of an Anisotropic Elastic Body, Mir Publishers, Moscow, 1981.

[86] K. Lichtenecker, Die Dielektrizitätkonstante naturlicher und kunstlicher Mischkorper, Physikalische Zeitschrift XXVII, 4/5 (1926) 115–158.

[87] J.E. Littlewood, A Mathematician's Miscellany, Methuen & Co., London, 1953.

[88] V.A. Lyubchak, L.A. Filshtinskii, Elastic equilibrium of an anisotropic shell with cracks, Issledovaniâ po Teorii Plastin i Obolochek. Kazan, Kazan State Univ. 19 (1985) 57–75 (in Russian).

[89] V.N. Maksimenko, L.A. Filshtinskii, Elastic behavior of anisotropic shells under the action of loads concentrated on lines, Mechanics of Solids 9 (5) (1974) 57–66.

[90] V.N. Maksimenko, L.A. Filshtinskii, Elastic equilibrium of anisotropic shells reinforced by stiffener ribs, Journal of Applied Mathematics and Mechanics 39 (5) (1975) 863–872.

[91] V.N. Maksimenko, L.A. Filshtinskii, On contact problems in the theory of anisotropic shells, in: Trudy X Vses. Konf. po Teorii Obolochek i Plastin, vol. 1, Tbilisi, Mezniereba, 1975, pp. 186–193 (in Russian).

[92] V.N. Maksimenko, L.A. Filshtinskii, Transmission of stresses from stringer of variable stiffness to reinforced shell, Soviet Applied Mechanics 12 (7) (1976) 641–647.

[93] V.N. Maksimenko, L.A. Filshtinskii, Contact of an anisotropic shell of revolution with rigid linear punch, Mechanics of Solids 12 (2) (1977) 89–94.

[94] V.N. Maksimenko, L.A. Filshtinskii, Calculation of anisotropic shells under the action of concentrated forces, Dinamika i Prochnost Mashin. Kharkov, Kharkov State Univ. 27 (1978) 22–26 (in Russian).

[95] V.N. Maksimenko, L.A. Filshtinskii, Load transmission from a stiffener to an anisotropic shell through a layer of adhesive between them, Soviet Applied Mechanics 14 (8) (1978) 835–840.

[96] L.I. Manevitch, I.V. Andrianov, V.O. Oshmyan, Mechanics of Periodically Heterogeneous Structures, Springer-Verlag, Berlin, 2002.

[97] J.C. Maxwell, Treatise on Electricity and Magnetism, Clarendon Press, Oxford, 1873.

[98] J.C. Maxwell Garnett, Colours in metal glasses and in metallic films, Philosophical Transactions of the Royal Society of London 203 (1904) 385–420.

[99] D.R. McKenzie, R.C. McPhedran, G.H. Derrick, The conductivity of lattices of spheres II. The body centred and face centred cubic lattices, Proceedings of the Royal Society of London. Series A 362 (1978) 211–232.

[100] R.C. McPhedran, D.R. McKenzie, The conductivity of lattices of spheres I. The simple cubic lattice, Proceedings of the Royal Society of London. Series A 359 (1978) 45–63.

[101] R.C. McPhedran, G.W. Milton, A review of anomalous resonance, its associated cloaking, and superlensing, Comptes Rendus. Physique 21 (2020) 409–423.

[102] R. McPhedran, S. Gluzman, V. Mityushev, N. Rylko (Eds.), 2D and Quasi-2D Composite and Nanocomposite Materials, Properties and Photonic Applications, Elsevier, Amsterdam, 2020.

[103] G.W. Milton, The Theory of Composites, Cambridge University Press, Cambridge, 2002.

[104] V. Mityushev, Functional equations and its applications in mechanics of composites, Demonstratio Mathematica 30 (1997) 64–70.

[105] V. Mityushev, Plane problem for the steady heat conduction of material with circular inclusions, Archives of Mechanics 45 (1993) 211–215.

[106] V. Mityushev, Transport properties of finite and infinite composite materials and Rayleigh's sum, Archives of Mechanics 49 (1997) 345–358.

Filshtinsky's contribution to Applied Mathematics & Mechanics Chapter | 1 **39**

[107] V. Mityushev, Transport properties of regular array of cylinders, ZAMM (Journal of Applied Mathematics and Mechanics) 177 (1997) 115–120.

[108] V. Mityushev, Transport properties of doubly periodic arrays of circular cylinders and optimal design problems, Applied Mathematics & Optimization 44 (2001) 17–31.

[109] V. Mityushev, Exact solution of the R-linear problem for a disk in a class of doubly periodic functions, Journal of Applied Functional Analysis 2 (2007) 115–127.

[110] V. Mityushev, Schwarz-Christoffel formula for multiply connected domains, Computational Methods and Function Theory 12 (2012) 449–463.

[111] V. Mityushev, S. Rogosin, Constructive Methods for Linear and Nonlinear. Boundary Value Problems for Analytic Functions: Theory and Applications, Monographs and Surveys in Pure and Applied Mathematics, Chapman & Hall / CRC, Boca Raton, 2000.

[112] V. Mityushev, N. Rylko, Optimal distribution of the non-overlapping conducting disks, Multiscale Modeling & Simulation. SIAM Interdisciplinary Journal 10 (2012) 180–190.

[113] V. Mityushev, N. Rylko, Maxwell's approach to effective conductivity and its limitations, Quarterly Journal of Mechanics and Applied Mathematics 66 (2) (2013) 241–251.

[114] V. Mityushev, Cluster method in composites and its convergence, Applied Mathematics Letters 77 (2018) 44–48.

[115] V. Mityushev, P. Drygaś, Effective properties of fibrous composites and cluster convergence, Multiscale Modeling & Simulation. SIAM Interdisciplinary Journal 17 (2) (2019) 696–715, https://doi.org/10.1137/18M1184278.

[116] V. Mityushev, W. Nawalaniec, Effective conductivity of a random suspension of highly conducting spherical particles, Applied Mathematical Modelling 72 (2019) 230–246.

[117] A.B. Movchan, N.A. Nicorovici, R.C. McPhedran, Green's tensors and lattice sums for electrostatics and elastodynamics, Proceedings of the Royal Society of London. Series A 453 (1997) 643–662.

[118] N.I. Muskhelishvili, Some Mathematical Problems of the Plane Theory of Elasticity, first Russian edition, Izd Akad Nauk SSSR, Moscow-Leningrad, 1933; English translation Dordrecht, Springer, 1977.

[119] V.Ya. Natanzon, On stresses in a tensioned plate with holes located in the chess order, Matematicheskii Sbornik 42 (1935) 617–636 (in Russian).

[120] J. von Neumann, The Mathematician, Works of the Mind I(1), University of Chicago Press, Chicago, 1947, pp. 180–196. See also Neumann's Collected Works.

[121] W. Nowacki, Thermoelasticity, second edition, revised and enlarged, Polish Scientific Publishers, Warsaw, 1986.

[122] N.A. Nicorovici, R.C. McPhedran, G.W. Milton, Transport properties of a three-phase composite material: the square array of coated cylinders, Proceedings of the Royal Society of London. Series A 442 (1993) 599–620.

[123] N.A. Nicorovici, R.C. McPhedran, G.W. Milton, Optical and dielectric properties of partially resonant composites, Physical Review B 49 (1994) 8479–8482.

[124] L. Poladian, R.C. McPhedran, Effective transport properties of periodic composite materials, Proceedings of the Royal Society of London. Series A 408 (1986) 45–59.

[125] J.M. De Ponti, A. Colombi, E. Riva, R. Ardito, F. Braghin, A. Corigliano, R.V. Craster, Experimental investigation of amplification, via a mechanical delay-line, in a rainbow-based metamaterial for energy harvesting, Applied Physics Letters 117 (2020) 143902.

[126] Rayleigh Lord, On the instability of cylindrical fluid surfaces, Philosophical Magazine Series 5 34 (207) (1892) 177–180.

[127] N. Rylko, Transport properties of a rectangular array of highly conducting cylinders, Journal of Engineering Mathematics 38 (2000) 1–12.

[128] N. Rylko, Effective anti-plane properties of piezoelectric fibrous composites, Acta Mechanica 224 (2013) 2719–2734.

[129] N. Rylko, Representative volume element in 2D for disks and in 3D for balls, Journal of Mechanics of Materials and Structures 9 (2014) 427–439.

40 Mechanics and Physics of Structured Media

[130] N. Rylko, A pair of perfectly conducting disks in an external field, Mathematical Modelling and Analysis 20 (2015) 273–288.

[131] L.I. Sedov, Plane Problems of Hydrodynamics and Aerodynamics, Gostekhizdat, Moscow, 1950.

[132] S. Timoshenko, S. Woinovsky-Krieger, Theory of Plates and Shells, McGraw-Hill, New York, 1959.

[133] T.C.T. Ting, Anisotropic Elasticity: Theory and Applications, Oxford University Press, New York, 1996.

[134] J. Turowski, Coupled fields, in: J.K. Sykulski (Ed.), Computational Magnetics, Springer, Dordrecht, 1995.

[135] G.A. Van Fo Fy, Theory of Reinforced Materials With Coatings, Naukova Dumka, Kiev, 1971.

[136] P. Vannucci, Anisotropic Elasticity, Springer, Singapore, 2018.

[137] K. Viertel, The development of the concept of uniform convergence in Karl Weierstrass's lectures and publications between 1861 and 1886, Archive for History of Exact Sciences 75 (2021) 455–490.

[138] A. Weil, Elliptic Functions According to Eisenstein and Kronecker, Springer-Verlag, Berlin, New York, 1976.

[139] P.T. Wootton, J. Kaplunov, D.J. Colquitt, An asymptotic hyperbolic-elliptic model for flexural-seismic metasurfaces, Proceedings of the Royal Society of London. Series A 475 (2019) 20190079.

[140] S. Yakubovich, P. Drygaś, V. Mityushev, Closed-form evaluation of 2D static lattice sums, Proceedings of the Royal Society of London. Series A 472 (2016) 20160510.

[141] E.I. Zverovich, Boundary value problems in the theory of analytic functions in Hölder classes on Riemann surfaces, Russian Mathematical Surveys 26 (1) (1971) 117–192.

Chapter 2

Cracks in two-dimensional magneto-electro-elastic medium

Dmytro Nosov[a,b], Leonid Filshtinsky[c], and Vladimir Mityushev[d]

[a]*Constantine the Philosopher University, Nitra, Slovakia,* [b]*Pedagogical University, Kraków, Poland,* [c]*Sumy State University, Sumy, Ukraine,* [d]*Research Group Materialica+, Faculty of Computer Science and Telecommunications, Cracow University of Technology, Kraków, Poland*

2.1 Introduction

Chapter 1 [10] of the present book contains a historical review on the plane problems of magneto-electro-elasticity (MEE) and the contribution of Filshtinsky in this topic. The preprint [6] summarizes investigations of Filshtinsky in the last years. Chapter 3 [11] concerns the extension of Lekhnitskii's formalism [8] for anisotropic plane elastic problems to coupled general MEE equations, and application of the vector-matrix problems to the boundary value problems of MEE. In the present chapter, we concentrate our attention to MEE materials with cracks. Following the preprint [6] we develop a method of integral equations. The regularity properties of the corresponding thermoelastic and electric fields near the crack edges were analyzed in [1,2].

Brittle and quasibrittle fracture can be modeled following the lines of the pure elastic model [3] by extension of the model to MEE. The Cherepanov-Rice integral [3] was extended to MEE medium in [17]. The path independent J_i-integral ($i = 1, 2$) and M-integral were introduced in [14]. These integrals express the total energy of the local fields near the tip of fracture and reduce the problem of fracture to the investigation of the corresponding stress intensity factors.

Using the generalized Lekhnitskii's representations in Section 2.2 we state the main boundary value problems for cracks on the plane in terms of complex potentials. Application of the Cauchy type integral representations of analytic functions yields a system of singular integral equations discussed in Section 2.3. Section 2.4 is devoted to computation of the stress intensity factors and energy flux at the tip of the crack.

Introduce the space coordinates x_j ($j = 1, 2, 3$). The third coordinate will be omitted in description of 2D problems. The coupled equations for MEE can be

42 Mechanics and Physics of Structured Media

written in form [16]

$$\sigma_{ij} = c_{ijks}\varepsilon_{ks} - e_{sij}E_s - h_{sij}H_s,$$
$$D_i = e_{iks}\varepsilon_{ks} + \epsilon_{is}E_s + \beta_{is}H_s, \qquad (2.1)$$
$$B_i = h_{iks}\varepsilon_{ks} + \beta_{is}E_s + \gamma_{is}H_s,$$

where σ_{ij}, ε_{ij}, D_i, B_i, E_i, H_i are the stress tensor components, the deformation tensor components, electrical and magnetic induction, and electrical and magnetic strains components, respectively, $c_{ijkl}, e_{iks}, h_{iks}, \beta_{is}$ are elastic, piezoelectric, piezomagnetic, and electromagnetic constants, $\epsilon_{is}, \gamma_{is}$ dielectrical and magnetic permitivities respectively. The equilibrium equations are added to the partial differential equations (2.1)

$$\sigma_{ij,i} = 0, \quad D_{i,i} = 0, \quad B_{i,i} = 0, \qquad (2.2)$$

where coma means partial derivative on the corresponding variable.

2D equations (2.1)–(2.2) correspond to the vector-matrix equations of magneto-electro-elasticity written in Chapter 1 [10] and to the same equation written in the expanded form in Chapter 3 [11].

2.2 Boundary-value problems for an unbounded domain

Consider the plane $\mathbb{R}^2$ of variables x_1 and x_2 weakened by cracks Γ_m ($m = \overline{1, M}$), $\Gamma = \bigcup\limits_{n=1}^{M} \Gamma_n$. The following assumptions are made in accordance with the general approach [12,15]

1. a crack is modeled by a slit Γ_m considered as a two-side Lyapunov's curve;
2. every crack pore is filled with air where constant electrical and magnetic fields D_1^0, D_2^0 and B_1^0, B_2^0 hold, respectively, in deformed state;
3. the sides of cracks do not contact;
4. the cracks are mutually disjoint slits, i.e., $\Gamma_i \cap \Gamma_j = \emptyset$ for $i \neq j$.

The sides of the cut $\Gamma_n = (a_n, b_n)$ are denoted by $\Gamma_n^{\pm}$. The sign plus corresponds to the left side of Γ_n while a point moves from its beginning a_n to the end b_n. The sign minus is assigned to the opposite side while a point moves from b_n to a_n. Let G denote the complement of Γ to the plane $\mathbb{R}^2$. Therefore, the boundary $\Gamma = \partial G$ of the domain G coincides with the union of two-sided slits

$$\gamma_n = \Gamma_n^+ \cup \Gamma_n^-. \qquad (2.3)$$

Let the considered domain G be occupied by magneto-electro-elastic material. Therefore, 2D equations (2.1)–(2.2) hold in G.

Let ψ denote the angle between the positive normal vector to the right side of Γ_m and the axis x_1. The normal stress is given as the vector $(X_{1n}^{\pm}, X_{2n}^{\pm})$ on

Cracks in two-dimensional magneto-electro-elastic medium **Chapter | 2 43**

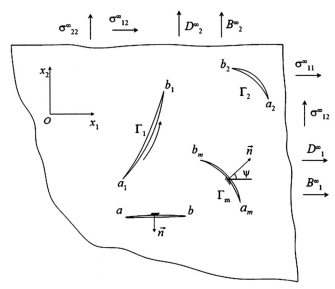

FIGURE 2.1 Crack in MEE plane. The arrow from a_1 to b_1 shows the orientation of Γ_1^+. The orientation of Γ_1^- is opposite.

the different sides of Γ_m. Consider the special case when the normal stress is expressed by the constant pressure $p = p_m$ in every crack

$$X_{1n}^{\pm} = -p_m \cos\psi, \quad X_{2n}^{\pm} = -p_m \sin\psi, \quad \text{on } \Gamma_m \ (m = \overline{1, M}). \tag{2.4}$$

This corresponds to the physical condition when the pressure air or liquid is given in a crack which is a void in real medium. (See Fig. 2.1.) The constant electrical and magnetic fields D_1^0, D_2^0 and B_1^0, B_2^0, respectively, are given on the sides of cracks.

$$D_n^{\pm} = D_n^0, \quad B_n^{\pm} = B_n^0, \quad \text{on } \Gamma_m. \tag{2.5}$$

Below, we use the complex potentials of the coupled fields and the corresponding constants explicitly described in Chapters 1 and 3 [10,11]. We do not repeat the formulas here except the ultimate representations of the local fields through the complex potentials. Let μ_k denote a root of characteristic equation with nonnegative Im μ_k. Introduce the transformed complex coordinate

$$z_k = \operatorname{Re} z + \mu_k \operatorname{Im} z. \tag{2.6}$$

The local fields have the form

$$\begin{pmatrix} \sigma_{11} \\ \sigma_{12} \\ \sigma_{22} \end{pmatrix} = 2\operatorname{Re} \sum_{k=1}^{4} A_{11}(1, \mu_k) \Phi_k(z_k) \begin{pmatrix} \mu_k^2 \\ -\mu_k \\ 1 \end{pmatrix}, \tag{2.7}$$

$$\begin{pmatrix} D_1 \\ D_2 \end{pmatrix} = 2\mathrm{Re} \sum_{k=1}^{4} A_{12}(1, \mu_k) \Phi_k(z_k) \begin{pmatrix} \mu_k \\ -1 \end{pmatrix}, \tag{2.8}$$

$$\begin{pmatrix} B_1 \\ B_2 \end{pmatrix} = 2\mathrm{Re} \sum_{k=1}^{4} A_{13}(1, \mu_k) \Phi_k(z_k) \begin{pmatrix} \mu_k \\ -1 \end{pmatrix}, \tag{2.9}$$

$$\begin{pmatrix} E_1 \\ E_2 \end{pmatrix} = 2\mathrm{Re} \sum_{k=1}^{4} \alpha_k^{(E)} \Phi_k(z_k) \begin{pmatrix} 1 \\ \mu_k \end{pmatrix}, \tag{2.10}$$

$$\begin{pmatrix} H_1 \\ H_2 \end{pmatrix} = 2\mathrm{Re} \sum_{k=1}^{4} \alpha_k^{(H)} \Phi_k(z_k) \begin{pmatrix} 1 \\ \mu_k \end{pmatrix}, \tag{2.11}$$

where the coefficients on A_{1j}, $\alpha_k^{(E)}$, $\alpha_k^{(H)}$ are given by the explicit formulas in Chapter 3 [11].

The $\mathbb{R}$-linear transformation (2.6) maps the slit Γ_m onto a slit $\Gamma_m^{(k)}$ on the complex plane z_k. Every function $\Phi_k(z_k)$ is analytic in the variable z_k in the domain G_k obtained from the domain G by the transformation (2.6). The function $\Phi_k(z_k)$ is continuous in the closure of G_k in the two sided topology for the slits $\Gamma_m^{(k)}$ except at the end-points where the complex potentials $\Phi_k(z_k)$ admit weak singularities. The weak singularity means that the function is locally integrated near the ends and the integral is almost bounded near the ends. The almost bounded means at most logarithmic singularity [12]. Such a class of complex potentials is consistent with the theory of boundary value problems in a class of piece-wise Hölder continuous functions on piece-wise Lyapunov's contours.

The boundary conditions can be written in the following form

$$2\mathrm{Re} \sum_{k=1}^{4} \mu_k A_{11}(1, \mu_k) a_k(\psi) \Phi_k^{\pm}(z_k) = -p \cos \psi,$$

$$2\mathrm{Re} \sum_{k=1}^{4} -A_{11}(1, \mu_k) a_k(\psi) \Phi_k^{\pm}(z_k) = -p \sin \psi,$$

$$2\mathrm{Re} \sum_{k=1}^{4} A_{12}(1, \mu_k) a_k(\psi) \Phi_k^{\pm}(z_k) = D_n^0, \tag{2.12}$$

$$2\mathrm{Re} \sum_{k=1}^{4} A_{13}(1, \mu_k) a_k(\psi) \Phi_k^{\pm}(z_k) = B_n^0, \quad z_k \in \Gamma_m^{(k)} \ (m = \overline{1, M}).$$

Adding and subtracting Eqs. (2.12) we obtain the boundary conditions

$$2\mathrm{Re}\sum_{k=1}^{4} R_{jk}a_k(\psi)\,[\Phi_k] = [F_j],$$

$$2\mathrm{Re}\sum_{k=1}^{4} R_{jk}a_k(\psi)\left(\Phi_k^+ + \Phi_k^-\right) = \left(F_j^+ + F_j^-\right), \quad j = \overline{1,4}, \qquad (2.13)$$

where

$$R_{1k} = \mu_k A_{11}(1,\mu_k), \quad R_{2k} = -A_{11}(1,\mu_k), \qquad (2.14)$$
$$R_{3k} = A_{12}(1,\mu_k), \quad R_{4k} = A_{13}(1,\mu_k),$$

and

$$F_1^\pm = -p\cos\psi, \quad F_2^\pm = -p\sin\psi, \quad F_3^\pm = D_n^0, \quad F_4^\pm = B_n^0.$$

The jumps of functions across the slits are introduced as follows

$$[\Phi_k] = \Phi_k^+(z_k) - \Phi_k^-(z_k), \quad [F_j] = F_j^+ - F_j^-. \qquad (2.15)$$

In the considered case, the jumps $[F_j]$ vanish and (2.13) becomes

$$2\mathrm{Re}\sum_{k=1}^{4} R_{jk}a_k(\psi)\,[\Phi_k] = 0, \quad 2\mathrm{Re}\sum_{k=1}^{4} R_{jk}a_k(\psi)\left(\Phi_k^+ + \Phi_k^-\right) = F_j, \quad j = \overline{1,4},$$

$$(2.16)$$

where $F_j = F_j^+ = F_j^-$.

2.3 Integral equations for an unbounded domain

A method of integral equation is developed in the present section in order to solve the boundary-value problem (2.12). It can be considered as an extension of Filshtinsky's method for pure elastic problems [4,5] to MEE.

The complex potentials $\Phi_k(z_k)$ are represented in terms of the Cauchy type integrals along the curve $\Gamma = \cup_{n=1}^{M}\Gamma_n$

$$\Phi_k(z_k) = b_k + \frac{1}{2\pi}\int_{\Gamma}\frac{\omega_k(\zeta)}{\zeta_k - z_k}ds. \qquad (2.17)$$

The function $\omega_k(\zeta)$ is Hölder continuous on the open slits Γ_n and almost bounded at its ends; the differential $ds = ds(\zeta)$. The constants b_k have to be

46 Mechanics and Physics of Structured Media

matched with the condition at infinity. The complex variable ζ_k is expressed through ζ by equation

$$\zeta_k = \operatorname{Re} \zeta + \mu_k \operatorname{Im} \zeta, \quad \zeta = x_1 + ix_2 \in \Gamma. \tag{2.18}$$

It is convenient to write Eqs. (2.17) in form

$$\Phi_k(z_k) = b_k + \frac{1}{2\pi} \int_\Gamma \frac{\omega_k(\zeta)d\zeta_k}{(\zeta_k - z_k)a_k(\psi)}, \quad z_k \in G_k. \tag{2.19}$$

Let $\omega_k^{(n)}(\zeta)$ denote the value of $\omega_k(\zeta)$ on Γ_n for a fixed n. In order to determine the limit values of $\Phi_k(z_k)$ on Γ_n we use the Sochocki formulas [7,12]

$$\{\Phi_k(z_k)\}^\pm_{z \to \zeta_0 \in \Gamma \ (z_k \to \zeta_{0k})} = b_k \pm \frac{i\omega_k^{(n)}(\zeta_0)}{2a_k(\psi_0)} + \frac{1}{2\pi} \int_\Gamma \frac{\omega_k(\zeta)}{\zeta_k - \zeta_{0k}} ds, \tag{2.20}$$

where

$$\psi_0 = \psi(\zeta_0), \ \zeta_{0k} = \operatorname{Re} \zeta_0 + \mu_k \operatorname{Im} \zeta_0, \ \zeta_0 \in \Gamma. \tag{2.21}$$

The integrals in (2.20) are considered by means of the Cauchy principal value. Substitution of the boundary values (2.20) into the boundary conditions (2.12) yields

$$2\operatorname{Re} \sum_{k=1}^4 R_{jk} a_k(\psi_0) \left\{ b_k \pm \frac{i\omega_k^{(n)}(\zeta_0)}{2a_k(\psi_0)} + \frac{1}{2\pi} \int_\Gamma \frac{\omega_k(\zeta)}{\zeta - \zeta_{0k}} ds \right\} = F_j \ (j = \overline{1,4}). \tag{2.22}$$

Taking into account (2.16) we obtain the following mixed system of algebraic and integral equations

$$\operatorname{Im} \sum_{k=1}^4 R_{jk}\omega_k^{(n)}(\zeta) = 0, \quad (j = \overline{1,4}; n = \overline{1,M}), \tag{2.23}$$

$$2\operatorname{Re} \sum_{k=1}^4 R_{jk} \frac{a_k(\psi_0)}{2\pi} \int_\Gamma \frac{\omega_k(\zeta)}{\zeta_k - \zeta_{0k}} ds = F_j - 2\operatorname{Re} \sum_{k=1}^4 R_{jk} a_k(\psi_0)b_k. \tag{2.24}$$

The right part of the system (2.23)–(2.24) depends on the external fields given at infinity through the constants C_k. Using (2.16) and the general representations of the fields through the complex potentials (2.7)–(2.11) we obtain

$$\sigma_{11}^{\infty} = 2\text{Re} \sum_{k=1}^{4} \mu_k^2 A_{11}(1, \mu_k) b_k,$$

$$\sigma_{12}^{\infty} = 2\text{Re} \sum_{k=1}^{4} -\mu_k A_{11}(1, \mu_k) b_k,$$

$$\sigma_{22}^{\infty} = 2\text{Re} \sum_{k=1}^{4} A_{11}(1, \mu_k) b_k,$$

$$D_1^{\infty} = 2\text{Re} \sum_{k=1}^{4} \mu_k A_{12}(1, \mu_k) b_k,$$

$$D_2^{\infty} = 2\text{Re} \sum_{k=1}^{4} -A_{12}(1, \mu_k) b_k,$$

$$B_1^{\infty} = 2\text{Re} \sum_{k=1}^{4} \mu_k A_{13}(1, \mu_k) b_k, \tag{2.25}$$

$$B_2^{\infty} = 2\text{Re} \sum_{k=1}^{4} -A_{13}(1, \mu_k) b_k.$$

Linear algebraic relations (2.25) can be considered as equations to determine the constants $b_k = b_k(\sigma_{11}^{\infty}, \ldots, B_2^{\infty})$, see for details Chapter 3 [11]. Therefore, one can assume that the constants b_k are known.

Taking into account (2.16) we obtain from (2.25) equations

$$-2\text{Re} \sum_{k=1}^{4} R_{1k} a_k(\psi_0) B_{1k} = -\left(\sigma_{11}^{\infty} \cos(\psi_0) + \sigma_{12}^{\infty} \sin \psi_0\right) = -N_1(\zeta_0),$$

$$-2\text{Re} \sum_{k=1}^{4} R_{2k} a_k(\psi_0) B_{1k} = -\left(\sigma_{12}^{\infty} \cos(\psi_0) + \sigma_{22}^{\infty} \sin \psi_0\right) = -N_2(\zeta_0),$$

$$-2\text{Re} \sum_{k=1}^{4} R_{3k} a_k(\psi_0) B_{1k} = -\left(D_1^{\infty} \cos(\psi_0) + D_2^{\infty} \sin \psi_0\right) = -N_3(\zeta_0),$$

$$-2\text{Re} \sum_{k=1}^{4} R_{4k} a_k(\psi_0) B_{1k} = -\left(B_1^{\infty} \cos(\psi_0) + B_2^{\infty} \sin \psi_0\right) = -N_4(\zeta_0),$$

$$\tag{2.26}$$

where for shortness the right part of the system (2.24) is denoted by $N_j(\zeta_0)$.

48 Mechanics and Physics of Structured Media

The algebraic system (2.23) can be excluded from Eqs. (2.23)–(2.24). First, write them in the form

$$\sum_{k=1}^{4} R_{jk} \omega_k^{(n)}(\zeta) = q_j^{(n)}, \quad (j = \overline{1,4}; \; n = \overline{1,M}), \tag{2.27}$$

where $\mathrm{Im}\, q_j^{(n)} = 0$.

Introduce the symmetric matrix

$$\mathcal{R} = \begin{bmatrix} R_{11} & R_{12} & R_{13} & R_{14} \\ R_{21} & R_{22} & R_{23} & R_{24} \\ R_{31} & R_{32} & R_{33} & R_{34} \\ R_{41} & R_{42} & R_{43} & R_{44} \end{bmatrix}, \tag{2.28}$$

where R_{jk} are given by (2.14) and $R_{kj} = R_{jk}$.

We have $\det \mathcal{R} \neq 0$ [6]. Introduce the vectors $q^{(n)} = \{q_1^{(n)}, q_2^{(n)}, q_3^{(n)}, q_4^{(n)}\}^T$ and

$$\omega^{(n)} = \mathcal{R}^{-1} q^{(n)}. \tag{2.29}$$

Then, Eqs. (2.24) yield the system of integral equations in the matrix form

$$\int_{\Gamma} K(\zeta, \zeta_0) q(\zeta) ds = \pi N(\zeta_0), \quad \zeta_0 \in \Gamma, \tag{2.30}$$

where

$$K(\zeta, \zeta_0) = \mathrm{Re}\, \{\mathcal{R} G(\zeta, \zeta_0) \mathcal{R}^{-1}\}, \quad q(\zeta) = \{q_1(\zeta), q_2(\zeta), q_3(\zeta), q_4(\zeta)\}^T,$$

$$G(\zeta, \zeta_0) = diag\, \left\{ \frac{a_1(\psi_0)}{\zeta_1 - \zeta_{01}}, \frac{a_2(\psi_0)}{\zeta_2 - \zeta_{02}}, \frac{a_3(\psi_0)}{\zeta_3 - \zeta_{03}}, \frac{a_4(\psi_0)}{\zeta_4 - \zeta_{04}} \right\},$$

$$N(\zeta_0) = \{-p \cos \psi_0 - N_1(\zeta_0), -p \sin \psi_0 - N_2(\zeta_0), D_n^0 - N_3(\zeta_0), B_n^0 - N_4(\zeta_0)\}^T.$$

Therefore, the boundary value problem (2.12) is reduced to the matrix singular integral equation of the first kind (2.30) on $q(\zeta)$. The unknown vector-function $q(\zeta)$ belongs to the class of functions Hölder continuous on Γ except at the ends where weak singularities are admissible [9,12].

The solution will be accomplished, when the conditions of single-valuedness for the physical fields determined by complex potentials, e.g. displacement, will be fulfilled. These conditions can be established in the following way. The mechanical displacements, electrical and magnetic potentials can be determined through the indefinite integrals of complex potentials (2.17)

$$\int \Phi_k(z_k) dz_k = b_k z_k - \frac{1}{2\pi} \int_{\Gamma} \omega_k(\zeta) \ln(\zeta_k - z_k) ds. \tag{2.31}$$

Cracks in two-dimensional magneto-electro-elastic medium Chapter | 2 **49**

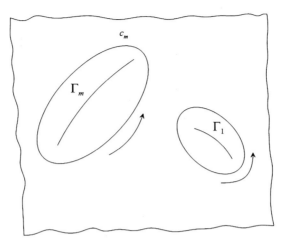

FIGURE 2.2 The curves c_m homotopic to the curve $(-\gamma_m) = -(\Gamma_m^+ \cup \Gamma_m^-)$.

Here, a branch of logarithm is arbitrarily fixed on the complex plane ζ_k without the cut connecting the points ζ_k and infinity.

Let a smooth closed curve c_n embraces the curve $(-\gamma_n)$ defined by (2.3) and does not intersect with others γ_m ($m \neq n$) as shown in Fig. 2.2. The increment of the logarithm $\ln(\zeta_k - z_k)$ over c_n is equal to $2\pi i$. Therefore, the increment of the function $\phi_k(z_k)$ has the form

$$[\Phi_k(z_k)]_{c_n} = i \int_{\Gamma_n} \omega_k^{(n)}(\zeta)ds, \quad (k = \overline{1,4}, \ n = \overline{1,M}). \tag{2.32}$$

Using (2.29) we rewrite the single-valuedness conditions in form

$$\mathrm{Im}(\mathcal{B}\mathcal{R}^{-1}) \int_{\Gamma_n} q_1^{(n)}(\zeta)ds = 0, \tag{2.33}$$

where

$$\mathcal{B} = \begin{bmatrix} r_1 & r_2 & r_3 & r_4 \\ q_1 & q_2 & q_3 & q_4 \\ \alpha_1^E & \alpha_2^E & \alpha_3^E & \alpha_4^E \\ \alpha_1^H & \alpha_2^H & \alpha_3^H & \alpha_4^H \end{bmatrix}.$$

Here, the coefficients on $r_k, q_k, \alpha_k^{(E)}, \alpha_k^{(H)}$ are given by the explicit formulas in Chapter 3 [11].

It can be shown, that

$$det(\mathcal{B}\mathcal{R}^{-1}) \neq 0.$$

Then, (2.33) can be written in the form

$$\int_{\Gamma_n} q_1^{(m)}(\zeta)ds = 0 \ (n = \overline{1, M}).$$ (2.34)

Thus, the problem is reduced to solution of the matrix integral equations (2.30) simultaneously with the additional conditions (2.34).

2.4 Asymptotic solution at the ends of cracks

Fixed a contour $\Gamma_n = (a_n, b_n)$. Omit the subscripts and write $\Gamma = (a, b)$ for shortness in the present section. Introduce the parameterization of Γ by means of the normalized parameters $-1 \le \beta \le 1$ in such a way that

$$\zeta = \zeta(\beta), \ a = \zeta(-1), \ b = \zeta(1), \ \zeta \in \Gamma.$$

The same designation $\zeta_0 = \zeta(\beta_0)$ is used for $\zeta_0 \in \Gamma$.

It follows from [9] that the densities $\omega_k(\zeta)$ of the integrals (2.17) have square root singularities which can be presented in the form

$$\omega_k(\zeta) = \frac{\omega_k^*(\zeta)}{\sqrt{(\zeta - a)(\zeta - b)}} = \frac{\omega_k(\beta)}{s'(\beta)\sqrt{1 - \beta^2}},$$ (2.35)

where $s'(\beta) = \frac{ds}{d\beta}$.

The function $\omega_k^*(\zeta)$ is continuous at the tips of the cracks and

$$\omega_k^*(c) = \frac{\omega_k(\pm 1)}{s'(\pm 1)\sqrt{2}}\sqrt{(a - b)\zeta'(\pm 1)},$$ (2.36)

where $\zeta' = \frac{d\zeta}{d\beta}$. The sign plus corresponds to the end of the crack $c = b$ and minus to $c = a$.

In order to study the asymptotic behavior of complex potentials at the tips of cracks, we use the general formula valid for the weak singularity of order $0 < \gamma < 1$ in the Cauchy type integral [12]

$$\frac{1}{2\pi i} \int_\Gamma \frac{f(\zeta)d\zeta}{(\zeta - c)^\gamma(\zeta - z)} = \pm \frac{e^{\pm i\pi\gamma}f(c)}{2i\sin\pi\gamma}(z - c)^{-\gamma} + F_c(z),$$ (2.37)

where the function $F_c(z)$ is analytic near $z = c$ except at the cut connecting $z = c$ and infinity defined by a branch of $(\zeta - c)^\gamma$. Moreover,

$$\lim_{z \to c}(z - c)^\gamma F_c(z) = 0.$$ (2.38)

Cracks in two-dimensional magneto-electro-elastic medium **Chapter | 2** **51**

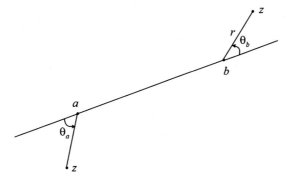

FIGURE 2.3 To the definition of the asymptotic behavior of function $\Phi_k(z_k)$ at the tips of the cracks. Angles θ_b and θ_a are measured from tangential line on continuation of the tip counter clock arrow.

Substituting $\gamma = \frac{1}{2}$ into (2.37) we arrive at the asymptotic relation

$$\frac{1}{2\pi i} \int_\Gamma \frac{f(\zeta)d\zeta}{\sqrt{\zeta - c}(\zeta - z)} = \frac{f(c)}{2}(z - c)^{-\frac{1}{2}} + F_c(z). \tag{2.39}$$

Application of (2.36), (2.39) to (2.19) yields

$$\Phi_k(z_k) = \frac{1}{2\pi} \int_\Gamma \frac{\omega_k(\zeta)d\zeta_k}{a_k(\psi)(\zeta_k - z_k)} = \frac{i\omega_k(\pm 1)}{2\sqrt{2\zeta_k'(\pm 1)}} [\pm(c_k - z_k)]^{-\frac{1}{2}}, \tag{2.40}$$

where

$$c_k = \operatorname{Re} c + \mu_k \operatorname{Im} c, \quad \zeta_k'(\pm 1) = \frac{d\zeta_k(\beta)}{d\beta}\bigg|_{\beta=\pm 1}. \tag{2.41}$$

As above the sign plus corresponds to the end $c = b$ and minus to $c = a$.

Introduce the local polar coordinates (r, θ_a) and (r, θ_b) near the points a and b, respectively

$$z - b = re^{i\theta_b}, \quad z - a = -re^{i\theta_a},$$
$$z_k - a_k = -r(\cos\theta_a + \mu_k \sin\theta_a), \tag{2.42}$$
$$b_k - z_k = -r(\cos\theta_b + \mu_k \sin\theta_b).$$

The choice of angles θ_b and θ_a is shown in Fig. 2.3. Now, the main asymptotic term of the function (2.40) can be written in the form

$$\Phi_k^c(z_k) = \frac{\omega_k(\pm 1)}{2\sqrt{2\zeta'(\pm 1)}} \frac{(\cos\theta_c + \mu_k \sin\theta_c)^{-\frac{1}{2}}}{\sqrt{r}} \equiv \frac{1}{2\sqrt{2r}}\Psi_k^c, \tag{2.43}$$

52 Mechanics and Physics of Structured Media

where the following constant is introduced for shortness

$$\Psi_k^c = \frac{\omega_k(\pm 1)}{\sqrt{\zeta_k'(\pm 1)}} (\cos\theta_c + \mu_k \sin\theta_c)^{-\frac{1}{2}}. \qquad (2.44)$$

Substituting the derived asymptotic formulas into (2.7)–(2.11) we obtain the main asymptotic terms of the local fields at the cracks tips

$$\begin{pmatrix} \sigma_{11}^{(1)} \\ \sigma_{12}^{(1)} \\ \sigma_{22}^{(1)} \end{pmatrix} = \frac{1}{\sqrt{2r}} \operatorname{Re} \sum_{k=1}^{4} A_{11}(1, \mu_k) \Psi_{1k}^c \begin{pmatrix} \mu_k^2 \\ -\mu_k \\ 1 \end{pmatrix} + O(1), \qquad (2.45)$$

$$\begin{pmatrix} D_1^{(1)} \\ D_2^{(1)} \end{pmatrix} = \frac{1}{\sqrt{2r}} \operatorname{Re} \sum_{k=1}^{4} A_{12}(1, \mu_k) \Psi_k^c \begin{pmatrix} \mu_k \\ -1 \end{pmatrix} + O(1), \qquad (2.46)$$

$$\begin{pmatrix} B_1^{(1)} \\ B_2^{(1)} \end{pmatrix} = \frac{1}{\sqrt{2r}} \operatorname{Re} \sum_{k=1}^{4} A_{13}(1, \mu_k) \Psi_{1k}^c \begin{pmatrix} \mu_k \\ -1 \end{pmatrix} + O(1), \qquad (2.47)$$

$$\begin{pmatrix} E_1^{(1)} \\ E_2^{(1)} \end{pmatrix} = \frac{1}{\sqrt{2r}} \operatorname{Re} \sum_{k=1}^{4} \alpha_k^{(E)} \Psi_{1k}^c \begin{pmatrix} 1 \\ \mu_k \end{pmatrix} + O(1), \qquad (2.48)$$

$$\begin{pmatrix} H_1^{(1)} \\ H_2^{(1)} \end{pmatrix} = \frac{1}{\sqrt{2r}} \operatorname{Re} \sum_{k=1}^{4} \alpha_k^{(H)} \Psi_{1k}^c \begin{pmatrix} 1 \\ \mu_k \end{pmatrix} + O(1). \qquad (2.49)$$

2.5 Stress intensity factors

The stress intensity factors K_I, K_{II}, and K_{III} are used as the main failure criterion of brittle fracture in fracture mechanics [3,13]. Following this theory we extend the notion of stress intensity factor to MEE restricting ourselves by the modes K_I and K_{II}.

Consider the vectors of normal σ_n and tangent stresses τ_{ns} near the tip crack c

$$\sigma_n = \sigma_{11} \cos^2\psi_c + \sigma_{12} \sin 2\psi_c + \sigma_{22} \sin^2\psi_c,$$

$$\tau_{ns} = \sigma_{12} \cos 2\psi_c + \frac{\sigma_{22} - \sigma_{11}}{2} \sin 2\psi_c. \qquad (2.50)$$

The stresses σ_n, τ_{ns} and the angle ψ_c are shown in Fig. 2.4.

Cracks in two-dimensional magneto-electro-elastic medium **Chapter | 2 53**

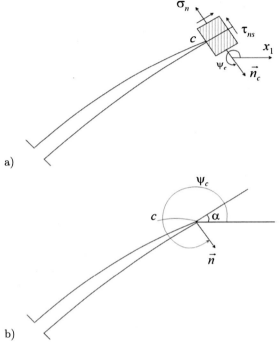

FIGURE 2.4 Illustration of normal σ_n and tangent stresses τ_{ns} near the point c. The normal vector to the crack is denoted by $\vec{n}_c$, the local angles by ψ_c and by $\alpha = \psi_c - \frac{3}{2}\pi$.

The stress intensity factors for the crack opening mode K_I and for the crack sliding mode K_{II} are defined by the limits [3,13]

$$K_I = \lim_{r \to 0} \left(\sqrt{2\pi r} \sigma_n \right), \quad K_{II} = \lim_{r \to 0} \left(\sqrt{2\pi r} \tau_{ns} \right). \tag{2.51}$$

In addition to them we introduce the stress intensity factors for the electrical and magnetic induction [16]

$$K_D = \lim_{r \to 0} \left(\sqrt{2\pi r} D_n \right), \quad K_B = \lim_{r \to 0} \left(\sqrt{2\pi r} B_n \right). \tag{2.52}$$

Put

$$a_k(\psi_c) = \cos \theta_c + \mu_k \sin \theta_c. \tag{2.53}$$

Then (2.43) becomes

$$\Psi^c_{1k} = \frac{\omega_k(\pm 1)}{a_k(\psi_c)\sqrt{s'(\pm 1)}}. \tag{2.54}$$

54 Mechanics and Physics of Structured Media

Using (2.54) and (2.45)–(2.49) write the stress intensity factors (2.51), (2.52) in the form

$$K_I = \sqrt{\frac{\pi}{s'(\pm 1)}} \operatorname{Re} \sum_{k=1}^{4} a_k(\psi_c) A_{11}(1, \mu_k) \omega_k(\pm 1),$$

$$K_{II} = \sqrt{\frac{\pi}{s'(\pm 1)}} \operatorname{Re} \sum_{k=1}^{4} b_k(\psi_c) A_{11}(1, \mu_k) \omega_k(\pm 1),$$

$$K_D = \sqrt{\frac{\pi}{s'(\pm 1)}} \operatorname{Re} \sum_{k=1}^{4} A_{12}(1, \mu_k) \omega_k(\pm 1),$$

$$K_B = \sqrt{\frac{\pi}{s'(\pm 1)}} \operatorname{Re} \sum_{k=1}^{4} A_{13}(1, \mu_k) \omega_k(\pm 1), \qquad (2.55)$$

where $b_k(\psi) = \frac{da_k(\psi)}{d\psi}$. Introduce the functions in β

$$q_{1j}(\beta) = \frac{Q_j(\beta)}{s'(\beta)\sqrt{1 - \beta^2}}, \quad \left(j = \overline{1, 4}\right). \qquad (2.56)$$

Application of (2.35) and (2.27) to (2.56) yields

$$\sum_{k=1}^{4} \mu_k A_{11}(1, \mu_k) \omega_k(\pm 1) = Q_1(\pm 1),$$

$$\sum_{k=1}^{4} A_{11}(1, \mu_k) \omega_k(\pm 1) = -Q_2(\pm 1),$$

$$\sum_{k=1}^{4} A_{12}(1, \mu_k) \omega_k(\pm 1) = Q_3(\pm 1), \qquad (2.57)$$

$$\sum_{k=1}^{4} A_{13}(1, \mu_k) \omega_k(\pm 1) = Q_4(\pm 1).$$

Taking into account (2.57) we find from (2.55) the stress intensity factors

$$K_I = \sqrt{\frac{\pi}{s'(\pm 1)}} \{Q_1(\pm 1) \cos \psi_c + Q_2(\pm 1) \sin \psi_c\},$$

$$K_{II} = \sqrt{\frac{\pi}{s'(\pm 1)}} \{-Q_1(\pm 1) \sin \psi_c + Q_2(\pm 1) \sin \psi_c\}, \qquad (2.58)$$

$$K_D = \sqrt{\frac{\pi}{s'(\pm 1)}} Q_3, \quad K_B = \sqrt{\frac{\pi}{s'(\pm 1)}} Q_4. \qquad (2.59)$$

Cracks in two-dimensional magneto-electro-elastic medium Chapter | 2 **55**

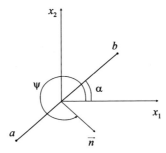

FIGURE 2.5 Straight-line crack with length $2l$.

Therefore, the crack opening mode K_I, the crack sliding mode K_{II}, the electrical and magnetic induction, K_D and K_B, respectively, are written by means of the functionals obtained from the solution to the matrix singular integral equation (2.30) and the conditions (2.34).

The stress intensity factors K_E and K_H can be introduced in the same way

$$K_E = \lim_{r \to 0} \left(\sqrt{2\pi r} E_S^{(1)} \right), \quad K_H = \lim_{r \to 0} \left(\sqrt{2\pi r} H_S^{(1)} \right). \tag{2.60}$$

Using (2.54) and (2.60) we ultimately find

$$K_E^{(1)} = \sqrt{\frac{\pi}{s'(\pm 1)}} \operatorname{Re} \sum_{k=1}^{4} \alpha_k^E \omega_k(\pm 1), \quad K_H^{(1)} = \sqrt{\frac{\pi}{s'(\pm 1)}} \operatorname{Re} \sum_{k=1}^{4} \alpha_k^H \omega_k(\pm 1).$$

A crack in MME plane

As a test example consider the plane (x_1, x_2) with a straight-line crack (a, b) displayed in Fig. 2.5. Take for definiteness $a = -le^{i\alpha}$ and $b = le^{i\alpha}$, where α denotes the inclination angle of the crack to the x_1-axis. Let the parametrization of the crack line Γ is given by equation

$$\zeta = \beta l e^{i\alpha}, \quad -1 \leq \beta \leq 1, \tag{2.61}$$

where $0 \leq \alpha \leq \pi$, $\alpha = \psi - \frac{3}{2}\pi$.

Introduce

$$\zeta_k(\beta) = \beta l a_k(\psi), \tag{2.62}$$

where a_k is given by (2.53).

Let the uniform MME field is given at infinity: the stresses σ_{ij}^∞, the constant electric and magnetic induction D_i^∞ and B_i^∞, respectively. Using (2.56)

56 Mechanics and Physics of Structured Media

we write the system of integral equations (2.30) in form

$$\frac{1}{\pi l} \int_{-1}^{1} \frac{Q_n(\beta)d\beta}{\sqrt{1 - \beta^2}(\beta - \beta_0)} = N_n, \quad (n = \overline{1, 4}). \tag{2.63}$$

One can see that the system of equations is decomposed onto scalar integral equations. Eq. (2.63) is called the Abel integral equation. Its solution has the form [7]

$$Q_n(\beta) = l\beta N_n, \tag{2.64}$$

where N_n are given by (2.26). Substitution of (2.64) into (2.59) yields the stress intensity factors

$$K_I = \sqrt{\pi l}\{p + \sigma_{11}^{\infty} \cos^2 \psi + \sigma_{12}^{\infty} \sin 2\psi + \sigma_{22}^{\infty} \sin^2 \psi\},$$

$$K_{II} = \sqrt{\pi l}\{\frac{\sigma_{22}^{\infty} - \sigma_{11}^{\infty}}{2} \sin 2\psi + \sigma_{12}^{\infty} \cos 2\psi\}, \tag{2.65}$$

$$K_D = \sqrt{\pi l}\{D_1^{\infty} \cos \psi + D_2^{\infty} \sin \psi\}, \tag{2.66}$$

$$K_B = \sqrt{\pi l}\{B_1^{\infty} \cos \psi + B_2^{\infty} \sin \psi\}.$$

2.6 Numerical example

Composites of cadmium selenide $CdSe$ and barium titanate $BaTiO_3$ are magneto-electro-elastic material [18]. The elastic and piezoelectric constants correspond to $CdSe$, piezomagnetic and magnetic constants to barium titanate $BaTiO_3$. The numerical constants for the coupled equations (2.1) are given below [6]

$$s_{11}/s_0 = 22.260, \quad s_{12}/s_0 = -6.437, \quad s_{22}/s_0 = 14.984, \quad s_{66}/s_0 = 47.481,$$

$$g_{16}/g_0 = 109.220, \quad g_{21}/g_0 = -4.333, \quad g_{22}/g_0 = 8.016,$$

$$p_{16}/p_0 = 268.318, \quad p_{21}/p_0 = 17.778, \quad p_{22}/p_0 = 31.206,$$

$$\beta_{11}/\beta_0 = 19.612, \quad \beta_{22}/\beta_0 = 10.612, \quad \nu_{11}/\nu_0 = 213.404, \quad \nu_{22}/\nu_0 = -5.534,$$

$$\chi_{11}/\chi_0 = 0.590, \quad \chi_{22}/\chi_0 = 0.575,$$

with

$$s_0 = 10^{-6} \text{ MPa}^{-1}, \quad g_0 = 10^{-2} \text{ MC}^{-1}\text{m}^2, \quad p_0 = 10^{-5} \text{ MT}^{-1},$$

$$\beta_0 = 10^3 \text{ MN·m}^2\text{·MC}^{-2}, \quad \nu_0 = 0.1 \text{ MC·m·MA}^{-1}, \quad \chi_0 = 0.1 \text{ MPa·MT}^{-2}.$$

Cracks in two-dimensional magneto-electro-elastic medium Chapter | 2 **57**

The next set of constants is given below

$$c_{11}/c_0 = 47.545, \ c_{12}/c_0 = 15.908, \ c_{22}/c_0 = 63.028, \ c_{66}/c_0 = 5.857,$$

$$e_{16}/e_0 = -5.992, \ e_{21}/e_0 = 6.182, \ e_{22}/e_0 = -43.176,$$

$$h_{16}/h_0 = -244.665, \ h_{21}/h_0 = -232.742, \ h_{22}/h_0 = -395.400,$$

$$\epsilon_{11}/\epsilon_0 = 8.347, \ \epsilon_{22}/\epsilon_0 = 9.065,$$

$$\beta_{11}/\beta_0 = -3.292, \ \beta_{22}/\beta_0 = -0.127, \ \gamma_{11}/\gamma_0 = 17.728, \ \gamma_{22}/\gamma_0 = 14.513$$

with

$$c_0 = 10^3 \ \text{MPa}, \ e_0 = 10^{-2} \ \text{MC·m}^{-2}, \ h_0 = 1 \ \text{MT}, \ \epsilon_0 = 10^{-5} \ \text{MN}^{-1}\text{·m}^{-2}\text{·MC}^2,$$
$$\beta_0 = 10^{-2} \ \text{MC}^{-1}\text{·m}^{-1}\text{·MA}, \ \gamma_0 = 1 \ \text{MPa}^{-1}\text{·MT}^2.$$

The coefficients $s_{ij}, g_{ij}, c_{ij}, \ldots$ do not vanish with the indicated subscripts written above. Otherwise, $s_{ij}, g_{ij}, c_{ij}, \ldots$ are equal to zero.

The roots of characteristic equation are $\mu_1 = 2.900i$, $\mu_2 = 1.254i$, $\mu_3 = -0.168 + 0.445i$, $\mu_4 = 0.168 + 0.445i$. The calculated values of $A_{1j}(1, \mu_k)$ are given below in the form of the matrix where j is the number of verse, k of column

$$\begin{bmatrix} 35283.825 & -450.502 & 312.227 - 229.664i & 312.227 + 229.664i \\ 0.441 & 0.087 & 0.005 + 0.009i & 0.005 - 0.009i \\ -2091.023 & -31.078 & 2.433 + 3.564i & 2.433 - 3.564i \end{bmatrix}.$$

$$(2.67)$$

The electromagnetic coefficients are given by numbers

$$\alpha_1^{(E)} = 895.8095i, \quad \alpha_2^{(E)} = -83.1886 + 453.1147i,$$
$$\alpha_3^{(E)} = 83.1886 + 453.1147i, \quad \alpha_4^{(E)} = 320.7244i$$

and

$$\alpha_1^{(H)} = 3.2173i, \quad \alpha_2^{(H)} = 0.3486 + 0.2210i,$$
$$\alpha_3^{(H)} = -0.3486 + 0.2210i, \quad \alpha_4^{(H)} = 0.4740i.$$

Consider two cracks (a_1, b_1) and (a_2, b_2) displayed in Figs. 2.9–2.10 with half-length $r_1 = 1$ m and $r_2 = 0.5$ m, respectively. The distance between the centers of cracks holds $h = 1.5$ m.

58 Mechanics and Physics of Structured Media

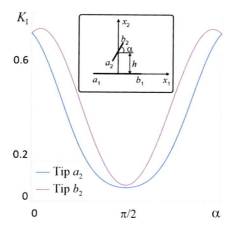

FIGURE 2.6 Stress intensity factors K_I at a_2 and b_2. The external stress holds $\sigma_{22}^\infty = 1$ MPa.

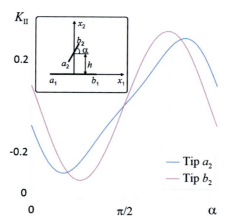

FIGURE 2.7 Stress intensity factors K_{II} at a_2 and b_2. The external stress holds $\sigma_{22}^\infty = 1$ MPa.

Let the external stress holds $\sigma_{22}^\infty = 1$ MPa. The stress intensity factor K_I at the tips of crack a_2 and b_2 depends on the angle $0 < \alpha < \pi$ as shown in Fig. 2.6. One can observe that the K_I at b_2 is greater than at a_2. The minimum value of K_I is attained at $\alpha = \frac{\pi}{2}$ for the both tips. The maximum values of K_I are attained for different angles near $\alpha = 0$.

The dependence of the stress intensity factor K_{II} at a_2 and b_2 on the angle $0 < \alpha < \pi$ is displayed in Fig. 2.7. One can observe a shift between K_I, K_{II} and different relations between K_{II} at a_2 and b_2 contrary to K_I.

The dependence of the magnetic stress intensity factor K_B at a_2 and b_2 on the angle $0 < \alpha < \pi$ is displayed in Fig. 2.8. The significant difference of K_B between the tips a_2 and b_2 is observed.

Cracks in two-dimensional magneto-electro-elastic medium Chapter | 2 59

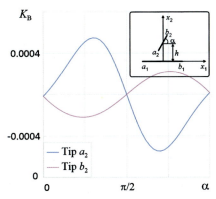

FIGURE 2.8 Stress intensity factors K_B at crack (a_2, b_2) tips. The stress field $\sigma_{22}^\infty = 1$ MPa is acting at infinity.

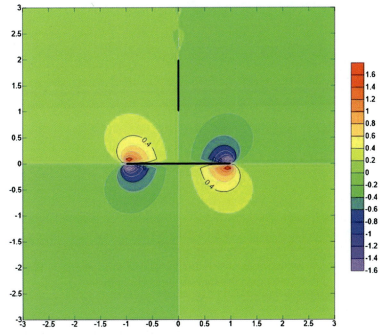

FIGURE 2.9 The local induction field D_1. The electric field $D_2^\infty = 1$ MC is given at infinity

The local electric induction fields D_1 and D_2 for two perpendicular cracks ($\alpha = \frac{\pi}{2}$) under the external field $D_2^\infty = 1$ MC are displayed in Figs. 2.9 and 2.10, respectively. The distance between the centers of cracks holds $h = 1.5$ m; the cracks half-length hold $r_1 = 1$ m, $r_2 = 0.5$ m for (a_1, b_1) and (a_2, b_2), respectively.

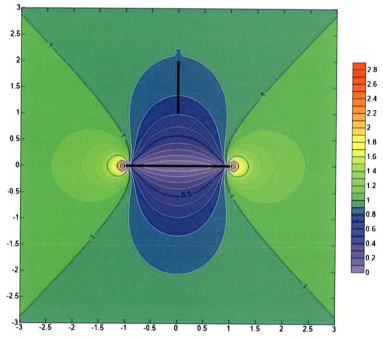

FIGURE 2.10 The local induction field D_2 under the same external field $D_2^\infty = 1$ MC given at infinity.

2.7 Conclusion

In the present chapter, the coupled mechanical, electrical, and magnetic fields in the body with cracks characteristics near their tips are developed in two-dimensional statement. The corresponding boundary value problem is stated in terms of the complex potentials which extend Lekhnitskii's complex potentials used for pure elastic fields. A method of singular integral equations is developed and implemented for the local MEE fields. The special attention is payed to computation of the stress intensity factors at the tip of the cracks.

References

[1] T. Buchukuri, O. Chkadua, R. Duduchava, D. Natroshvili, Interface crack problems for metallic-piezoelectric composite structures, Memoirs on Differential Equations and Mathematical Physics 55 (2012) 1–150.

[2] T. Buchukuri, O. Chkadua, D. Natroshvili, Mixed boundary value problems of thermopiezo-electricity for solids with interior cracks, Integral Equations and Operator Theory 64 (2009) 495–537.

[3] G.P. Cherepanov, Mechanics of Brittle Fracture, McGraw-Hill, New York, 1979.

[4] L.A. Filshtinskii (Filshtinsky), Elastic equilibrium of a plane anisotropic medium weakened by arbitrary curvilinear cracks. Limit transition to an isotropic medium, Izvestiya Akademii Nauk SSSR. Mehanika Tverdogo Tela 5 (1976) 91–97 (in Russian).

Cracks in two-dimensional magneto-electro-elastic medium Chapter | 2 **61**

[5] L.A. Filshtinskii (Filshtinsky), Boundary-value problems of theory of elasticity for anisotropic half-plane weakened by hole or cut, Izvestiya Akademii Nauk SSSR. Mehanika Tverdogo Tela 6 (1982) 72–79 (in Russian).

[6] L.A. Filshtinskii (Filshtinsky), Tutorial to the implementation of master's theses on the topic: "Boundary value problems of fracture mechanics of new composite materials", Sumy University Publ., 2015 (in Russian).

[7] F.D. Gakhov, Boundary Value Problems, Pergamon Press, Oxford, etc., 1966.

[8] S.G. Lekhnitskii, Theory of Elasticity of an Anisotropic Elastic Body, Mir Publishers, Moscow, 1981.

[9] I.K. Lifanov, Method of Singular Integral Equations and Numerical Experiment, Yanus, Moscow, 1995 (in Russian).

[10] V. Mityushev, I. Andrianov, S. Gluzman, L.A. Filshtinsky's contribution to Applied Mathematics and Mechanics of Solids, Chapter 1 of the present book.

[11] V. Mityushev, D. Nosov, R. Wojnar, Two-dimensional equations of magneto-electro-elasticity, Chapter 3 of the present book.

[12] N.I. Muskhelishvili, Some Mathematical Problems of the Plane Theory of Elasticity, Springer, Dordrecht, 1977.

[13] V.V. Panasyuk, M.P. Savruk, A.P. Datsyshyn, Stress Distribution Around the Crack in Plates and Shells, Naukova Dumka, Kyiv, 1976 (in Russian).

[14] W.Y. Tian, R.K.N.D. Rajapakse, Fracture analysis of magnetoelectroelastic solids by using path independent integrals, International Journal of Fracture 131 (2005) 311–335.

[15] V.S. Vladimirov, Equations of Mathematical Physics, 2nd ed., Mir Publishers, Moscow, 1983.

[16] B.L. Wang, Y.W. Mai, Fracture of piezoelectromagnetic materials, Mechanics Research Communications 31 (2004) 65–73.

[17] B.L. Wang, Y.W. Mai, Crack tip field in piezoelectric/piezomagnetic media, European Journal of Mechanics. A, Solids 22 (2003) 591–602.

[18] Y. Yamamoto, K. Miya, Electromagnetomechanical Interactions in Deformable Solids and Structures, Elsevier, Amsterdam, 1987.

Chapter 3

Two-dimensional equations of magneto-electro-elasticity

Vladimir Mityushev[a], Dmytro Nosov[b,c], and Ryszard Wojnar[d]

[a]*Research Group Materialica+, Faculty of Computer Science and Telecommunications, Cracow University of Technology, Kraków, Poland,* [b]*Constantine the Philosopher University, Nitra, Slovakia,* [c]*Pedagogical University, Kraków, Poland,* [d]*Institute of Fundamental Technological Research PAS, Warszawa, Poland*

3.1 Introduction

Magneto-electro-elastic (MEE) composites extend the physical notion of piezoelectric composites when the electric and elastic fields are coupled with the magnetic field. MEE composites and multiferroic phases belong to a new class of materials which find wide applications in modern technology. Their peculiar property is the cross-coupling between electric polarization, magnetization and elastic fields, what creates new possibilities for electronic devices (sensors, actuators, transducers, memories). The introduction of composites offers a practical method for realizing strong interactions of magnetic and electric fields at room temperature.

Piezoelectric materials are anisotropic [54]. Therefore, the complex potentials by Kolosov-Muskhelishvili do not fit to model the piezoelectric fields in two-dimensional (2D) composites except at the special cases discussed below. Fortunately, Lekhnitskii extended the method of complex potentials in 1939 to anisotropic 2D elastic problems and summarized the results in his fundamental book [35]. The extension of Lekhnitskii's formalism to the 2D piezoelectric materials was first made perhaps in 1975 by Kosmodamianskii and Lozhkin [30]. In the same time, Kudryavtsev, Parton, and Rakitin [33] investigated a crack on the boundary of a two-phase piezoelectrics. Later, this and other results on the electroelastic waves and the heat conduction in piezoelectrics were summarized in the book [45]. Kosmodamianskii and his coauthors solved the plane problems with holes, cracks and inclusions in 1975–1984. They considered piezoelectrics with an elliptical hole, for a point source in half-plane, etc. [31,32]. Guz et al. selected the results on the coupled fields up to 1988 in five volumes of the treatise [16]. Unfortunately, these works were published in local hardly accessible journals and publishing companies in Russian.

Mechanics and Physics of Structured Media. https://doi.org/10.1016/B978-0-32-390543-5.00008-6
Copyright © 2022 Elsevier Inc. All rights reserved.

A short introduction to the general 2D MEE fields and contribution by Filshtinsky are described in Chapter [43]. One may divide the complex potentials methods in 2D MEE problems onto the analytical method and the method of integral equations. The present chapter is devoted to an analytical approach. Filshtinsky included into group of scientists worked in piezoelectricity in 1979 [5,22]. He prepared reprint in 2015 on the method of integral equations for MEE composites discussed in the separate chapter [44].

It is worth noting that there is the "parallel Universe" of the same investigations beginning from the paper by D.M. Barnett and J. Lothe [4] published in 1975. This paper [4] extended Stroh's eigenvalue equation (1958) equivalent to Lekhnitskii's characteristic (1939) equation to dislocated piezoelectric insulators in the whole piezoelectric plane. This corresponds to Kosmodamianskii's result for a point source. A part of results by Kosmodamianskii, Filshtinsky and others were repeated in terms of Stroh's formalism later, see [13,28] and references therein. New results on the mathematical problems of piezoelectricity obtained in the last years besides Filshtinsky can be found in [13, Chapter 11]. The special attention deserves the exact solution in series of the plane elastic problem for an elliptic anisotropic inclusion embedded in matrix having other type of anisotropy [13, Chapter 8]. In the previous works, an anisotropic media with holes were successfully discussed by conformal mappings and by application of the Riemann-Hilbert type problem. The study of inclusions instead of holes essentially complicates the corresponding boundary value problem for analytic functions, since the $\mathbb{R}$-linear condition contains different $\mathbb{R}$-linear shifts on the opposite sides of contour. Below, we write such a problem in the form (3.66). It is a new boundary value problem in the theory of analytic functions called by $\mathbb{R}\mathbb{R}$-linear problem in the present chapter.

In Section 3.2 of the present paper, we develop Lekhnitskii's formalism to 2D MEE problems following Filshtinsky's preprint [21]. The MEE fields are written in terms of complex potentials. It is worth noting that Filshtinsky's extension of Lekhnitskii's formalism does not reduce to a formal increasing of analytic functions from two to four. Section 3.3 is devoted to statement of boundary value problems for complex potentials in the vector-matrix form. In Section 3.4, we formally present Filshtinsky's method for dielectrics, i.e., for a problem for one complex potential. Section 3.4 does not contain a new result, but clearly demonstrate the difference in these approaches and the powerful perspectives of Filshtinsky's method. In particular, the $\mathbb{R}$-linear problem is stated similar to the antiplane piezoelectric problems discussed in [46] and to the problems solved in [14,47,48].

One of the widely used mathematical methods to describe the effective properties of heterogeneous bodies is the method of two-scale asymptotic homogenization described in monographs whose authors are Bensoussans, Lions, and Papanicolaou [6], Sanchez-Palencia [49], Bakhvalov and Panasenko [3], and Jikov et al. [29]. Section 3.6 is devoted to homogenization of MME composites. In [52] the formulas were derived by Telega for the effective moduli

Two-dimensional equations of magneto-electro-elasticity Chapter | 3 **65**

of a piezoelectric composite exhibiting fine periodic structure. The static case considered was studied by the method of Γ-convergence, [2]. In contribution [23] the preliminary results concerning homogenization of the equations of the linear thermopiezoelectricity were presented. To such a problem the method of Γ-convergence is not applicable. Therefore the relevant analysis was performed by using the method of two-scale asymptotic expansions. On the basis of results [23] a two-component thermopiezoelectric material was discussed. By applying the Ritz method *local* problems were numerically solved in [24]. In [53] the effective properties of physically nonlinear piezoelectric composites were investigated. Gambin [26] gave an in-depth analysis of influence of microstructure on properties of elastic, piezoelectric, and thermoelastic composites.

The constructive homogenization formulas for fibrous composites are outlined in Section 3.7. We concern Hill's type universal relations for fibrous composites following [7,37,50,51]. Such a relation minimizes the number of the special independent problems needed for determination of the effective constants. The set of problems can be optimized by the reduction of the coupled problem to uncoupled equations when the coupled constitutive matrix is symmetric [37,51], see Section 3.7.1.

The Natanzon-Filshtinsky method was extended in [9] to the square and hexagonal arrays of piezoelectric composites in the antiplane statement. The problem was reduced to an infinite system of linear algebraic equations solved numerically by the truncation method.[1] A series of exact formulas known for the antiplane statement of composites was automatically extended to piezoelectric composites by a matrix transformation of the $\mathbb{R}$-problem introduced in [46]. However, the approach [46] had a technical restriction related to a condition on matrix transformations in the real space. This restriction is removed in Section 3.7.2. This result completes the constructive theory of the piezoelectric longitudinal transversely isotropic fibrous composites. More precisely, any formula for the scalar conductivity of 2D composite can be transformed by a matrix formalism to a formula for the effective tensor of piezoelectric fibrous composite.

3.2 2D equations of magneto-electro-elasticity

Consider fibrous composites formed by unidirectional infinite cylinders shown in Fig. 3.1. It is assumed that the cylinder axis coincides with the x_3-axis and a section parallel to the plane (x_1, x_2) represents the fibrous composite.

Introduce the complex variable $z = x_1 + ix_2$, where $i = \sqrt{-1}$ denotes the imaginary unit. Consider a domain G on the extended complex plane $\widehat{\mathbb{C}} \cup \{\infty\}$ with the Lyapunov boundary ∂G. The domain G can be unbounded, multiply connected and even not connected. We will also discuss the continuous boundary ∂G consisting of a finite number of the closed Lyapunov arcs.

---

[1] It should be noted that the term *closed form solution* is used misleadingly in [9], see the discussion in review [1].

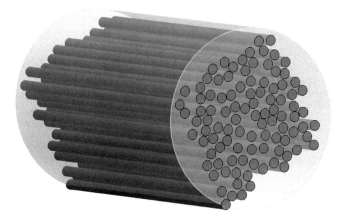

FIGURE 3.1 Fibrous composite.

In the present section, we introduce complex potentials for the local linear MEE equations in a plane domain G following Filshtinsky's approach [17–21].

3.2.1 Linear equations of magneto-electro-elasticity and potentials

Let $\partial_j \equiv \frac{\partial}{\partial x_j}$ denote the partial derivative ($i, j = 1, 2$). Introduce the following designations for the local fields. Let σ_{ij} ($i, j = 1, 2$) denote the stresses, $\epsilon_{ij} = \frac{1}{2}(\partial_i u_j + \partial_j u_i)$ the strains expressed by means of the displacement u_j ($i, j = 1, 2$), D_j and E_j ($j = 1, 2$) electric displacement and electric fields, B_j and H_j ($j = 1, 2$) magnetic induction and magnetic displacement fields. The equilibrium differential equations of elastic, electric, and magnetic fields have the classic form

$$\partial_1 \sigma_{11} + \partial_2 \sigma_{12} = 0, \quad \partial_1 \sigma_{12} + \partial_2 \sigma_{22} = 0, \tag{3.1}$$

$$\partial_1 B_1 + \partial_2 B_2 = 0, \quad \partial_1 D_1 + \partial_2 D_2 = 0, \tag{3.2}$$

$$\partial_2 H_1 - \partial_1 H_2 = 0, \quad \partial_2 E_1 - \partial_1 E_2 = 0. \tag{3.3}$$

The compatibility condition for strains has the form

$$\partial_1^2 \epsilon_{22} + \partial_2^2 \epsilon_{11} = 2\partial_1 \partial_2 \epsilon_{12}. \tag{3.4}$$

The local fields are related by the linear constitutive equations

$$\epsilon_{11} = s_{11}\sigma_{11} + s_{12}\sigma_{22} + s_{16}\sigma_{12} + g_{11}D_1 + g_{21}D_2 + p_{11}B_1 + p_{21}B_2,$$
$$\epsilon_{22} = s_{12}\sigma_{11} + s_{22}\sigma_{22} + s_{26}\sigma_{12} + g_{12}D_1 + g_{22}D_2 + p_{12}B_1 + p_{22}B_2,$$
$$2\epsilon_{12} = s_{16}\sigma_{11} + s_{26}\sigma_{22} + s_{66}\sigma_{12} + g_{16}D_1 + g_{26}D_2 + p_{16}B_1 + p_{26}B_2,$$
$$E_1 = -g_{11}\sigma_{11} - g_{12}\sigma_{22} - g_{16}\sigma_{12} + \beta_{11}D_1 + \beta_{12}D_2 + \nu_{11}B_1 + \nu_{12}B_2,$$
$$E_2 = -g_{21}\sigma_{11} - g_{22}\sigma_{22} - g_{26}\sigma_{12} + \beta_{12}D_1 + \beta_{22}D_2 + \nu_{12}B_1 + \nu_{22}B_2,$$

$$H_1 = -p_{11}\sigma_{11} - p_{12}\sigma_{22} - p_{16}\sigma_{12} + \nu_{11}D_1 + \nu_{12}D_2 + \chi_{11}B_1 + \chi_{12}B_2,$$
$$H_2 = -p_{21}\sigma_{11} - p_{22}\sigma_{22} - p_{26}\sigma_{12} + \nu_{12}D_1 + \nu_{22}D_2 + \chi_{12}B_1 + \chi_{22}B_2.$$
$$(3.5)$$

Here, the coefficients s_{ij} describe Hook's elastic law; g_{ij} and p_{ij} piezoelectric and piezomagnetic effects; β_{ij}, ν_{ij}, and χ_{ij} dielectric, magnetic, and electromagnetic effects. The differential equations (3.1)–(3.4) separately describe elastic, electric, and magnetic fields. The constitutive equations (3.5) express linear interactions between the considered fields.

3.2.2 Complex representation of field values

Using the elasticity equations (3.1) we introduce the standard Airy function $F_1(x_1, x_2)$ in such a way that

$$\sigma_{11} = \partial_2^2 F_1, \quad \sigma_{12} = -\partial_1\partial_2 F_1, \quad \sigma_{22} = \partial_1^2 F_1. \quad (3.6)$$

The electric and magnetic potentials U and $\varphi^{(H)}$, respectively, satisfy the relations

$$E_1 = -\partial_1 U, \quad E_2 = -\partial_2 U, \quad H_1 = -\partial_1\varphi^{(H)}, \quad H_2 = -\partial_2\varphi^{(H)}. \quad (3.7)$$

Then, Eqs. (3.3) are fulfilled identically. Along similar lines, the magnetic induction and magnetic displacement are expressed by means of the functions $F_2(x_1, x_2)$ and $F_3(x_1, x_2)$ as follows

$$D_1 = \partial_2 F_2, \quad D_2 = -\partial_1 F_2, \quad B_1 = \partial_2 F_3, \quad B_2 = -\partial_1 F_3. \quad (3.8)$$

Then, Eqs. (3.2) are fulfilled.

Substitute (3.7) and (3.8) into the right part of (3.5). As a result we obtain an expression of ϵ_{ij}, E_i and H_i in terms of the functions $F_j(x_1, x_2)$ not explicitly written here. Substituting these expressions into three equations (3.3) and (3.4) we arrive at the system of differential equations on $F_j(x_1, x_2)$

$$\sum_{j=1,2,3} L_{ij}(\partial_1, \partial_2)F_j = 0, \quad i = 1, 2, 3, \quad (3.9)$$

where the differential operators L_{ij} have the form

$$L_{11}(\partial_1, \partial_2) = s_{11}\partial_2^4 - 2s_{16}\partial_1\partial_2^3 + (2s_{12} + s_{66})\partial_1^2\partial_2^2 - 2s_{26}\partial_1^3\partial_2 + s_{22}\partial_1^4,$$
$$L_{12}(\partial_1, \partial_2) = g_{11}\partial_2^3 - g_{22}\partial_1^3 - (g_{21} + g_{16})\partial_1\partial_2^2 + (g_{12} + g_{26})\partial_1^2\partial_2,$$
$$L_{13}(\partial_1, \partial_2) = p_{11}\partial_2^3 - p_{22}\partial_1^3 - (p_{21} + p_{16})\partial_1\partial_2^2 + (p_{12} + p_{26})\partial_1^2\partial_2,$$
$$L_{22}(\partial_1, \partial_2) = -\beta_{11}\partial_2^2 - \beta_{22}\partial_1^2 + 2\beta_{12}\partial_1\partial_2, \quad (3.10)$$
$$L_{23}(\partial_1, \partial_2) = -\nu_{11}\partial_2^2 - \nu_{22}\partial_1^2 + 2\nu_{12}\partial_1\partial_2,$$
$$L_{33}(\partial_1, \partial_2) = -\chi_{11}\partial_2^2 - \chi_{22}\partial_1^2 + 2\chi_{12}\partial_1\partial_2.$$

68 Mechanics and Physics of Structured Media

We now proceed to derive a representation of $F_j(x_1, x_2)$ in terms of the complex potentials to identically fulfill the differential equations (3.9). Let $A = \{A_{ij}\}$ denote the matrix of cofactors to the 3×3 matrix $L = \{L_{ij}\}$. The components A_{1j} are explicitly written below

$$
\begin{aligned}
A_{11}(\partial_1, \partial_2) &= (\beta_{22}\chi_{22} - \nu_{22}^2)\partial_1^4 + (\beta_{11}\chi_{22} - 2\nu_{11}\nu_{12} + \beta_{22}\chi_{11})\partial_1^2\partial_2^2 \\
&\quad + (\beta_{11}\chi_{11} - \nu_{11}^2)\partial_2^4, \hspace{3cm} (3.11) \\
A_{12}(\partial_1, \partial_2) &= (g_{22}\chi_{22} + p_{22}\nu_{22})\partial_1^5 + (g_{22}\chi_{11} + p_{22}\nu_{11} \\
&\quad + g_{62}\chi_{22} + p_{62}\nu_{22})\partial_1^3\partial_2^2 + (g_{62}\chi_{11} + p_{62}\nu_{11})\partial_1\partial_2^4, \\
A_{13}(\partial_1, \partial_2) &= -(g_{22}\nu_{22} + p_{22}\beta_{22})\partial_1^5 - (g_{22}\nu_{11} + p_{22}\beta_{11} \\
&\quad + g_{62}\nu_{22} + p_{62}\beta_{22})\partial_1^3\partial_2^2 - (g_{62}\nu_{11} + p_{62}\beta_{11})\partial_1\partial_2^4.
\end{aligned}
$$

We shall need only these components A_{1j}.

Let $\Psi(x_1, x_2)$ be a sufficiently smooth function. Substitute the expressions

$$
F_j = A_{1j}\Psi, \quad j = 1, 2, 3, \tag{3.12}
$$

into (3.9). The first equation (3.9) becomes

$$
\sum_{j=1,2,3} L_{1j} A_{1j}\Psi \equiv \det(L)\,\Psi = 0, \tag{3.13}
$$

where the Laplace expansion formula for the determinant $\det(L)$ is used. The second and third equations (3.9) are fulfilled identically because of Cramer's formulas. One can consider the transformation of the system (3.9) to Eq. (3.13) as elimination of F_2, F_3 and the introduction of the new unknown function $F_1 = A_{11}\Psi$.

The differential equation (3.13) can be solved by the characteristic equation method [35] based on the representation

$$
\Psi(x_1, x_2) = f(x_1 + \mu x_2), \tag{3.14}
$$

where $f(z)$ is sufficiently smooth function and μ is an undetermined constant. Substitution of (3.14) into (3.13) yields the algebraic equation of 8th power with real coefficients

$$
\det L(1, \mu) = 0. \tag{3.15}
$$

The positively defined matrices $\{s_{ij}\}$, $\{\beta_{ij}\}$, and $\{\chi_{ij}\}$ imply that the roots of (3.15) are complex [21,35]. Below, we consider only the case of different roots. Let us take four roots μ_k ($k = 1, 2, 3, 4$) having the positive imaginary parts and introduce the corresponding complex variables

$$
z_k = x_1 + \mu_k x_2 \quad (k = 1, 2, 3, 4). \tag{3.16}
$$

Eq. (3.16) can be considered as the $\mathbb{R}$-linear transformations of the complex plane

$$z_k = \frac{1 - i\mu_k}{2}z + \frac{1 + i\mu_k}{2}\overline{z} \quad (k = 1, 2, 3, 4). \tag{3.17}$$

Remark. If one of the values μ_k is equal to i, say for definiteness $\mu_1 = i$, Eq. (3.17) becomes $z_1 = z$. This case does not simplify the study, since all the other roots are not equal to i by the assumption. More precisely, they can be equal to i, but then, the representation (3.12) must be replaced by another which holds in the theory of differential equations with multiple roots μ_k.

Introduce the domain $G_k = \{z_k = x_1 + \mu_k x_2 \in \mathbb{C} : (x_1, x_2) \in G\}$. The general solution to the differential equation (3.13) has the form

$$\Psi(x_1, x_2) = 2\mathrm{Re} \sum_{k=1,2,3,4} c_k \varphi_k(z_k), \tag{3.18}$$

where c_k are complex constants, the function $\varphi_k(z_k)$ is analytic in the domain G_k in the variable z_k.

Let $\Phi_k = \frac{d^6\varphi_k}{dz_k^6}$ denote the sixth order derivative of the function $\varphi_k(z_k)$. Substitution of (3.18), (3.12) into (3.6), (3.8) yields the complex representation of local fields

$$\begin{pmatrix} \sigma_{11} \\ \sigma_{12} \\ \sigma_{22} \end{pmatrix} = 2\mathrm{Re} \sum_{k=1}^{4} A_{11}(1, \mu_k)\Phi_k(z_k) \begin{pmatrix} \mu_k^2 \\ -\mu_k \\ 1 \end{pmatrix}, \tag{3.19}$$

$$\begin{pmatrix} D_1 \\ D_2 \end{pmatrix} = 2\mathrm{Re} \sum_{k=1}^{4} A_{12}(1, \mu_k)\Phi_k(z_k) \begin{pmatrix} \mu_k \\ -1 \end{pmatrix}, \tag{3.20}$$

$$\begin{pmatrix} B_1 \\ B_2 \end{pmatrix} = 2\mathrm{Re} \sum_{k=1}^{4} A_{13}(1, \mu_k)\Phi_k(z_k) \begin{pmatrix} \mu_k \\ -1 \end{pmatrix}. \tag{3.21}$$

Here, $A_{1j}(1, \mu_k)$ are polynomial expressions in μ_k. The left parts of (3.19)–(3.21) depend on $(x_1, x_2) \in G$ through the complex variables (3.16) in the right parts.

Substitution of (3.19)–(3.21) into (3.5) yields

$$\begin{pmatrix} E_1 \\ E_2 \end{pmatrix} = 2\mathrm{Re} \sum_{k=1}^{4} \alpha_k^{(E)}\Phi_k(z_k) \begin{pmatrix} 1 \\ \mu_k \end{pmatrix} \tag{3.22}$$

70 Mechanics and Physics of Structured Media

and

$$\begin{pmatrix} H_1 \\ H_2 \end{pmatrix} = 2\mathrm{Re} \sum_{k=1}^{4} \alpha_k^{(H)} \Phi_k(z_k) \begin{pmatrix} 1 \\ \mu_k \end{pmatrix}, \qquad (3.23)$$

where

$$\alpha_k^{(E)} = \mu_k[g_{16}A_{11}(1,\mu_k) + \beta_{11}A_{12}(1,\mu_k) + \nu_{11}A_{13}(1,\mu_k)] \qquad (3.24)$$

and

$$\alpha_k^{(H)} = \mu_k[p_{16}A_{11}(1,\mu_k) + \nu_{11}A_{12}(1,\mu_k) + \chi_{11}A_{13}(1,\mu_k)]. \qquad (3.25)$$

The coefficients on A_{1j} are taken from the constitutive equations (3.5). The strains are expressed by means of the complex potentials as follows

$$\partial_1 u_1 = \epsilon_{11} = 2\mathrm{Re} \sum_{k=1}^{4} r_k \Phi_k(z_k), \quad \partial_2 u_2 = \epsilon_{22} = 2\mathrm{Re} \sum_{k=1}^{4} \mu_k q_k \Phi_k(z_k),$$

$$(3.26)$$

where

$$r_k = (s_{11}\mu_k^2 + s_{12})A_{11}(1,\mu_k) - g_{21}A_{12}(1,\mu_k) - p_{21}A_{13}(1,\mu_k)$$

and

$$q_k = (s_{12}\mu_k + \frac{s_{22}}{\mu_k})A_{11}(1,\mu_k) - \frac{g_{22}}{\mu_k}A_{12}(1,\mu_k) - \frac{p_{22}}{\mu_k}A_{13}(1,\mu_k).$$

After integration of (3.26) we obtain the elastic displacement up to an additive constant vector

$$\begin{pmatrix} u_1 \\ u_2 \end{pmatrix} = 2\mathrm{Re} \sum_{k=1}^{4} \varphi_k(z_k) \begin{pmatrix} r_k \\ q_k \end{pmatrix}. \qquad (3.27)$$

3.3 Boundary value problem

In the present section, we consider the problem with one hole or inclusion embedded in the MEE host material. It is assumed that the complex plane is divided by a simple closed Lyapunov curve Γ onto the simply connected bounded domain G^+ and its complement $G \equiv G^-$ to the extended complex plane $\widehat{\mathbb{C}} = \mathbb{C} \cup \{\infty\}$.

The problem for a hole is stated in the following way. Let a MEE material occupy the domain G. The domain G^+ is considered as a hole. The field is given on the boundary of the hole $\Gamma = \partial G^+$ oriented in the positive direction, i.e., G^+ is on the left side of Γ. It is necessary to find the field in the domain G. The boundary field on Γ can be given in various forms as it is in the classic

Two-dimensional equations of magneto-electro-elasticity **Chapter | 3 71**

theories of separated electric, magnetic and elastic fields. We consider one of the possible statements.

Let θ denote the angle between the x_1-axis and the normal outward unit vector $\mathbf{n} = (\cos\theta, \sin\theta)$ to Γ. Then, the arc length vector along Γ has the form $\mathbf{s} = (-\sin\theta, \cos\theta)$. The differential $dz_k = dx_1 + \mu_k dx_2$ and the infinitesimal arc length ds on Γ are related by equation $dz_k = g_k(\theta)ds$ where $g_k(\theta) = \mu_k \cos\theta - \sin\theta$. The function $g_k(\theta)$ can be written as a function in z and $\bar{z}$ as it is done for a circle in Section 3.5. The normal stresses vector applied to Γ are given by formula

$$\begin{pmatrix} X_1 \\ X_2 \end{pmatrix} = \begin{pmatrix} \sigma_{11} & \sigma_{12} \\ \sigma_{12} & \sigma_{22} \end{pmatrix} \begin{pmatrix} \cos\theta \\ \sin\theta \end{pmatrix}. \tag{3.28}$$

Using (3.19) we obtain

$$\begin{pmatrix} X_1 \\ X_2 \end{pmatrix} = 2\text{Re} \sum_{k=1}^{4} A_{11}(1, \mu_k) g_k \Phi_k(z_k) \begin{pmatrix} \mu_k \\ -1 \end{pmatrix}. \tag{3.29}$$

The normal components of the vectors (3.20)–(3.21) are determined by formulas

$$D_\mathbf{n} = 2\text{Re} \sum_{k=1}^{4} A_{12}(1, \mu_k) g_k \Phi_k(z_k) \tag{3.30}$$

and

$$B_\mathbf{n} = 2\text{Re} \sum_{k=1}^{4} A_{13}(1, \mu_k) g_k \Phi_k(z_k). \tag{3.31}$$

The boundary values of the stress vector and of the scalars $D_\mathbf{n}$ and $B_\mathbf{n}$ on Γ can be given in the form of the vector-function $\mathbf{X} = (X_1, X_2, D_\mathbf{n}, B_\mathbf{n})^T$ depending on $(x_1, x_2) \in \Gamma$. Introduce the matrices

$$\mathbf{A} = \begin{pmatrix} \mu_1 A_{11}(1, \mu_1) & \mu_2 A_{11}(1, \mu_2) & \mu_3 A_{11}(1, \mu_3) & \mu_4 A_{11}(1, \mu_4) \\ -A_{11}(1, \mu_1) & -A_{11}(1, \mu_2) & -A_{11}(1, \mu_3) & -A_{11}(1, \mu_4) \\ A_{12}(1, \mu_1) & A_{12}(1, \mu_2) & A_{12}(1, \mu_3) & A_{12}(1, \mu_4) \\ A_{13}(1, \mu_1) & A_{13}(1, \mu_2) & A_{13}(1, \mu_3) & A_{13}(1, \mu_4) \end{pmatrix} \tag{3.32}$$

and

$$\mathbf{g} = \begin{pmatrix} g_1 & 0 & 0 & 0 \\ 0 & g_2 & 0 & 0 \\ 0 & 0 & g_3 & 0 \\ 0 & 0 & 0 & g_4 \end{pmatrix}. \tag{3.33}$$

72 Mechanics and Physics of Structured Media

Then, the boundary conditions on Γ can be written in the form of the vector-matrix Riemann-Hilbert problem with the shifts (3.17)

$$2\text{Re } \mathbf{Ag } \mathbf{\Phi} \circ \boldsymbol{\gamma}(z) = \mathbf{X}(z), \quad z \in \Gamma, \tag{3.34}$$

where the matrix-function $\mathbf{A}$ and $\mathbf{g}$ are defined on Γ. The shift $\boldsymbol{\gamma}(z)$ of the curve Γ is introduced in every component of $\mathbf{\Phi}$

$$\mathbf{\Phi} \circ \boldsymbol{\gamma}(z) := (\Phi_1(z_1), \Phi_2(z_2), \Phi_3(z_3), \Phi_4(z_4))^T. \tag{3.35}$$

Assume that the stresses σ_{ij}^∞, the electric displacement D_i^∞ and the magnetic induction B_i^∞ are given at infinity. Substituting the infinite point into (3.19)–(3.21) we obtain seven real equations on four complex constants $C_k = \Phi_k(\infty)$ $(k = 1, 2, 3, 4)$

$$\begin{pmatrix} \sigma_{11}^\infty \\ \sigma_{12}^\infty \\ \sigma_{22}^\infty \end{pmatrix} = 2\text{Re} \sum_{k=1}^{4} A_{11}(1, \mu_k) C_k \begin{pmatrix} \mu_k^2 \\ -\mu_k \\ 1 \end{pmatrix},$$

$$\begin{pmatrix} D_1^\infty \\ D_2^\infty \end{pmatrix} = 2\text{Re} \sum_{k=1}^{4} A_{12}(1, \mu_k) C_k \begin{pmatrix} \mu_k \\ -1 \end{pmatrix}, \tag{3.36}$$

$$\begin{pmatrix} B_1^\infty \\ B_2^\infty \end{pmatrix} = 2\text{Re} \sum_{k=1}^{4} A_{13}(1, \mu_k) C_k \begin{pmatrix} \mu_k \\ -1 \end{pmatrix}.$$

One of the complex constant, say C_4, can be taken real, i.e., Im $C_4 = 0$. Therefore, the number of real unknowns in the system (3.36) holds seven. We do not investigate general equations on C_k in the present paper and assume that a set of constants $\{C_k, \ k = 1, 2, 3, 4\}$ is determined. Anyway, the existence and uniqueness of solution for the given field at infinity [21] implies the unique existence of the constants C_k satisfying the system (3.36).

3.4 Dielectrics

In the present section, we consider a particular case of (3.5) when all the effects vanish besides the electric displacement $\mathbf{D} = (D_1, D_2)$ and electric field $\mathbf{E} = (E_1, E_2)$. The following equations are considered in a simply connected domain G

$$\partial_1 D_1 + \partial_2 D_2 = 0, \quad \partial_2 E_1 - \partial_1 E_2 = 0, \tag{3.37}$$

and

$$E_1 = \beta_{11} D_1 + \beta_{12} D_2, \quad E_2 = \beta_{12} D_1 + \beta_{22} D_2. \tag{3.38}$$

It is assumed for definiteness that infinity belongs to G. The complement of the closure $G \cup \Gamma$ to the extended complex plane is considered as a hole. We formally follow the general scheme described in the previous section in order to explicitly demonstrate Filshtinsky's method and to compare it to the classic method of complex potential in the considered simple case.

The Airy function $F \equiv F_2$ from Eqs. (3.8) is introduced in such a way that

$$D_1 = \partial_2 F, \quad D_2 = -\partial_1 F. \tag{3.39}$$

Substitute (3.39) into (3.38)

$$E_1 = \beta_{11}\partial_2 F - \beta_{12}\partial_1 F, \quad E_2 = \beta_{12}\partial_2 F - \beta_{22}\partial_1 F. \tag{3.40}$$

Substituting these expressions into the second equation (3.37) and using the symmetry of the dielectric tensor we arrive at the differential equation

$$(\beta_{11}\partial_2^2 - 2\beta_{12}\partial_1\partial_2 + \beta_{22}\partial_1^2)F = 0. \tag{3.41}$$

The matrix operator (3.11) is introduced in order to reduce the system of differential equations (3.9) to the differential equation (3.13). This step is skipped for the differential equation (3.41). The function $F(x_1, x_2)$ is found from the representation

$$F(x_1, x_2) = f(x_1 + \mu x_2) \tag{3.42}$$

which yields the second order algebraic equation [35]

$$\beta_{11}\mu^2 - 2\beta_{12}\mu + \beta_{22} = 0. \tag{3.43}$$

The quadratic equation (3.43) has two complex conjugated roots since its discriminant $4(\beta_{12}^2 - \beta_{11}\beta_{22})$ is negative due to the positiveness of the dielectric tensor [35]. Therefore, the function (3.42) can be expressed through a function $\varphi(z_1)$ analytic in the variable $z_1 = x_1 + \mu x_2$

$$F(x_1, x_2) = 2\mathrm{Re}\,\varphi(z_1). \tag{3.44}$$

Here, for definiteness the root $\mu = \frac{1}{\beta_{11}}(\beta_{12} + i\Delta_\beta)$ with the positive imaginary part is taken. The designation $\Delta_\beta = \sqrt{\beta_{11}\beta_{22} - \beta_{12}^2}$ is introduced for shortness.

The local fields in G are expressed in terms of the complex potential $\varphi(z_1)$. Let $z_1 = x + iy$, where

$$x = x_1 + x_2\frac{\beta_{12}}{\beta_{11}}, \quad y = x_2\frac{\Delta_\beta}{\beta_{11}} \quad \Leftrightarrow \quad x_1 = x - \frac{\beta_{12}}{\Delta_\beta}y, \quad x_2 = \frac{\beta_{11}}{\Delta_\beta}y. \tag{3.45}$$

Calculate the derivative of the analytic function $\varphi(z_1)$ through the partial derivatives of its real part $U(x, y) = \frac{1}{2}F(x_1, x_2)$

$$\varphi'(z_1) = \partial_x U - i\partial_y U = \frac{1}{2}\partial_1 F - \frac{i}{2\Delta_\beta}(-\beta_{12}\partial_1 F + \beta_{11}\partial_2 F). \tag{3.46}$$

74 Mechanics and Physics of Structured Media

This formula can be written in the form

$$\text{Re }\varphi'(z_1) = \frac{1}{2}\partial_1 F, \quad \text{Im }\varphi'(z_1) = \frac{1}{2\Delta_\beta}(\beta_{12}\partial_1 F - \beta_{11}\partial_2 F). \qquad (3.47)$$

Express the derivatives $\partial_1 F$ and $\partial_2 F$ in terms of $\text{Re }\varphi'(z_1)$ and $\text{Im }\varphi'(z_1)$

$$\partial_1 F = 2\text{Re }\varphi'(z_1), \quad \partial_2 F = 2\left(\frac{\beta_{12}}{\beta_{11}}\text{Re }\varphi'(z_1) - \frac{\Delta_\beta}{\beta_{11}}\text{Im }\varphi'(z_1)\right). \qquad (3.48)$$

The second equation (3.48) is equivalent to the following one

$$\partial_2 F = 2\text{Re}[\mu\varphi'(z_1)]. \qquad (3.49)$$

Eqs. (3.48) and (3.39) imply that

$$D_1 = 2\left(\frac{\beta_{12}}{\beta_{11}}\text{Re }\varphi'(z_1) - \frac{\Delta_\beta}{\beta_{11}}\text{Im }\varphi'(z_1)\right), \quad D_2 = -2\text{Re }\varphi'(z_1). \qquad (3.50)$$

Eqs. (3.38) yield

$$E_1 = -2\Delta_\beta\text{Im }\varphi'(z_1), \quad E_2 = -\frac{2\Delta_\beta}{\beta_{11}}\left(\beta_{12}\text{Im }\varphi'(z_1) + \Delta_\beta\text{Re }\varphi'(z_1)\right). \qquad (3.51)$$

One can check that the second equation (3.51) can be written in the form $E_2 = 2\Delta_\beta\text{Im}\,[\mu\varphi'(z_1)]$. Eqs. (3.50)–(3.51) express the components of dielectric field in terms of the complex potential $\varphi'(z_1)$.

Remark. The potential U can be used instead of the function F_2 in the following way. Using (3.38) express the displacement (D_1, D_2) in terms of (E_1, E_2)

$$D_1 = \frac{1}{\Delta_\beta^2}(\beta_{22}E_1 - \beta_{12}E_2), \quad D_2 = \frac{1}{\Delta_\beta^2}(-\beta_{12}E_1 + \beta_{11}E_2). \qquad (3.52)$$

Substitute the relations

$$E_1 = -\partial_1 U, \quad E_2 = -\partial_2 U \qquad (3.53)$$

into (3.52). Using the first equation (3.37) we arrive at the same differential equation (3.41) for U. It follows from the theory of differential equations that a complex potential $\varphi^{(E)}$ real part of which coincides with U is a linear combination of analytic functions in z_1 and $\overline{z_1}$, i.e., $\varphi^{(E)}(z_1) = C_1\varphi_1(z_1) + C_2\varphi_2(\overline{z_1})$.

The domain G is mapped to the domain G_1 by the transformation $z_1 = x_1 + \mu x_2$. Therefore, a boundary value problem of one complex variable z_1 arises. This essentially simplifies the constructive study of the problems. For instance, the boundary value problem for a hole with the prescribed normal component of the displacement can be written as the classic boundary value

Two-dimensional equations of magneto-electro-elasticity **Chapter | 3 75**

problem of complex analysis. Let Γ be oriented in the counterclockwise sense and $\mathbf{n} = (n_1, n_2)$ denote the outward unit vector to Γ and $\mathbf{s} = (n_2, -n_1)$ the tangent vector to Γ. We have

$$D_n = -\mathbf{D} \cdot \mathbf{n} = \mathbf{s} \cdot \nabla F \quad \text{on } \Gamma. \tag{3.54}$$

Integrating (3.54) along Γ and using (3.44) we arrive at the boundary value problem

$$2\text{Re } \varphi(z_1) = \int D_n \, ds, \quad z_1 \in \Gamma', \tag{3.55}$$

where Γ' is obtained from Γ by the shift $z_1 = x_1 + \mu x_2$. The indefinite integral in (3.55) contains an arbitrary additive constant.

Consider the problem with $D_n = 0$ and the normalized electric field prescribed at infinity $\mathbf{E}_0 = (\cos\theta_0, \sin\theta_0) \equiv \exp(i\theta_0)$. The value $\varphi'(\infty)$ is determined by means of (3.51)

$$\varphi'(\infty) = \frac{1}{2\Delta_\beta^2}[(\beta_{12} + i)\cos\theta_0 - \beta_{11}\sin\theta_0]. \tag{3.56}$$

The condition (3.56) is equivalent to the Laurent series at infinity

$$\varphi(z_1) = \frac{z_1}{2\Delta_\beta^2}[(\beta_{12} + i)\cos\theta_0 - \beta_{11}\sin\theta_0] + \alpha_0 + \frac{\alpha_1}{z_1} + \dots. \tag{3.57}$$

The boundary condition (3.55) becomes

$$\text{Re } \varphi(z_1) = 0, \quad z_1 \in \Gamma'. \tag{3.58}$$

The arbitrary constant is fixed as zero without loss of generality, since a complex potential is defined up to an additive constant. Thus, we arrive at the classic boundary value problem of the analytic function theory [25]. Let a simply connected domain G' contain infinity and bounded by a smooth curve Γ'. It is necessary to find a function $\varphi(z_1)$ analytic in G' and continuously differentiable in its closure except at infinity where $\varphi(z_1)$ has a simple pole described by (3.57) with the boundary condition (3.58). The problem (3.57)–(3.58) can be solved in terms of the conformal mapping of G' onto a disk. Section 3.5 contains example of the general MEE medium with a circular hole. The general vector-matrix problem for a circular hole is decomposed onto scalar problems such as (3.57)–(3.58).

Consider now another boundary condition when instead of D_n the tangent electric field E_s is given on the boundary of hole

$$E_s = \mathbf{E} \cdot \mathbf{s} \quad \text{on } \Gamma. \tag{3.59}$$

The method of solution can be based on the potential U defined by (3.53) instead of the Airy function F. Such an approach repeats the above method.

76 Mechanics and Physics of Structured Media

The situation becomes complicated when the both fields $\mathbf{E}$ and $\mathbf{D}$ have to be involved in the boundary condition. Such a problem arises when instead of the hole we consider an inclusion occupied by another medium. Let G^+ denote the inclusion domain and $G^- \equiv G$ the host domain. We use the above presentations for the local fields and potentials adding the superscripts $\pm$ to indicate the domain under consideration. Let the contact between G^+ and G^- be perfect. Hence,

$$E_s^+ = E_s^-, \quad D_n^+ = D_n^- \quad \text{on } \Gamma. \tag{3.60}$$

Following (3.54) we write the second equality (3.60) in the form

$$\mathbf{s} \cdot \nabla F^+ = \mathbf{s} \cdot \nabla F^- \quad \text{on } \Gamma. \tag{3.61}$$

The Airy functions $F^\pm$ in $G^\pm$ are expressed by means of complex potentials. Using (3.48) we write (3.61) in terms of the derivatives of complex potentials

$$\left(s_1 + s_2 \frac{\beta_{12}^+}{\beta_{11}^+} \right) \operatorname{Re} \varphi'^+(z_1^+) - s_2 \frac{\Delta_{\beta^+}}{\beta_{11}^+} \operatorname{Im} \varphi'^+(z_1^+) = \tag{3.62}$$

$$\left(s_1 + s_2 \frac{\beta_{12}^-}{\beta_{11}^-} \right) \operatorname{Re} \varphi'^-(z_1^-) - s_2 \frac{\Delta_{\beta^-}}{\beta_{11}^-} \operatorname{Im} \varphi'^-(z_1^-) \quad \text{on } \Gamma.$$

One can integrate Eq. (3.61) along Γ and write it in terms of the complex potentials

$$\operatorname{Re} \varphi^+(z_1^+) = \operatorname{Re} \varphi^-(z_1^-) \quad \text{on } \Gamma. \tag{3.63}$$

Here, $(x_1, x_2) \in \Gamma$, the transformations $z_1^+ = x_1 + \mu^+ x_2$ and $z_1^- = x_1 + \mu^- x_2$ map the curves Γ onto Γ^+ and Γ^-, respectively. The curves Γ^+ and $(-\Gamma^-)$ are oriented in the counterclockwise sense and bound the domains G'^+ and G'^-. One can consider the transformation of Eq. (3.63) into (3.62) as the differentiation along the curve Γ.

Consider now the first relation (3.60) in the form $\mathbf{s} \cdot \mathbf{E}^+ = \mathbf{s} \cdot \mathbf{E}^-$ on Γ. Using (3.51) we can write it as follows

$$\Delta_{\beta^+} \left[\left(s_1 + s_2 \frac{\beta_{12}^+}{\beta_{11}^+} \right) \operatorname{Im} \varphi'^+(z_1^+) + s_2 \frac{\Delta_{\beta^+}}{\beta_{11}^+} \operatorname{Re} \varphi'^+(z_1^+) \right] = \tag{3.64}$$

$$\Delta_{\beta^-} \left[\left(s_1 + s_2 \frac{\beta_{12}^-}{\beta_{11}^-} \right) \operatorname{Im} \varphi'^-(z_1^-) + s_2 \frac{\Delta_{\beta^-}}{\beta_{11}^-} \operatorname{Re} \varphi'^-(z_1^-) \right] \quad \text{on } \Gamma.$$

Integration of (3.64) along Γ yields

$$\operatorname{Im} \varphi^+(z_1^+) = \gamma \operatorname{Im} \varphi^-(z_1^-) \quad \text{on } \Gamma, \tag{3.65}$$

Two-dimensional equations of magneto-electro-elasticity **Chapter | 3 77**

where $\gamma = \frac{\Delta_{\beta-}}{\Delta_{\beta+}}$. The transformation of (3.65) into (3.64) can be considered as the differentiation along the curve Γ.

Multiply (3.65) by i and add (3.63). As a result we obtain one complex equation equivalent to two real equations (3.63) and (3.65)

$$\varphi^+[z_1^+(t)] = \frac{\gamma+1}{2}\varphi^-[z_1^-(t)] - \frac{\gamma-1}{2}\overline{\varphi^-[z_1^-(t)]}, \quad t \in \Gamma. \qquad (3.66)$$

Following [38] and [27, Chapter 2] the condition (3.66) in the particular case $z_1^+(t) \equiv z_1^-(t) \equiv t$ is called by the $\mathbb{R}$-linear conjugation problem. In the case with the shifts defined by (3.17), i.e. by the $\mathbb{R}$-linear shifts

$$z_1^\pm(t) = \frac{1-i\mu^\pm}{2}t + \frac{1+i\mu^\pm}{2}\bar{t} \qquad (3.67)$$

we call it by the $\mathbb{RR}$-linear problem. The exact solution of this problem in the case of ellipse Γ follows from [13, Chapter 8].

We have $\mu^\pm = i$ in the case of locally isotropic constitutes. Then, the shifts become identical transformations and we arrive at the classic $\mathbb{R}$-linear problem [38] and [27, Chapter 2]. The method of functional equations and the generalized alternating method of Schwarz were applied in the above books and works cited therein to solve the problem for locally isotropic media. In all other cases including $\mu^+ = \mu^-$, the method of integral equations due to Filshtinsky were developed, see for details Chapter "Cracks in two-dimensional magneto-electro-elastic composites" of the present book.

3.5 Circular hole

We now consider the case when Γ is the unit circle. This case is useful to understand the structure of the boundary value problem (3.35). We now fix the first coordinate of (3.35) and write it in the extended form

$$2\text{Re}[a_{11}g_1\Phi_1(z_1) + a_{12}g_2\Phi_2(z_2) + a_{13}g_3\Phi_3(z_3) + a_{14}g_4\Phi_4(z_4)] = X_1(z),$$
$$z \in \Gamma, \qquad (3.68)$$

where a_{1j} are the elements of the matrix (3.32). Using the parametric equation $z = \exp(i\theta)$ of the unit circle we write the function $g_k(\theta)$ as the function of z and $\bar{z}$

$$g_k(z) = \frac{\mu_k+i}{2}z + \frac{\mu_k-i}{2}\bar{z} \quad (k=1,2,3,4). \qquad (3.69)$$

Using the relations $\bar{z} = z^{-1}$ on the unit circle, (3.69) and (3.16), we can write the boundary condition (3.68) in the form

$$\text{Re}\sum_{k=1}^{4}a_{1k}\left((\mu_k+i)z + \frac{\mu_k-i}{z}\right)\Phi_k[\gamma_k(z)] = X_1(z), \quad |z|=1, \qquad (3.70)$$

78 Mechanics and Physics of Structured Media

where

$$z_k = \gamma_k(z) \equiv \frac{1 - i\mu_k}{2}z + \frac{1 + i\mu_k}{2z}. \tag{3.71}$$

The shift (3.71) conformally transforms the domain $|z| > 1$ onto the domain G_k, the exterior of the ellipse defined by the parametric equations on the plane $z_k = x_k + iy_k$

$$x_k = \cos\theta - \operatorname{Im}\mu_k \sin\theta, \quad y_k = \operatorname{Re}\mu_k \sin\theta \quad (0 \le \theta < 2\pi). \tag{3.72}$$

The infinite point $z = \infty$ is mapped onto infinity. The ellipse degenerates to the slit $(-1, 1)$ on the real axis, when $\operatorname{Re}\mu_k = 0$.

Introduce the function analytic in $|z| > 1$ and continuously differentiable in $|z| \ge 1$

$$\Omega_1(z) = \sum_{k=1}^{4} a_{1k}\left((\mu_k + i)z + \frac{\mu_k - i}{z}\right)\Phi_k[\gamma_k(z)] - \omega_1 z, \tag{3.73}$$

where $\omega_1 = \sum_{k=1}^{4} a_{1k}(\mu_k + i)C_k$, the constants C_k are found from the system (3.36). The boundary condition (3.70) can be written in the form

$$\operatorname{Re}\left[\overline{\Omega_1\left(\frac{1}{\bar{z}}\right)} + \omega_1 z\right] = X_1(z), \quad |z| = 1, \tag{3.74}$$

where the function in brackets is analytic in the unit disk. The Schwarz operator for the unit disk [25], [38, Section 2.7.1] solves the problem (3.74). After application of the transformation $z \longmapsto \frac{1}{\bar{z}}$ and the complex conjugation we get

$$\Omega_1(z) = -\frac{\overline{\omega_1}}{z} - \frac{1}{2\pi i}\int_{|\zeta|=1}X_1(\zeta)\frac{\zeta + z}{\zeta - z}\frac{d\zeta}{\zeta}, \quad |z| > 1. \tag{3.75}$$

The same arguments can be applied to other coordinates of the boundary value problem (3.35). We obtain

$$\Omega_m(z) = -\frac{\overline{\omega_m}}{z} - \frac{1}{2\pi i}\int_{|\zeta|=1}X_m(\zeta)\frac{\zeta + z}{\zeta - z}\frac{d\zeta}{\zeta}, \quad |z| > 1, \tag{3.76}$$

where

$$\omega_m = \sum_{k=1}^{4} a_{mk}(\mu_k + i)C_k \quad (m = 1, 2, 3, 4). \tag{3.77}$$

The complex potentials $\Phi_k(z_k)$ satisfy the system of linear algebraic equations

$$\sum_{k=1}^{4} a_{mk}\left((\mu_k + i)z + \frac{\mu_k - i}{z}\right)\Phi_k(z_k) = \Omega_m(z) + \omega_m z \quad (m = 1, 2, 3, 4). \tag{3.78}$$

Two-dimensional equations of magneto-electro-elasticity **Chapter | 3 79**

After its solution the local fields are calculated by the formulas (3.19)–(3.21).

Consider the case when the forces on the boundary $|z| = 1$ are absent and $B_n = D_n = 0$. This implies that $X_m(\zeta) = 0$ $(m = 1, 2, 3, 4)$, hence,

$$\Omega_m(z) = -\frac{\overline{\omega_m}}{z}. \tag{3.79}$$

We have to find the constants C_k from the system of Eqs. (3.36). Let these constants be found. Calculate ω_m by (3.77) and substitute (3.79) in (3.78)

$$\sum_{k=1}^{4} a_{mk} \left((\mu_k + i)z + \frac{\mu_k - i}{z} \right) \Phi_k(z_k) = \omega_m z - \frac{\overline{\omega_m}}{z} \quad (m = 1, 2, 3, 4). \tag{3.80}$$

It is convenient to write (3.80) the result in the vector-matrix form

$$\mathbf{Ax} = \boldsymbol{\omega}, \tag{3.81}$$

where $\mathbf{A} = \{a_{mk}\}$, the vector $\mathbf{x}$ has the coordinates $\left((\mu_k + i)z + \frac{\mu_k - i}{z} \right) \Phi_k(z_k)$ and the vector $\boldsymbol{\omega}$ the coordinates $\omega_m z - \frac{\overline{\omega_m}}{z}$. Let the system (3.81) is solved $\mathbf{x} = \mathbf{A}^{-1}\boldsymbol{\omega}$. Hence, we obtain the expressions

$$\left((\mu_k + i)z + \frac{\mu_k - i}{z} \right) \Phi_k(z_k) = (\mathbf{A}^{-1}\boldsymbol{\omega})_k(z), \quad k = 1, 2, 3, 4, \tag{3.82}$$

where $(\mathbf{A}^{-1}\boldsymbol{\omega})_k(z)$ denotes the kth coordinate of the vector $\mathbf{A}^{-1}\boldsymbol{\omega}(z)$. Ultimately, the formula (3.82) yields

$$\Phi_k(z_k) = \frac{(\mathbf{A}^{-1}\boldsymbol{\omega})_k(z)}{(\mu_k + i)z + \frac{\mu_k - i}{z}}, \quad k = 1, 2, 3, 4, \tag{3.83}$$

where $z \in G$ is expressed in terms of $z_k \in G_k$ by the function inverse to (3.71)

$$z = \frac{z_k - \sqrt{z_k^2 - \mu_k^2 - 1}}{1 - i\mu_k}, \quad (k = 1, 2, 3, 4), \tag{3.84}$$

where the branch of square root is selected in the complex plane with the excluded cut connecting the points $z_k = \pm\sqrt{\mu_k^2 + 1}$. The local fields are calculated by (3.19)–(3.24).

80 Mechanics and Physics of Structured Media

Numerical example

Consider a numerical example of the matrix (3.32) when

$$
\mathbf{A} =
\begin{pmatrix}
0.00714i & 219.2 - 327.6i & -219.2 - 327.6i & -0.0016i \\
-0.0049 & 396.1 + 180.2i & 396.1 - 180.2i & 0.0025 \\
0.1188 & 0.0413 + 0.0154i & 0.0414 - 0.0154i & 0.0196 \\
-80.399 & 6.0673 - 4.483i & 6.067 + 4.483i & 5.4626
\end{pmatrix}.
$$

$$(3.85)$$

Four roots of the characteristic equation (3.15) have the form

$$
\mu_1 = 1.4573i, \quad \mu_2 = -0.1468 + 0.8938i, \tag{3.86}
$$
$$
\mu_3 = 0.1468 + 0.8938i, \quad \mu_4 = 0.6397i.
$$

Let the dimensionless external electric displacement $\mathbf{D}_{ext} = (1, 0)$ is given at infinity. The external forces and the external magnetic induction are absent at infinity, i.e., $\sigma_{ij}^{\infty} = 0$ $(i, j = 1, 2)$, $\mathbf{B}_{ext} = (0, 0)$. The constants C_k are found from the system of linear algebraic equations (3.36)

$$
C_1 = 0.0043691 - 2.19037i, \quad C_2 = 17.8583 - 11.9491i, \tag{3.87}
$$
$$
C_3 = -17.8583 - 11.9491i, \quad C_4 = 0.0646326.
$$

The constants (3.77) are calculated by the formulas

$$
\omega_1 = -3.72 \times 10^{-6} + 2486.56i, \quad \omega_2 = 3030.8 - 3.72 \times 10^{-6}i, \tag{3.88}
$$
$$
\omega_3 = 1.19837 - 0.0002048i, \quad \omega_4 = 129.02 - 0.2878i.
$$

The complex potentials are calculated by (3.83)–(3.84). As an example, we write the first function

$$
\Phi_1(z_1) = \frac{P(z_1)}{Q(z_1)}, \tag{3.89}
$$

where

$$
P(z_1) = (2.19037 + 0.0043691\,i)z_1\sqrt{z_1^2 + 1.12373}
$$
$$
+ 0.555387 - 0.00490971\,i - (2.19037 + 0.0043691\,i)z_1^2, \tag{3.90}
$$
$$
Q(z_1) = -i(z_1^2 - z_1\sqrt{z_1^2 + 1.12373} + 1.12372).
$$

The local fields are displayed in Figs. 3.2–3.5 for the considered data in the form of vector and scalar fields.

Two-dimensional equations of magneto-electro-elasticity Chapter | 3 | 81

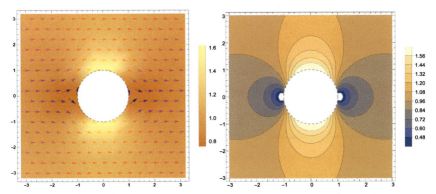

FIGURE 3.2 The vector field **D** (left picture). The intensity of **D** is displayed by the color legend. The scalar field |**D**| is presented by contour plot (right picture). The values of level lines are displayed by the color legend.

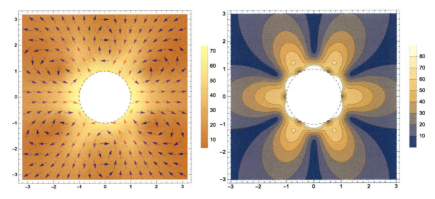

FIGURE 3.3 The vector field **B** (left picture). The scalar field |**B**| (right picture).

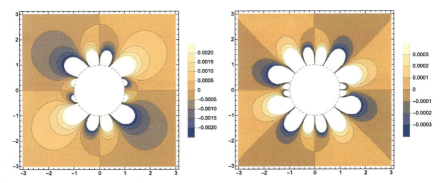

FIGURE 3.4 The components σ_{11} (left) and σ_{22} (right) of the stress tensor.

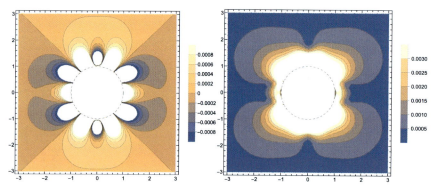

FIGURE 3.5 The component σ_{12} of the stress tensor (left). The scalar field $\sqrt{\sigma_{11}^2 - \sigma_{11}\sigma_{22} + \sigma_{22}^2 + 3\sigma_{12}^2}$ by von Mises (right).

3.6 MEE equations and homogenization

In the present section, we outline the main MEE equations and discuss the corresponding homogenized equations. Equations are considered in a smooth domain $G \subset \mathbb{R}^3$. The same results hold in $G \subset \mathbb{R}^2$.

The field equations of our steady state problem are

- equation of equilibrium

$$\frac{\partial \sigma_{ij}}{\partial x_j} = 0 \quad \text{with } i, j = 1, 2, 3, \tag{3.91}$$

- Maxwell equations

$$\frac{\partial B_i}{\partial x_i} = 0 \quad \text{and} \quad \frac{\partial D_i}{\partial x_i} = 0, \tag{3.92}$$

$$\varepsilon_{kij} \frac{\partial H_j}{\partial x_i} = 0 \quad \text{and} \quad \varepsilon_{kij} \frac{\partial E_j}{\partial x_i} = 0. \tag{3.93}$$

The ε_{kij} with $i, j, k = 1, 2, 3$, is known as the antisymmetric symbol, alternating symbol or the Levi-Civita symbol.

The appropriate constitutive relations read

$$\begin{aligned}
\epsilon_{ij} &= s_{ijmn}\sigma_{mn} + g_{kij}D_k + p_{kij}B_k, \\
E_i &= -g_{imn}\sigma_{mn} + \beta_{ik}D_k + \nu_{ik}B_k, \\
H_i &= -p_{imn}\sigma_{mn} + \nu_{ik}D_k + \chi_{ik}B_k.
\end{aligned} \tag{3.94}$$

Eqs. (3.94) in two dimensions coincide with Eqs. (3.5). The inverse transformation of tensor relation (3.94) can be obtained by the equivalent inversion of

Two-dimensional equations of magneto-electro-elasticity Chapter | 3 **83**

matrix relation. The inverse tensor form to (3.94) is given by equations

$$\sigma_{ij} = c_{ijmn}\epsilon_{mn} - e_{kij}E_k - h_{kij}H_k,$$

$$D_i = e_{imn}\epsilon_{mn} + \epsilon_{ik}^{(D)}E_k + \mu_{ik}^{(D)}H_k, \qquad (3.95)$$

$$B_i = h_{imn}\epsilon_{mn} + \epsilon_{ik}^{(B)}E_k + \mu_{ik}H_k.$$

There are also geometrical relationships

$$\epsilon_{ij} = \frac{1}{2}\left(\frac{\partial u_i}{\partial x_j} + \frac{\partial u_j}{\partial x_i}\right) \qquad (3.96)$$

and the compatibility equations

$$\varepsilon_{imn}\,\varepsilon_{jab}\,\frac{\partial^2 \epsilon_{nb}}{\partial x_a \partial x_m} = 0. \qquad (3.97)$$

Here σ_{ij}, ϵ_{ij}, u_i, E_i, D_i, H_i, and B_i are the stress tensor, the strain tensor, the displacement vector, the electric field vector, the electric displacement vector, magnetizing vector, and magnetic vector, respectively. The electric U and magnetic $\varphi^{(H)}$ potentials satisfy the relations

$$E_i = -\frac{\partial U}{\partial x_i} \quad \text{and} \quad H_i = -\frac{\partial \varphi^{(H)}}{\partial x_i}. \qquad (3.98)$$

The following assumption is made: there exists a constant $c > 0$ such that for almost every $x \in G$ the following conditions are satisfied

$$c_{ij\mu\nu}(x)e_{ij}e_{\mu\nu} \geq c|\mathbf{e}|^2,$$

$$\epsilon_{ij}^{(D)}(x)a_i a_j \geq c|\mathbf{a}|^2, \qquad \epsilon_{ij}^{(B)}(x)a_i a_j \geq c|\mathbf{a}|^2,$$

$$\mu_{ij}^{(D)}(x)a_i a_j \geq c|\mathbf{a}|^2, \qquad \mu_{ij}(x)a_i a_j \geq c|\mathbf{a}|^2$$

for each $\mathbf{e} \in I\!E^3$ and $\mathbf{a} \in I\!R^3$; here $I\!E^3$ is the space of symmetric 3×3 matrices. The tensors of material coefficients satisfy (usual) symmetry conditions

$$s_{ijmn} = s_{mnij} = s_{ijnm} = s_{jimn},$$

$$g_{imn} = g_{inm}, \quad h_{imn} = h_{inm}, \quad v_{ij} = v_{ji}, \quad \text{and} \quad \chi_{ij} = \chi_{ji}.$$

84 Mechanics and Physics of Structured Media

Substituting relations (3.95)–(3.98) into Eqs. (3.91)–(3.93) we get

$$\frac{\partial}{\partial x_j}\left(c_{ijmn}\frac{\partial u_m}{\partial x_n} - e_{kij}\frac{\partial U}{\partial x_k} - h_{kij}\frac{\partial \varphi^{(H)}}{\partial x_k}\right) = 0,$$

$$\frac{\partial}{\partial x_i}\left(e_{imn}\frac{\partial u_m}{\partial x_n} - \epsilon_{ik}^{(D)}\frac{\partial U}{\partial x_k} - \mu_{ik}^{(D)}\frac{\partial \varphi^{(H)}}{\partial x_k}\right) = 0, \qquad (3.99)$$

$$\frac{\partial}{\partial x_i}\left(h_{imn}\frac{\partial u_m}{\partial x_n} - \epsilon_{ik}^{(B)}\frac{\partial U}{\partial x_k} - \mu_{ik}\frac{\partial \varphi^{(H)}}{\partial x_k}\right) = 0.$$

Homogenization uses the idea of extending the area of the body to be homogenized in such a way that the solution would be presented by an expansion around a small parameter.

Let $\Gamma = \partial G$ denote the boundary of the domain G. We introduce a parameter

$$\lambda = l/L,$$

where l and L are typical length scales associated with micro-inhomogeneities and the region G, respectively. According to the asymptotic two-scale method, instead of one spatial variable x, we introduce two variables, the macroscopic x and the microscopic y, where $y = x/\lambda$, and instead of the function $f(x)$ we consider the function $f(x, y)$.

Consequently, instead of the domain G, we consider the domain $G \times Y$, where Y is a basic cell of micro-periodicity. We use the formula for the total derivative

$$\frac{\partial f(x, y)}{\partial x} \rightsquigarrow \frac{\partial f(x, y)}{\partial x} + \frac{1}{\lambda}\frac{\partial f(x, y)}{\partial y} \quad \text{where } y = \frac{x}{\lambda}.$$

In line with the two-scale asymptotic expansions method we assume

$$f^{\lambda} = f^{\lambda}(x) = f^{(0)}(x, y) + \lambda^1 f^{(1)}(x, y) + \lambda^2 f^{(2)}(x, y) + \ldots,$$

where the functions $f^{(i)}(x, y)$, $i = 0, 1, 2, \ldots$ are Y-periodic. The superscript λ indicates the micro-periodicity of the respective quantities.

It is tacitly assumed that all derivatives appearing in the procedure of asymptotic homogenization make sense. The effect of micro-structural heterogeneity is described by periodic functions, the so-called *local* functions on the cell.

Appropriate expansions in our case are

$$u_i^{\lambda} = u_i^{(0)}(x, y) + \lambda u_i^{(1)}(x, y) + \lambda^2 u_i^{(2)}(x, y) + \ldots,$$

$$U^{\lambda} = U^{(0)}(x, y) + \lambda U^{(1)}(x, y) + \lambda^2 U^{(2)}(x, y) + \ldots, \qquad (3.100)$$

$$\varphi^{(H)\lambda} = \varphi^{(0)}(x, y) + \lambda \varphi^{(1)}(x, y) + \lambda^2 \varphi^{(2)}(x, y) + \ldots,$$

and after substitution into Eqs. (3.99) one obtains

$$\left(\frac{\partial}{\partial x_j} + \frac{1}{\lambda}\frac{\partial}{\partial y_j}\right)$$

$$\left[c_{ijmn}\left(\frac{\partial}{\partial x_n} + \frac{1}{\lambda}\frac{\partial}{\partial y_n}\right)(u_m^{(0)}(x, y) + \lambda u_m^{(1)}(x, y) + \lambda^2 u_m^{(2)}(x, y) + \ldots)\right.$$

$$- e_{kij}\left(\frac{\partial}{\partial x_k} + \frac{1}{\lambda}\frac{\partial}{\partial x_k}\right)(U^{(0)}(x, y) + \lambda U^{(1)}(x, y) + \lambda^2 U^{(2)}(x, y) + \ldots)$$

$$\left. - h_{kij}\left(\frac{\partial}{\partial x_k} + \frac{1}{\lambda}\frac{\partial}{\partial x_k}\right)(\varphi^{(0)}(x, y) + \lambda\varphi^{(1)}(x, y) + \lambda^2\varphi^{(2)}(x, y) + \ldots)\right] = 0,$$

$$\left(\frac{\partial}{\partial x_i} + \frac{1}{\lambda}\frac{\partial}{\partial y_i}\right)$$

$$\left[e_{imn}\left(\frac{\partial}{\partial x_n} + \frac{1}{\lambda}\frac{\partial}{\partial y_n}\right)(u_m^{(0)}(x, y) + \lambda u_m^{(1)}(x, y) + \lambda^2 u_m^{(2)}(x, y) + \ldots)\right.$$

$$- \epsilon_{ik}^{(D)}\left(\frac{\partial}{\partial x_k} + \frac{1}{\lambda}\frac{\partial}{\partial y_k}\right)(U^{(0)}(x, y) + \lambda U^{(1)}(x, y) + \lambda^2 U^{(2)}(x, y) + \ldots)$$

$$\left. - \mu_{ik}^{(D)}\left(\frac{\partial}{\partial x_k} + \frac{1}{\lambda}\frac{\partial}{\partial y_k}\right)(\varphi^{(0)}(x, y) + \lambda\varphi^{(1)}(x, y) + \lambda^2\varphi^{(2)}(x, y) + \ldots)\right] = 0,$$

$$\left(\frac{\partial}{\partial x_i} + \frac{1}{\lambda}\frac{\partial}{\partial y_i}\right)$$

$$\left[h_{imn}\left(\frac{\partial}{\partial x_n} + \frac{1}{\lambda}\frac{\partial}{\partial y_n}\right)(u_m^{(0)}(x, y) + \lambda u_m^{(1)}(x, y) + \lambda^2 u_m^{(2)}(x, y) + \ldots)\right.$$

$$- \epsilon_{ik}^{(B)}\left(\frac{\partial}{\partial x_k} + \frac{1}{\lambda}\frac{\partial}{\partial y_k}\right)(U^{(0)}(x, y) + \lambda U^{(1)}(x, y) + \lambda^2 U^{(2)}(x, y) + \ldots)$$

$$\left. - \mu_{ik}\left(\frac{\partial}{\partial x_k} + \frac{1}{\lambda}\frac{\partial}{\partial y_k}\right)(\varphi^{(0)}(x, y) + \lambda\varphi^{(1)}(x, y) + \lambda^2\varphi^{(2)}(x, y) + \ldots)\right] = 0.$$

Comparing to zero the coefficients at successive negative powers of λ one finds: at λ^{-2}

$$\frac{\partial}{\partial y_j}\left[c_{ij\mu\nu}\frac{\partial u_m^{(0)}(x, y)}{\partial y_\nu} - e_{kij}\frac{\partial U^{(0)}(x, y)}{\partial x_k} - h_{kij}\frac{\partial \varphi^{(0)}(x, y)}{\partial x_k}\right] = 0,$$

$$\frac{\partial}{\partial y_i}\left[e_{imn}\frac{\partial u_m^{(0)}(x, y)}{\partial y_n} - \epsilon_{ik}^{(D)}\frac{\partial U^{(0)}(x, y)}{\partial y_k} - \mu_{ik}^{(D)}\frac{\partial \varphi^{(0)}(x, y)}{\partial y_k}\right] = 0, \quad (3.101)$$

$$\frac{\partial}{\partial y_i}\left[h_{imn}\frac{\partial u_m^{(0)}(x, y)}{\partial y_n} - \epsilon_{ik}^{(B)}\frac{\partial U^{(0)}(x, y)}{\partial y_k} - \mu_{ik}\frac{\partial \varphi^{(0)}(x, y)}{\partial y_k}\right] = 0.$$

Hence

$$u_m^{(0)}(x, y) = u_m^{(0)}(x), \quad U^{(0)}(x, y) = U^{(0)}(x), \quad \varphi^{(0)}(x, y) = \varphi^{(0)}(x). \quad (3.102)$$

86 Mechanics and Physics of Structured Media

The zeroth terms of expansions do not depend on the microperiodic variable y. We have at λ^{-1}

$$
\frac{\partial}{\partial y_j}\left[c_{ijmn}\left(\frac{\partial u_m^{(0)}(x)}{\partial x_n} + \frac{\partial u_m^{(1)}(x,y)}{\partial y_n}\right) - e_{kij}\left(\frac{\partial U^{(0)}(x)}{\partial x_k} + \frac{\partial U^{(1)}(x,y)}{\partial y_k}\right)\right.
$$
$$
\left. - h_{kij}\left(\frac{\partial \varphi^{(0)}(x)}{\partial x_k} + \frac{\partial \varphi^{(1)}(x,y)}{\partial y_k}\right)\right] = 0,
$$
$$
\frac{\partial}{\partial y_i}\left[e_{imn}\left(\frac{\partial u_m^{(0)}(x)}{\partial x_n} + \frac{\partial u_m^{(1)}(x,y)}{\partial y_n}\right) - \epsilon_{ik}^{(D)}\left(\frac{\partial U^{(0)}(x)}{\partial x_k} + \frac{\partial U^{(1)}(x,y)}{\partial y_k}\right)\right.
$$
$$
\left. - \mu_{ik}^{(D)}\left(\frac{\partial \varphi^{(0)}(x)}{\partial x_k} + \frac{\partial \varphi^{(1)}(x,y)}{\partial y_k}\right)\right] = 0,
$$
$$
\frac{\partial}{\partial y_i}\left[h_{imn}\left(\frac{\partial u_m^{(0)}(x)}{\partial x_n} + \frac{\partial u_m^{(1)}(x,y)}{\partial y_n}\right) - \epsilon_{ik}^{(B)}\left(\frac{\partial U^{(0)}(x)}{\partial x_k} + \frac{\partial U^{(1)}(x,y)}{\partial y_k}\right)\right.
$$
$$
\left. - \mu_{ik}\left(\frac{\partial \varphi^{(0)}(x)}{\partial x_k} + \frac{\partial \varphi^{(1)}(x,y)}{\partial y_k}\right)\right] = 0.
$$

$$(3.103)$$

To satisfy the last system we make an assumption

$$
u_i^{(1)}(x,y) = Y_{imn}(y)\frac{\partial u_m^{(0)}(x)}{\partial x_n} + F_{im}(y)\frac{\partial U^{(0)}(x)}{\partial x_m} + G_{im}(y)\frac{\partial \varphi^{(0)}(x)}{\partial x_m},
$$
$$
U^{(1)}(x,y) = \Psi_{mn}(y)\frac{\partial u_m^{(0)}(x)}{\partial x_n} + P_m(y)\frac{\partial U^{(0)}(x)}{\partial x_m} + Q_m(y)\frac{\partial \varphi^{(0)}(x)}{\partial x_m},
$$
$$
\varphi^{(1)}(x,y) = \Phi_{mn}(y)\frac{\partial u_m^{(0)}(x)}{\partial x_n} + R_m(y)\frac{\partial U^{(0)}(x)}{\partial x_m} + S_m(y)\frac{\partial \varphi^{(0)}(x)}{\partial x_m}.
$$

$$(3.104)$$

Substituting assumptions (3.104) into Eqs. (3.103) we find that the following *local* relations should be satisfied

from Eq. $(3.103)_1$

$$
\frac{\partial}{\partial y_j}\left[c_{ijmn}\left(\delta_{mr}\delta_{ns} + \frac{\partial Y_{mrs}}{\partial y_n}\right) - e_{kij}\frac{\partial \Psi_{rs}}{\partial y_k} - h_{kij}\frac{\partial \Phi_{rs}}{\partial y_k}\right] = 0,
$$
$$
\frac{\partial}{\partial y_j}\left[c_{ijmn}\frac{\partial F_{mr}}{\partial y_n} - e_{kij}\left(\delta_{rk} + \frac{\partial P_r}{\partial y_k}\right) - h_{kij}\frac{\partial R_r}{\partial y_k}\right] = 0, \qquad (3.105)
$$
$$
\frac{\partial}{\partial y_j}\left[c_{ijmn}\frac{\partial G_{mr}}{\partial y_n} - e_{kij}\frac{\partial Q_r}{\partial y_k} - h_{kij}\left(\delta_{rk} + \frac{\partial S_r}{\partial y_k}\right)\right] = 0,
$$

Two-dimensional equations of magneto-electro-elasticity **Chapter | 3 87**

from Eq. (3.103)$_2$

$$\frac{\partial}{\partial y_i}\left[e_{imn}\left(\delta_{mr}\delta_{ns}+\frac{\partial Y_{mrs}}{\partial y_n}\right)-\epsilon_{ik}^{(D)}\frac{\partial\Psi_{rs}}{\partial y_k}-\mu_{ik}^{(D)}\frac{\partial\Phi_{rs}}{\partial y_k}\right]=0,$$

$$\frac{\partial}{\partial y_i}\left[e_{imn}\frac{\partial F_{ir}}{\partial y_n}-\epsilon_{ik}^{(D)}\left(\delta_{rk}+\frac{\partial P_r}{\partial y_k}\right)-\mu_{ik}^{(D)}\frac{\partial R_r}{\partial y_k}\right]=0, \qquad (3.106)$$

$$\frac{\partial}{\partial y_i}\left[e_{imn}\frac{\partial F_{ir}}{\partial y_n}-\epsilon_{ik}^{(D)}\frac{\partial Q_r}{\partial y_k}-\mu_{ik}^{(D)}\left(\delta_{rk}+\frac{\partial S_r}{\partial y_k}\right)\right]=0,$$

and from Eq. (3.103)$_3$

$$\frac{\partial}{\partial y_i}\left[h_{imn}\left(\delta_{mr}\delta_{ns}+\frac{\partial Y_{\mu rs}}{\partial y_n}\right)-\epsilon_{ik}^{(B)}\frac{\partial\Psi_{rs}}{\partial y_k}-\mu_{ik}\frac{\partial\Phi_{rs}}{\partial y_k}\right]=0,$$

$$\frac{\partial}{\partial y_i}\left[h_{imn}\frac{\partial F_{mr}}{\partial y_n}-\epsilon_{ik}^{(B)}\left(\delta_{rk}+\frac{\partial P_r}{\partial y_k}\right)-\mu_{ik}\frac{\partial R_r}{\partial y_k}\right]=0, \qquad (3.107)$$

$$\frac{\partial}{\partial y_i}\left[h_{imn}\frac{\partial G_{mr}}{\partial y_n}-\epsilon_{ik}^{(B)}\frac{\partial Q_r}{\partial y_k}-\mu_{ik}\left(\delta_{rk}+\frac{\partial S_r}{\partial y_k}\right)\right]=0.$$

Finally, at the power λ^0 we get

$$\frac{\partial}{\partial x_j}\left[c_{ijmn}\left(\frac{\partial u_m^{(0)}(x)}{\partial x_n}+\frac{\partial u_m^{(1)}(x,y)}{\partial y_n}\right)-e_{kij}\left(\frac{\partial U^{(0)}(x)}{\partial x_k}+\frac{\partial U^{(1)}(x,y)}{\partial y_k}\right)\right.$$
$$\left.-h_{kij}\left(\frac{\partial\varphi^{(0)}(x)}{\partial x_k}+\frac{\partial\varphi^{(1)}(x,y)}{\partial y_k}\right)\right]+\frac{\partial}{\partial y_j}\left[\cdots\right]=0,$$

$$\frac{\partial}{\partial x_i}\left[e_{imn}\left(\frac{\partial u_m^{(0)}(x)}{\partial x_n}+\frac{\partial u_m^{(1)}(x,y)}{\partial y_n}\right)-\epsilon_{ik}^{(D)}\left(\frac{\partial U^{(0)}(x)}{\partial x_k}+\frac{\partial U^{(1)}(x,y)}{\partial y_k}\right)\right.$$
$$\left.-\mu_{ik}^{(D)}\left(\frac{\partial\varphi^{(0)}(x)}{\partial x_k}+\frac{\partial\varphi^{(1)}(x,y)}{\partial y_k}\right)\right]+\frac{\partial}{\partial y_i}\left[\cdots\right]=0,$$

$$\frac{\partial}{\partial x_i}\left[h_{imn}\left(\frac{\partial u_m^{(0)}(x)}{\partial x_n}+\frac{\partial u_m^{(1)}(x,y)}{\partial y_n}\right)-\epsilon_{ik}^{(B)}\left(\frac{\partial U^{(0)}(x)}{\partial x_k}+\frac{\partial U^{(1)}(x,y)}{\partial y_k}\right)\right.$$
$$\left.-\mu_{ik}\left(\frac{\partial\varphi^{(0)}(x)}{\partial x_k}+\frac{\partial\varphi^{(1)}(x,y)}{\partial y_k}\right)\right]+\frac{\partial}{\partial y_i}\left[\cdots\right]=0.$$

$$(3.108)$$

To both sides of Eq. (3.108) we apply the averaging operator

$$\langle(\cdots)\rangle\equiv\frac{1}{Y}\int_Y d\mathbf{y}\left[(\cdots)\right]$$

88 Mechanics and Physics of Structured Media

and, since by periodic boundary conditions

$$\frac{1}{Y} \int_Y d\mathbf{y} \frac{\partial}{\partial y_j} \Big[\cdots \Big] = 0$$

after using expressions (3.104) and appropriate rearrangement of terms we get a system with homogenized coefficients

$$
\begin{aligned}
c_{ijmn}^{\mathrm{h}} \frac{\partial^2 u_m^{(0)}}{\partial x_j \partial x_n} - e_{kij}^{\mathrm{h}} \frac{\partial^2 U^{(0)}}{\partial x_j \partial x_k} - h_{kij}^{\mathrm{h}} \frac{\partial \varphi^{(H)(0)}}{\partial x_j \partial x_k} &= 0, \\
e_{imn}^{\mathrm{h}} \frac{\partial^2 u_m^{(0)}}{\partial x_i \partial x_n} - \epsilon_{ik}^{(D)\mathrm{h}} \frac{\partial^2 U^{(0)}}{\partial x_i \partial x_k} - \mu_{ik}^{(D)\mathrm{h}} \frac{\partial \varphi^{(H)(0)}}{\partial x_i \partial x_k} &= 0, \\
h_{imn}^{\mathrm{h}} \frac{\partial^2 u_m^{(0)}}{\partial x_i \partial x_n} - \epsilon_{ik}^{(B)\mathrm{h}} \frac{\partial U^{(0)}}{\partial x_i \partial x_k} - \mu_{ik}^{\mathrm{h}} \frac{\partial^2 \varphi^{(H)(0)}}{\partial x_i \partial x_k} &= 0.
\end{aligned}
\tag{3.109}
$$

Here

$$
\begin{aligned}
c_{ijrs}^{\mathrm{h}} &= \Big\langle c_{ijmn} \Big(\delta_{mr}\delta_{ns} + \frac{\partial Y_{mrs}}{\partial y_n} \Big) - e_{kij} \frac{\partial \Psi_{rs}}{\partial y_k} - h_{kij} \frac{\partial \Phi_{rs}}{\partial y_k} \Big\rangle, \\
e_{rij}^{\mathrm{h}} &= \Big\langle c_{ijmn} \frac{\partial F_{mr}}{\partial y_n} - e_{kij} \Big(\delta_{kr} + \frac{\partial P_r}{\partial y_k} \Big) - h_{kij} \frac{\partial R_r}{\partial y_k} \Big\rangle, \\
h_{rij}^{\mathrm{h}} &= \Big\langle c_{ijmn} \frac{\partial G_{mr}}{\partial y_n} - e_{kij} \frac{\partial Q_r}{\partial y_k} - h_{kij} \Big(\delta_{kr} + \frac{\partial S_r}{\partial y_k} \Big) \Big\rangle, \\
e_{irs}^{\mathrm{h}} &= \Big\langle e_{imn} \Big(\delta_{mr}\delta_{ns} + \frac{\partial Y_{mrs}}{\partial y_n} \Big) - \epsilon_{ik}^{(D)} \frac{\partial \Psi_{rs}}{\partial y_k} - \mu_{ik}^{(D)} \frac{\partial \Phi_{rs}}{\partial y_k} \Big\rangle, \\
\epsilon_{ir}^{(D)\mathrm{h}} &= \Big\langle e_{imn} \frac{\partial F_{mr}}{\partial y_n} - \epsilon_{ik}^{(D)} \Big(\delta_{kr} + \frac{\partial P_r}{\partial y_k} \Big) - \mu_{ik}^{(D)} \frac{\partial R_r}{\partial y_k} \Big\rangle, \\
\mu_{ir}^{(D)\mathrm{h}} &= \Big\langle e_{imn} \frac{\partial G_{mr}}{\partial y_n} - \epsilon_{ik}^{(D)} \frac{\partial Q_r}{\partial y_k} - \mu_{ik}^{(D)} \Big(\delta_{kr} + \frac{\partial S_r}{\partial y_k} \Big) \Big\rangle, \\
h_{irs}^{\mathrm{h}} &= \Big\langle h_{imn} \Big(\delta_{\mu r}\delta_{ns} + \frac{\partial Y_{mrs}}{\partial y_n} \Big) - \epsilon_{ik}^{(B)} \frac{\partial \Psi_{rs}}{\partial y_k} - \mu_{ik} \frac{\partial \Phi_{rs}}{\partial y_k} \Big\rangle, \\
\epsilon_{ir}^{(B)\mathrm{h}} &= \Big\langle h_{imn} \frac{\partial F_{mr}}{\partial y_n} - \epsilon_{ik}^{(B)} \Big(\delta_{kr} + \frac{\partial P_r}{\partial y_k} \Big) - \mu_{ik} \frac{\partial R_r}{\partial y_k} \Big\rangle, \\
\mu_{ir}^{\mathrm{h}} &= \Big\langle h_{imn} \frac{\partial G_{mr}}{\partial y_n} - \epsilon_{ik}^{(B)} \frac{\partial Q_r}{\partial y_k} - \mu_{ik} \Big(\delta_{kr} + \frac{\partial S_r}{\partial y_k} \Big) \Big\rangle.
\end{aligned}
\tag{3.110}
$$

In the list (3.110), there are duplicate definitions of the coefficients e_{rij}^{h} and h_{rij}^{h}. They are equivalent to each other, as it can be shown by the use of the *local* equations (3.105)–(3.108) and multiple integration by parts.

3.7 Homogenization of 2D composites by decomposition of coupled fields

We now proceed to discuss homogenization of fibrous composites formed by unidirectional infinite cylinders shown in Fig. 3.1 (plane strain). An analogous approach holds for the plane stress problem concerning thin plates and shells. The constructive homogenization should follow the lines of [27, Chapter 2] where statistically homogeneous (deterministic or random) composites were investigated by introduction of the representative cell and solution to a boundary value problem. The considered problem for a MEE composite has to be stated in the same way as for a scalar conductive composite, but with a vector-matrix extension. In the present section, the attention is paid to decomposition of the coupled equations onto the scalar ones by a matrix transform T. After this transformation one may use the known exact or approximate formulas for the scalar effective conductivity and by application of T^{-1} to get formulas for MEE composites.

Let the tensors s_{ijmn}, g_{kij}, p_{kij}, ... from Eqs. (3.94) are constant but different in the host and in the fibers. Let the contact between constituents be perfect. Consider the uniform external field. Despite the same geometry and the same physical properties in every section S, the local fields in the considered composite are essentially 3D and can be decomposed onto the field$_\perp$ (transversal) and the field$_\parallel$ (longitudinal) in the special cases. This question is important for calculations of the effective constants (3.110). More precisely, how many and which problems for a cell one should solve in order to determine the effective constants of MEE composites.

The above question was considered in main for piezoelectric fibrous composites. Following [12,50] we consider transversely isotropic fibrous composites which corresponds to the geometrical type 3-1. The composite exhibits the plane isotropy in every section S perpendicular to the axis of fibers. Such a composite is described by seven constants including four effective elastic moduli, two piezoelectric constants and one permittivity. Five independent relationships on four effective elastic constants and three on the effective physical constants were derived in [12,50]. These five universal relationships include the concentration of fibers.

It is worth noting that the complete set of the effective piezoelectric constants of transversely isotropic fibrous composites can be obtained from the independent transverse (in-plane) and the longitudinal problems. The in-plane isotropy implies that the transverse coupled piezoelectric effects vanish and we arrive at the pure plane elastic problem. The analytical formulas for regular and random transversely isotropic composites were established in [15]. The second longitudinal problem consists in the antiplane elastic field coupled with the electric field. Below in Sections 3.7.1 and 3.7.2, we will see that the piezoelectric longitudinal problem can be decomposed onto two auxiliary scalar problems.

90 Mechanics and Physics of Structured Media

The longitudinal piezoelectric effects are governed by equations [46] following from (3.95). It is convenient to write them in the symmetric form

$$\sigma_{i3} = d_{11}\frac{\partial u_3}{\partial x_i} + d_{12}(-E_i),$$

$$D_i = d_{12}\frac{\partial u_3}{\partial x_i} + d_{22}(-E_i), \quad i = 1, 2, \tag{3.111}$$

where for instance $d_{22} = -\epsilon_{11}^{(D)}$. First, consider a porous medium described by Eqs. (3.111) in a multiply connected domain G. Following [37,51] we introduce the symmetric matrix

$$\mathbf{d} = \begin{pmatrix} d_{11} & d_{12} \\ d_{12} & d_{22} \end{pmatrix}. \tag{3.112}$$

For simplicity consider macroscopically isotropic effective tensors d_{ij}^{h} which are reduced to scalars. Therefore, three effective constants d_{ij}^{h} determine the longitudinal piezoelectric tensor of fibrous composite

$$\mathbf{d}^{\mathrm{h}} = \begin{pmatrix} d_{11}^{\mathrm{h}} & d_{12}^{\mathrm{h}} \\ d_{12}^{\mathrm{h}} & d_{22}^{\mathrm{h}} \end{pmatrix} \tag{3.113}$$

3.7.1 Straley-Milgrom decomposition

The matrix (3.112) is reduced to a diagonal matrix

$$\Lambda = \begin{pmatrix} \lambda_1 & 0 \\ 0 & \lambda_2 \end{pmatrix} \tag{3.114}$$

by the transformation $\mathbf{d} = T\Lambda T^{-1}$. Here, T is a nondegenerate matrix and

$$\lambda_{1,2} = \frac{1}{2}\left(d_{11} + d_{22} \pm \sqrt{(d_{11} - d_{22})^2 + 4d_{12}^2}\right). \tag{3.115}$$

Four equations (3.111) can be considered in the vector-matrix form

$$\sigma_i = \mathbf{d}\frac{\partial \mathbf{u}}{\partial x_i}, \quad i = 1, 2, \tag{3.116}$$

Two-dimensional equations of magneto-electro-elasticity Chapter | 3 **91**

where $\sigma_i = (\sigma_{i3}, D_i) \equiv (\mathcal{D}_{i1}, \mathcal{D}_{i2})$ and $\mathbf{u} = (u_3, U)^T \equiv (\mathcal{U}_{i1}, \mathcal{U}_{i2})^T$, the potential U is determined by (3.7). Introduce the auxiliary values

$$\sigma_i' = T^{-1}\sigma_i \equiv (\mathcal{D}_{i1}', \mathcal{D}_{i2}'), \quad \mathbf{u}' = T^{-1}\mathbf{u} \equiv (\mathcal{U}_1', \mathcal{U}_2')^T. \tag{3.117}$$

Then, Eqs. (3.116) are decomposed onto two vector equations (four scalar equations)

$$\sigma_i' = \lambda_i \, \nabla \mathcal{U}_i', \quad i = 1, 2. \tag{3.118}$$

The conductivity λ_i of the auxiliary composites determines the auxiliary effective conductivity λ_i^h ($i = 1, 2$). The effective tensor of the considered piezoelectric porous medium has the form

$$\mathbf{d}^h = T \begin{pmatrix} \lambda_1^h & 0 \\ & \\ 0 & \lambda_2^h \end{pmatrix}. \tag{3.119}$$

3.7.2 Rylko decomposition

The above Straley-Milgrom [37,51] approach admits the generalization to two-phase composites developed by Rylko [46]. Let the constitutive law is given by Eqs. (3.111) in $G = G^+ \sqcup \Gamma \sqcup G^-$, where the tensor (3.112) is equal to $\mathbf{d}_\pm$ in the domains $G^\pm$, respectively. It is assumed that a smooth curve Γ divides the domains $G^\pm$. For definiteness, we consider a polydispersed composite with the connected host G^+ and the set of inclusions G^-. The developed below approach can be extended to other types of composite after the corresponding description. The condition of perfect contact can be written in terms of the complex potential analogously to (3.66). In the considered case, the complex potential are vector-functions $\varphi^+(z)$ and $\varphi^-(z)$ analytic in G^+ and G^-, respectively, and continuously differentiable in the closures of the considered domains. If G is a representative cell, the function $\varphi^+(z)$ has the prescribed constant jumps per cell corresponding to the external field. The perfect contact condition has the form [46]

$$\varphi^+(t) = \varphi^-(t) - \mathbf{R}\,\overline{\varphi^-(t)}, \quad t \in \Gamma, \tag{3.120}$$

where

$$\mathbf{R} = -(\mathbf{I} - \mathbf{d}_+^{-1}\mathbf{d}_-)(\mathbf{I} + \mathbf{d}_+^{-1}\mathbf{d}_-)^{-1}, \tag{3.121}$$

$\mathbf{I}$ is the identity 2×2 matrix. The matrix (3.121) can be considered as the contrast parameter matrix.

Let the matrix $\mathbf{R}$ is decomposed as follows

$$\mathbf{R} = T\lambda T^{-1}, \tag{3.122}$$

where

$$\mathbf{T} = \begin{pmatrix} \frac{h(\varrho_1-\varrho_2)-\sqrt{\Delta}}{2c(\varrho_1-\varrho_3)} & \frac{h(\varrho_1-\varrho_2)+\sqrt{\Delta}}{2c(\varrho_1-\varrho_3)} \\ 1 & 1 \end{pmatrix} \tag{3.123}$$

and

$$\lambda = \begin{pmatrix} \lambda_1 & 0 \\ 0 & \lambda_2 \end{pmatrix} \tag{3.124}$$

with

$$\lambda_{1,2} = \frac{1}{2(1+h)}[h(\varrho_1+\varrho_2)+2\varrho_3\pm\sqrt{\Delta}]. \tag{3.125}$$

Here, the constants are calculated by means of the components of $\mathbf{d}_\pm$ as follows

$$\varrho_1 = \frac{d_{11}^- - d_{11}^+}{d_{11}^- + d_{11}^+}, \quad \varrho_2 = \frac{d_{22}^- - d_{22}^+}{d_{22}^- + d_{22}^+}, \quad \varrho_3 = \frac{d_{12}^- - d_{12}^+}{d_{12}^- + d_{12}^+}, \quad c = \frac{d_{11}^- + d_{11}^+}{d_{12}^- + d_{12}^+}, \tag{3.126}$$

$$e = -\frac{d_{22}^- + d_{22}^+}{d_{12}^- + d_{12}^+}, \quad h = ce, \quad \Delta = h[h(\varrho_1-\varrho_2)^2 - 4(\varrho_1-\varrho_3)(\varrho_2-\varrho_3)].$$

The matrix $\mathbf{T}$ by Rylko [46] and the matrix T by Straley-Milgrom [37,51] are different. The matrix $\mathbf{T}$ is real, if Δ is positive. It is not always fulfilled since the matrix $\mathbf{R}$ as well as $\mathbf{T}$ is not necessary symmetric.

For simplicity, we consider a macroscopically isotropic structure when the components d_{ij}^h ($i, j = 1, 2$) of the symmetric effective tensor $\mathbf{d}^h$ are scalars. Consider now an auxiliary scalar conductivity problem for the same geometrical structure. Introduce the auxiliary complex potentials $\psi^\pm(z)$ satisfying the $\mathbb{R}$-linear problem

$$\psi^+(t) = \psi^-(t) - \varrho\,\overline{\psi^-(t)}, \quad t \in \Gamma, \tag{3.127}$$

where

$$\varrho = \frac{\lambda-1}{\lambda+1} \tag{3.128}$$

denotes the contrast parameters in an auxiliary composite with the ratio conductivity of inclusions to matrix λ. The generalized alternating method of Schwarz was developed for the traditional polydispersed composites in [39,41,42]. The method of Schwarz can be written in the form of integral equations [8,27] different from the integral equations of potential theory. In the case of 2D conductive polydispersed composites, the method of Schwarz can be applied in the form of contrast parameter expansion. Let f denote the concentration of inclusions and

ϱ the contrast parameter (3.128). Below, such an expansion is written for the normalized effective conductivity of macroscopically isotropic composites

$$\lambda^{h} = 1 + 2\varrho f P(\varrho), \qquad (3.129)$$

where the conductivity of matrix is normalized to unity. The function $P(\varrho)$ is analytic in the unit disk $|\varrho| < 1$ and represented by the series

$$P(\varrho) = \sum_{k=0}^{\infty} \mathcal{A}_k \varrho^k. \qquad (3.130)$$

The coefficients $\mathcal{A}_k$ are determined by the location and shapes of inclusions. In the case of circular inclusions, $\mathcal{A}_k$ are exactly written in [27]. It was proved [8,27,41] that the series (3.129) converges absolutely in the unit disk $|\varrho| < 1$. The coefficients $\mathcal{A}_k$ are real for a macroscopically isotropic composite.

Consider the case $\Delta > 0$. Then, the effective tensor of the considered piezo-electric composite has the form [46]

$$\mathbf{d}^{h} = \mathbf{d}_{+} (\mathbf{I} + 2f\mathbf{R}\boldsymbol{\psi}), \qquad (3.131)$$

where the elements of the matrix $\boldsymbol{\psi}$ are given by the expressions

$$\psi_{11} = \frac{1}{2}[P(\lambda_1) + P(\lambda_2)] - \frac{h(\varrho_1 - \varrho_2)}{2\sqrt{\Delta}}[P(\lambda_1) - P(\lambda_2)], \qquad (3.132)$$

$$\psi_{12} = \frac{e}{\sqrt{\Delta}}(\varrho_2 - \varrho_3)[P(\lambda_1) - P(\lambda_2)],$$

$$\psi_{21} = \frac{c}{\sqrt{\Delta}}(\varrho_3 - \varrho_1)[P(\lambda_1) - P(\lambda_2)],$$

$$\psi_{22} = \frac{1}{2}[P(\lambda_1) + P(\lambda_2)] + \frac{h(\varrho_1 - \varrho_2)}{2\sqrt{\Delta}}[P(\lambda_1) - P(\lambda_2)].$$

Eqs. (3.132) are written in the slightly another form than in [46]. It is done in order to study the case $\Delta < 0$ which was omitted in the investigation [46]. We now assume that $\Delta < 0$. The matrices (3.123) and (3.124) are complex. Nevertheless, the formulas (3.131)–(3.132) will be modified and justified below in the general case.

First, we note that for $\Delta < 0$ the values (3.125) are conjugated and satisfy the quadratic equation with real coefficients

$$\lambda^2 - \frac{h(\varrho_1 + \varrho_2) + 2\varrho_3}{1 + h}\lambda + \frac{h\varrho_1\varrho_2 + \varrho_3^2}{1 + h} = 0. \qquad (3.133)$$

Below, these roots are denoted by λ and $\overline{\lambda}$. It follows from Vieta's formula and inequalities $|\varrho_i| < 1$, $h > 0$ that $|\lambda|^2 = \left|\frac{h\varrho_1\varrho_2 + \varrho_3^2}{1 + h}\right| < 1$. Therefore, the complex

94 Mechanics and Physics of Structured Media

roots λ and $\overline{\lambda}$ lie in the unit disk, the convergence domain of the series (3.130). The function $P(\lambda)$ satisfies the relations

$$P(\lambda) + P(\overline{\lambda}) = 2\mathrm{Re}\,[P(\lambda)], \quad P(\lambda) - P(\overline{\lambda}) = 2\mathrm{i}\,\mathrm{Im}\,[P(\lambda)], \qquad (3.134)$$

since its coefficients $\mathcal{A}_k$ are real. Substitute $\lambda_1 = \lambda$ and $\lambda_2 = \overline{\lambda}$ into (3.132) and use the relations (3.134)

$$\psi_{11} = \mathrm{Re}\,[P(\lambda)] - \frac{h(\varrho_1 - \varrho_2)}{\sqrt{|\Delta|}}\mathrm{Im}\,[P(\lambda)], \qquad (3.135)$$

$$\psi_{12} = \frac{2e}{\sqrt{|\Delta|}}(\varrho_2 - \varrho_3)\mathrm{Re}\,[P(\lambda)], \quad \psi_{21} = \frac{2c}{\sqrt{|\Delta|}}(\varrho_3 - \varrho_1)\mathrm{Im}\,[P(\lambda)],$$

$$\psi_{22} = \mathrm{Re}\,[P(\lambda)] + \frac{h(\varrho_1 - \varrho_2)}{|\Delta|}\mathrm{Im}\,[P(\lambda)].$$

One can see that the components of the matrix $\boldsymbol{\psi}$ are real. We assert that the formula (3.132) holds for the effective tensor $\mathbf{d}^h$ in the case of complex representation (3.122)–(3.124), where the matrix $\boldsymbol{\psi}$ is given by (3.135).

The arguments [46] cannot be applied to justify the above statement, since the restriction $\Delta \geq 0$ is essential in transformation of the vector-matrix $\mathbb{R}$-linear problem to a pair of scalar problems. We now apply the arguments of analytic continuation on the parameters (3.126). The formulas (3.131)–(3.132) hold for $\Delta > 0$ with the corresponding values (3.125) substituted in the argument of the function (3.130), more precisely for the arguments $\varrho = \lambda_1$ and $\varrho = \lambda_2$. Consider the ball $\mathbb{B} = \{(\lambda_1, \lambda_2) \in \mathbb{C}^2 : |\lambda_1| \leq q, |\lambda_2| \leq q\}$ in the two-dimensional complex space $\mathbb{C}^2$ of the arbitrarily fixed radius $q < 1$. One can suppose that the formulas (3.131)–(3.132) hold in the subset $\mathbb{B}_+ = \{(\lambda_1, \lambda_2) \in \mathbb{B} : \Delta > 0\}$. The absolute convergence of the series (3.130) in the disk $|\varrho| \leq q$ on the complex plane implies its uniform convergence on the parameters (3.126), hence, the analytic continuation on these parameters into the whole ball $\mathbb{B}$. Therefore, the formula (3.131) for positive Δ with (3.132) is analytically continued on the parameters to the set $\Delta < 0$ of $\mathbb{B}$. As a result of the continuation we arrive at the same formula (3.131) but with $\boldsymbol{\psi}$ given by (3.135). It is worth noting that the case $\Delta = 0$ requires usage of the Jordan normal form instead of the diagonal matrix (3.124) [46].

The justification of (3.131), (3.135) will be completed by the statement that the effective tensor $\mathbf{d}^h$ is an analytic function of the parameters in the analyticity domain $\mathbb{B}$. Such an analyticity was established in [34,36] for a scalar problem. It is suggested that the same result is true for the considered piezoelectric problem.

3.7.3 Example

The described method allows to write exact formulas for the effective piezoelectric tensor $\mathbf{d}^h$ through the known exact formulas for the scalar effective conductivity. Before to give an example we have, unfortunately, to say that our

Two-dimensional equations of magneto-electro-elasticity **Chapter | 3 95**

formulas are *exact* in the sense explained in [1] contrary to other *exact* or *closed-form* formulas.

As an example consider the square array of circular cylinders. Let f stand for the concentrations of disks and S_{2m} the lattice sums and

$$s_k^{(m)} = \frac{(2m + 2k - 3)!}{(2m - 1)!(2k - 2)!} S_{2(m+k-1)}. \tag{3.136}$$

The function $P(\varrho)$ is written explicitly by the series [27,40]

$$P(\varrho) = 1 + \varrho f$$
$$+ \frac{\varrho f}{\pi} \sum_{k=1}^{\infty} \varrho^k \sum_{m_1=1}^{\infty} \cdots \sum_{m_k=1}^{\infty} s_{m_1}^{(1)} s_{m_2}^{(m_1)} \cdots s_{m_k}^{(m_{k-1})} s_1^{(m_k)} \left(\frac{f}{\pi} \right)^{2(m_1+\ldots+m_k)-k}. \tag{3.137}$$

The details of computations of S_{2m} can be found in [27].

The effective piezoelectric tensor $\mathbf{d}^h$ is given by the exact formulas (3.131)–(3.132). The exact formula (3.137) can be transformed to the following asymptotic formula including the percolation regimes $(|\varrho|f) \sim \frac{\pi}{4}$

$$2\varrho f P(\varrho) \simeq \sqrt{\frac{\frac{\pi}{4} + \varrho f}{\frac{\pi}{4} - \varrho f}} \sqrt[4]{\frac{1 + 1.454\varrho f + 0.388(\varrho f)^2}{1 - 1.454\varrho f + 0.388(\varrho f)^2}} - 1. \tag{3.138}$$

Ultimately, the longitudinal effective piezoelectric tensor of the square array is estimated by the formula (3.131), where $\mathbf{d}_{\pm}$ denotes the tensors of phases, $\mathbf{R}$ is given by (3.121), ψ by (3.132) or by (3.135), the function $P(\varrho)$ is calculated exactly by (3.137) or approximately by (3.138).

3.8 Conclusion

In the present chapter, we follow Filshtinsky's extension of Lekhnitskii's formalism to 2D MEE problems for composites. The difference between Filshtinsky's and Lekhnitskii's approach is clearly demonstrated on the example of dielectric problem in Section 3.4. Using Filshtinsky's complex potentials we reduce a boundary value problem for a hole and for an inclusion embedded in the MEE host material to the vector-matrix Riemann-Hilbert and to the $\mathbb{R}$-linear problems with a shift, respectively. A problem for circular hole in the MEE bulk material is solved in closed form by the method of conformal mapping and Sochocki's formulas [25]. Numerical examples are considered. The theoretical investigation of the considered integral equations can be found in [10,11].

The general homogenization approach is applied to MEE composites. The special attention is paid to the antiplane coupled problem for the skew-symmetric piezoelectric media. The considered problem can be reduced to uncoupled conductivity equations. First, the Straley-Milgrom decomposition for porous MME and thermoelastic media is outlined. Next, the Rylko decomposition for two-phase composites MME is extended. As a result, we derive exact and analytical approximate formulas for the longitudinal effective piezoelectric tensor and complete the constructive theory by Rylko [46]. We demonstrate that any formula for the scalar conductivity of 2D composite can be transformed by a matrix formalism to a formula for the effective tensor of piezoelectric fibrous composite.

Therefore, we derive the following scheme to determine the effective constants of piezoelectric transversely isotropic fibrous composites. The in-plane piezoelectric effects vanish. Hence, it is sufficient to find the plane elastic effective constants of macroscopically isotropic 2D composite. The corresponding analytical formulas for regular and random locations of circular inclusions can be taken from the book [15]. The longitudinal piezoelectric constants can be found by Rylko's decomposition by using the standard matrix calculus. After this decomposition we can directly use the wide set of exact and approximate formulas valid for 2D scalar conductivity. An example of application is given in Section 3.7.3. The obtained decomposition is useful for numerical applications too, since it is sufficient to solve uncoupled one plane elastic problem an two scalar problems instead of the vector-matrix problem expensive in computations.

References

[1] I. Andrianov, V. Mityushev, Exact and "exact" formulae in the theory of composites, in: P. Drygaś, S. Rogosin (Eds.), Modern Problems in Applied Analysis, Birkhäuzer, Springer International Publishing AG, Cham, 2018, pp. 15–34.

[2] H. Attouch, Variational Convergence for Functions and Operators, Pitman, London, 1984.

[3] N.S. Bakhvalov, G.P. Panasenko, Homogenisation: Averaging Processes in Periodic Media: Mathematical Problems in the Mechanics of Composite Materials, Kluwer, Dordrecht - Boston - London, 1989.

[4] D.M. Barnett, J. Lothe, Dislocations and line charges in anisotropic piezoelectric insulators, Physica Status Solidi B 67 (1975) 105–111.

[5] L.V. Belokopytova, L.A. Fil'shtinskii, Two-dimensional boundary value problem of electro-elasticity for a piezoelectric medium with cuts, Journal of Applied Mathematics and Mechanics 43 (1979) 147–153.

[6] A. Bensoussans, J.-L. Lions, G. Papanicolaou, Asymptotic Analysis of Periodic Structures, North Holland Publishing Company, Amsterdam - New York - Oxford, 1980.

[7] Y. Benveniste, Magnetoelectric effect in fibrous composites with piezoelectric and piezomagnetic phases, Physical Review. B, Condensed Matter 51 (1995) 16424–16427.

[8] B. Bojarski, V. Mityushev, $\mathbb{R}$-linear problem for multiply connected domains and alternating method of Schwarz, Journal of Mathematical Sciences 189 (2013) 68–77.

[9] J. Bravo-Castillero, R. Rodriguez-Ramos, R. Guinovart-Diaz, F.J. Sabina, A.R. Aguiar, Uziel Paulo da Silva, J. Gomez-Munoz, Analytical formulae for electromechanical effective properties of 3–1 longitudinally porous piezoelectric materials, Acta Materialia 57 (2009) 795–803.

[10] T. Buchukuri, O. Chkadua, R. Duduchava, D. Natroshvili, Interface crack problems for metallic-piezoelectric composite structures, Memoirs on Differential Equations and Mathematical Physics 55 (2012) 1–150.

[11] T. Buchukuri, O. Chkadua, D. Natroshvili, Mixed boundary value problems of thermopiezoelectricity for solids with interior cracks, Integral Equations and Operator Theory 64 (2009) 495–537.

[12] Ce-Wen Nan, Comment on "relationships between the effective properties of transversely isotropic piezoelectric composites": K. Schulgasser (1992) J. Mech. Phys. Solids 40, 473-479, Journal of the Mechanics and Physics of Solids 41 (1993) 1567–1570.

[13] Chyanbin Hwu, Anisotropic Elastic Plates, Springer, Berlin, 2010.

[14] R. Czapla, V. Mityushev, N. Rylko, Conformal mapping of circular multiply connected domains onto slit domains, Electronic Transactions on Numerical Analysis 39 (2012) 286–297.

[15] P. Drygaś, S. Gluzman, V. Mityushev, W. Nawalaniec, Applied Analysis of Composite Media. Analytical and Computational Results for Materials Scientists and Engineers, Elsevier, Oxford, 2020.

[16] A.N. Guz, et al., The Mechanics of Coupled Fields in Structural Elements, Naukova Dumka, Kiev, 1988.

[17] L.A. Fil'shtinskii (Filshtinsky), Yu.D. Kovalyov, Concentration of mechanical stresses near a hole in a piezoceramic layer, Mechanics of Composite Materials 38 (2002) 121–124.

[18] L.A. Fil'shtinskii (Filshtinsky), D.I. Bardzokas, L.V. Shramko, Homogeneous solutions of the electroelasticity equations for the piezoceramic layer into $\mathbb{R}^3$, Acta Mechanica 209 (2010).

[19] L.A. Fil'shtyns'kyi (Filshtinsky), Yu.V. Shramko, D.S. Kovalenko, Averaging of the magnetic properties of fibrous ferromagnetic composites, Materials Science 46 (2011) 808–818.

[20] L.A. Filshtinsky, V. Mityushev, Mathematical models of elastic and piezoelectric fields in two-dimensional composites, in: Panos Pardalos, Themistocles M. Rassias (Eds.), Mathematics Without Boundaries, Springer New York, 2014, pp. 217–262.

[21] L.A. Filshtinskii (Filshtinsky), Tutorial to the implementation of master's theses on the topic: "Boundary value problems of fracture mechanics of new composite materials", Sumy University Publ., 2015 (in Russian).

[22] L.A. Filshtinskii (Filshtinsky), L.V. Volkova, O.A. Ivanenko, V.A. Lyubchak, Some static and dynamic boundary value problem of elasticity and electro-elasticity for media with cuts and elastic inclusions, in: Proceedings of All-Soviet Union Conference on the Theory of Elasticity, Erevan, 1979, pp. 56–58.

[23] A. Gałka, J.J. Telega, R. Wojnar, Homogenization and thermopiezolectricity, Mechanics Research Communications 19 (1992) 315–324.

[24] A. Gałka, J.J. Telega, R. Wojnar, Some computational aspects of homogenization of thermopiezoelectric composites, Computer Assisted Mechanics and Engineering Sciences 3 (1996) 133–154.

[25] F.D. Gakhov, Boundary Value Problems, Pergamon Press, Oxford, 1966.

[26] B. Gambin, Influence of microstructure on properties of elastic, piezoelectric and thermoelastic composites, in: Prace IPPT - IFTR Reports, Warszawa, 2006.

[27] S. Gluzman, V. Mityushev, W. Nawalaniec, Computational Analysis of Structured Media, Elsevier, 2018.

[28] Jiashi Yang, Special Topics in the Theory of Piezoelectricity, Springer, Berlin, 2009.

[29] V.V. Jikov, S.M. Kozlov, O.A. Oleinik, Homogenization of Differential Operators and Integral Functionals, Springer, Berlin, Heidelberg, 1994.

[30] A.S. Kosmodamianskii, V.N. Lozhkin, General two-dimensional stressed state of thin piezoelectric slabs, Soviet Applied Mechanics 11 (5) (1975) 495–501.

[31] A.S. Kosmodamianskii, V.N. Lozhkin, An action of electrical point charge on the boundary of piezoelectrical half-plane weakened by an elliptic hole, Izvestiâ Akademii Nauk Armânskoj SSR, XXX 1 (1977) 13–20.

[32] A.S. Kosmodamianskii, A.P. Kravchenko, Distribution of electroelastic fields of direct piezoeffect in crystal half-plane with an elliptic hole, Teoreticheskaya i Prikladnaya Mekhanika (Kiev, Donetsk) 16 (1985) 66–70.

[33] B.A. Kudriavtsev, V.Z. Parton, V.I. Rakitin, Fracture mechanics of piezoelectric materials. Rectilinear tunnel crack on the boundary with a conductor, Journal of Applied Mathematics and Mechanics 39 (1975) 136–146.

98 Mechanics and Physics of Structured Media

[34] M. Lanza de Cristoforis, P. Musolino, Analytic dependence of a periodic analog of a fundamental solution upon the periodicity parameters, Annali Di Matematica Pura Ed Applicata 197 (2018) 1089–1116.

[35] S.G. Lekhnitskii, Theory of Elasticity of an Anisotropic Elastic Body, Mir Publishers, Moscow, 1981.

[36] P. Luzzini, P. Musolino, Perturbation analysis of the effective conductivity of a periodic composite, Networks and Heterogeneous Media 15 (2020) 581–603.

[37] M. Milgrom, S. Shtrikman, Linear response of two-phase composites with cross moduli: exact universal relations, Physical Review. A, General Physics 40 (1989) 1568–1575.

[38] V. Mityushev, S. Rogosin, Constructive Methods for Linear and Nonlinear Boundary Value Problems for Analytic Functions: Theory and Applications, Monographs and Surveys in Pure and Applied Mathematics, Chapman & Hall / CRC, Boca Raton, 2000.

[39] V. Mityushev, Generalized method of Schwarz and addition theorems in mechanics of materials containing cavities, Archives of Mechanics 47 (1995) 1169–1181.

[40] V. Mityushev, Exact solution of the R-linear problem for a disk in a class of doubly periodic functions, Journal of Applied Functional Analysis 2 (2007) 115–127.

[41] V. Mityushev, Random 2D composites and the generalized method of Schwarz, Advances in Mathematical Physics 2015 (2015) 535128.

[42] V. Mityushev, W. Nawalaniec, D. Nosov, E. Pesetskaya, Schwarz's alternating method in a matrix form and its applications to composites, Applied Mathematics and Computation 356 (2019) 144–156.

[43] V. Mityushev, I. Andrianov, S. Gluzman, L.A. Filshtinsky's contribution to Applied Mathematics and Mechanics of Solids, Chapter 1 of the present book.

[44] D. Nosov, L.A. Filshtinsky, V. Mityushev, Cracks in two-dimensional magneto-electro-elastic medium, Chapter 2 of the present book.

[45] V.Z. Parton, B.A. Kudryavtsev, Electro-Magneto-Elasticity of Piezoelectric and Conducting Materials, Nauka, Moscow, 1988.

[46] N. Rylko, Effective anti-plane properties of piezoelectric fibrous composites, Acta Mechanica 224 (2013) 2719–2734.

[47] N. Rylko, Representative volume element in 2D for disks and in 3D for balls, Journal of Mechanics of Materials and Structures 9 (2014) 427–439.

[48] N. Rylko, A pair of perfectly conducting disks in an external field, Mathematical Modelling and Analysis 20 (2015) 273–288.

[49] E. Sanchez-Palencia, Non-Homogeneous Media and Vibration Theory, Springer Verlag, Berlin, Heidelberg, New York, 1980.

[50] K. Schulgasser, Relationships between the effective properties of transversely isotropic piezoelectric composites, Journal of the Mechanics and Physics of Solids 40 (1992) 473–479.

[51] J.P. Straley, Thermoelectric properties of inhomogeneous materials, Journal of Physics. D, Applied Physics 14 (1981) 2101–2105.

[52] J.J. Telega, Piezoelectricity and homogenization. Application to biomechanics, in: G.A. Maugin (Ed.), Continuum Models and Discrete Systems, vol. 2, 220, Longman, Essex, 1991.

[53] J.J. Telega, A. Gałka, B. Gambin, Effective properties of physically nonlinear piezoelectric composites, Archives of Mechanics 59 (1998) 321–340.

[54] V.Yu. Topolov, Ch.R. Bowen, P. Bisegna, Piezo-Active Composites: Microgeometry-Sensitivity Relations, Springer Series in Materials Science, Springer International Publishing, Berlin, 2018.

Chapter 4

Hashin-Shtrikman assemblage of inhomogeneous spheres

Andrej Cherkaev[a] and Vladimir Mityushev[b]
[a]*Department of Mathematics, University of Utah, Salt Lake City, UT, United States,*
[b]*Research Group Materialica+, Faculty of Computer Science and Telecommunications,
Cracow University of Technology, Kraków, Poland*

4.1 Introduction

The Hashin-Shtrikman assemblage (HSA) of coated spheres with the same radii ratio constitutes the exceptional class of composites [1,2,4,6] due to the "invisible" properties in mutual interactions of inclusions. The coated spheres of different sizes fill the whole space. The uniform external field produces the same local field up to scaling in each cell. The effective properties of such a composite are calculated by exact formulas discussed below. The HSA for the polycrystalline structure was considered in [9]. The HSA were extended in [1] to various two-dimensional (2D) structures such as wheel-shaped phase, three-material composite laminates and to a class of 2D random macroscopically isotropic composites assemblages. In the present chapter, we develop one of the model discussed in [1] to 3D structures.

Structures similar to the HSA were observed in the experimental pictures presented in [5,10,11] for $Al/nano - TiC$ and other composites. The formation in $Al/nano - TiC$ composites occurs due to addition of moderator which leads to the pushing nanoparticles (called below by particles for shortness) to spontaneously generated clusters. The location and size of a cluster depends on the fluctuation of moderator, i.e., on its local concentration. The clusters form a partition of the space onto cells. The boundaries of clusters are clearly observed and can be approximated by spheres. It is worth noting that the 2D pictures are in our disposal where the sections of clusters are displayed. The considered composite is macroscopically isotropic because of the isotropic external parameters of manufacturing process and due to the Pierre Curie principle "the symmetries of the causes are to be found in the effects" [3]. Therefore, any plane section is representative, and the 2D pictures display the traces of boundaries similar to spherical.

The geometry with which the real composite is modeled does not perfectly coincide with the geometry of the Hashin-Strikman ensemble. It is rather similar

100 Mechanics and Physics of Structured Media

to the geometry of randomly shaped cells, but with "rounded" interiors. It is assumed that clusters can be approximated by a coating sphere cells filing the whole space.

In the present chapter, we use a radially inhomogeneous approximation of clusters. The inhomogeneity is modeled by a power function which describes the local density of particles in a spherical cell, see Fig. 4.4. This implies that the local conductivity K is a function of the radial coordinate r. The control parameter s is introduced in (4.4) and its impact on K is shown in Fig. 4.5. We do not discuss there the adequate approximation or the optimal choice of s because we do not have accurate pictures of the real $Al/nano - TiC$ structures in our disposal yet.

The classic HSA is described in Section 4.2. The new model with radially inhomogeneous sphere cell is developed in Section 4.3. The chapter ends by Conclusion concerning the effective properties of the considered model of composites.

4.2 The classic Hashin-Shtrikman assemblage

Consider a two-phase 3D macroscopically isotropic composite. Let σ_1 and σ_2 be conductivities of components, f and $1 - f$ their concentrations, respectively. It is assumed that $\sigma = \frac{\sigma_1}{\sigma_2} < 1$ for definiteness throughout the present paper. This case corresponds to $Al/nano - TiC$ composites with TiC particles when the conductivity of Al and TiC hold $237\,(\mathrm{W/(m \cdot K)})$ and 21, respectively. Hence, their ratio holds $\sigma = 0.0886$.

The effective conductivity σ_e of macroscopically isotropic two-phase composites satisfies the Hashin-Shtrikman bounds [4]

$$\sigma^- \le \frac{\sigma_e}{\sigma_2} \le \sigma^+, \tag{4.1}$$

where

$$\frac{\sigma^-}{\sigma_2} = \sigma \left[1 + \frac{3(1 - f)(1 - \sigma)}{f(1 - \sigma) + 3\sigma} \right], \quad \frac{\sigma^+}{\sigma_2} = 1 + \frac{3f(\sigma - 1)}{(1 - f)(\sigma - 1) + 3}. \tag{4.2}$$

The bound σ^+ coincides with the Clausius-Mossotti approximation (CMA)

$$\frac{\sigma_e}{\sigma_2} = \frac{1 + 2f\frac{\sigma - 1}{\sigma + 2}}{1 - f\frac{\sigma - 1}{\sigma + 2}}. \tag{4.3}$$

The CMA holds for spheres of conductivity σ_1 diluted ($f \ll 1$) in the host of conductivity σ_2. The same formula (4.3) is fulfilled exactly for the classic HSA of coated spheres. The coated sphere cell is displayed in Fig. 4.1. Every cell has the same structure up to a scale factor. In particular, the concentration of coating holds $f = 1 - r_c^2$, where r_c denotes the ratio of the core radius to the external radius. The HSA consists of the mutually disjoint coated sphere cells of different sizes filling the whole space. Every cell of HSA does not interact with others and can be considered as an "invisible" inclusion. [1]. The Hashin-Shtrikman bounds (4.1) are displayed in Fig. 4.2.

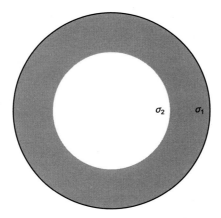

FIGURE 4.1 The normalized coated sphere of the external unit radius and the core radius $r_c < 1$.

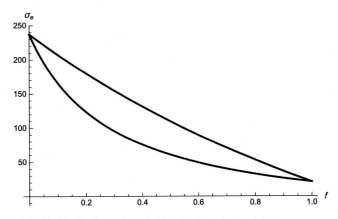

FIGURE 4.2 The Hashin-Shtrikman bounds (4.1) for the ratio $\sigma = 0.0886$.

The bound σ^+ coinciding with the CMA (4.3) can be applied to estimation of the effective conductivity of composites consisting of spherical cell filled by particles of conductivity σ_1. The coated sphere in Fig. 4.1 approximates the structure with particles densely packed near the boundary. In the considered $Al/nano - TiC$ composite, the concentration f is fixed and the bounds in Fig. 4.2 show the attainable domain of σ_e. The distribution of particles is the main factor which impacts on its macroscopic properties. The distribution can be roughly described by the parameter r_c^2, see the dependence of σ_e on r_c^2 in

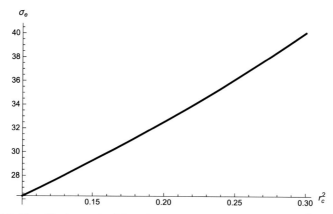

FIGURE 4.3 The effective conductivity calculated by (4.3) as a function of $r_c^2 = 1 - f$; $\sigma = 0.0886$.

Fig. 4.3. One may go ahead and develop a multi-coating model in order to get more accurate geometrical approximations. However, the multi-coating model contains too many parameters which complicate the investigation. A radially inhomogeneous model will be developed in the next section in order to model the distribution of particles.

4.3 HSA-type structure

We proceed to discuss a class of structures resembling the perfect HSA. Consider a set of particles distributed in the space $\mathbb{R}^3$. The original distribution of particles is supposed to be uniform. A moderator is added during the manufacturing process. Let an infinite set of points $\mathbf{x}_k \in \mathbb{R}^3$ model the points (regions) of the moderator concentrations in the space with the assigned dimensionless degree of concentration d_k. It is assumed that near each point $\mathbf{x}_k$ a large number of particles is located. In the real composites, this number is of order 10^4. Each point $\mathbf{x}_k$ is a center of radially pushing particles with the intensity proportional to d_k. The competition of neighbor opposite pushing points leads to formation of cells with the local concentration of particles near the boundaries of cells. It is assumed that the following geometrical assumptions are fulfilled:

- The space $\mathbb{R}^3$ is filled by different sphere cells except a set of measure zero.
- The interior structure of each sphere, i.e., the distribution of particles is the same up to scaling.

These conditions determine a HSA-type structure of composed spheres. Each sphere after normalization can be considered as a unit sphere in the local coordinates where the origin is the center of sphere. It is convenient to consider the standard spherical coordinates (r, θ, ϕ), where $r \geq 0$, $-\pi < \phi \leq \pi$ and $0 \leq \theta \leq \pi$.

We consider the 3D extension of the classic HSA and 2D theory [1, Section 3.3 Effective medium theory]. First, the discrete distribution of particles shown in Fig. 4.4a is replaced by the continuous density distribution $0 \leq f(r) \leq 1$ ($0 < r < 1$) displayed in Fig. 4.4b. The spherical cell displayed in Fig. 4.3a is replaced by the radially inhomogeneous cell in Fig. 4.3b by heuristic evaluation based on the observation.

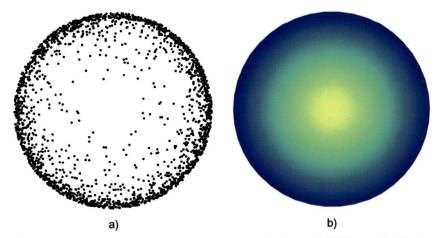

FIGURE 4.4 a) A typical distribution of particles in the unit sphere cell. b) The smoothed density plot radial distribution of particles in the unit sphere cell $\mathbb{U}$. The colors correspond to the local density of particles.

The constructed sphere with the fuzzy core and coating is considered as the cell used for generation of HSA. Assume that an infinite space is filled with a material σ_e and has a spherical inclusion of unit radius $\mathbb{U}$. Let the conductivity $K(r)$ of the sphere change with the radius r in the local spherical coordinates with the center of the spheres. Let this local conductivity be approximated by the power function

$$K(r, s) = \sigma_2 + (\sigma_1 - \sigma_2)r^s, \tag{4.4}$$

where the positive parameter s corresponds to different distributions of particles. One can see that

$$K(0, s) = \sigma_2, \quad K(1, s) = \sigma_1. \tag{4.5}$$

The shape of (4.4) depends on the distribution parameter s; it is displayed in Fig. 4.5. Various shapes of $K(r, s)$ are chosen by means of the preliminary observations of pictures and by estimations outlined in [7]. A more precise analysis of the observed local concentration of particles with further estimations of the local conductivity can be developed after preparation of the corresponding 3D pictures.

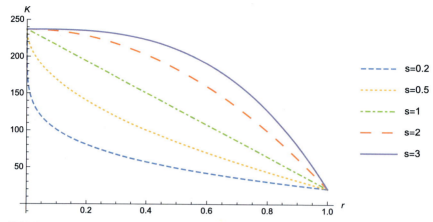

FIGURE 4.5 The local conductivity (4.4) for different distribution parameter s corresponding to different density distributions of particles.

The potential U satisfies the partial elliptic differential equation

$$\nabla \cdot (K \nabla U) = 0, \quad 0 < r < 1. \tag{4.6}$$

Eq. (4.6) will be used in the spherical coordinates (r, ϕ, θ) with $U = U(r, \theta, \phi)$, where the radial distance $r \geq 0$, the azimuthal angle $-\pi < \phi \leq \pi$, and the polar angle $0 \leq \theta \leq \pi$ are used.

Following [1] we assume that the potential has the form

$$U = u(r) \cos \theta, \quad r \leq 1. \tag{4.7}$$

Substitution of (4.7) into Eq. (4.6) written in the spherical coordinates yields the ordinary differential equation for the auxiliary function $u(r)$

$$K \frac{d^2 u}{dr^2} + \left(\frac{dK}{dr} + \frac{2K}{r} \right) \frac{du}{dr} - \frac{2K}{r^2} u = 0, \quad 0 < r < 1. \tag{4.8}$$

The following boundary condition can be added to the axially symmetric equation (4.8)

$$u(0) = 0. \tag{4.9}$$

Following the effective medium approximation for the "invisible" inclusion [1, Section 3.3] we assume that the exterior of the unit sphere $\mathbb{U}$ is occupied by the macroscopically isotropic material of conductivity σ_e which has to be determined. Let the potential outside the inclusion be given by the formula

$$U_{ext}(r) = r \cos \theta. \tag{4.10}$$

The function satisfies Laplace equation and the condition at infinity

$$\lim_{r \to \infty} \frac{U_{ext}(r)}{r} = \cos\theta. \quad (4.11)$$

This condition guarantees that the inclusion is invisible to an outside observer.

The perfect contact expresses the continuity of the normal current and the potential at the boundary

$$U(1) = U_{ext}(1), \quad \sigma_1 \frac{dU}{dr}(1) = \sigma_e \frac{dU_{ext}}{dr}(1). \quad (4.12)$$

Substituting (4.7) and (4.10) into (4.12) and adding the condition (4.9) we obtain

$$u(0) = 0, \quad u(1) = 1, \quad \sigma_1 \frac{du}{dr}(1) = \sigma_e. \quad (4.13)$$

The first and second equations (4.13) can be considered as the boundary conditions. The third equation yields the formula for the calculation of effective conductivity after solution to the boundary value problem

$$\sigma_e = \sigma_1 \frac{du}{dr}(1). \quad (4.14)$$

The boundary value problem is solved numerically for the different parameters s. The polynomial fitting can be applied to the obtained numerical solution. For instance, the function $u(r)$ for $s = 2$ can be approximated by the polynomial

$$u(r) \approx 0.45r + 1.46r^2 - 3.04r^3 + 2.13r^4. \quad (4.15)$$

This formula is based on the measured properties of constitutes $\sigma_1 = 21$, $\sigma_2 = 237$ and on the suggested particles distribution given by (4.4) with $s = 2$. The approximation (4.15) is compared with the numerical calculations in Fig. 4.6. Few plots $u(r)$ for various s are displayed in Fig. 4.7.

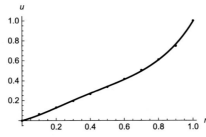

FIGURE 4.6 The polynomial approximation (4.15) of the function $u(r)$ for $s = 2$. The numerical results are shown by points.

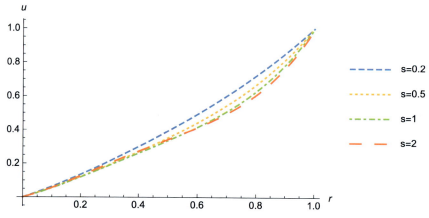

FIGURE 4.7 The function $u(r)$ for various s.

The results for the effective conductivity σ_e are presented in Fig. 4.8 with the polynomial fitting

$$\sigma_e(s) = 25.09 + 44.36s - 20.0s^2 + 2.98s^3. \qquad (4.16)$$

The strong dependence of the local conductivity in a spherical cell on the ef-

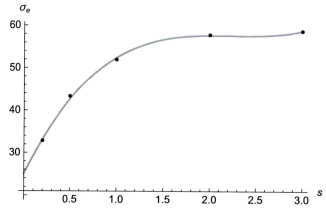

FIGURE 4.8 The effective conductivity σ_e for different parameters s. The numerical results are shown by points and its polynomial fitting (4.16) by the continuous line.

fective conductivity is demonstrated in Fig. 4.8. This observation can be used in the manufacturing process of $Al/nano - TiC$ composites where a moderator impacts on the distribution of nanoparticles.

4.4 Conclusion

In the present chapter, the Hashin-Shtrikman approach is generalized. The considered model extends the classic theory of HSA to more complicated geometric structures. Instead of coated spheres in HSA, we consider the radially inhomogeneous spherical inclusions. The considered model is a 3D extension of one of the 2D models considered in [1]. The effective conductivity σ_e is given by the formula (4.14), where the function $u(r)$ is obtained by numerical solution to the boundary value problem for the ordinary differential equation (4.8). A fitting procedure in the distribution parameter s is applied to the numerical solution. An example of polynomial approximation is given by Eq. (4.16).

The obtained results demonstrate the dependence of σ_e on the distribution of particles through the parameter s in Fig. 4.8. Figs. 4.5 and 4.8 can be used to optimize the structure of $Al/nano - TiC$ composites. The results shown in Figs. 4.8 and 4.3 cannot be directly compared, since it is difficult to properly investigate the dependence of σ_e on the distribution of particles within the Hashin-Shtrikman model. The usage of the parameter s seems to be more preferable than r_c.

The reiterated homogenization can be carry out when the homogenization over the sphere can be taken as the first step where two different scales are considered. The first step includes the heuristic replacement of the spherical cell displayed in Fig. 4.3a by the radially inhomogeneous cell in Fig. 4.3b. The second step performed in Section 4.3 is to solve the problem of inserting a radially inhomogeneous sphere into a homogeneous matrix with unknown properties solution of which implicitly determines these properties. The simulations demonstrate the shortage of the simplified reiterated homogenization outlined above because of highly volatile results strongly depending on the gradient geometrical location of particles near the boundary. We suppose that the simplified reiterated homogenization should be essentially modified because of the boundary effects, see application of Adler's resurgence flow to the heat conduction [8].

The obtained results are valid for (the numerically fixed conductivity $\sigma_1 = 21$ (W/m/K) and $\sigma_2 = 237$) of components Al and TiC. The radial distribution of inclusions is modeled by the power function (4.4). We expect to fit the model with experimental data when they become available.

Acknowledgments

We are grateful to Barbara Gambin (Warsaw) for useful remarks and fruitful discussion. This work was partially supported by National Science Centre, Poland, Research Project No. 2016/21/B/ST8/01181.

Andrej Cherkaev in thankful for the support by NSF DMS, Grant No. 1515125.

References

[1] A. Cherkaev, Optimal three-material wheel assemblage of conducting and elastic composites, International Journal of Engineering Science 59 (2012) 27–39.

108 Mechanics and Physics of Structured Media

[2] A. Cherkaev, A.D. Pruss, Effective conductivity of spiral and other radial symmetric assemblages, Mechanics of Materials 65 (2013) 103–109.

[3] P. Curie, Sur la symétrie dans les phénomènes physiques, symétrie d'un champ électrique et d'un champ magnétique, Journal de Physique Théorique et Appliquée, EDP Sciences 3 (1894) 393–415.

[4] Z. Hashin, S. Shtrikman, A variational approach to the theory of the effective magnetic permeability of multiphase materials, Journal of Applied Physics 33 (10) (1962) 3125–3131.

[5] W. Maziarz, A. Wojcik, P. Bobrowski, A. Bigos, L. Szymanski, P. Kurtyka, N. Rylko, E. Olejnik, SEM and TEM studies on in-situ cast Al-TiC composites, Materials Transactions 60 (2019) 714–717.

[6] G.W. Milton, The Theory of Composites, Cambridge University Press, 2002.

[7] N. Rylko, Representative volume element in 2D for disks and in 3D for balls, Journal of Mechanics of Materials and Structures 9 (2014) 427–439.

[8] N. Rylko, R. Wojnar, Resurgence edge effects in composites: fortuity and geometry, in: M. Mladenov, Mariana Hadzhilazova, Vasyl Kovalchuk (Eds.), Geometry, Integrability, Mechanics and Quantization, Ivailo Avangard Prima, Sofia, 2015, pp. 342–349.

[9] K. Schulgasser, Sphere assemblage model for polycrystals and symmetric materials, Journal of Applied Physics 54 (1983) 1380–1382.

[10] M. Stawiarz, P. Kurtyka, N. Rylko, S. Gluzman, Influence of FSP process modification on selected properties of Al-Si-Cu/SiCp composite surface layer, Composites Theory and Practice 19 (2019) 161–168.

[11] A. Wojcik, E. Olejnik, A. Bigos, R. Chulist, P. Bobrowski, P. Kurtyka, A. Tarasek, N. Rylko, L. Szymanski, W. Maziarz, Microstructural characterization and mechanical properties of in situ cast nanocomposites Al/TiC type, Journal of Materials Research and Technology 9 (2020) 12707–12715.

Chapter 5

Inverse conductivity problem for spherical particles

Vladimir Mityushev[a], Natalia Rylko[a], Zhanat Zhunussova[b,c], and Yeskendyr Ashimov[b,c]

[a]*Research Group Materialica+, Faculty of Computer Science and Telecommunications, Cracow University of Technology, Kraków, Poland,* [b]*Institute of Mathematics and Mathematical Modeling, Almaty, Kazakhstan,* [c]*al-Farabi Kazakh National University, Almaty, Kazakhstan*

5.1 Introduction

Inverse problems for random composites with nonoverlapping spherical inclusions are of considerable interest in a number of fields. Detection of locations of inclusions by means of the measurement data on the boundary of a sample can be performed by a method of integral equations and the far field patterns [2,3,6,10]. Application of conformal mappings yielded effective computational methods for two-dimensional problems [11]. The main attention was paid to determination of shapes of one or few inclusions. Various numerical reconstructions by the time reversal method [9] by acousto-electric tomography [13], thermoacoustic and photoacoustic tomography, etc., [1] were developed.

The considered inverse problem can be studied via a perturbation problem when the boundary field induced by a homogeneous body is compared to the field caused by inclusions embedded in the host material. Such a problem can be stated by the Dirichlet-to-Neumann operator for fields governed by Laplace's equation when the Dirichlet data (potential) is fixed on the boundary, and the difference of the normal fluxes called the perturbation term is used to determine the physical properties of inclusions and their location. Let the shapes, sizes of inclusions, and their physical properties be given. The inverse problem is reduced to determination of the location of inclusions by the perturbation term.

In the present paper, we propose a new statement of inverse problem when the shape of each inclusion is fixed as a sphere of radius r_k ($k = 1, 2, \ldots, n$). The centers of spheres $\mathbf{a}_k$ ($k = 1, 2, \ldots, n$) have to be found during solution to the problem. The radii r_k and the number of spheres n may be also considered as unknown parameters. Here, we consider the case of given n and r_k, in particular, the case of the same radius $R = r_k$. This problem yields an approximate solution to the shape problem, since any shape can be arbitrarily approximated by packing spheres. Though the statement of the previously discussed inverse shape problem formally includes an arbitrary number n, the existing publications actually contain numerical results at most for three inclusions. We develop an analytical method of the detection of spheres with a fixed concentration in the unit sphere.

The considered problem is called the problem with passive interior inclusions contrary to the problem with active interior points. The external flux (energy) is given on the known external boundary and the passive interior inclusions do not degenerate to points. A problem with active interior points is stated with energy prescribed to the unknown points, see for instance [8]. The problem with active interior inclusions cannot be used as an approximation to the problem with passive inclusions, since inclusions have nonvanishing concentration.

The present paper concerns stationary problems for dispersed media. It differs from the inverse Stefan problem and the reconstruction problem of heat transfer coefficient [21,22], where a continuous curve (surface) is determined.

In the present paper, we derive the analytical approximate formula (5.41) for the perturbation term of highly conducting equal small spherical nonoverlapping inclusions embedded in a large sphere displayed in Fig. 5.1. Such perturbation terms were studied implicitly by integral equations. A numerical example with ten unknown spheres is discussed.

FIGURE 5.1 Randomly located nonoverlapping spheres in the unit sphere.

5.2 Modified Dirichlet problem

5.2.1 Reduction to functional equations

Let $U = \{\mathbf{x} \in \mathbb{R}^3 : |\mathbf{x}| < 1\}$ denote the unit ball. Consider mutually disjoint balls $D_k = \{\mathbf{x} \in U : |\mathbf{x} - \mathbf{a}_k| < r_k\}$ $(k = 1, 2, \ldots, n)$ in U. Introduce the complement of all the balls $|\mathbf{x} - \mathbf{a}_k| \leq r_k$ to the unit ball $D = U \setminus \cup_{k=1}^{n} (D_k \cup \partial D_k)$. Let a continuously differentiable function $f(\mathbf{x})$ be given on the unit sphere $|\mathbf{x}| = 1$.

We find a function $u(\mathbf{x})$ harmonic in D and continuously differentiable in the closure of D with the boundary conditions

$$u(\mathbf{x}) = f(\mathbf{x}), \quad |\mathbf{x}| = 1, \tag{5.1}$$

$$u = c_k, \quad \mathbf{x} \in \partial D_k \quad (k = 1, 2, \ldots, n). \tag{5.2}$$

Here, c_k are undetermined constants which should be found during solution to the boundary value problem. Moreover, the following conditions are fulfilled

$$\int_{\partial D_k} \frac{\partial u}{\partial \mathbf{n}} ds = 0 \iff \int_{\partial D_k} \frac{\partial u}{\partial r} ds = 0 \quad (k = 1, 2, \ldots, n), \tag{5.3}$$

where $\frac{\partial u}{\partial \mathbf{n}}$ stands for the normal derivative to the sphere ∂D_k and $\frac{\partial}{\partial r}$ the radial derivative in the local spherical coordinates (r, θ, ϕ) centered at $\mathbf{a}_k$, hence, $r = |\mathbf{x} - \mathbf{a}_k|$.

The inversion with respect to a sphere ∂D_k is introduced by formula

$$\mathbf{x}_{(k)}^* = \frac{r_k^2}{|\mathbf{x} - \mathbf{a}_k|^2} (\mathbf{x} - \mathbf{a}_k) + \mathbf{a}_k. \tag{5.4}$$

The Kelvin transform with respect to ∂D_k has the form [23]

$$\mathcal{K}_k w(\mathbf{x}) = \frac{r_k}{r} w(\mathbf{x}_{(k)}^*). \tag{5.5}$$

If a function $w(\mathbf{x})$ is harmonic in $|\mathbf{x} - \mathbf{a}_k| < r_k$, the function $\mathcal{K}_k w(\mathbf{x})$ is harmonic in $|\mathbf{x} - \mathbf{a}_k| > r_k$ and vanishes at infinity. Moreover, $\mathcal{K}_k w(\mathbf{x}) = w(\mathbf{x})$ on the sphere $r = r_k$.

Let $v(\mathbf{x})$ be a given Hölder continuous function on a fixed sphere $|\mathbf{x} - \mathbf{a}_k| = r_k$. Consider the auxiliary Robin problem

$$\frac{1}{r_k} u_k + 2 \frac{\partial u_k}{\partial r} = v, \quad r = r_k, \tag{5.6}$$

on the function $u_k(\mathbf{x})$ harmonic in the ball $|\mathbf{x} - \mathbf{a}_k| < r_k$. The analogous problem on $u_0(\mathbf{x})$ harmonic in the exterior of the unit sphere $D_0 = \{\mathbf{x} \in \mathbb{R}^3 : |\mathbf{x}| > 1\}$ is also considered

$$u_0 + 2 \frac{\partial u_0}{\partial r} = v, \quad r = 1. \tag{5.7}$$

The local coordinate r is introduced separately for every problem (5.6), (5.7). Every problem (5.6), (5.7) has a unique solution [14,15]. Let $u_k(\mathbf{x})$ be a solution of the problem (5.6) with $v = \frac{\partial u}{\partial r}$, i.e.,

$$\frac{1}{r_k} u_k + 2 \frac{\partial u_k}{\partial r} = \frac{\partial u}{\partial r}, \quad r = r_k. \tag{5.8}$$

112 Mechanics and Physics of Structured Media

One can harmonically continue the function $f(\mathbf{x})$ into $|\mathbf{x}| < 1$ by solution to the Dirichlet problem for the unit sphere. Consider the following problem (5.7) on $u_0(\mathbf{x})$ harmonic in $|\mathbf{x}| > 1$ and continuously differentiable in $|\mathbf{x}| \geq 1$

$$u_0 + 2\frac{\partial u_0}{\partial r} = \frac{\partial u}{\partial r} - \frac{\partial f}{\partial r}, \quad r = 1. \tag{5.9}$$

It is worth noting that the right hand part of (5.9) does not vanish despite of the boundary condition (5.1).

Introduce the function piecewise harmonic in $\mathbb{R}^3$ including the infinite point

$$U(\mathbf{x}) = \begin{cases} u_k(\mathbf{x}) + \sum_{m \neq k} \frac{r_m}{|\mathbf{x} - \mathbf{a}_m|} u_m(\mathbf{x}^*_{(m)}) + \frac{u_0(\mathbf{x}^*_{(0)})}{|\mathbf{x}|} + c_k - f(\mathbf{x}), \\ \qquad |\mathbf{x} - \mathbf{a}_m| \leq r_k, \quad k = 1, 2, \ldots, n, \\ u(\mathbf{x}) + \sum_{m=1}^{n} \frac{r_m}{|\mathbf{x} - \mathbf{a}_m|} u_m(\mathbf{x}^*_{(m)}) + \frac{u_0(\mathbf{x}^*_{(0)})}{|\mathbf{x}|} - f(\mathbf{x}), \quad \mathbf{x} \in D, \\ u_0(\mathbf{x}) + \sum_{m=1}^{n} \frac{r_m}{|\mathbf{x} - \mathbf{a}_m|} u_m(\mathbf{x}^*_{(m)}), \qquad |\mathbf{x}| \geq 1. \end{cases} \tag{5.10}$$

Let $U^+(\mathbf{x}) = \lim_{\cup_{k=1}^{n} D_k \ni \mathbf{y} \to \mathbf{x}} U(\mathbf{y})$ and $U^-(\mathbf{x}) = \lim_{D \ni \mathbf{y} \to \mathbf{x}} U(\mathbf{y})$. Calculate the jump $\Delta_k(\mathbf{x}) = U^+(\mathbf{x}) - U^-(\mathbf{x})$ across ∂D_k $(k = 1, 2, \ldots, n)$. Using the definition of $U(\mathbf{x})$ and the boundary condition (5.2) we get

$$\Delta_k = u_k + c_k - u - \frac{r_k}{|\mathbf{x} - \mathbf{a}_k|} u_k = 0. \tag{5.11}$$

The jump $\Delta'_k = \frac{\partial U^+}{\partial r} - \frac{\partial U^-}{\partial r}$ across ∂D_k vanishes as it is demonstrated in [7, Chapter 8]. Analogously the jumps Δ_0 and Δ'_0 across the unit sphere vanish. Then, Theorem 15 from [7, Chapter 8] implies that $U(\mathbf{x}) \equiv 0$. This equation written in the considered domains gives the system of functional equations

$$u_k(\mathbf{x}) = -\sum_{m \neq k} \frac{r_m}{|\mathbf{x} - \mathbf{a}_m|} u_m(\mathbf{x}^*_{(m)}) - \frac{1}{|\mathbf{x}|} u_0(\mathbf{x}^*_{(0)}) + f(\mathbf{x}) - c_k, \tag{5.12}$$

$$|\mathbf{x} - \mathbf{a}_m| \leq r_k, \quad (k = 1, 2, \ldots, n),$$

$$u_0(\mathbf{x}) = -\sum_{m=1}^{n} \frac{r_m}{|\mathbf{x} - \mathbf{a}_m|} u_m(\mathbf{x}^*_{(m)}), \quad |\mathbf{x}| \geq 1. \tag{5.13}$$

It is convenient to write the system (5.12)–(5.13) in the form

$$u_k(\mathbf{x}) = -\sum_{m \neq k} \frac{r_m}{|\mathbf{x} - \mathbf{a}_m|} u_m(\mathbf{x}^*_{(m)}) + h(\mathbf{x}), \tag{5.14}$$

$$\mathbf{x} \in D_k, \quad (k = 0, 1, 2, \ldots, n),$$

where $h(\mathbf{x}) = f(\mathbf{x}) - c_k$ in D_k ($k = 1, 2, \ldots, n$) and $h(\mathbf{x}) = 0$ in D_0. The integer number m runs from 0 to n except at $m = k$ in the sum $\sum_{m \neq k}$. After solution to the system (5.14) the auxiliary functions $u_k(\mathbf{x})$ are substituted into (5.14) to get the required function

$$u(\mathbf{x}) = -\sum_{m=0}^{n} \frac{r_m}{|\mathbf{x} - \mathbf{a}_m|} u_m(\mathbf{x}^*_{(m)}) + f(\mathbf{x}), \quad \mathbf{x} \in D. \tag{5.15}$$

Introduce the space $\mathcal{C}(D_k)$ of functions harmonic in all D_k and continuous in their closures endowed with the norm $\|h\|_k = \max_{\mathbf{x} \in \partial D_k} |h(\mathbf{x})|$. Let $\mathcal{C} = \mathcal{C}\left(\cup_{k=0}^{n} D_k\right)$ denote the space of functions harmonic in all D_k and continuous in their closures. The norm in $\mathcal{C}$ has the form

$$\|h\| = \max_k \|h\|_k = \max_k \max_{\mathbf{x} \in \partial D_k} |h(\mathbf{x})|. \tag{5.16}$$

The system of functional equations (5.14) can be considered as an equation in the space $\mathcal{C}$ on a function equal to $u_k(\mathbf{x})$ in each closed ball $D_k \cup \partial D_k$. The operator defined by the right part of (5.14) is compact in $\mathcal{C}$. It follows from [15] that the system of functional equations (5.14) has a unique solution in $\mathcal{C}$. This solution can be found by successive approximations converging in $\mathcal{C}$, i.e., uniformly in $\cup_{k=0}^{n}(D_k \cup \partial D_k)$.

Let k_s run over $0, 1, 2, \ldots, n$. Consider the sequence of inversions with respect to the spheres $\partial D_{k_1}, \partial D_{k_2}, \ldots, \partial D_{k_m}$ determined by the recurrence formula

$$x^*_{(k_m k_{m-1} \ldots k_1)} := \left(x^*_{(k_{m-1} k_{m-2} \ldots k_1)} \right)^*_{k_m}. \tag{5.17}$$

It is supposed that no equal neighbor numbers in the sequence $k_1, k_2, \ldots, k_m$. The transformations (5.17) for $m = 0, 1, 2, \ldots$ with the identity map form the Schottky group $\mathcal{S}$ of maps acting in $\mathbb{R}^3$, see the 2D theory [5,16]. Straightforward application of the successive approximations gives an exact formula in terms of the uniformly (not necessary absolutely) convergent Poincaré type series

$$u_k(\mathbf{x}) = h(\mathbf{x}) - \sum_{m \neq k} \frac{r_m}{|\mathbf{x} - \mathbf{a}_m|} h(\mathbf{x}^*_{(m)}) \tag{5.18}$$

$$+ \sum_{m \neq k} \sum_{k_1 \neq k} \frac{r_m}{|\mathbf{x} - \mathbf{a}_m|} \frac{r_{k_1}}{|\mathbf{x}^*_{(m)} - \mathbf{a}_{k_1}|} h(\mathbf{x}^*_{(k_1 m)})$$

$$- \sum_{m \neq k} \sum_{\substack{k_1 \neq m \\ k_2 \neq k_1}} \frac{r_m}{|\mathbf{x} - \mathbf{a}_m|} \frac{r_{k_1}}{|\mathbf{x}^*_{(m)} - \mathbf{a}_{k_1}|} \frac{r_{k_2}}{|\mathbf{x}^*_{(k_1 m)} - \mathbf{a}_{k_2}|} h(\mathbf{x}^*_{(k_2 k_1 m)})$$

$$+ \ldots,$$

114 Mechanics and Physics of Structured Media

where for instance $\sum_{\substack{k_1 \neq m \\ k_2 \neq k_1}} := \sum_{k_1 \neq m} \sum_{k_2 \neq k_1}$. Every sum $\sum_{k_s \neq k_{s-1}}$ contains terms with $k_s = 0, 1, \ldots, n$ except $k_s = k_{s-1}$.

Substitution of (5.18) into (5.15) yields

$$u(\mathbf{x}) = f(\mathbf{x}) - \sum_{k=0}^{n} \frac{r_k}{|\mathbf{x} - \mathbf{a}_k|} h(\mathbf{x}_{(k)}^*) \qquad (5.19)$$

$$+ \sum_{k=0}^{n} \sum_{k_1 \neq k} \frac{r_k}{|\mathbf{x} - \mathbf{a}_k|} \frac{r_{k_1}}{|\mathbf{x}_{(k)}^* - \mathbf{a}_{k_1}|} h(\mathbf{x}_{(k_1 k)}^*)$$

$$- \sum_{k=0}^{n} \sum_{\substack{k_1 \neq k \\ k_2 \neq k_1}} \frac{r_k}{|\mathbf{x} - \mathbf{a}_k|} \frac{r_{k_1}}{|\mathbf{x}_{(k)}^* - \mathbf{a}_{k_1}|} \frac{r_{k_2}}{|\mathbf{x}_{(k_1 k)}^* - \mathbf{a}_{k_2}|} h(\mathbf{x}_{(k_2 k_1 k)}^*)$$

$$+ \sum_{k=0}^{n} \sum_{\substack{k_1 \neq k \\ k_2 \neq k_1 \\ k_3 \neq k_2}} \frac{r_k}{|\mathbf{x} - \mathbf{a}_k|} \frac{r_{k_1}}{|\mathbf{x}_{(k)}^* - \mathbf{a}_{k_1}|} \frac{r_{k_2}}{|\mathbf{x}_{(k_1 k)}^* - \mathbf{a}_{k_2}|} \frac{r_{k_3}}{|\mathbf{x}_{(k_2 k_1 k)}^* - \mathbf{a}_{k_3}|} h(\mathbf{x}_{(k_3 k_2 k_1 k)}^*)$$

$$+ \ldots, \quad \mathbf{x} \in D.$$

The constants c_k ($k = 1, 2, \ldots, n$) can be found from the conditions (5.3). Using (5.8) we reduce (5.3) to

$$\int_{\partial D_k} u_k ds = 0, \quad k = 1, 2, \ldots, n. \qquad (5.20)$$

Here, the relation $\int_{\partial D_k} \frac{\partial u_k}{\partial \mathbf{n}} ds = 0$ for u_k harmonic in D_k is used. The integral from (5.20) is calculated by the mean value theorem for harmonic functions [23, p. 294]. Then, (5.20) becomes

$$u_k(\mathbf{a}_k) = 0, \quad k = 1, 2, \ldots, n. \qquad (5.21)$$

Substituting $\mathbf{x} = \mathbf{a}_k$ into (5.18) and using (5.21) we obtain a system of linear algebraic equations on c_k ($k = 1, 2, \ldots, n$) having a unique solution [19].

5.2.2 Explicit asymptotic formulas

In the present section, we consider the third order approximation in the linear size of inclusions. For simplicity, let $R = r_k$ ($k = 1, 2, \ldots, n$) and the boundary function $f(\mathbf{x}) = x_1$. It follows from (5.13)–(5.15) that the precision $O(R^3)$ for $u(\mathbf{x})$ will be achieved if we determine $u_k(\mathbf{x})$ and c_k up to $O(R^4)$. The series (5.18) for $k \neq 0$ can be approximated by the expression

Inverse conductivity problem for spherical particles **Chapter | 5 115**

$$u_k(\mathbf{x}) = x_1 - c_k - R \sum_{m \neq 0, k} \frac{\frac{R^2}{|\mathbf{x}-\mathbf{a}_m|^2}(x_1 - a_{m1}) + a_{m1} - c_m}{|\mathbf{x} - \mathbf{a}_m|} \tag{5.22}$$

$$+ R \sum_{m \neq k, 0} \frac{\frac{R^2}{|\mathbf{x}-\mathbf{a}_m|^2}(x_1 - a_{m1}) + a_{m1} - c_m}{|\mathbf{x}||\mathbf{x}^*_{(0)} - \mathbf{a}_m|}$$

$$+ R^2 \sum_{m \neq k, 0} \sum_{k_1 \neq k, 0} \frac{1}{|\mathbf{x} - \mathbf{a}_m|} \frac{a_{k_1,1} - c_{k_1}}{|\mathbf{a}_m - \mathbf{a}_{k_1}|}$$

$$- R^2 \sum_{m \neq k, 0} \sum_{k_1 \neq m, 0} \frac{1}{|\mathbf{x} - \mathbf{a}_m|} \frac{1}{|\mathbf{a}_m|} \frac{a_{k_1,1} - c_{k_1}}{|\mathbf{a}_m - \mathbf{a}_{k_1}|} + O(R^3),$$

where $a_{k_1,1}$ denotes the x_1-coordinate of $\mathbf{a}_{k_1}$.

Substituting $\mathbf{x} = \mathbf{a}_k$ into (5.22) and using (5.21) we obtain a system of linear algebraic equations on c_k $(k = 1, 2, \ldots, n)$

$$c_k = a_{k1} - R \sum_{m \neq 0, k} \frac{\frac{R^2}{|\mathbf{a}_k-\mathbf{a}_m|^2}(a_{k1} - a_{m1}) + a_{m1} - c_m}{|\mathbf{a}_k - \mathbf{a}_m|} \tag{5.23}$$

$$+ R|\mathbf{a}_k| \sum_{m \neq k, 0} \frac{\frac{R^2}{|\mathbf{a}_k-\mathbf{a}_m|^2}(a_{k1} - a_{m1}) + a_{m1} - c_m}{|\mathbf{a}_k - |\mathbf{a}_k|^2\mathbf{a}_m|}$$

$$+ R^2 \sum_{m \neq k, 0} \sum_{k_1 \neq k, 0} \frac{1}{|\mathbf{a}_k - \mathbf{a}_m|} \frac{a_{k_1,1} - c_{k_1}}{|\mathbf{a}_m - \mathbf{a}_{k_1}|}$$

$$- R^2 \sum_{m \neq k, 0} \sum_{k_1 \neq m, 0} \frac{1}{|\mathbf{a}_k - \mathbf{a}_m|} \frac{1}{|\mathbf{a}_m|} \frac{a_{k_1,1} - c_{k_1}}{|\mathbf{a}_m - \mathbf{a}_{k_1}|} + O(R^4).$$

Applying two iterations to (5.23) one can see that $c_k = a_{k1} + O(R^3)$. This simplifies (5.22) to

$$u_k(\mathbf{x}) = x_1 - a_{k1} + O(R^3). \tag{5.24}$$

Remark 5.1. The simple form (5.24) holds within the considered precision. The terms beginning from R^3 have much more complicated form.

It follows from (5.24) that

$$u_m(\mathbf{x}^*_{(m)}) = \frac{R^2}{|\mathbf{x} - \mathbf{a}_m|^2}(x_1 - a_{m1}) + O(R^3). \tag{5.25}$$

Substitute (5.25) into (5.13)

$$u_0(\mathbf{x}) = -R^3 \sum_{m=1}^n \frac{x_1 - a_{m1}}{|\mathbf{x} - \mathbf{a}_m|^2} + O(R^4), \quad |\mathbf{x}| \geq 1, \tag{5.26}$$

116 Mechanics and Physics of Structured Media

and apply the inversion on the unit sphere

$$\frac{1}{|\mathbf{x}|}u_0(\mathbf{x}^*_{(0)}) = -R^3|\mathbf{x}|^3 \sum_{m=1}^{n} \frac{x_1 - |\mathbf{x}|^2 a_{m1}}{|\mathbf{x} - |\mathbf{x}|^2 \mathbf{a}_m|^3} + O(R^4), \quad |\mathbf{x}| \le 1. \tag{5.27}$$

Substituting (5.25) and (5.27) into (5.15) we obtain the required third order approximation

$$u(\mathbf{x}) = x_1 + R^3 \sum_{m=1}^{n} \left[|\mathbf{x}|^3 \frac{x_1 - |\mathbf{x}|^2 a_{m1}}{|\mathbf{x} - |\mathbf{x}|^2 \mathbf{a}_m|^3} - \frac{x_1 - a_{m1}}{|\mathbf{x} - \mathbf{a}_m|^3} \right] + O(R^4), \quad \mathbf{x} \in D. \tag{5.28}$$

Explanations on the precision of (5.28) are required when $\mathbf{x}$ tends to a boundary point $\mathbf{z}$ lying on the fixed sphere $|\mathbf{z} - \mathbf{a}_k| = R$. We have

$$u(\mathbf{z}) = a_{k1} + R^3 \sum_{m \ne k} \left[|\mathbf{z}|^3 \frac{z_1 - |\mathbf{z}|^2 a_{m1}}{|\mathbf{z} - |\mathbf{z}|^2 \mathbf{a}_m|^3} - \frac{z_1 - a_{m1}}{|\mathbf{z} - \mathbf{a}_m|^3} \right] \tag{5.29}$$

$$+ R^3 |\mathbf{z}|^3 \frac{z_1 - |\mathbf{z}|^2 a_{k1}}{|\mathbf{z} - |\mathbf{z}|^2 \mathbf{a}_k|^3} + O(R^4).$$

One can see that (5.28) with $\mathbf{x} = \mathbf{z}$ and (5.29) coincides on $|\mathbf{z} - \mathbf{a}_k| = R$ up to $O(R^2)$, not up to $O(R^3)$. This divergence in precision is explained by the tacit assumption in derivation of (5.28) that $\mathbf{x}$ is sufficiently far away from the small spheres. More precisely, (5.28) holds up to $O(R^2)$ for any $\mathbf{x}$ in the closure of the domain D. In the same time, (5.28) holds up to $O(R^4)$ in a compact subset of D away from the small spheres. This case is important for further consideration of the perturbation term on the unit sphere for the remote inclusions.

5.3 Inverse boundary value problem

Using (5.28) introduce the perturbation term of order $O(nR^3)$ equal to the order of concentration of small balls in the unit ball

$$U(\mathbf{x}) = \sum_{m=1}^{n} \left[|\mathbf{x}|^3 \frac{x_1 - |\mathbf{x}|^2 a_{m1}}{|\mathbf{x} - |\mathbf{x}|^2 \mathbf{a}_m|^3} - \frac{x_1 - a_{m1}}{|\mathbf{x} - \mathbf{a}_m|^3} \right], \quad \mathbf{x} \in D. \tag{5.30}$$

Let (r, θ, ϕ) be the spherical coordinates (radial, azimuthal, polar) with the origin at $\mathbf{x} = \mathbf{0}$ and $\mathbf{x} = r\mathbf{y}$ where

$$\mathbf{y} = (\cos\theta \sin\phi, \sin\theta \sin\phi, \cos\phi), \quad 0 \le \theta < 2\pi, \ 0 \le \phi \le \pi. \tag{5.31}$$

Rewrite (5.30) in the form

$$U(r, \mathbf{y}) = \sum_{m=1}^{n} U_m(r, \mathbf{y}), \quad |\mathbf{y}| = 1, \tag{5.32}$$

where

$$U_m(r, \mathbf{y}) = r \frac{y_1 - r a_{m1}}{|\mathbf{y} - r \mathbf{a}_m|^3} - \frac{r y_1 - a_{m1}}{|r \mathbf{y} - \mathbf{a}_m|^3}, \qquad |\mathbf{y}| = 1. \tag{5.33}$$

Calculate the normal derivative $\frac{\partial U_m}{\partial r}$ on the unit sphere $r = 1$

$$g_m(\mathbf{y}) := \frac{\partial U_m}{\partial r}(1, \mathbf{y}) = 3 \frac{(y_1 - a_{m1})(1 - |\mathbf{a}_m|^2)}{|\mathbf{y} - \mathbf{a}_m|^5}. \tag{5.34}$$

Then, (5.32) yields

$$g(\mathbf{y}) := \frac{\partial U}{\partial r}(1, \mathbf{y}) = 3 \sum_{m=1}^{n} \frac{(y_1 - a_{m1})(1 - |\mathbf{a}_m|^2)}{|\mathbf{y} - \mathbf{a}_m|^5}, \qquad |\mathbf{y}| = 1. \tag{5.35}$$

The following useful formula can be applied to compute (5.35)

$$|\mathbf{y} - \mathbf{a}_m| = (1 + |\mathbf{a}_m|^2 - 2\mathbf{a}_m \cdot \mathbf{y})^{1/2}, \tag{5.36}$$

where $\mathbf{a}_m \cdot \mathbf{y} = a_{m1} y_1 + a_{m2} y_2 + a_{m3} y_3$.

The generating function for the Legendre polynomials $P_l(x)$ has the form

$$\frac{1}{\sqrt{1 + t^2 - 2xt}} = \sum_{l=0}^{\infty} P_l(x) t^l. \tag{5.37}$$

The series converges absolutely in the disk $|t| < 1$. The following formula takes place [24]

$$\frac{1 - t^2}{(1 + t^2 - 2xt)^{3/2}} = \sum_{l=0}^{\infty} (2l + 1) P_l(x) t^l. \tag{5.38}$$

Its derivative on x has the form

$$\frac{3t(1 - t^2)}{(1 + t^2 - 2xt)^{5/2}} = \sum_{l=1}^{\infty} (2l + 1) P_l'(x) t^l. \tag{5.39}$$

Substitute $t = |\mathbf{a}_m|$ and $x = \frac{1}{|\mathbf{a}_m|} \mathbf{a}_m \cdot \mathbf{y}$ into (5.39) using (5.36)

$$3 \frac{1 - |\mathbf{a}_m|^2}{|\mathbf{y} - \mathbf{a}_m|^5} = \sum_{l=1}^{\infty} (2l + 1) P_l' \left(\frac{\mathbf{a}_m \cdot \mathbf{y}}{|\mathbf{a}_m|} \right) |\mathbf{a}_m|^{l-1}. \tag{5.40}$$

Transform (5.34) using (5.40). Ultimately, the third order perturbation of the normal derivative on the unit sphere takes the form

$$g(\mathbf{y}) = \sum_{l=1}^{\infty} (2l + 1) \sum_{m=1}^{n} (y_1 - a_{m1}) P_l' \left(\frac{\mathbf{a}_m \cdot \mathbf{y}}{|\mathbf{a}_m|} \right) |\mathbf{a}_m|^{l-1}, \qquad |\mathbf{y}| = 1. \tag{5.41}$$

118 Mechanics and Physics of Structured Media

The explicit forms (5.35) and (5.41) of the perturbation term yield various statements of the inverse boundary value problem and constructive methods of their solutions. Let the measurement data be given on the whole unit sphere, i.e., a continuous function $g(\mathbf{y})$ be known on the unit sphere $|\mathbf{y}| = 1$. It is required to determine the points $\mathbf{a}_m$ ($m = 1, 2, \ldots, n$) in the unit ball from Eq. (5.35) (or from the equivalent equation (5.41)). At the present time, existence and uniqueness problems are not resolved.

Substitute (5.31) into (5.35) in order to get the explicit dependence of the perturbation term on θ and ϕ

$$g(\mathbf{y}) \equiv \widetilde{g}(\theta, \phi) = 3 \sum_{m=1}^{n} \frac{(\cos \theta \sin \phi - a_{m1})(1 - |\mathbf{a}_m|^2)}{|\mathbf{y} - \mathbf{a}_m|^5}, \qquad (5.42)$$

where

$$|\mathbf{y} - \mathbf{a}_m|^2 = 1 + |\mathbf{a}_m|^2 - 2(a_{m1} \cos \theta \sin \phi + a_{m2} \sin \theta \sin \phi + a_{m3} \cos \phi). \qquad (5.43)$$

Let the function (5.42) be given at the set of points $\Theta = \{(\theta_p, \phi_q) : p, q = 1, 2, \ldots, M\}$ for some integer M, and $g_{pq} := \widetilde{g}(\theta_p, \phi_q)$. The inverse problem can be reduced to determination of the set $\mathbf{A} = \{\mathbf{a}_m, m = 1, 2, \ldots, n\}$ from the following minimization problem

$$\min_{\mathbf{A}} \sum_{p,q} [\widetilde{g}(\theta_p, \phi_q) - g_{pq}]^2. \qquad (5.44)$$

The choice of the set Θ is important. For instance, numerical examples demonstrate that in the case $\phi_q = 0$ the number of solution $\mathbf{A}$ can be infinite. We address to the interpolation problem on the unit sphere to take a poised set (θ_p, ϕ_q) when the interpolation problem has a unique solution [4,25]. For definiteness, let n be even. Then, the set of points lying on $(n + 1)$ latitudes (parallel circles on the unit sphere), each of them contain $(n + 1)$ equally spaced nodes is a poised set.

Consider a numerical example with $n = 10$ small inclusions of radius $R = 0.15$. First, the function $\widetilde{g}(\theta, \phi)$ is constructed with (5.42) with the randomly prescribed centers shown in the first column of Table 5.1. Next, the latitudes $\phi_q = \frac{\pi q}{11}$ and $\theta_p = \frac{2\pi p}{11} - 0.2$ ($p, q = 1, 2, \ldots, 10$), and the data g_{pq} are computed and substituted into (5.44). The solution of the minimization problem is in the second column of Table 5.1. The third column contains the distances between the given and calculated points. One can see that all the points almost coincide except at the third point for which the given and calculated results are different. These points are displayed in Fig. 5.2. The third given point is shown in black as the rightmost center of sphere. The calculated third point is shown in gray, see the leftmost point in Fig. 5.2.

Inverse conductivity problem for spherical particles **Chapter | 5 119**

TABLE 5.1 The coordinates of the prescribed and calculated points are given with three digits after comma. The distances between them, the error in the calculation of points, are given in the third column.

Prescribed points	Calculated points	Distance between them
$(-0.782, -0.132, -0.162)$	$(-0.780, -0.132, -0.162)$	0.0014
$(-0.176, -0.352, 0.801)$	$(-0.176, -0.352, 0.801)$	0.00017
$(0.848, -0.220, 0.130)$	$(-0.959, -0.148, -0.242)$	**1.846**
$(-0.311, -0.642, -0.124)$	$(-0.310, -0.647, -0.124)$	0.0050
$(-0.267, 0.109, -0.345)$	$(-0.267, 0.111, -0.347)$	0.0026
$(0.050, -0.643, 0.032)$	$(0.038, -0.646, 0.029)$	0.013
$(0.348, 0.231, 0.0620)$	$(0.360, 0.234, 0.060)$	0.012
$(-0.014, 0.419, 0.667)$	$(-0.014, 0.419, 0.667)$	0.0003
$(-0.104, -0.009, 0.615)$	$(-0.104, -0.008, 0.615)$	0.0011
$(-0.457, 0.120, 0.161)$	$(-0.458, 0.122, 0.161)$	0.002

Other examples demonstrate excellent coincidence of the given and calculated points when n is about 5. The error increases with n and for $n \geq 10$ the results for some points significantly diverge as in the presented example.

5.4 Discussion and conclusion

In the present paper, we derive the perturbed field (5.28) induced by randomly located nonoverlapping perfectly conducting balls in analytical form. The explicit dependence of the perturbation term (5.30) on the location of inclusions is established. The obtained results can be applied to electrical, magnetic, diffusive, and other physical fields, in particular, to electrical impedance tomography [12].

It is worth noting that the perturbation term (5.28) cannot be obtained from dipoles [20] by the limit $R \to 0$. It is explicitly shown in the 2D case [17] that the field around two charged disks is described by dipoles and around not charged perfectly conducting disks by the elliptic functions.

The results are applied to the inverse problem when the potential and the normal flux are given on the exterior boundary, and the location of inclusions is determined.

The inverse problem can be stated in terms of distributions of spheres [18] when the error in detection of spheres may be considered as a statistical error. This question requires a separate numerical investigation.

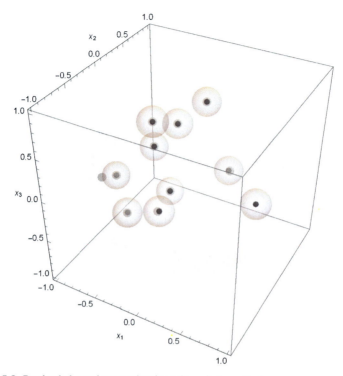

FIGURE 5.2 Randomly located nonoverlapping spheres in the unit sphere.

Acknowledgments

V. Mityushev, Zh. Zhunussova, and Y. Ashimov appreciate the support from the Science Committee of the Ministry of Education and Science of the Republic of Kazakhstan (Grant No. AP08856381).

References

[1] M. Agranovsky, P. Kuchment, Uniqueness of reconstruction and an inversion procedure for thermoacoustic and photoacoustic tomography with variable sound speed, Inverse Problems 23 (2007) 2089.
[2] H. Ammari, H. Kang, Reconstruction of Small Inhomogeneities From Boundary Measurements, Lecture Notes in Mathematics, vol. 1846, Springer-Verlag, Berlin, 2004.
[3] H. Ammari, H. Kang, Polarization and Moment Tensors, Applied Mathematical Sciences, Springer-Verlag, Berlin, 2007.
[4] W. Castell, N.L. Fernandez, Yuan Xu, Polynomial interpolation on the unit sphere II, arXiv: math/0407448v1, 2004.
[5] R. Czapla, V. Mityushev, N. Rylko, Conformal mapping of circular multiply connected domains onto slit domains, Electronic Transactions on Numerical Analysis 39 (2012) 286–297.
[6] D. Colton, R. Kress, Inverse Acoustic and Electromagnetic Scattering Theory, Springer-Verlag, Berlin, 2013.
[7] S. Gluzman, V. Mityushev, W. Nawalaniec, Computational Analysis of Structured Media, Elsevier, Amsterdam, 2017.

Inverse conductivity problem for spherical particles **Chapter | 5 121**

[8] D.P. Hardin, E.B. Saff, Minimal Riesz energy point configurations for rectifiable d-dimensional manifolds, Advances in Mathematics 193 (2005) 174–204.

[9] Yu. Hristova, P. Kuchment, Linh Nguyen, Reconstruction and time reversal in thermoacoustic tomography in acoustically homogeneous and inhomogeneous media, Inverse Problems 24 (2008) 055006.

[10] V. Isakov, Inverse Problems for Partial Differential Equations, Springer-Verlag, Berlin, 2017.

[11] R. Kress, Inverse problems and conformal mapping, Complex Variables and Elliptic Equations 57 (2012) 301–316.

[12] P. Kuchment, The Radon Transform and Medical Imaging, CBMS-NSF Regional Conference Series in Applied Mathematics, 2013.

[13] P. Kuchment, L. Kunyansky, 2D and 3D reconstructions in acousto-electric tomography, Inverse Problems 27 (2011) 055013.

[14] L. Lanzani, Z. Shen, On the Robin boundary condition for Laplace's equation in Lipschitz domains, Communications in Partial Differential Equations 29 (2004) 91–109.

[15] V.V. Mityushev, S.V. Rogosin, Constructive Methods for Linear and Nonlinear Boundary Value Problems for Analytic Functions, Chapman & Hall / CRC, Boca Raton, 2000.

[16] V. Mityushev, Riemann-Hilbert problems for multiply connected domains and circular slit maps, Computational Methods and Function Theory 11 (2011) 575–590.

[17] N. Rylko, A pair of perfectly conducting disks in an external field, Mathematical Modelling and Analysis 20 (2015) 273–288.

[18] N. Rylko, Representative volume element in 2D for disks and in 3D for balls, Journal of Mechanics of Materials and Structures 9 (2014) 427–439.

[19] S.G. Mikhlin, Integral Equations and Their Applications to Certain Problems in Mechanics, Mathematical Physics and Technology, International Series of Monographs in Pure and Applied Mathematics, vol. 5, Pergamon Press, Oxford etc., 1957.

[20] W.K.H. Panofsky, M. Phillips, Classical Electricity and Magnetism, Addison-Wesley Publ., London, 1962.

[21] E. Hetmaniok, D. Slota, R. Witula, A. Zielonka, An analytical method for solving the two-phase inverse Stefan problem, Bulletin of the Polish Academy of Sciences. Technical Sciences 63 (2015) 583–590.

[22] E. Hetmaniok, D. Slota, A. Zielonka, Using the swarm intelligence algorithms in solution of the two-dimensional inverse Stefan problem, Computers & Mathematics with Applications 69 (2015) 347–361.

[23] A.N. Tikhonov, A.A. Samarskii, Equations of Mathematical Physics, Dover Books on Physics, 1963.

[24] Eric W. Weisstein, Legendre polynomial, From MathWorld–A Wolfram Web Resource, https://mathworld.wolfram.com/LegendrePolynomial.html.

[25] Yuan Xu, Polynomial interpolation on the unit ball and on the unit sphere, Advances in Computational Mathematics 20 (2004) 247–260.

Chapter 6

Compatibility conditions: number of independent equations and boundary conditions

Igor Andrianov and Heiko Topol
Institute of General Mechanics, RWTH Aachen University, Aachen, Germany

6.1 Introduction

The *compatibility conditions* of continuum mechanics define the relationship between strains and displacements. These conditions are necessary and sufficient for solutions under a self-balanced external loading to exist, and they reflect the fact that the strain field corresponds to some displacement field. Mathematically, the compatibility conditions are reduced to the requirement that the components of the Riemann-Christoffel tensor are equal to zero. In other words, the compatibility conditions are mathematical conditions that determine whether a particular deformation will leave a body in a compatible state.

The compatibility conditions have been in the focus of numerous works, and the historical background of these conditions has been reviewed in [1–5]. Apparently Kirchhoff [6] was the first to write down these conditions (in incomplete form) in 1859. In 1860 Barré de Saint-Venant [7] proposed a complete system of the compatibility conditions, and in 1864 he presented a detailed description of the compatibility conditions in the appendix of Navier's book [8]. The necessity of compatibility conditions is obvious. De Saint-Venant [8] and Boussinesq [9] asserted sufficiency of compatibility conditions, but it is now believed [2,4] that the first rigorous proof of sufficiency was given by Beltrami in 1886 [10]. Beltrami also obtained the *Beltrami stress compatibility conditions*. They are also denoted as the *Beltrami-Michell compatibility conditions* or *Beltrami-Donati-Michell compatibility conditions* (see [11,12]). The stress compatibility conditions in the absence of external forces were proposed by Beltrami in 1892 [10], and for the general case by Donati in 1894 [11] and Michell in 1900 [12] (for further comments on the paper by Michell see [13]).

Mechanics and Physics of Structured Media. https://doi.org/10.1016/B978-0-32-390543-5.00011-6
Copyright © 2022 Elsevier Inc. All rights reserved.

124 Mechanics and Physics of Structured Media

The fact that the integral of some functions of strain must vanish over any closed path in a body can also be used in order to derive the compatibility conditions [14].

"The conditions of compatibility necessary and sufficient condition that the space of which the continuum part be locally Euclidean. They are supposed to hold everywhere upon certain singular points, curves, or surfaces, where the material may divide or join" [15].

One can find different opinions concerning the independence of the compatibility conditions in the literature. Many books [16–24] state that compatibility conditions are independent. Often the authors of these monographs do not explain why they consider the compatibility conditions to be independent. It seems that only Barber [22] clearly states that he considers these conditions to be independent *"in the sense that no one of them can be derived from the other five"*. However, the presence of three dependencies between the six equations of compatibility of deformation was indicated by Beltrami yet in 1892 [10].

It is interesting to note that the differential Bianchi identity was discovered by Beltrami in 1892 [10], whereas Bianchi's paper was published in 1902 [25]. On the other hand, Voss discovered the differential Bianchi identity in 1880 [26]. This once again confirms *Stigler's law of eponymy*, which states that no scientific discovery is named after its original discoverer [27].

Many books that treat the compatibility conditions do not address the dependency of these conditions [28–57]. It is most likely that these books do not consider the issue of the dependence of the compatibility conditions to be important, at least for beginners. This is somewhat surprising for Southwell's monograph [47], since it was Southwell [58,59] who showed that only three of the six Barré de Saint-Venant compatibility conditions follow from the principle of the minimum complementary energy.

Lurie [60] refers to the results by Krutkov [61] that show that homogeneous problems of statics, which are formulated in terms of stresses, can be reduced to a system of six compatibility conditions for six stress functions.

Hahn [62] presents the Beltrami relations [10] (i.e. Bianchi differential identity) in his book, but he does not comment further on them in any way.

Rabotnov [63] presents the differential Bianchi identity for deformations. He noted that the overdetermination of the problem is related to the order of the compatibility conditions system, which was artificially increased in the process of derivation. As a result, some "parasitic" additional solutions can be obtained. For example, stresses that linearly depend on the coordinates satisfy the compatibility conditions without in general satisfying equilibrium equations (also see [64]).

Trefftz [65] noticed the overdetermination of the problem in the form of six equations of the second order and three of the first order with six boundary con-

Compatibility conditions Chapter | 6 **125**

ditions. However, he makes a remark that three first order equilibrium equations can be treated in some sense as boundary conditions.

According to Donnell [66] there are nine relations, equilibrium equations, and compatibility conditions that are not independent, but there is no acceptable way of reducing them to six relations.

Christensen [67] writes that only three compatibility conditions are independent without discussing any criteria for choosing them.

According to Zubchaninov [68] any combination of three stress compatibility conditions can be chosen as independent.

Leybenzon [69,70] treats the work by Southwell [59] on the derivation of the de Saint-Venant compatibility conditions from Castigliano's variational principle. In [70] Leybenzon points out that the three de Saint-Venant equations were obtained by using Castigliano's variational principle in the paper by Haar and von Kármán [71]. The fact that all de Saint-Venant identities can be obtained by using Castigliano's variational principle was first noted in [72, Chapter IV, §12, subsection 12].

Slaughter [73] points out the interdependence of the compatibility condition, citing the work of Malvern [74]. Malvern [74] in turn treats Southwell's paradox [58,59] and the results by Washizu [75,76].

Gould and Feng [77] mentioned: *"It has been shown that these equations are not strictly independent. A further differential combination of the six equations produces three identities connecting the equations which are known as the Bianchi formulas. However, one may not choose any three of the six, so the usual approach is to include all six in the formulation."* [74].

Sadd [78] indicates the interdependence of the different compatibility condition and presents the differential Bianchi identity.

Ionov and Ogibalov [79] describe the reduction of six compatibility conditions in differential form to three integro-differential equations and three stress functions.

Betten [80] discussed the differential Bianchi identity and results by Washizu [75,76].

Chandrasekharaiah and Debnath [81] mentioned that the compatibility equations and Cauchy's equations serve as a closed system of governing equations for elastostatic problems.

Wilmanski [82] mentioned that the Beltrami-Michell stress equations together with the boundary conditions for tractions form the well-posed problem for the determination of stresses.

A meaningful discussion of compatibility conditions can be found in [5].

Temam and Miranville [83] analyze the relationship between the Schwarz conditions integrability and compatibility conditions. The same issue is discussed in detail in [84].

In [85], the compatibility conditions are investigated for planar network structures, which are assembled of nodes which are connected by bars.

126 Mechanics and Physics of Structured Media

This survey is devoted to the formation of the correct boundary conditions in the application of the compatibility conditions for a bounded domain. We assume simply-connected bodies, and in order to simplify the treatment, we restrict our considerations to linear problems in rectangular Cartesian coordinate systems.

This article is organized as follows. The statement of the problem and Southwell's paradox are described in Section 6.2. The formulations of correct boundary value problems are described in Section 6.3. Some counterexamples that have been proposed by Pobedrya and Georgievskii are analyzed in Section 6.4. Statements of the problem in terms of stresses are described in Section 6.5. Section 6.6 is devoted to some approaches which differ from the ones in the previous section. Some generalizations of the classical compatibility conditions are analyzed in Section 6.7. Finally, Section 6.8 presents the concluding remarks.

6.2 Governing relations and Southwell's paradox

The compatibility conditions define the geometric conditions for the continuity of deformations. If the compatibility conditions are fulfilled, then the deformations can be represented as single-valued functions of the displacements. In a Cartesian coordinate system x, y, and z, the six de Saint-Venant compatibility conditions can be written in the form [60]:

$$e^{xx} \equiv \frac{\partial^2 \varepsilon_x}{\partial y^2} + \frac{\partial^2 \varepsilon_y}{\partial z^2} - \frac{\partial^2 \varepsilon_{xy}}{\partial y \partial z} = 0, \qquad (x, y, z), \qquad (6.1a)$$

$$e^{xy} \equiv -\frac{\partial^2 \varepsilon_{zz}}{\partial x \partial y} + \frac{1}{2} \frac{\partial}{\partial z} \left(\frac{\partial \varepsilon_{zx}}{\partial y} + \frac{\partial \varepsilon_{yz}}{\partial z} - \frac{\partial \varepsilon_{xy}}{\partial y} \right) = 0, \qquad (x, y, z), \qquad (6.1b)$$

where $\varepsilon_{xx} = \frac{\partial u_x}{\partial x}$, $\varepsilon_{yy} = \frac{\partial u_y}{\partial y}$, and $\varepsilon_{zz} = \frac{\partial u_z}{\partial z}$ are the normal components of the strain tensor, $\varepsilon_{xy} = \varepsilon_{yx} = \frac{1}{2} \left(\frac{\partial u_x}{\partial y} + \frac{\partial u_y}{\partial x} \right)$, $\varepsilon_{xz} = \varepsilon_{zx} = \frac{1}{2} \left(\frac{\partial u_x}{\partial z} + \frac{\partial u_z}{\partial x} \right)$, and $\varepsilon_{yz} = \varepsilon_{zy} = \frac{1}{2} \left(\frac{\partial u_y}{\partial z} + \frac{\partial u_z}{\partial y} \right)$ are shear components of the strain tensor, here presented in terms of the derivatives of the components u_x, u_y, and u_z of the displacements vector. The symbol (x, y, z) indicates a circular permutation of the variables, with the help of which two more sets of relations can be obtained from relation (6.1).

The system of partial differential equations in (6.1) is overdetermined in the sense that the numbers of equations exceed the number of the unknowns (see, e.g. [86]). Ostrosablin [87] has shown that the matrix of the de Saint-Venant operators is of the rank three. However, the statement of the boundary value problem is correct, since there are no partial differential equations in (6.1) that would contradict each other.

The differential Bianchi identity for the system (6.1) is

$$\frac{\partial e^{xx}}{\partial x} + \frac{\partial e^{xy}}{\partial y} + \frac{\partial e^{xz}}{\partial z} = 0, \quad (x, y, z). \tag{6.2}$$

In a similar fashion, the equilibrium equations for the components σ_{xx}, σ_{xy}, σ_{xz}, ..., of the stress tensor can be presented in the form:

$$\frac{\partial \sigma_{xx}}{\partial x} + \frac{\partial \sigma_{xy}}{\partial y} + \frac{\partial \sigma_{xz}}{\partial z} = 0, \quad (x, y, z). \tag{6.3}$$

Beltrami's conditions are

$$E^{xx} \equiv \frac{\partial^2 \sigma_x}{\partial y^2} + \frac{\partial^2 \sigma_y}{\partial x^2} - \frac{\nu}{\nu+1}\left(\frac{\partial^2 s}{\partial x^2} + \frac{\partial^2 s}{\partial y^2}\right) - 2\frac{\partial^2 \tau_{yz}}{\partial x \partial y} = 0, \quad (x, y, z),$$

$$\tag{6.4a}$$

$$E^{xy} \equiv \frac{\partial}{\partial z}\left(\frac{\partial \tau_{zx}}{\partial y} + \frac{\partial \tau_{yz}}{\partial z} - \frac{\partial \tau_{xy}}{\partial y}\right) = \frac{\partial^2}{\partial x \partial y}\left(\sigma_z - \frac{\nu}{\nu+1}s\right), \quad (x, y, z),$$

$$\tag{6.4b}$$

where ν is the Poisson's ratio and $s = \sigma_x + \sigma_y + \sigma_z$.

Southwell [58] applied Castigliano's theorem for the de Saint-Venant compatibility conditions and showed that the state granting the continuity of deformations corresponds to the minimum value of the accumulated potential energy.

The solution of the equilibrium equations can be represented in the form of Maxwell's stress functions φ_{xx}, φ_{yy}, and φ_{zz} [88–91],

$$\sigma_{xx} = \frac{\partial^2 \varphi_{yy}}{\partial y^2}, \quad \sigma_{xy} = -\frac{\partial^2 \varphi_{zz}}{\partial x \partial y}, \quad (x, y, z), \tag{6.5}$$

or in term of Morera's stress functions φ_{xy}, φ_{xz}, φ_{yz} [88,89,91],

$$\sigma_{xx} = -2\frac{\partial^2 \varphi_{yz}}{\partial x \partial y}, \quad \sigma_{xy} = \frac{\partial^2 \varphi_{xz}}{\partial y \partial z} + \frac{\partial^2 \varphi_{yz}}{\partial x \partial z} - \frac{\partial^2 \varphi_{xy}}{\partial z^2}, \quad (x, y, z). \tag{6.6}$$

Southwell [58] has shown that by using either the expressions (6.5) or (6.6) the theorem of minimum complementary energy results in only three compatibility conditions.

In his further paper [59] Southwell mentioned that some investigators drew the wrong conclusions from the results. In particular, "...*Professor G.I. Taylor remarked to me that by them three of the six conditions [compatibility conditions] would seem to be made redundant; ...*".

However, Marguerre [92] noted that despite the differential Bianchi identity, the correct statement of the problem must contain six stress functions. This contradiction was named *Southwell's paradox*.

128 Mechanics and Physics of Structured Media

Southwell proposed the following solution to this paradox: When the restrictions imposed by the conditions of equilibrium at the boundary are taken into account then all six compatibility equations can be treated as a consequence of Castigliano's principle [59].

Washizu [75,76] pointed out that some conditions of compatibility of deformations can be used as the missing boundary conditions. In the paper [75] Washizu notes (in Washizu's monograph [76] this remark is missing):

"Now, as pointed out by S.H. Crandall in private discussion, there would arise a problem as to what are the necessary and sufficient boundary conditions for compatibility. [...] The solution of this problem could be obtained by the careful treatment of the boundary conditions in the variational procedure."

Stickforth [93] treated Southwell's paradox in detail for all possibilities by setting three Beltrami stress functions equal to zero. In this approach Stickforth used the Castigliano-Colonetti's principle and Friedrichs' involutory transformation, which convert the natural conditions of one problem into the constraints of the other problem [86,94]. Stickforth [93] showed that there are 17 possibilities of setting three Beltrami stress functions equal to zero. The same results were previously obtained by Blokh [88,89]. Similar results have been obtained by Ostrosablin [95,96], Sachenkov and Saïfullin [97], and Saïfullin et al. [98].

6.3 System of ninth order

Malyi [99,100] reduced the original system that consists of the compatibility conditions and the equilibrium equations to a ninth-order system with nine boundary conditions. For the problem in deformation in [100] the compatibility conditions (6.1b) and the equilibrium conditions (6.2) must be satisfied everywhere. This gives a system of partial differential equations of the ninth order with six original boundary conditions. Some combinations of the compatibility equations (6.1a) on the boundaries give three additional initial conditions. In what follows there will be explained why the obtained boundary conditions can be treated as initial ones.

It has been shown in [101] that the application of the operator method [86] to the system of equilibrium conditions (6.2) and the compatibility conditions (6.4a) leads to the initial-boundary value problem

$$\frac{\partial^3}{\partial x \partial y \partial z} \nabla^2 \nabla^2 \nabla^2 \Phi = 0, \tag{6.7}$$

were Φ is the potential function.

Eq. (6.7) has a mixed form and requires an odd number of boundary conditions, seven conditions concerning each variable. This confirms the conclusion of papers [99,100] about an odd number of boundary conditions for the problem in stresses.

In order to clarify the problem, let us consider an elastic cube of side lengths $2L$ with the origin of the Cartesian coordinate system in its center and edges parallel to the axes of the coordinate system. We suppose that the compatibility conditions (6.1b) are satisfied in the cube's domain. We set the boundary conditions

$$e^{xx} = 0 \quad \text{at} \quad x = \pm L, \quad (x, y, z). \tag{6.8}$$

From (6.2) one obtains

$$\frac{\partial e^{xx}}{\partial x} = 0, \quad (x, y, z). \tag{6.9}$$

The fulfillment of the conditions at $x = L$, (x, y, z) leads automatically to the fulfillment of the conditions on the opposite face of the cube at $x = -L$, (x, y, z). That is why here we are dealing with initial condition, but not with boundary conditions. Thus, to the six initial boundary conditions, three initial conditions

$$e^{xx} = 0 \quad \text{at} \quad x = L, \quad (x, y, z), \tag{6.10}$$

are added.

Malyi [100] and Vlasov [102] have proven that when the compatibility conditions (6.1b) and the initial conditions (6.10) are satisfied, the compatibility conditions (6.1a) are satisfied inside the domain of a body.

Problems in stresses with Beltrami conditions (6.4) and equilibrium equations (6.2) can be treated similarly [99]. The compatibility conditions (6.4b) and the equilibrium conditions (6.2) must be satisfied everywhere. Some combinations of the compatibility equations (6.4a) on the boundaries give three additional initial conditions.

6.4 Counterexamples proposed by Pobedrya and Georgievskii

Pobedrya and Georgievskii wrote in their article [103]:

> "For $n = 3$, two counterexamples are presented that show the impossibility to transfer three "diagonal" or three "nondiagonal" Beltrami-Michell equations from the domain of an elastic solid to its boundary."

We would like to show that these counterexamples are the result of a misunderstanding.

The first example in [103] studies the deformation of an elastic sphere under uniform normal pressure at the boundary, where the solution depends only on

130 Mechanics and Physics of Structured Media

the radius and the material parameters. The problem is described by one equilibrium equation and three (instead of six) compatibility conditions. The solution that has been proposed in [103] satisfies the equilibrium equations, but the compatibility conditions are only satisfied at the boundaries. Then the authors show that the obtained expression (which is not a solution to the original problem!) does not satisfy the compatibility conditions inside the sphere, because the compatibility conditions have been previously ignored.

Generally, the solution depends on the coordinates r, φ, θ. In the notation of [103], we have the equilibrium equations:

$$S_r = 0, \tag{6.11a}$$
$$S_\phi = 0, \tag{6.11b}$$
$$S_\theta = 0, \tag{6.11c}$$

as well as two sets of compatibility conditions

$$H_{r\phi} = H_{\theta\phi} = H_{r\theta} = 0, \tag{6.12a}$$
$$H_{rr} = H_{\phi\phi} = H_{\theta\theta} = 0. \tag{6.12b}$$

The problem solved in [103] is reduced to Eqs. (6.11a) and (6.12b). The authors assume that conditions (6.12b) can be replaced by conditions (6.12a). However, conditions (6.12a) are satisfied identically; therefore, such a replacement is incorrect.

For the correct solution to the original problem, see, for example, Blokh [89] and Lurie [60]. The second "counterexample" presented in article [103] can be studied in a similar way.

6.5 Various formulations of the linear theory of elasticity problems in stresses

The equilibrium equations (6.3) have the same structure as the Bianchi identity (6.2). This makes it possible to reduce the problem in terms of stresses to the system of six equations for the compatibility conditions. The solution is then sought in terms of functions that satisfy the system of equilibrium equations. Apparently, Krutkov [61] (also see [60]) drew attention to this matter for the first time in 1949 (see the comments regarding [61] in [104]).

Vasil'yev and Fedorov [105] rediscovered Southwell's paradox and formulated their point of view as to its resolutions. They underlined that for the complete description of the theory of elasticity problem six stress functions are needed. Three stress functions allow to satisfy the equilibrium equations, and three others are necessary to determine the displacements. Using the analogy between equilibrium equations (6.3) and differential Bianchi identity (6.2), in [105] it has been shown that the tensor of the stress functions φ_{ab} can be determined as follows

$$\sigma_{xx} = \frac{\partial \varphi_{yy}}{\partial z^2} + \frac{\partial \varphi_{zz}}{\partial y^2} - 2\frac{\partial \varphi_{yz}}{\partial y \partial z}, \qquad (x, y, z), \qquad (6.13a)$$

$$\sigma_{xx} = -2\frac{\partial \varphi_{yz}}{\partial y \partial z}, \qquad (x, y, z), \qquad (6.13b)$$

$$\sigma_{xy} = \frac{\partial \varphi_{xz}}{\partial y \partial z} + \frac{\partial \varphi_{yz}}{\partial x \partial z} - \left(\frac{\partial^2 \varphi_{xy}}{\partial z^2} + \frac{\partial^2 \varphi_{zz}}{\partial x \partial y} \right), \qquad (x, y, z). \qquad (6.13c)$$

It can be noticed that equilibrium equations (6.3) are satisfied for any choice of the functions φ_{ab}. For the general case, gauge transformations are applied in [105].

A different approach to the stress formulation was proposed in [106–110], which is based on the following idea: the operator of the theory of elasticity both in displacements and in stresses has a factorized form ACA^*, where the symbol $(\cdot)^*$ refers to the conjugation operation.

The variational formulation allows to use the Lagrange multiplier method and its generalizations such as Kuhn-Tucker's conditions [111] or the Friedrichs' transformation [112–118]. The use of Friedrichs' transformation was first discussed, apparently, by Washizu [75] in 1957, and then later by Stickforth [93].

In another form, Friedrichs' transformation has been used in a series of publications [112–118]. In these works, the equilibrium equations are considered as some additional restrictions. The functions that must provide a minimum of the functional must satisfy these constraints. Then they are introduced into the functional using Lagrange multipliers. Application of the Friedrichs' transformation procedure [72,94] makes it possible to reduce the original problem to the 12th order system for six equations, while the equilibrium equations are transformed into boundary conditions. In a certain sense, this is a realization of Trefftz's idea [65] – equilibrium conditions play the role of missing boundary conditions. The mathematical analysis of the resulting system is described in [119,120].

6.6 Other approximations

Different techniques and approximations have been applied in order to analyze the compatibility conditions. Klyushnikov [121], Grycz [122], and Vlasov [123] applied Castigliano's principle. Kozák [124–127] used the principle of minimum complementary energy as well as the dual forms of principle of virtual work.

Borodachev [128,129] applied a three-dimensional Fourier integral transformation in order to show that between the six compatibility conditions only three are independent. The author comes to the conclusion that in practice only Eq. (6.1a) or Eq. (6.1b) needs to be used. This conclusion should be treated with caution. In [128,129] an infinite elastic medium is considered. As stated in the articles, *"for simplicity"*. In reality, this is a fundamental simplification.

132 Mechanics and Physics of Structured Media

As shown above, solution of problems in the bounded domain requires all six compatibility conditions. Later Borodachev [130] obtained nine compatibility conditions solely in terms of first-order partial derivatives, and he proved that of these nine equations only six are independent.

In the articles [131–134] (also see [79]) the original six compatibility conditions in differential form are deduced to the three equations in integro-differential form and three stress functions. This approach is then applied in order to find solutions to particular problems.

In article [135] it is proposed to satisfy three compatibility equations everywhere in the domain, and some relations are obtained as boundary conditions that combine other compatibility conditions and equilibrium equations. Unfortunately, it is not clear from the text of the article whether the three discarded compatibility equations in the domain are satisfied. Perhaps this approach is more justified for plane problems of the theory of elasticity [136].

6.7 Generalization

Let us briefly describe the generalizations of the problems considered in the article. These generalizations include complication of physical models (e.g., taking into account multi-connectedness of regions, nonlinearity) or weaking of mathematical constraints imposed on the original relations.

Compatibility conditions that result from the principle of minimum complementary energy in the three-dimensional multiply-connected bodies are treated in [137].

De Saint-Venant tensor for multiply-connected domain is also studied in [138]. Compatibility conditions for micropolar elasticity have been analyzed by Szeidl [139].

Different works [127,140–143] focus on compatibility conditions for nonlinear problems. A detailed description of the compatibility conditions in the nonlinear theory of elasticity can be found in [144].

De Saint-Venant's theorem constitutes a classical characterization of smooth matrix fields considered as linearized strain tensor fields. This theorem has been extended to matrix fields with components in L^2 in [145,146].

In their simplest form, the compatibility conditions obtained for the twice continuously differentiable functions, while the article [147] presents results, which are more general. It has been shown that de Saint-Venant's theorem is the matrix analogue of Poincaré's lemma [148].

The classical Donati theorem has been used to characterize smooth matrix fields as linearized strain tensor fields, and several generalizations of this theorem are presented in [147]. These extensions of Donati's theorem allow to reformulate linearized 3D elasticity problems as quadratic minimization problems with the strains in a novel fashion.

The work [149] presents a new method for approximating two- and three-dimensional linear elasticity problems. This approach approximated strains

Compatibility conditions Chapter | 6 **133**

directly without a simultaneous approximation of the displacements in finite element spaces where the Barré de Saint-Venant compatibility conditions are satisfied in a weak form.

We note with regret that, as in many other fields [150], the cooperation of pure mathematicians and specialists in the field of elasticity theory is very weak. The results of pure mathematicians are little used by practicing mechanical engineers. These results are too mathematically complicated and general, and pure mathematicians do not strive to make them available to applied mathematicians.

6.8 Concluding remarks

This review highlights, in particular, the influence of the Iron Curtain on the development of science in the USSR and the Western world (this influence is also discussed in the chapter devoted to the work of L.A. Filshtinsky in this book [151]).

The assessment of the scientific development is objective, and researchers on both sides of the Iron Curtain often obtained similar results. But they often did not refer to each other, because it was difficult to obtain information "from the other side" in the pre-Internet era.

Another observation is the loss of information that is caused by multiple citations and recitations skipping the true creators. In any case, it is preferable to cite the original sources instead of secondary literature whenever it is possible.

Let us dwell on the question of the role of the compatibility conditions in various theories and the possibilities of formulating the problem in stresses. For the dynamic theory of elasticity, it is impossible to formulate the problem in stresses, although the equations of motion can be written with respect to the components of the stress tensor [2]. In the two-dimensional theory of elasticity for simply connected bodies, only one of the de Saint-Venant equations remains. For an isotropic homogeneous body this equation in the absence of mass forces is reduced to the M. Lévy equation [18]. After substitution of the Airy stress function into this equation, we obtain a biharmonic equation for these functions. Formulation of the boundary conditions is not difficult. In the theory of plate bending, there is also only one consistency equation. In the nonlinear case, the introduction of the Airy stress function allows to reduce the problem to the system of two fourth order Föppl-Kármán equations with respect to the normal displacement and the Airy function [152].

In the theory of shells, with the accuracy accepted in this theory, the problem in forces and moments is a system of six PDEs with six unknowns. Three fourth-order equations correspond to equilibrium equations, and three fourth-order compatibility condition equations are, in fact, Gauss-Codazzi equations. For this system of the eighth order, eight boundary conditions are posed [153].

Compatibility conditions are an important subject of research in both mechanics of solids and in differential geometry [13,154–157]. In recent years, it is mathematicians working in the field of differential geometry that have been

134 Mechanics and Physics of Structured Media

investigating such important issues as compatibility conditions for nonsimply-connected bodies [13], compatibility conditions in the presence of boundary conditions [154,156], etc.

This review was written by experts in the field of solid mechanics and for researchers working in this field. Several purely mathematical works are only mentioned in it. It would be interesting and useful that some pure mathematician to write a detailed survey of the results concerning compatibility conditions obtained by specialists in the field of differential geometry. And, of course, this review should be written in a language understandable to mechanics and applied mathematicians.

In practice most 3D theory of elasticity problems are treated in terms of displacements instead of stress. But the issues under consideration in this review continue to be of interest to the specialists in the theory of elasticity, computational mathematics, and pure mathematicians. We hope that from this point of view our review provides some contributions to this topic.

Conflict of interest

The authors declare that they have no conflict of interest.

Acknowledgments

The authors thank to Prof. V. Mityushev, Prof. J. Kaplunov, Prof. V. Malyi, and Prof. T.J. Pence for their valuable comments and suggestions.

References

[1] D. Capecchi, G. Ruta, Strength of Materials and Theory of Elasticity in 19th Century Italy – A Brief Account of the History of Mechanics of Solids and Structures, Springer, 2015.

[2] M.E. Gurtin, The linear theory of elasticity, in: S. Flügge (Ed.), Encyclopedia of Physics, vol. VIa/2, Mechanics of Solids II (editor: C. Truesdell), Springer, 1972, pp. 1–296.

[3] G.A. Maugin, Continuum Mechanics Through the Eighteenth and Nineteenth Centuries – Historical Perspectives from John Bernoulli (1727) to Ernst Hellinger (1914), Springer, 2014.

[4] I. Todhunter, A History of the Theory of Elasticity and of the Strength of Materials: From Galilei to the Present Time. Volume II. Saint-Venant to Lord Kelvin. Part II (editor K. Pearson), Cambridge University Press, 1893 (reprinted in 2014).

[5] R.W. Soutas-Little, Elasticity, Dover, 1999.

[6] G. Kirchhoff, Über das Gleichgewicht und die Bewegung eines unendlich dünnen elastischen Stabes, Journal für die Reine und Angewandte Mathematik 56 (1859) 285–313.

[7] A.J.C. Barré de Saint-Venant, Des conditions pour que six fonctions des coordonnées x, y, z des points d'un corps élastiques représentent des composantes de pression s'exercant sur trois plans rectangulaires à l'intérieur de ce corps, par suite de petits changements de distance de ses parties, L'Institut, Journal Universel des Scienses et des Sociétés Savantes en France et de L'étranger. Section 1. Sciences Mathématiques, Physiques et Naturelles XXYIII (1860) 294.

[8] A.J.C. Barré de Saint-Venant, Établissement élémentaire des formules et équations générales de la théorie de l'élasticité des corps solides. Appendice III to: Navier C.-L.-M.-H. Résumé des Leçons Données à l'École des Ponts et Chaussées sur l'Application de la Mécanique A l'Établissement des Constructions Et des Machines, Vol. 1: Premiére Partie, Contenant les

Compatibility conditions Chapter | 6 **135**

Leçons sur la Résistance des Matériaux et sur l'Établissement des Constructions en Terre, en Maçonnerie et en Charpente; Premiere Section, de la Resistance des Corps Solides, par 3rd ed., Paris, Dunod, 1864, pp. 541–617 (compatibility conditions 597–600). Reprinted: Forgotten Books, London, 2018.

[9] J. Boussinesq, Étude nouvelle sur l'Équilibre et le mouvement des corps solides élastiques dont certaines dimensions sont très petites per rapport à d'autres. Premier Mémoire (Des tiges), Journal de Mathématiques Pures et Appliquées, Ser. II 16 (1871) 125–240, compatibility conditions 132–134.

[10] E. Beltrami, Osservazioni della nota precedente. Rend. R. Acc. Lincei, serie V, 1(1), I semestre (1892) 141–142. Reprinted as: E. Beltrami, Osservazioni alla nota Prof. Morera, in: Opere Matematiche Di Eugenio Beltrami, Pubblicate Per Cura Della Facolta Di Scienze Della R. Universita Di Roma 4, Milano, Ulriko Hoeply (1920) 510–512.

[11] L. Donati, Ulteriori osservazioni intorno al teorema del Menabrea, Memorie dell'Accademia delle Scienze di Bologna 5 (4) (1894) 449–474.

[12] J.H. Michell, On the direct determination of stress in an elastic solid, with application to the theory of plates, Proceedings of the London Mathematical Society 31 (1899) 100–124.

[13] A. Yavari, Compatibility equations of nonlinear elasticity for non-simply-connected bodies, Archive for Rational Mechanics and Analysis 209 (2013) 237–253.

[14] L. Brillouin, Les tenseurs en mécanique et en élasticité, Dover, 1946.

[15] C. Truesdell, The mechanical foundations of elasticity and fluid dynamics, Journal of Rational Mechanics and Analysis 1 (1952) 125–171, 173–230.

[16] P. Germain, Cours de Mécanique des milieux continus. Tome 1, Masson, 1973.

[17] I.N. Sneddon, D.S. Berry, The Classical Theory of Elasticity, Springer, 1958.

[18] Y.A. Amenzade, Theory of Elasticity, Mir, 1979.

[19] W. Nowacki, Teoria sprężystości, PWN, 1973 (in Polish).

[20] M.E. Eglit, Lectures on the Basics of Continuum Mechanics, 2nd edition, Librokom, 2010 (in Russian).

[21] V.V. Novozhilov, Theory of Elasticity, Pergamon Press, 1961.

[22] J.R. Barber, Elasticity, 2nd ed., Kluwer, 2002.

[23] L.I. Sedov, A Course in Continuum Mechanics, vol. I, Wolters-Noordhoff, 1971.

[24] B.A. Boley, J.H. Weiner, Theory of Thermal Stresses, Wiley, 1960.

[25] L. Bianchi, Sui simboli a quattro indici e sulla curvatura di Riemann, Rendiconti dell'Accademia Nazionale dei Lincei 11 (1902) 3–7.

[26] A. Voss, Zur Theorie der Transformation quadratischer Differentialausdrücke und der Krümmung höherer Mannigfaltigkeiten, Mathematische Annalen 16 (1880) 129–179.

[27] S.M. Stigler, Stigler's law of eponymy, Transactions of the New York Academy of Sciences 39 (1980) 147–157.

[28] O.I. Terebushko, Foundations of the Theory of Elasticity and Plasticity, Nauka, 1984 (in Russian).

[29] I.S. Sokolnikoff, Mathematical Theory of Elasticity, McGraw-Hill, 1946.

[30] S.P. Timoshenko, J.N. Goodier, Theory of Elasticity, McGraw-Hill, 1954.

[31] L.D. Landau, E.M. Lifshitz, Theory of Elasticity, Pergamon Press, 1989.

[32] M.M. Filonenko-Borodich, Theory of Elasticity, Dover, 1965.

[33] L.P. Lebedev, M.J. Cloud, Introduction to Mathematical Elasticity, World Scientific, 2009.

[34] A.E.H. Love, A Treatise on the Mathematical Theory of Elasticity, 4th ed., Dover, 1944 (reissued in 2011).

[35] A.E. Green, W. Zerna, Theoretical Elasticity, Clarendon Press, 1968.

[36] P.F. Papkovich, Theory of Elasticity, Oborongiz, 1939 (in Russian).

[37] N.I. Muskhelishvili, Some Basic Problems of the Mathematical Theory of Elasticity: Fundamental Equations, Plane Theory of Elasticity, Torsion and Bending, Springer, 1977, pp. 28–51, Ch. Analysis of Strain.

[38] P.G. Ciarlet, Mathematical Elasticity, Vol. I: Three-Dimensional Elasticity, North-Holland, 1988.

136 Mechanics and Physics of Structured Media

[39] J.E. Marsden, T.J.R. Hughes, Mathematical Foundations of Elasticity, Dover, 1994.

[40] R.B. Hetnarski, J. Ignaczak, The Mathematical Theory of Elasticity, CRC Press, 2013.

[41] A.P. Boresi, K.P. Chong, J.D. Lee, The Mathematical Theory of Elasticity, John Wiley & Sons, 2011.

[42] R.J. Atkin, N. Fox, An Introduction to the Theory of Elasticity, Dover, 2005.

[43] C.-C. Wang, C. Truesdell, Introduction to Rational Elasticity, Noordhoff, 1973.

[44] P. Podio-Guidugli, A. Favata, Elasticity for Geotechnicians: A Modern Exposition of Kelvin, Boussinesq, Flamant, Cerruti, Melan, and Mindlin Problems, Springer, 2014.

[45] G.T. Mase, R.E. Smelser, G.E. Mase, Continuum Mechanics for Engineers, 3rd ed., CRC Press, 2010.

[46] A.P. Filin, Applied Mechanics of Rigid Deformed Body, vol. 1, Nauka, 1975 (in Russian).

[47] R.V. Southwell, An Introduction to the Theory of Elasticity for Engineers and Physicists, 2nd ed., Oxford University Press, 1941.

[48] Ye.I. Shemyakin, Introduction to the Theory of Elasticity, Moscow University Publishing, 1993 (in Russian).

[49] H.-C. Wu, Continuum Mechanics and Plasticity, Chapman & Hall/CRC Press, 2005.

[50] F. Irgens, Continuum Mechanics, Springer, 2008.

[51] W.M. Lai, D. Rubin, E. Krempl, Introduction to Continuum Mechanics, 3rd edition, Butterworth-Heinemann, 1999.

[52] M.E. Gurtin, E. Fried, L. Anand, The Mechanics and Thermodynamics of Continua, Cambridge University Press, 2012.

[53] A. Bertram, Elasticity and Plasticity of Large Deformations – An Introduction, Springer, 2012.

[54] M.E. Gurtin, An Introduction to Continuum Mechanics, Academic Press, 1981.

[55] X. Oliver, C. Agelet de Saracibar, Continuum Mechanics for Engineers – Theory and Problems, 2nd edition, 2017.

[56] J.H. Heinbockel, Introduction to Tensor Calculus and Continuum Mechanics, Department of Mathematics and Statistics, Old Dominion University, 1996.

[57] H.F. Wang, Theory of Linear Poroelasticity with Applications to Geomechanics and Hydrogeology, Princeton University Press, 2000.

[58] R.V. Southwell, Castigliano's principle of minimum strain-energy, Proceedings of the Royal Society A 154 (1936) 4–21.

[59] R.V. Southwell, Castigliano's principle of minimum strain energy, and the conditions of compatibility for strain, in: Stephen Timoshenko 60th Anniversary Volume, The Macmillan Co., New York, 1938, pp. 211–217, reprinted in: London, Edinburgh & Dublin Philosophical Magazine 30 (1940) 252–258.

[60] A.I. Lurie, Theory of Elasticity, Springer, 2005.

[61] Y.A. Krutkov, Tensor of Stress Functions and General Solutions in Statics of the Theory of Elasticity, Izd. AN SSSR, 1949 (in Russian).

[62] H.G. Hahn, Elastizitätstheorie. Grundlagen der Linearen Theorie und Anwendungen auf eindimensionale, ebene und räumliche Probleme, Vieweg+Teubner Verlag, 1985.

[63] Y.N. Rabotnov, Mechanics of a Deformable Rigid Body, Nauka, 1988 (in Russian).

[64] G.V. Kolosov, The Use of a Complex Variable in the Theory of Elasticity, ONTI, 1935 (in Russian).

[65] E. Trefftz, Mathematische Elastizitätstheorie, in: G. Angenheister, A. Busemann, O. Föppl, J.W. Geckeler, A. Nadai, F. Pfeiffer, T. Pöschl, P. Riekert, E. Trefftz, R. Grammel (Eds.), Mechanik der Elastischen Körper, Springer, Berlin, Heidelberg, 1928, pp. 47–140.

[66] L.H. Donnell, Beams, Plates, and Shells, McGraw-Hill, 1976.

[67] R.M. Christensen, Mechanics of Composite Materials, Dover Publications, 2005.

[68] V.P. Zubchaninov, Foundation of the Theory of Elasticity and Plasticity, Visshaya Shkola, 1990 (in Russian).

[69] L.S. Leybenzon, Elasticity Theory Course, 2nd ed., Gostekhizdat (State Publishing of Technical and Theoretical Literature), 1947 (in Russian).

Compatibility conditions Chapter | 6 **137**

[70] L.S. Leybenzon, Variational Methods for Solving Problems of Elasticity Theory, Gostekhiz-dat (State Publishing of Technical and Theoretical Literature), 1943 (in Russian).

[71] A. Haar, T. von Kármán, Zur Theorie der Spannungszustände in plastischen und sandartigen Medien, Nachrichten von der Gesellschaft der Wissenschaften zu Göttingen, Mathematisch-Physikalische Klasse 1909 (1909) 204–218.

[72] R. Courant, D. Hilbert, Methods of Mathematical Physics, Vol. I, Interscience Publishers, 1953 (first edition in 1924, in German).

[73] W.S. Slaughter, The Linearized Theory of Elasticity, Birkhäuser, 2002.

[74] L.E. Malvern, Introduction to the Mechanics of a Continuous Medium, Prentice Hall, 1969.

[75] K.A. Washizu, A note on the conditions of compatibility, Journal of Mathematical Physics 36 (1957) 306–312.

[76] K.A. Washizu, Variational Methods in Elasticity and Plasticity, 3rd edition, Pergamon Press, 1982.

[77] P.L. Gould, Y. Feng, Introduction to Linear Elasticity, 4th ed., Springer, 2018.

[78] M. Sadd, Introduction to Linear Elasticity, 4th ed., Academic Press, 2020.

[79] V.N. Ionov, P.M. Ogibalov, Strength of Spatial Structure Elements, 2. Statics and Oscillations, Vysshaya Shkola, 1979 (in Russian).

[80] J. Betten, Kontinuumsmechanik: Elasto-, Plasto- und Kriechmechanik, Springer, 1993.

[81] D.S. Chandrasekharaiah, L. Debnath, Continuum Mechanic, Elsevier, 1994.

[82] K. Wilmanski, Fundamentals of Solid Mechanics, IUSS Press, 2010.

[83] R. Temam, A. Miranville, Mathematical Modeling in Continuum Mechanics, Cambridge University Press, 2005.

[84] E.H. Brown, On the most general form of the compatibility equations and the conditions of integrability of strain rate and strain, Journal of Research of the National Bureau of Standards 59 (1957) 421–426.

[85] A. Treibergs, A. Cherkaev, P. Krtolica, Compatibility conditions for discrete planar structures, International Journal of Solids and Structures 184 (2020) 248–278.

[86] R. Courant, D. Hilbert, Methods of Mathematical Physics, Vol. II: Partial Differential Equations, Interscience Publishers, 1962.

[87] N.I. Ostrosablin, On Beltrami-Michell equations and the Saint-Venant operator, Dinamika Splošnoj Sredy 116 (2000) 211–216 (in Russian).

[88] V.I. Blokh, The stress functions in the theory of elasticity, Prikladnaya Matematika i Mekhanika 14 (1950) 415–422 (in Russian).

[89] V.I. Blokh, Theory of Elasticity, Kharkov University, 1964 (in Russian).

[90] J.C. Maxwell, On reciprocal figures, frames, and diagrams of forces, Transactions of the Royal Society of Edinburgh 26 (1870) 1–40.

[91] B. Morera, Soluzione generale delle equazioni indefinite dell'equilibrio di un corpo continuo, Atti Della Accademia Dei Lincei. Rendiconti, Serie V 1 (1) (1892) 137–141, I semestre.

[92] K. Marguerre, Supplementary note to the author's abstract of Southwell's paper [59], Zentralblatt für Mechanik 10 (1940/41) 5.

[93] J. Stickforth, On the derivation of the conditions of compatibility from Castigliano's principle by means of three-dimensional stress functions, Studies in Applied Mathematics 44 (1965) 214–226.

[94] V. Slivker, Mechanics of Structural Elements. Theory and Applications, Springer, 2007.

[95] N.I. Ostrosablin, Compatibility conditions of small deformations and stress functions, Journal of Applied Mechanics and Technical Physics 38 (1997) 774–783.

[96] N.I. Ostrosablin, Comment on the publication "Compatibility conditions of small deformations and stress functions", Journal of Applied Mechanics and Technical Physics 40 (1999) 549.

[97] A.V. Sachenkov, È.G. Saïfullin, Stress functions and displacement functions in three-dimensional elasticity theory, Issledovaniya po Teorii Plastin i Oboloček 16 (1981) 36–41 (in Russian).

138 Mechanics and Physics of Structured Media

[98] È.G. Saĭfullin, A.V. Sachenkov, P.M. Timerbaev, Basic equations of the theory of elasticity in stresses and displacements, Issledovaniya po Teorii Plastin i Oboloček 18 (1985) 66–79 (in Russian).

[99] V.I. Malyi, Independent conditions of stress compatibility for elastic isotropic body, Doklady Akademii Nauk Ukrainskoj SSR. Seriya A 7 (1987) 43–46 (in Russian).

[100] V.I. Malyi, One representation of the conditions of the compatibility of deformations, Journal of Applied Mathematics and Mechanics 50 (1986) 679–681.

[101] I.V. Andrianov, J. Awrejcewicz, Compatibility equations in the theory of elasticity, Journal of Vibration and Acoustics 125 (2003) 244–245.

[102] B.F. Vlasov, Integration of the strain-continuity equations in St. Venant form, Soviet Applied Mechanics 5 (1969) 1283–1285.

[103] B.E. Pobedrya, D.V. Georgievskii, Equivalence of formulations for problems in elasticity theory in terms of stresses, Russian Journal of Mathematical Physics 13 (2006) 203–209.

[104] B.F. Vlasov, Equations for the determination of the Monera and Maxwell stress functions, Soviet Physics Doklady 16 (1971) 252–254.

[105] V.V. Vasil'yev, L.V. Fedorov, To the problem of the theory of elasticity, formulated in stresses, Mechanics of Solids 31 (1996) 82–92.

[106] A.N. Konovalov, Problems of the theory of elasticity in terms of stresses, in: J.-L. Lions, G.I. Marchuk (Eds.), Vychislitel'nye Metody v Matematicheskoĭ Fizike, Geofizike i Optimal'nom Upravlenii (Computational Methods in Mathematics, Geophysics and Optimal Control), Nauka (SO), 1978, pp. 121–124 (in Russian).

[107] A.N. Konovalov, Solution of the Theory of Elasticity Problems in Terms of Stresses, Novosibirsk State University, 1979 (in Russian).

[108] A.N. Konovalov, S.B. Sorokin, Structure of equation of elasticity theory. Static problem, Preprint 665, Vichisl. Zentr Sib. Otd. AN SSSR, 1986 (in Russian).

[109] A.N. Konovalov, S.B. Sorokin, Numerical methods in elasticity problems, Bulletin of the Novosibirsk Computing Center. Series Numerical Analysis 5 (1994) 27–34.

[110] A.N. Konovalov, Numerical methods for static problems of elasticity, Siberian Mathematical Journal 36 (1995) 491–505.

[111] E. Chraptovič, J. Atkočiūnas, Role of Kuhn-Tucker conditions in elasticity equations in terms of stresses, Statyba 6 (2000) 104–112 (in Russian).

[112] B.E. Pobedrya, Some general theorems of the mechanics of a deformable solid, Journal of Applied Mathematics and Mechanics 43 (1979) 531–541.

[113] B.E. Pobedrya, On the problem in stresses, Soviet Mathematics Doklady 23 (1978) 351–353.

[114] B.E. Pobedrya, A new formulation of the problem in mechanics of a deformable solid body under stress, Soviet Mathematics Doklady 22 (1980) 88–91.

[115] B.E. Pobedrya, Numerical Methods in the Theory of Elasticity and Plasticity, MGU, 1981 (in Russian).

[116] B.E. Pobedrya, A new formulation of the problem in mechanics of a deformable solid body under stress, Moscow University Mechanics Bulletin 58 (2003) 6–12.

[117] B.E. Pobedrya, S.V. Sheshenin, T. Kholmatov, Problem in Terms of Stresses, Fan, 1988 (in Russian).

[118] D.V. Georgievskii, General solutions of weakened equations in terms of stresses in the theory of elasticity, Moscow University Mechanics Bulletin 68 (2013) 1–7.

[119] V. Kucher, X. Markenscoff, M. Paukshto, Some properties of the boundary value problems of linear elasticity in terms of stresses, Journal of Elasticity 74 (2004) 135–145.

[120] S. Li, A. Gupta, X. Markenscoff, Conservation laws of linear elasticity in stress formulation, Proceedings of the Royal Society A 461 (2005) 99–116.

[121] V.D. Klyushnikov, Derivation of the Beltrami-Michell equations from the Castigliano's variational equations, Prikladnaya Matematika i Mekhanika 18 (1958) 171–174 (in Russian).

[122] J. Grycz, On the compatibility conditions in the classical theory of elasticity, Archiwum Mechaniki Stosowanej 19 (1967) 883–891.

Compatibility conditions Chapter | 6 **139**

[123] B.F. Vlasov, On equations of continuity of deformations, Soviet Applied Mechanics 6 (1970) 1227–1231.

[124] I. Kozák, Notes on the field equations with stresses and on the boundary conditions in the linearized theory of elastostatics, Acta Technica Academiae Scientiarum Hungaricae 90 (1980) 221–245.

[125] I. Kozák, Determination of compatibility boundary conditions in linear electrostatics with the aid of the principle of minimum complementary energy, Publications of the Technical University for Heavy Industry. Series D, Natural Sciences 34 (1980) 83–98.

[126] I. Kozák, Principle of virtual work in terms of stress functions, Publications of the Technical University for Heavy Industry. Series D, Natural Sciences 34 (1980) 147–163.

[127] I. Kozák, Principle of the complementary virtual work and the Riemann-Christoffel curvature tensor as compatibility condition, Journal of Computational and Applied Mechanics 1 (2000) 71–79.

[128] N.M. Borodachev, About one approach in the solution of 3D problem of the elasticity in stresses, International Applied Mechanics 31 (1995) 38–44.

[129] N.M. Borodachev, Three-dimensional problem of the theory of elasticity in strains, Strength of Materials 27 (1995) 296–299.

[130] N.M. Borodachev, The equations of compatibility of deformation, Journal of Applied Mathematics and Mechanics 65 (2001) 1021–1024.

[131] B.F. Vlasov, About number of independent equations of continuity, Trudy Univ. Družby Narodov 34 (1968) 171–174 (in Russian).

[132] S.Ya. Makovenko, Integration of the equations of continuity of deformations in an arbitrary coordinate system, Prikladnaya Mekhanika 16 (1980) 122–124 (in Russian).

[133] S.Ya. Makovenko, On the equivalent transformation of the Cesaro displacement formula, Stroit. Mekh. Inzh. Konstr. Soor., Ros. Univ. Družby Narodov 1 (1980) 3–9 (in Russian).

[134] S.Ya. Makovenko, On one scheme of deriving of the deformation compatibility equations in the integro-differential form, Stroit. Mech. Inzh. Konstr. Soor., Ros. Univ. Družby Narodov 2 (2006) 15–17 (in Russian).

[135] S. Patnaik, D. Hopkins, Stress formulation in three dimensional elasticity, NASA/TP-2001-210515, 2001.

[136] S.N. Patnaik, S.S. Pai, D.A. Hopkins, Compatibility condition in theory of solid mechanics (elasticity, structures, and design optimization), Archives of Computational Methods in Engineering 14 (2001) 431–457.

[137] Gy. Szeidl, On compatibility conditions for mixed boundary value problems, Technische Mechanik 17 (1997) 245–262.

[138] G. Geymonat, F. Krasucki, Hodge decomposition for symmetric matrix fields and the elasticity complex in Lipschitz domains, Communications on Pure and Applied Analysis 8 (2009) 295–309.

[139] Gy. Szeidl, On compatibility conditions for mixed boundary value problems in micropolar theory of elasticity, Publications of the University of Miskolc. Series D. Natural Sciences. Mathematics 37 (1997) 105–116.

[140] J. Grycz, On the conditions of compatibility in the theory of elasticity, Archives of Civil Engineering 47 (2001) 271–289.

[141] S.Ya. Makovenko, About problem in stresses in nonlinear theory of elasticity, Vestnik Ros. Univ. Družby Narodov ser. Inzh. Issl. 2 (2003) 23–28 (in Russian).

[142] S.Ia. Makovenko, Certain versions of the formulations of problems of non-liner elasticity in terms of stresses, Journal of Applied Mathematics and Mechanics 47 (1983) 775–782.

[143] J. Blume, Compatibility conditions for a left Cauchy-Green strain field, Journal of Elasticity 21 (1989) 271–308.

[144] S.K. Godunov, E. Romenskii, Elements of Continuum Mechanics and Conservation Laws, Kluwer, 2003.

[145] G. Geymonat, F. Krasucki, Some remarks on the compatibility conditions in elasticity, Accademia Nazionale delle Scienze XL 123 (2005) 175–182.

140 Mechanics and Physics of Structured Media

[146] P.G. Ciarlet, P. Ciarlet Jr., Another approach to linearized elasticity and a new proof of Korn's inequality, Mathematical Models and Methods in Applied Sciences 15 (2005) 259–271.

[147] C. Amrouche, P.G. Ciarlet, L. Gratie, S. Kesavan, New formulations of linearized elasticity problems, based on extensions of Donati's theorem, Comptes Rendus de L'Académie Des Sciences. Série I 342 (2006) 785–789.

[148] C. Amrouche, P.G. Ciarlet, L. Gratie, S. Kesavan, On Saint Venant's compatibility conditions and Poincaré's lemma, Comptes Rendus de l'Académie des Sciences. Série I 342 (2006) 888–891.

[149] P.G. Ciarlet, P. Ciarlet Jr., A new approach for approximating linear elasticity problems, Comptes Rendus de l'Académie des Sciences. Série I 346 (2008) 351–356.

[150] I.V. Andrianov, Mathematical models in pure and applied mathematics, in: A. Abramyan, I. Andrianov, V. Gaiko (Eds.), Nonlinear Dynamics of Discrete and Continuous Systems, in: Advanced Structured Materials, vol. 139, Springer Nature, 2021, pp. 15–30.

[151] V. Mityushev, I. Andrianov, S. Gluzman, L.A. Filshtinsky's contribution to applied mathematics and mechanics of solids, this book.

[152] A.S. Volmir, Stability of Elastic Systems, Foreign Technology Division, Air Force Systems Command, Wright-Patterson Air Force Base, OH, 1967.

[153] V.V. Novozhilov, The Theory of Thin Shells, Noordhoff, 1963.

[154] A. Angoshtari, A. Yavari, On the compatibility equations of nonlinear and linear elasticity in the presence of boundary conditions, Zeitschrift für Angewandte Mathematik und Physik 66 (2015) 3627–3644.

[155] C. Vallée, Compatibility equations for large deformations, International Journal of Engineering Science 30 (1992) 1753–1757.

[156] V. Georgescu, On the operator of symmetric differentiation on a compact Riemannian manifold with boundary, Archive for Rational Mechanics and Analysis 74 (1980) 143–165.

[157] C.G. Böhmer, Y. Lee, Compatibility conditions of continua using Riemann-Cartan geometry, Mathematics and Mechanics of Solids 26 (2021) 513–529.

Chapter 7

Critical index for conductivity, elasticity, superconductivity. Results and methods

Simon Gluzman
Research Group Materialica+, ON, Toronto, Canada

7.1 Introduction

Critical phenomena are ubiquitous, ranging from the field theory to hydrodynamics [28]. And it is challenging to explain related critical indices theoretically. The function $\Phi(x)$ of a real variable x exhibits critical behavior, with a critical index α, at a finite critical point x_c, when

$$\Phi(x) \simeq A(x_c - x)^\alpha, \text{ as } x \to x_c - 0. \tag{7.1}$$

The definition covers the case of negative index when function can tend to infinity, or the sought function can tend to zero if the index is positive. The variable $x > 0$ can represent, e.g., a coupling constant, inverse temperature or concentration of particles. Sometimes, the values of critical index and critical point are known from some sources, and the problem consists in finding the critical amplitude A, as extensively exemplified in [25].

The case when critical behavior occurs at infinity,

$$\Phi(x) \simeq Ax^\alpha, \text{ as } x \to \infty, \tag{7.2}$$

can be analyzed similarly. It can be understood as the particular case with the critical point positioned at infinity.

As a rule, for realistic physical systems one can learn only its behavior at small variable, which follows from some perturbation theory. And the function $\Phi(x)$ is approximated by its expansion. Most often one finds that such expansions are divergent, and valid only for very small or very large x. Sometimes, theoretically, one even has a convergent series, resulting in a rather good numerically convergent, truncated polynomial approximations. But there is still a problem of extrapolating outside of the region of numerical convergence, where the critical behavior sets in.

Mechanics and Physics of Structured Media. https://doi.org/10.1016/B978-0-32-390543-5.00012-8
Copyright © 2022 Elsevier Inc. All rights reserved.

142 Mechanics and Physics of Structured Media

Let as assume that some kind of perturbation theory is possible to develop, so that for a smooth function $\Phi(x)$, we have the asymptotic power series

$$\Phi(x) \sim \sum_{n=0}^{\infty} a_n x^n. \tag{7.3}$$

Our task is to recast the series (7.3) into some convergent expressions by means of a nonlinear analytical construct, the so-called approximants. When literally all of the terms in divergent series are known, one can invoke Euler of Borel summation. Even for convergent series there is still a problem of how to accelerate convergence and (or) continue the expansion outside of radius of convergence, where the approximants could be useful. In practice we are dealing with (normalized to unity) truncation

$$\Phi_k(x) = 1 + \sum_{n=1}^{k} a_n x^n. \tag{7.4}$$

One can always express the critical index directly by using its definition, and find it as the limit of explicitly expressed approximants. For instance, critical index can be expressed as the following derivative

$$\mathcal{B}_a(x) = \partial_x \log(\Phi(x)) \simeq -\frac{\alpha}{x_c - x}, \tag{7.5}$$

as $x \to x_c$, thus defining the critical index as the residue in the corresponding single pole. The pole corresponds to the critical point x_c. The critical index corresponds to the residue

$$\alpha = \lim_{x \to x_c} (x - x_c)\mathcal{B}_a(x).$$

To the $DLog$-transformed series $\mathcal{B}_a(x)$ one is bound to apply the Padé approximants [1]. Moreover, the whole table of Padé approximants can be constructed [2]. I.e., the $DLog$ Padé method does not lead to a unique algorithm for finding critical indices. Basically, different values are produced by different Padé approximants. Then it is not clear which of these estimates to prefer. Standard approach consists in applying a diagonal Padé approximants [1].

When a function, at asymptotically large variable, behaves as in (7.2), then the critical exponent can be defined similarly, by means of the $DLog$ transformation. It is represented by the limit

$$\alpha = \lim_{x \to \infty} x \mathcal{B}_a(x). \tag{7.6}$$

Assume that the small-variable expansion for the function $\mathcal{B}_a(x)$ is given. In order for the critical index to be finite it is necessary to take only the approximants behaving as x^{-1} as $x \to \infty$ [29]. It leaves us no choice but to select the nondiagonal $P_{n,n+1}(x)$ approximants, so that the corresponding approximation α_n is finite. Some examples of application of the method can be found in [19,22,25]. Such approach can be applied when the critical point is known, in conjunction with transformation

$$z = \frac{x}{x_c - x} \Leftrightarrow x = \frac{z x_c}{z + 1},$$

applied to the original series. Instead of the Padé approximants one can also apply some different approximants or their combinations [19,23,25].

One can calculate critical index directly from the explicit expressions, such as root and factor approximants. The self-similar root approximant has the following general form [26],

$$\mathcal{R}_k^*(x, m_k) = \left(\left((1 + \mathcal{P}_1 x)^{s_1} + \mathcal{P}_2 x^2 \right)^{s_2} + \ldots + \mathcal{P}_k x^k \right)^{s_k} . \qquad (7.7)$$

In principle, all the parameters $\mathcal{P}_i$ and s_i may be found from asymptotic equivalence with a given power series. Root approximants can be applied to the critical index calculations [19,22], and will be employed for such tasks throughout the current paper.

The singular solutions emerge also from factor approximants, and they correspond to critical points and phase transitions [30], including also the case of singularity located at ∞. Factor approximants can do the same work as Padé and $DLog$ Padé approximants.

All methods discussed above are direct methods of extracting critical properties and related dependencies from various series without invoking any further assumptions about systems behavior in the critical region. Such methods are universally applicable as soon as the series are available.

When the truncated series is long, one would expect that the accuracy is going to improve with increasing numbers of terms. Sometimes, an optimum is achieved for some finite number of terms, reflecting the asymptotic nature of the underlying series. It is very difficult to improve the quality of results when the series are short. In the current paper we choose the strategy of increasing the number of methods applied to the short series, create a sample of plausible estimates for the index and apply traditional, simple statistical measures, such as average and variance to produce the final estimate [19,28]. Weighted averages over all estimates [19,23,54,55], can be considered as well.

In Section 7.1 we briefly discussed the main methods of calculating critical index form the asymptotic series.

In Sections 7.2, 7.3, 7.5, 7.6 we apply root approximants in conjunction with various (five) optimization techniques, and calculate critical index for key cases of random media. The critical indices are calculated uniformly for the conductivity, elasticity, sedimentation, and 2D Ising model.

In Section 7.7 we present the so-called unbiased estimations of the critical index for three-dimensional Ising model.

144 Mechanics and Physics of Structured Media

In Section 7.8 we calculate critical indices for the three-dimensional super-conductivity problem.

In Section 7.9 we consider the critical index of the graphene-type composite and analyze its evolution with increasing concentration of vacancies.

In Section 7.10 we consider the critical index for the expansion factor of the three-dimensional polymer coil.

7.2 Critical index in 2D percolation. Root approximants

The problem of conductivity for site percolation is studied within the framework of a minimal model for transport of classical particles through a random medium [46]. This minimal model, known as the Lorenz 2D gas, is a particularly simple statistic hopping model allowing both for analytical consideration and numerical simulations [21,46]. It can be realized on a square lattice with a fraction of sites being excluded at random. The test particle, or tracer, walks randomly with Poisson-distributed waiting times between the moves. At every move the tracer attempts to jump on to one of the neighboring sites also selected at random. The move is accepted if the site is not excluded. Through the diffusion coefficient for the tracer one can express the macroscopic conductivity [46]. The diffusion ceases to exist at the critical density of the excluded sites.

If f stands for the concentration of conducting or not excluded sites in the Lorenz model, then $x = 1 - f$ is the concentration of excluded sites. In the vicinity of the site percolation threshold the conductivity behaves as

$$\sigma(x) \sim (x_c - x)^t, \text{ as } x \to x_c - 0, \tag{7.8}$$

with $x_c = 0.4073$, $t = 1.310$ [31,56]. Perturbation theory in powers of the variable $x = 1 - f$ gives [46] for the two-dimensional square lattice the expansion

$$\sigma(x) \simeq 1 - \pi x + 1.28588x^2, \text{ as } x \to 0. \tag{7.9}$$

In the paper [28], we obtained the critical index $t = 1.388 \pm 0.036$. The scatterers should not necessarily be restricted to the regular lattice positions. They could be distributed randomly and independently in the whole space [36]. In this case there is an analogy of the Lorenz gas and continuum percolation when particles can overlap. In 2D the value of critical index remains the same as in the lattice model [5] despite overlaps. But in 3D the value of the critical index $t = 2.88$ is considerably larger than its lattice counterpart [36].

All methods based on extraction of critical properties only from second-order truncated series should be treated with caution since physical predictions based only on two starting coefficients can not be always true. But in the particular cases studied below they can be verified against experimental evidence and can be obtained at least by a few methods. In fact, five methods will be applied for estimations of the critical index. They all differ in the way how the condition of convergence is expressed, with the critical parameters playing the role

Critical index for conductivity, elasticity, superconductivity Chapter | 7 **145**

of controls. Let us employ various special resummation techniques and estimate the critical index. The strategy consists in employing various conditions on the fixed point in order to compensate for a shortage of the coefficients, as discussed first in [28,51,52].

7.2.1 Minimal difference condition according to original

Let us first get back to the original suggestion of [51,52]. In this case one applies resummation technique to the original truncated series. First, we should construct a two different approximants, for instance the simplest pair is given by the lowest-order expressions with s being a control parameter,

$$\sigma_1^*(x) = \left(1 + \frac{a_1}{s}x\right)^s, \quad \sigma_2^*(x) = 1 + a_1 x \left[1 + \frac{a_2 x}{a_1(s-1)}\right]^{(s-1)}. \quad (7.10)$$

From the first-order approximant we estimate threshold as a function of s,

$$x_{c,1}(s) = -\frac{s}{a_1}. \quad (7.11)$$

One would like to have the two solutions to differ from each other minimally. The minimal difference condition between the two approximations is reduced to the condition on stabilizer s. It will be determined as a minimizer of the following expression

$$\left| 1 + a_1 x_{c,1}(s) \left[1 + \frac{a_2 x_{c,1}(s)}{a_1(s-1)}\right]^{(s-1)} \right|, \quad (7.12)$$

and is located at $s = 1.287$, leading to the critical index $t = s$.

7.2.2 Iterated roots. Conditions imposed on thresholds

One can also select the iterated root for the second approximant, so that,

$$\sigma_1^*(x) = \left(1 + \frac{a_1}{s}x\right)^s, \quad \sigma_2^*(x) = ((A_1(s)x + 1)^2 + A_2(s)x^2)^s, \quad (7.13)$$

where

$$A_1(s) = \frac{a_1}{2s}, \quad A_2(s) = \frac{-2a_1^2 s + a_1^2 + 4a_2 s}{4s^2}.$$

We also require that control parameters $s_1 = s_2 = s$, ignoring the difference between critical indices of different approximations. The parameter s is the sought critical index.

We would like to find the approximants to be virtually the same,

$$\sigma_1^*(x) = \sigma_2^*(x), \quad (7.14)$$

146 Mechanics and Physics of Structured Media

meaning that the sequence of iterated approximants is supposed to converge in one single step, in the fastest possible manner. The control parameter s can be found uniquely in this case to guarantee the equality. The solution can be obtained by requiring

$$\left(\sigma_2^* \left(x_{c,1}\left(s\right)\right)\right)^{1/s} = 0. \tag{7.15}$$

One can express the second order approximant in another form, allowing for a comparison with first-order approximant,

$$\sigma_2^*(x) = \left(1 + \frac{a_1}{s}x + b(s)x^2\right)^s,$$

where

$$b(s) = \frac{\left(-a_1^2(s-1) + 2a_2 s\right)}{2s^2}.$$

Let us set $b(s) = 0$, and find the sought control parameter

$$s = \frac{a_1^2}{a_1^2 - 2a_2}. \tag{7.16}$$

The critical index for the 2D conductivity is $t = s = 1.352$.

Yet, one can come to the same solution for the index and threshold differently. The second-order approximant is still simple enough to allow for an explicit expression for the threshold,

$$x_{c,2}(s) = \frac{\sqrt{2a_1^2 s^3 - a_1^2 s^2 - 4a_2 s^3} + a_1 s}{a_1^2 s - a_1^2 - 2a_2 s}. \tag{7.17}$$

The second branch of the solution is irrelevant for our study. The expression (7.17) can be useful for construction of higher-order approximate conditions on critical index, extending equation (7.15). Such variants of an unbiased estimation of the critical properties will be studied below, in Section 7.7.

We can also consider the threshold to be independent on approximation, i.e. impose the condition

$$x_{c,2}(s) = x_{c,1}(s), \tag{7.18}$$

and find from here the stabilizer s. The values of the critical index appear to be the same for the two schemes, explaining why the title of subsection was chosen.

7.2.3 Conditions imposed on the critical index

It seems even more natural to insist that the critical index by itself should not depend on the approximation. Of course in this case we would need an additional parameter to be determined from this requirement. E.g., one can use the value

Critical index for conductivity, elasticity, superconductivity **Chapter | 7 147**

of threshold as a parameter x_0, and introduce it to the original series through the transformation

$$z = \frac{x}{x_0 - x} \Leftrightarrow x = \frac{z x_0}{z + 1}.$$

The two root-approximants in two different lowest orders are given as follows,

$$\sigma_1^*(z) = (1 + b_1(x_0)z)^{s_1(x_0)},$$
$$\sigma_2^*(z) = \left((1 + B_1(x_0)z)^2 + B_2(x_0)z^2\right)^{s_2(x_0)/2}. \tag{7.19}$$

Let us find explicitly and then minimize the difference between the two approximations for the critical index

$$|s_2(x_0) - s_1(x_0)|,$$

with respect to the parameter x_0.

The parameters of the first-order approximant are defined uniquely, e.g.

$$s_1(x_0) = \frac{a_1^2 x_0}{a_1^2 x_0 + 2a_1 - 2a_2 x_0}.$$

There are two solutions for the parameters of the second order approximant and we bring only the branch which turns to be relevant,

$$s_2(x_0) =$$
$$\frac{3a_1^3 x_0^2 + 6a_1^2 x_0 - a_1 x_0 \sqrt{a_1^4 x_0^2 + 12a_1^3 x_0 - 12a_1^2 a_2 x_0^2 + 12a_1^2 - 24a_1 a_2 x_0 + 36a_2^2 x_0^2 - 6a_1 a_2 x_0^2}}{2\left(a_1^3 x_0^2 + 3a_1^2 x_0 - 3a_1 a_2 x_0^2 + 3a_1 - 6a_2 x_0\right)}. \tag{7.20}$$

The difference possesses a minimum at a reasonable value of $x_0 = 0.39$. From the first order approximant one can find the critical index $\mathbf{t} = 1.121$.

7.2.4 Conditions imposed on amplitudes

Assume that we do know the correct value of threshold $x_c = 0.4073$ in advance, and try to use the knowledge. Let us apply the transformation, $z = \frac{x}{x_c - x} \Leftrightarrow x = \frac{z x_c}{z + 1}$ to the original series (7.29), with the resulting inverse series

$$\sigma(z) = 1 + b_1 z + b_2 z^2 + O(z^3). \tag{7.21}$$

The set of approximations including just the two starting terms from (7.21), can be written down and the expression for the renormalized quantity σ_1^* can be readily obtained:

$$\sigma_1^* = \left(1 - \frac{b_1 z}{s_1}\right)^{-s_1} \simeq \left(\frac{s_1}{-b_1}\right)^{s_1} z^{-s_1}, \quad z \to +\infty. \tag{7.22}$$

148 Mechanics and Physics of Structured Media

The control parameter/stabilizer s_1 should be positive, if we want to reproduce in the limit of $z \to \infty$, the correct power-law behavior of effective conductivity.

For comparison one also needs to construct a different set of approximations. It can be accomplished simply by leaving the constant term outside of the renormalization procedure. There are two approximations

$$\overline{\sigma_1} = b_1 z, \quad \overline{\sigma_2} = b_1 z + b_2 z^2, \tag{7.23}$$

which are resummed as follows [51,52],

$$\sigma_2^* = 1 + b_1 z \left[1 - \frac{b_2 z}{b_1 (1 + s_2)} \right]^{-(1+s_2)} \simeq \left(-\frac{b_2}{1 + s_2} \right)^{-(1+s_2)} b_1^{2+s_2} z^{-s_2},$$

$$z \to +\infty. \tag{7.24}$$

Let us demand now that both expressions (7.22) and (7.24) have the same power-law behavior at $z \to \infty$, while

$$s_2 = s_1 \equiv s.$$

And let us require now the fulfillment of the convergence or else, stability criteria for the two available approximations in the form of the minimal-difference condition for critical amplitude at infinity. Thus one obtains the condition on stabilizer s. It is to be determined from the minimum of the expression:

$$\left| \left(\frac{-b_2}{1 + s} \right)^{-(1+s)} b_1^{(2+s)} - \left(\frac{s}{-b_1} \right)^{s} \right|. \tag{7.25}$$

Generally speaking, it is sufficient to ask for an extremum of this difference.

The critical index $t = s = 1.272$.

7.2.5 Minimal derivative (sensitivity) condition

We can also select the iterated root for the second-order approximant, so that

$$\sigma_2^*(z) = \left((1 + B_1(s) z)^2 + B_2(s) z^2 \right)^{s}, \tag{7.26}$$

where

$$B_1(s) = \frac{b_1}{2s}, \quad B_2(s) = \frac{-2b_1^2 s + b_1^2 + 4b_2 s}{4s^2}.$$

The minimal difference condition leads to the simplest $DLog$ Padé approximant.

There is some other way to proceed. Based on the form of Eq. (7.26) the control parameter s can be interpreted as the "critical" index in the vicinity of

Critical index for conductivity, elasticity, superconductivity **Chapter | 7 149**

the (quasi)threshold $X_e(s)$, given explicitly as follows

$$X_e(s) = \frac{\sqrt{-B_2(s)} - B_1(s)}{B_1(s)^2 + B_2(s)}. \tag{7.27}$$

It seems natural to require that the critical index at infinity would not depend on the position of a quasithreshold. Without actually inverting (7.27), one can find the value of s from the minimal sensitivity condition imposed on the quasithreshold,

$$\frac{\partial X_e(s)}{\partial s} = 0,$$

and the solution leads to the critical index $t = 2s = 1.423$.

Thus there are five various but still close enough estimates for the critical index t

$$1.121; \quad 1.272; \quad 1.287; \quad 1.352; \quad 1.423.$$

Such "framing" of the solution is what one could realistically expect from the methodology based on a short series. Their average equals $t^* = 1.291$, and the margins of error can be estimated through the variance, so that $t = 1.291 \pm 0.1$.

Pruning the lowest and highest extremes, we find the three solutions centered around the value of 1.304 ± 0.037. Note also that the first three methods of subsections 7.2.1, 7.2.2, 7.2.3, are not reliant on any preknowledge of the threshold value, but still produce reasonable estimates. Current estimates are better than our previous result $t = 1.388 \pm 0.36$ [28].

7.3 3D Conductivity and elasticity

For 3D site percolation on the cubic lattice, perturbation theory gives the following expansion for conductivity [38],

$$\sigma(f) \simeq f - 1.52 f(1 - f).$$

This expression can be rewritten in powers of variable $x = 1 - f$, in the form

$$\sigma(x) \simeq 1 + a_1 x + a_2 x^2,$$

$$a_1 = -2.52, \quad a_2 = 1.52. \tag{7.28}$$

It is also known that the three-dimensional site percolation conductivity problem, similar to the two-dimensional case, exhibits the critical behavior [5,36,38] as in the formula (7.8), with $x_c = 0.688$, $t = 1.9 \pm 0.1$ [14,56]. One can extract critical properties of the 3D case from the expansion (7.28). Technical part is accomplished exactly the same way as above.

Using method of subsection 7.2.1, we get for the critical index $t_1 = 1.877$.

By applying method of subsection 7.2.2, we calculate $t_2 = 1.918$.

150 Mechanics and Physics of Structured Media

The method of subsection 7.2.3 gives $t_3 = 1.789$.

By imposing minimal difference condition on the critical amplitude as in subsection 7.2.4, we obtain $t_4 = 1.659$.

The minimal derivative condition of subsection 7.2.5 gives $t_5 = 1.855$.

The average over all solutions and the variance give us $t^* = 1.82 \pm 0.09$. Discarding the lowest and highest extremes, we find the three solutions centered around the value of 1.84 ± 0.04. Again, the three methods not exploiting the known threshold, give rather good estimates for the index. Good estimate $t = 1.887 \pm 0.032$ was obtained in our previous work [28].

7.3.1 3D elasticity, or high-frequency viscosity

Consider the problem of a perfectly rigid spherical inclusions embedded at random into an incompressible matrix. In the case of isotropic components we have two shear moduli, G_1, G, for the particles and matrix respectively. The limiting jamming regime when $\frac{G_1}{G}$ is close to infinity and f tends to the random close packing fraction $f_c \approx 0.637$, can be hardly investigated numerically. The elasticity problem is analogous to the problem of high-frequency effective viscosity of a suspension [4]. The suspension consists of hard spheres immersed into a Newtonian fluid of viscosity μ, when only hydrodynamic interactions between pairs of suspended particles are considered. Brownian motion of a suspension thus is not taken into account. Such suspension (composite) has a Newtonian elastic behavior and the ratio of the effective shear modulus $G_e \equiv \mu_e$ to that of the matrix is [50]

$$G_e(f) = 1 + a_1 f + a_2 f^2 + O(f^3),$$

$$a_1 = \frac{5}{2}, \quad a_2 = 5.0022.$$

(7.29)

Cluster expansion is involved in derivation, and the second order term takes into account interactions within the two-particle clusters. The central point of cluster approach is to account precisely for the interactions among all the particles inside the progressively larger clusters that are supposed to represent the composite material with greater accuracy. For the coefficient a_2 Batchelor and Green give slightly higher estimate, $a_2 = 5.2$ [4]. The estimates for the critical index presented below, appear to be weakly sensitive to such variation.

It is expected that in the vicinity of the 3D threshold that

$$G_e(f) \simeq A(f_c - f)^{-S},$$

where [28,41] $f_c \approx 0.637$, $S \approx 1.7$. Here S is the critical index for 3D elasticity (high-frequency viscosity). Let us employ various special resummation techniques and estimate the critical index S. The strategy once gain consists in employing various conditions on the fixed point in order to compensate for a

Critical index for conductivity, elasticity, superconductivity Chapter | 7 **151**

shortage of the coefficients. Of course, there is a tacit assumption involved, that the metastable branch of the suspension is considered. Thus, smooth interpolation between the low-and-high-concentration regimes can be performed.

The method of subsection 7.2.1 finds the critical index $S = 1.57646$.

The method of subsection 7.2.2 gives the critical index $S = 1.66471$.

The technique of subsection 7.2.3 brings the result for the critical index $S = 1.61913$.

Minimal difference condition of subsection 7.2.4 gives the critical index $S = 1.60483$.

Minimal derivative condition of subsection 7.2.5 leads to the critical index $S = 1.788$.

The solutions are centered around the average value of 1.651. Such "framing" of the unknown solution is what one should realistically expect from the methodology based on a short series. Margins of error could be found by calculating variance, so that $S = 1.651 \pm 0.074$. Discarding the lowest and highest extremes, we find the three solutions centered around the value of 1.63 ± 0.033. Such estimate is based solely on the three methods not exploiting the known threshold. Mind that slightly different estimate, 1.726 ± 0.06, was obtained in [28].

7.4 Compressibility factor of hard-disks fluids

The state of hard-disks fluid is described by the compressibility factor

$$Z = \frac{P}{\rho k_B T} = Z(f) \qquad \left(f \equiv \frac{\pi \rho}{4} a_s^2 \right) , \qquad (7.30)$$

in which P is pressure, ρ is density, T is temperature, a_s is the disk diameter, and f is called packing fraction. The compressibility factor exhibits critical behavior at a finite critical point. This behavior has been found from phenomenological equations as

$$Z(f) \sim (f_c - f)^{-\alpha} \qquad (f \to f_c - 0) , \qquad (7.31)$$

with the fitted parameters $f_c = 1$ and $\alpha = 2$ [43,48], although these are not asymptotically exact values. For low packing fraction, the compressibility factor is represented by the virial expansion

$$Z(f) \simeq 1 + 2f + 3.12802 f^2 , \qquad (7.32)$$

and much more terms are available [16,42]. Using the same combination of methods as above, we find with $f_c = 1$, the following estimates.

The method of subsection 7.2.1 finds the critical index $\alpha = 1.69403$.

The method of subsection 7.2.2 gives the critical index $\alpha = 1.77302$.

The technique of subsection 7.2.3 brings the value $\alpha = 1.66355$.

152 Mechanics and Physics of Structured Media

Minimal difference condition of subsection 7.2.4 gives the index $\alpha = 2.19518$.

Minimal derivative condition of subsection 7.2.5 leads to $\alpha = 2.68716$. Therefore our prediction for the critical index is their average

$$\alpha = 2.003 \pm 0.392\,.$$

Discarding the lowest and highest extremes, we find the three solutions centered around the value of $\alpha = 1.887 \pm 0.248$.

7.5 Sedimentation coefficient of rigid spheres

Sedimentation is a fundamental problem of studying how a suspension moves under gravity. The dispersion considered is build of small rigid spheres with random positions falling through Newtonian fluid under gravity. The mixture of solid particles and the fluid in a container is assumed to be homogeneous. The particles settle out under gravity at a rate depending, in particular, on concentration controlled by the hydrodynamic interactions between particles.

The basic quantity of interest is the sedimentation velocity U, which is the averaged velocity of suspended particles, measured with respect to the velocity U_0 with which a single particle would move in the suspending fluid under the given force field in the absence of any other particles. This ratio is termed the collective mobility or sedimentation coefficient.

The dependence of the collective mobility at low packing fractions is similar to the single-particle mobility, but quickly decreases at high packing fractions. The problem of sedimentation reminds that of a flow in a porous medium, although their relation is not simple, because the physics of particle interactions are rather different. More details on the physics of sedimentation can be found in the paper by Batchelor [6]. The dimensionless sedimentation velocity $u \equiv U/U_0$ is considered as a function of the packing fraction f. This velocity exhibits the critical behavior

$$u(f) \propto (1 - f)^{\alpha} \qquad (f \to 1 - 0) \tag{7.33}$$

at the pole $f_c = 1$. The critical index, however, has been defined differently by different authors. Thus Batchelor [6] gives $\alpha = 5$. While other authors [8, 9,34,39] suggest $\alpha = 3$. Below we find the critical index by extrapolating the expansion derived in [13],

$$u(f) \simeq 1 - 6.546\,f + 21.918\,f^2, \qquad f \to 0\,. \tag{7.34}$$

Before, in [28] we estimated the critical index $\alpha = 3.3049 \pm 0.26$.

Let us again employ the same various special resummation technique as above and estimate the critical index α. Mind that the strategy consists in employing various conditions on the fixed point in order to compensate for a shortage of the coefficients.

The method of subsection 7.2.1 gives the critical index $\alpha = 2.966$.

The method of subsection 7.2.2 produces unreasonably large value and is discarded.

The method of subsection 7.2.3 brings the critical index $\alpha = 2.863$.

Minimal difference condition of subsection 7.2.4 gives the index $\alpha = 2.577$.

Minimal derivative condition of subsection 7.2.5 leads to the critical index $\alpha = 3.566$.

Let us also find the critical index by some other method, say the $DLog$ Padé method as explained in the preceding section, resulting in $\alpha = 3.044$.

There are now five estimates for the critical index α:

$$2.577; \quad 2.863; \quad 2.966; \quad 3.044; \quad 3.566.$$

We find that the solutions are neatly centered around their average, giving the evalue of 3.003 ± 0.323, very close to the expected number. The current estimate is more certain about he value of index than our previous result from [28] already quuoted above. Discarding the lowest and highest extremes, we find three solutions neatly centered around the value of 2.957 ± 0.087. In reverse, when the critical index is known one can estimate the value of a_2 independently on the method of [13], as was accomplished in [19]. Such estimated value appeared to be close to the number given in [13].

7.6 Susceptibility of 2D Ising model

Consider spin-1/2 Ising model characterized by the Hamiltonian [12]

$$\hat{H} = -\frac{J}{2} \sum_{\langle ij \rangle} s_i^z s_j^z \quad \left(s_j^z \equiv \frac{S_j^z}{S} \right) \tag{7.35}$$

on a lattice, with the ferromagnetic interaction of nearest neighbors J, for spins $S_j^z = \pm 1/2$. The dimensionless interaction parameter is defined as

$$g \equiv \frac{J}{k_B T} . \tag{7.36}$$

A high-temperature expansion of the susceptibility χ on the square lattice, in the powers of inverse temperature g could be obtained in rather high orders [12]. Its starting terms are given as follows,

$$\chi(g) = 1 + a_1 g + a_2 g^2 + O(g^3),$$
$$a_1 = 4, \quad a_2 = 12, \tag{7.37}$$

154 Mechanics and Physics of Structured Media

and $a_3 = \frac{104}{3}$. It is expected [7], that in the vicinity of the 2D-threshold

$$g_c = \left(\frac{2}{\log(1 + \sqrt{2})}\right)^{-1} \approx 0.440687,$$

the susceptibility diverges as

$$\chi(g) \sim (g_c - g)^{-\gamma},$$

with critical index $\gamma = \frac{7}{4}$.

Let us again employ the same various special resummation technique as above and estimate the index γ. The strategy remains the same and consists in employing various conditions on the fixed point in order to compensate for a shortage of the coefficients.

The method of subsection 7.2.1 finds the critical index $\gamma = 2$.

The method of subsection 7.2.2 gives the same value for the critical index $\gamma = 2$.

The method of subsection 7.2.3 brings the critical index $\gamma = -1.65177$.

Minimal difference condition of subsection 7.2.4 gives the critical index $\gamma = 1.74659$. In this case we find that the difference between approximated indices possesses minimum at the very reasonable value of threshold $g_0 = 0.466181$. The index is found from the first order approximant $\chi_1^*(g) = \left(1 + \frac{1.06764g}{0.466181 - g}\right)^{1.74659}$.

Minimal derivative condition of subsection 7.2.5 leads to the critical index $\gamma = 1.84618$.

Let us also find the critical index by some other method. To this end let us apply the transformation, $z = \frac{g}{g_c - g} \Leftrightarrow g = \frac{zg_c}{z+1}$ to the original series and calculate the index by the $DLog$ Padé method as explained in the Introduction, resulting in $\gamma = 1.57581$.

There are now five different estimates for the critical index

$$1.57581; \quad 1.65177; \quad 1.74659; \quad 1.84618; \quad 2.$$

Such "framing" of the solution is typical for the methodology based on a short series. Average and margins of error could be found, so that $\gamma = 1.76 \pm 0.15$. Discarding the two most extreme solutions we arrive to an almost perfect estimate, $\gamma = 1.748 \pm 0.111$. Although in this particular case there is an practically exact hit on the solution, such situation should be considered as exception. The current estimates are much better than our previous result $\gamma = 1.923 \pm 0.077$ [28].

Critical index for conductivity, elasticity, superconductivity Chapter | 7 **155**

7.7 Susceptibility of three-dimensional Ising model. Root approximants of higher orders

Consider the three-dimensional Ising model on a simple cubic lattice, with the ferromagnetic interaction J of nearest neighbors, for spins $S_j^z = \pm 1/2$. The dimensionless interaction parameter is defined as $g \equiv \frac{J}{k_B T}$. The susceptibility is expected to diverge at a critical point $g_c = 0.22165463(8)$ [17,35], as

$$\chi(g) \simeq B(g_c - g)^{-\gamma} , \qquad (7.38)$$

with the critical index $\gamma = 1.237 - 1.244$ [47,53,57]. The weak-interaction or high-temperature expansion of the susceptibility yields [12] the series in powers of g,

$$\chi(g) = 1 + 6g + 30g^2 + 148g^3 + 706g^4 + \frac{16804}{6}g^5 + \frac{42760}{3}g^6$$
$$+ \frac{7744136}{105}g^7 + \frac{35975026}{105}g^8 + \dots \qquad (g \to 0) . \qquad (7.39)$$

Another thermodynamic characteristic, $\chi_0^{(2)}$, the second derivative of the susceptibility with respect to an external (weak) magnetic field,

$$\chi_0^{(2)} = 1 + 24g + 318g^2 + 3232g^3 + 27946g^4 + \frac{1087216g^5}{5} + \frac{23504732g^6}{15}$$
$$+ \frac{1119734144g^7}{105} + \frac{7285355386g^8}{105} + \dots \qquad (g \to 0) . \qquad (7.40)$$

It is expected to diverge at critical point as

$$\chi_0^{(2)} \sim (g_c - g)^{-\gamma_4}$$

with the critical index $\gamma_4 = \gamma + 2\Delta$, where is Δ is the so-called gap exponent. From the estimates of Δ, one can find yet different critical indices.

E.g., in three-dimensional case the critical index for correlation length $\nu = \frac{2\Delta - \gamma}{3}$. While the critical index for magnetization $\beta = \Delta - \gamma$. Guttmann found that $\gamma_4 = 4.37$, and estimated $\Delta = 1.565 \pm 0.004$ [33].

7.7.1 Comment on unbiased estimates. Iterated roots

How to find the so-called unbiased estimates for the critical index when the exact threshold is not known? As was explained above, one can establish an approximate, analytical connection between the threshold and index, using the low-order iterated root approximants. After that the threshold is not needed and there is only single equation to be solved for the critical index as unknown.

156 Mechanics and Physics of Structured Media

The simplest expression relating the threshold

$$x_{c,1} = -\frac{s}{a_1},$$

is found form the first-order root approximant, see subsection 7.2.1. The sequence of approximate solutions for the control parameter s_i, leading to the estimates for the critical index, can be obtained by requiring

$$\left(\mathcal{R}_i^* \left(x_{c,1} \left(s_i\right)\right)\right)^{1/s_i} = 0, \quad i = 2, ...N, \tag{7.41}$$

where N is the maximal number of terms form the expansion employed in the construction of the iterated root.

Consider the critical behavior of the ratio $\frac{\chi_0^{(2)}(g)}{\chi(g)} \sim (g_c - g)^{-2\Delta}$, with $N = 10$.

The following subsequence of uniquely defined approximate values can be found from (7.41):

$$2\Delta_2 = 9, \ 2\Delta_3 = 3.40232, \ 2\Delta_5 = 3.26098, \ 2\Delta_7 = 3.18217, \ 2\Delta_9 = 3.13208.$$

The instances not shown appear to deliver only a minimum to the LHS of (7.41), with corresponding estimates being vastly inferior.

The second-order approximant also allows for an explicit relation between threshold and control parameter (index)

$$x_{c,2}(s) = \frac{\sqrt{2a_1^2 s^3 - a_1^2 s^2 - 4a_2 s^3} + a_1 s}{a_1^2 s - a_1^2 - 2a_2 s},$$

as explained in subsection 7.2.2.

Yet different sequence of approximate solutions for s_i can be obtained by requiring

$$\left(\mathcal{R}_i^* \left(x_{c,2} \left(s_i\right)\right)\right)^{1/s_i} = 0, \quad i = 3, ..., N. \tag{7.42}$$

Consider the critical behavior of the susceptibility $\chi(g) \sim (g_c - g)^{-\gamma}$, with $N = 10$. The following subsequence of uniquely defined approximate values can be found from (7.42):

$$\gamma_4 = 1.34589, \ \gamma_6 = 1.27884, \ \gamma_8 = 1.23935.$$

The instances not shown appear to be either nonuniquely defined, or deliver just a minimum to the LHS of (7.42). The latter case corresponds to $i = 9$. Such disappearance of zero-solutions serves as a natural stopping signal.

Both results for $2\Delta = 3.13208$ and $\gamma = 1.23935$, appear to be good. One can reasonably accurately estimate more familiar critical indices $\nu = 0.630911$ and $\beta = 0.326693$.

7.8 3D Superconductivity critical index of random composite

It was demonstrated in [19] that the classical Jeffrey formula contains the wrong f^2 term. The proper expansion in volume fraction of inclusions f for the random composite with superconducting (perfectly conducting) spherical inclusions was obtained in [19]. The terms f^2 and f^3 are written explicitly. In particular, the f^3 term depends on the deterministic or random locations of inclusions. In the limiting case of a perfectly conducting spherical inclusions, the effective conductivity σ_e is expected to tend to infinity as a power-law, as the concentration of inclusions f tends to f_c, the maximal value in 3D.

General methodology of [19] had been applied to the numerical estimation of the effective conductivity of random macroscopically isotropic random composites. For samples generation the Random Sequential Adsorption (RSA) protocol was employed. The consecutive objects/spheres were placed randomly in the cell, rejecting those that overlap with previously adsorbed one. For macroscopically isotropic composites the expansion for scalar effective conductivity was found at small f as follows

$$\sigma_e = 1 + 3f + 3f^2 + 4.80654f^3 + O(f^{\frac{10}{3}}). \tag{7.43}$$

Jeffrey [37] in this case gives a different result, $4.51f^2$. It was concluded in [19], that it is certainly methodological, not a computational error, related to an intuitive physical treatment of the conditionally convergent integral discussed in [37]. It is possible to extrapolate (7.43) to all f, and find the critical index and amplitude from the truncated series.

Assume that the threshold is known and corresponds to the random close packing (RCP), $f_c = 0.637$. Then in the vicinity of RCP it is assumed that

$$\sigma_e \simeq A(f_c - f)^{-s}, \tag{7.44}$$

where the superconductivity critical index s is expected to have the value of 0.73 ± 0.01 [14]. There is also a slightly larger estimate, $s \approx 0.76$ [10].

Let us estimate the value of s based on asymptotic information encapsulated in (7.43). There is a possibility to obtain for σ_e, the simplest factor approximant with fixed position of singularity and floating critical index, from the requirement of asymptotic equivalence with (7.43). I.e., we obtain the following result

$$\sigma_e^*(f) = \mathcal{F}_4^*(f) = \frac{(2.48123f + 1)^{0.766996}}{(1 - 1.56986f)^{0.698732}}. \tag{7.45}$$

Equation (7.45) suggests the value of 0.7 for the superconductivity critical index s.

Assume that in the vicinity of threshold,

$$\sigma(f) \sim (0.637 - f)^{-(1+s')},$$

158 Mechanics and Physics of Structured Media

with unity to be expected from the usual contribution from the radial distribution function at the particles contact $G(2, f)$, [11,41], and the value of s' coming from the particle interactions in the composite. One can suggest a simple root approximant for the effective conductivity,

$$r(z) = 1 + b_1 z (1 + b_2 z)^{s'}, \quad z = \frac{f}{f_c - f},$$

which explicitly takes unity-contribution into account. After imposing the asymptotic equivalence with the truncated series, we find all three parameters with the final result

$$\sigma_e^*(f) = 1 + \frac{1.911 f}{\left(\frac{0.709259 f + 0.637}{0.637 - f}\right)^{0.212373} (0.637 - f)}. \tag{7.46}$$

The expression (7.46) allows us to estimate $s' \approx -0.212$. Total value of the critical index $s \approx 0.788$, is still close enough to the expected values.

Let us employ a more systematic methodology used to construct the table of indices. The method of construction is explained below. It is based on considering iterated root approximants as functions of the critical index by itself. The index is to be found by imposing optimization condition in the form of minimal difference on critical amplitudes [28].

For convenience let us again the variable

$$z(f) = \frac{f}{f_c - f}, \quad f(z) = \frac{f_c z}{1 + z},$$

and construct the iterated root approximants [25],

$$\mathcal{R}_k^*(z) = \left(\left(\left((1 + \mathcal{P}_1 z)^2 + \mathcal{P}_2 z^2\right)^{3/2} + \mathcal{P}_3 z^3\right)^{4/3} + \ldots + \mathcal{P}_k z^k\right)^{s_k/k}, \tag{7.47}$$

defining the parameters $\mathcal{P}_j$ from the asymptotic equivalence with the truncated series for the effective conductivity. This gives the large-z asymptotic form

$$\mathcal{R}_k^*(z) \simeq A_k z^{s_k} \quad (z \to \infty), \tag{7.48}$$

where the amplitudes $A_k = A_k(s_k)$ are

$$A_k = \left(\left((\mathcal{P}_1^2 + \mathcal{P}_2)^{3/2} + \mathcal{P}_3\right)^{4/3} + \ldots + \mathcal{P}_k\right)^{s_k/k}. \tag{7.49}$$

In order to define the critical index s_k, we analyze the differences [19,22,28]

$$\Delta_{kn}(s_k) = A_k(s_k) - A_n(s_k). \tag{7.50}$$

Critical index for conductivity, elasticity, superconductivity **Chapter | 7 159**

TABLE 7.1 Critical indices for the superconductivity s_k obtained from the optimization conditions $\Delta_{kn}(s_k) = 0$.

s_k	$\Delta_{k,k+1}(s_k) = 0$	$\Delta_{k3}(s_k) = 0$
s_1	0.725	0.721
s_2	0.715	0.715

Composing the sequences $\Delta_{kn} = 0$, we find the related approximate values s_k for the critical indices. It is possible to investigate different sequences of the conditions $\Delta_{kn} = 0$, such as $\Delta_{k,k+1} = 0$ and $\Delta_{k3} = 0$, with $k = 1, 2$.

The series (7.43) allows to get only three independent estimates for the critical index s, see Table 7.1. All three are fairly close to 0.72 and to the expected value of 0.73.

For possible applications, one can simply adjust the iterated root approximants to the plausible value of 0.73 for the critical exponent,

$$
\begin{aligned}
\mathcal{R}_2^*(f) &= \left(\frac{2.56706 f^2 + 2.06109 f + 0.405769}{(0.637-f)^2} \right)^{0.365}, \\
\mathcal{R}_3^*(f) &= \left(\left(\frac{2.56706 f^2 + 2.06109 f + 0.405769}{(0.637-f)^2} \right)^{3/2} - \frac{0.372504 f^3}{(0.637-f)^3} \right)^{0.243333}.
\end{aligned}
\tag{7.51}
$$

The two expressions are very close numerically, and the critical amplitude can be found from (7.51). From the former approximant we obtain $A \approx 1.449$, and from the latter $A \approx 1.456$, giving close results. The 4th order coefficient found from $\mathcal{R}_3^*(f)$ equals 7.48.

Application of various techniques brings again very close results, especially for the critical amplitude. For instance, one can simply extract the known singularity first, and then apply to the remainder the Padé technique, obtaining the following approximant

$$
\sigma_e^*(f) = \frac{22.1332(f + 0.439227)}{(1 - 1.56986 f)^{0.73} \left(-f^2 + 4.1095 f + 9.72153 \right)},
\tag{7.52}
$$

which brings the value of $A \approx 1.44$ close to other estimates, but the projected 4th order coefficient appears to be different from the estimate found above and equals 6.59.

7.9 Effective conductivity of graphene-type composites

We extend here the study of [20], to a larger number of particles/larger samples.

Consider a regular honeycomb array of perfectly conducting (superconducting) disks as their volume fraction $f \to f_c$, and effective conductivity of the array goes to infinity, as a continuum analog of graphene. Mind that graphene

160 Mechanics and Physics of Structured Media

consists simply of carbon atoms tightly packed into the honeycomb lattice, with lattice as a whole having very good electric conductivity, at par with silver and copper.

Let us now study the role of vacancy defects. In graphene, even with some vacancies, one would still expect good conductivity properties.

But what will happen if some vacancies are introduced at random within the continuum model? The threshold f_c, or maximal possible volume fraction of disks in regular case, will be lowered to the value $f_c(p)$, where the parameter p is proportional to the number of vacancies per cell and $f_c(0) = f_c$. The conductivity is expected to diverge with some critical index $s(p)$. The value of the index could become unconventional, $s(p) < 1/2$, where $s(0) = 1/2$ corresponds to the regular array of highly conducting (superconducting) disks. The quantity $\frac{f_c - f_c(p)}{f_c}$ gives the fraction of atoms removed from the lattice, and for graphene it only makes sense to consider the limit $f \rightarrow f_c(p)$. The expansion for small f is possible to obtain though, by means of some well understood technique described in [20], leading to polynomials representing effective conductivity at small and moderate f. By some careful extrapolation to $f_c(p)$ we can get back to the graphene with vacancies. But for the problem of regular honeycomb array of perfectly conducting (superconducting) disks considered by itself, concentration f has physical meaning even far away from the threshold, and the critical index could quantify the behavior of conductivity as the threshold is approached.

The complete methodology of deriving the effective conductivity of finite samples is described in [20]. The original complete honeycomb location contains $864 = 24 \times 36$ disks. We delete randomly $24, 24 \times 2, 24 \times 3, ..., 24 \times 35$ disks, and take randomly 10 samples for every structure. The mean values of 35 polynomials for every fixed $p = 1, 2, ..., 36$ in this case up to $O(f^7)$ have the form presented in Appendix 7.B.

Further statistical analysis of the polynomials (7.B.1) can be performed following [20].

Is it possible to extrapolate polynomials given by the set (7.B.1) to all f, and find the critical index from the truncated series. Consider the case when the threshold is known and given by the formula (7.54).

In the vicinity of threshold we can assume that

$$\sigma_e(f, p) \sim (f_c(p) - f)^{-s(p)}, \qquad (7.53)$$

where the superconductivity critical index s is expected to depend on p, and have the value of $1/2$ as $p = 0$. Below, we are going to calculate s by extrapolating the small-f polynomials to $f \rightarrow f_c(p)$.

The threshold $f_c(p)$ also depends on p. It is given by the compact formula

$$f_c(p) = \frac{\pi}{3\sqrt{3}} \frac{p_{max} - p}{p_{max}}, \qquad (7.54)$$

Critical index for conductivity, elasticity, superconductivity Chapter | 7 **161**

TABLE 7.2 Critical indices for the superconductivity $s_k(p)$ as $p = 0$, obtained from the optimization conditions $\Delta_{kn}(s_k) = 0$. The two sequences demonstrate numerical convergence to the value $s(0) = 0.519$.

s_k	$\Delta_{k,k+1}(s_k) = 0$	$\Delta_{k6}(s_k) = 0$
s_1	0.6046	0.54501
s_2	0.53667	0.52341
s_3	0.51652	0.51633
s_4	0.5141	0.51619
s_5	0.51911	0.51911

where $f_c(0) = \frac{\pi}{3\sqrt{3}}$ is the maximum volume fraction attainable for the honeycomb lattice, and the maximum value of p is denoted as p_{max}. We have $p_{max} = 36$.

Since the threshold is known, we would first apply the following transformation,

$$z = \frac{f}{f_c(p) - f} \Leftrightarrow f = \frac{z f_c(p)}{z + 1}$$

to the original series. The transformation maps the segment to a half-line.

The method of construction of the table for the critical index was explained in detail in Section 14 of [28] and in the Introduction to the book [19]. It is based on considering iterated root approximants as functions of the critical index by itself. The index is to be found by imposing optimization condition in the form of minimal difference $\Delta_{kn}(s_k) = 0$ (where k stands for the order of approximation), imposed on the critical amplitudes [19,28].

It is instructive to study first the regular case with no defects. The small-f polynomial has the following form:

$$\sigma_e \simeq 1 + 2f + 2f^2 + 2f^3 + 4.14933 f^4 + 6.29865 f^5 + 8.44798 f^6. \quad (7.55)$$

The truncated series (7.55) allows us to get several estimates for the critical index s, as presented in Table 7.2. The sequences demonstrate good numerical convergence to the value $s(0) \approx 0.519$, with plausibly best result $s(0) \approx 0.514$.

Applying the *DLog* Padé technique for $p = 0$ [20], and imposing the asymptotic equivalence with corresponding truncated series, one readily calculates $s(0) = 0.4989$ with four nontrivial terms from (7.55) involved, and $s(0) = 0.65355$, with all six terms employed. The 6th order approximant gives a compact expression for the effective conductivity with practically correct index [20]. The method of optimization gives more consistent results, slightly overestimating the exact value of 0.5.

162 Mechanics and Physics of Structured Media

TABLE 7.3 Critical indices for the superconductivity $s_k(p)$ as $p = 2$, obtained from the optimization conditions $\Delta_{kn}(s_k) = 0$. The two sequences demonstrate numerical convergence to the value $s(0) = 0.48138$.

s_k	$\Delta_{k,k+1}(s_k) = 0$	$\Delta_{k6}(s_k) = 0$
s_1	0.57101	0.51168
s_2	0.50704	0.4912
s_3	0.48614	0.48323
s_4	0.48109	0.48122
s_5	0.48138	0.48138

TABLE 7.4 Critical indices for the superconductivity $s_k(p)$ as $p = 8$, obtained from the optimization conditions $\Delta_{kn}(s_k) = 0$. The starting two sequences demonstrate numerical convergence to the value $s(0) \approx 0.38$.

s_k	$\Delta_{k,k+1}(s_k) = 0$	$\Delta_{k6}(s_k) = 0$
s_1	0.47024	0.41373
s_2	0.41724	0.0.39665
s_3	0.39555	0.38771
s_4	0.38596	0.38301
s_5	0.37964	0.37964

Consider the case of $p = 2$. The truncated series (7.B.1) allows us to get several estimates for the critical index s, as presented in Table 7.3. The sequences demonstrate good numerical convergence to the value $s_5(2) \approx 0.48138$.

Consider also the case of $p = 8$. The truncated series (7.B.1) allows us to get several estimates for the critical index s, as presented in Table 7.4. The sequences demonstrate good numerical convergence to the value $s_5(8) \approx 0.38$.

Another way to calculate the index follows from the general idea of [20,25]. Let one think about a generalization of polynomials (7.B.1), i.e., the transition formula from the regular honeycomb array to the array with vacancies. We expect to obtain a dependence of the critical index on the degree of randomness introduced by vacancies. For "zero"-randomness the formula should lead to the regular honeycomb array. For "maximum" possible concentration of vacancies it leads to the vanishingly small index, the case discussed below. All cases with intermediate degrees of randomness are expected to fall in between the two cases.

Critical index for conductivity, elasticity, superconductivity Chapter | 7 **163**

Let us select the initial approximation to be corrected, as describing a regular honeycomb array of inclusions, namely

$$\sigma_0(p) = \left(1 - \frac{f}{f_c(p)}\right)^{-1/2}. \qquad (7.56)$$

This formula incorporates the critical index $1/2$ of the regular array as the starting approximation, and the threshold for the honeycomb array. Let us divide the original series (7.B.1) by $\sigma_0(p)$, thus extracting the part corresponding to the random vacancies effects only. Then express the newly found series in terms of variable z, then apply $DLog$ transformation and call the transformed series $L(z, p)$. Applying the Padé approximants $P_{n,n+1}$ to the transformed series $L(z, p)$ one can obtain the sequence of corrected approximations to the critical index,

$$s_n(p) = \frac{1}{2} + \lim_{z \to \infty} (z\, P_{n,n+1}(z, p)), \qquad (7.57)$$

where $P_{n,n+1}(z, p)$ stands for the Padé approximants [2,29], constructed for the series $L(z, p)$ with such a power at infinity that defines p-dependent correction to $s(0) = 1/2$.

Applying the corrected $DLog$ Padé technique [25], say, for $p = 3$ and corresponding asymptotic series, one calculates

$$s_1(3) = 0.48924, \quad s_2(3) = 0.48808.$$

We can also obtain after integration [25], the following compact result for the approximant for the effective conductivity valid for all f,

$$\sigma_6^*(f, 3) =$$

$$\frac{0.418952 e^{-0.817205 \tan^{-1}\left(1.46637 - \frac{1.81741}{0.554216 - f}\right)}}{\sqrt{1 - 1.80435 f}\left(-\frac{0.554216}{f - 0.554216} - 0.781666\right)^{0.0000636243}\left(\frac{f(f + 0.583499) + 0.417947}{(0.554216 - f)^2}\right)^{0.00593016}}, \qquad (7.58)$$

employing all six terms from the expansion at small f. The approximation suggests the final value of 0.48808 for the superconductivity critical index. Similar approach can be applied for all p. E.g., for $p = 12$, we find

$$s_1(12) = 0.30343, \quad s_2(12) = 0.2491,$$

and

$$\sigma_6^*(f, 12) = \frac{0.511747 e^{-0.72912 \tan^{-1}\left(1.36506 - \frac{1.43819}{0.403067 - f}\right)}}{\sqrt{1 - 2.48098 f}\left(1 - \frac{0.391343}{1.f - 0.403067}\right)^{0.233618}\left(\frac{f(f + 0.565123) + 0.332114}{(0.403067 - f)^2}\right)^{0.00865001}} \cdot$$

$$(7.59)$$

164 Mechanics and Physics of Structured Media

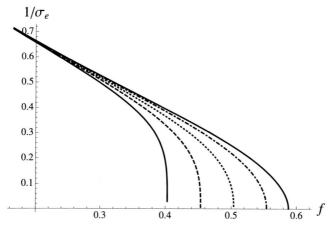

FIGURE 7.1 The case $p_{max} = 36$. The dependence of inverse effective conductivity on volume fraction f is shown for a few values of discrete parameter p: $p = 1$ (solid), $p = 3$ (dot-dashed), $p = 6$ (dotted), $p = 9$ (dashed), $p = 12$ (solid, leftmost). The steepening of curves is clearly seen. It is reflected in the critical index decreasing with more vacancies being randomly placed on a honeycomb lattice.

The dependence of inverse effective conductivity on volume fraction f is shown in Fig. 7.1 for a few values of the discrete parameter p: $p = 1$ (solid), $p = 3$ (dot-dashed), $p = 6$ (dotted), $p = 9$ (dashed), $p = 12$ (solid, leftmost). The steepening of curves is clearly seen. It is reflected in the critical index decrease with more vacancies being randomly placed on a honeycomb lattice. The dependence of s on the degree of disorder measured in steps of randomization managed through random placement of vacancies on a honeycomb lattice, is displayed in Fig. 7.2. The results given by optimization procedure are shown with squares. The results found by applying the corrected $DLog$ Padé method are shown with disks. We see that up to $p = 10$ both methods give rather close results for the index. They also agree qualitatively for larger p, up to $p \to p_{max}$, where a very small, tending to zero, index is anticipated. The optimization method of [28] seems to be the most consistent, since it gives $0 < s(p) < 1/2$. Previously, in [25] we discussed the case of random conductors with $1/2 < s(p) < 1.3$, with conservation of total number of particles with degree of randomness increasing. In the present setup the number of particles in the cell is not conserved with increasing randomness.

Corresponding curves for the inverse effective conductivity demonstrate steepening effect with increasing fraction of vacancies, with threshold simultaneously moving to smaller values. The index by itself is nonuniversal, smaller than $1/2$, and decreases with increasing concentration of vacancies. We have also conducted calculations for $p_{max} = 12$, with qualitatively similar conclusions. Moreover, there is a good qualitative agreement with the cases of

Critical index for conductivity, elasticity, superconductivity Chapter | 7 **165**

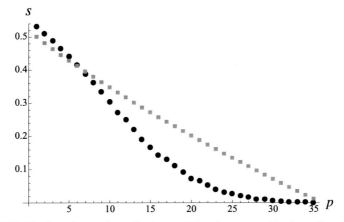

FIGURE 7.2 The dependence of s on the degree of disorder measured in steps of randomization through randomly placing vacancies on a honeycomb lattice, is displayed. The results given by optimization procedure are shown with squares. Corrected $DLog$ Padé results are shown with disks.

$p_{max} = 12$, $p_{max} = 24$ [20], when approximately equivalent vacancies fractions are considered.

We conclude that the honeycomb array of highly conducting disks easily adapts to randomly placed vacancies, by adjusting (lowering) the threshold and critical index, effectively remaining quasiregular. One can think that analogous graphene can also withstand such defects and retain its good conductivity.

The dependence of the critical index on the degree of disorder resembles the behavior of critical index in the so-called eight-vertex model and equivalent to it, Ising model with four-spin interaction [7], where the critical index similarly depends on temperature [49]. The two models defy the universality hypothesis in regard to the critical indices of many equilibrium physical systems [7].

7.10 Expansion factor of three-dimensional polymer chain

A polymer is a chain of molecular monomers attached at random angles to the end of the chain. A monomer cannot be attached at an already occupied spot. For many monomers, the molecular chain will be modeled as a random walk, which cannot cross its own path. This is applicable to macromolecules such as proteins or DNA, but here the monomers should be treated as not identical.

Polymer can serve as an example of self-organized criticality, since it demonstrates power-law feature without any external tuning. The expansion (swelling) factor of a 3D polymer chain, as a function of the dimensionless coupling parameter g, can be expressed with sufficient accuracy by the phenomenological equation [44,45]

$$\Upsilon(g) = (1 + 7.52g + 11.06g^2)^{0.1772}. \tag{7.60}$$

166 Mechanics and Physics of Structured Media

At large g, this gives

$$\Upsilon(g) \simeq 1.531 g^{\alpha}, \text{ as } g \to \infty, \tag{7.61}$$

with the critical index at infinity $\alpha = 0.3544$. In the theory of polymers, one also considers the critical index

$$\nu \equiv \frac{1}{2}\left(1 + \frac{\alpha}{2}\right), \tag{7.62}$$

which characterizes power-law increase

$$\sqrt{\langle R^2 \rangle} \sim N^{\nu},$$

of the typical chain radius $\sqrt{\langle R^2 \rangle}$ with number of monomers N composing the chain, so that $\nu = 0.5886$. Other numerical calculations [40] give slightly lower results, $\nu = 0.5877$, or 0.5876 [15]. In [28], we found a rather good estimate $\nu = 0.5814 \pm 0.006$.

At small g, perturbation theory yields [44,45] the following expansion

$$\Upsilon(g) \simeq 1 + \frac{4}{3}g - 2.07539g^2 + 6.29688g^3 - 25.0573g^4$$
$$+ 116.135g^5 - 594.717g^6, \tag{7.63}$$

as $g \to 0$. Let us construct a simplified root approximants, in the form often employed in the theory of polymers [47],

$$\mathcal{R}_k^*(x) = \left(1 + A_1 x + A_2 x^2 + A_3 x^3 + \ldots + A_k x^k\right)^{\alpha_k / k}, \tag{7.64}$$

defining the parameters A_j from the asymptotic equivalence with the truncated series (7.63). The same optimization technique as already employed in the case of superconductivity index, gives reasonable results, shown in Table 7.5. The second column seems the most reasonable, with the best estimate is achieved with $\alpha \approx 0.3503$, giving $\nu \approx 0.5876$. The $DLog$ Padé method gives smaller value, $\nu = 0.5869$.

One can easily reconstruct the 4th-order approximant

$$R_4^*(g) = \left(144.155g^4 + 185.44g^3 + 82.067g^2 + 15.2262g + 1\right)^{0.0875686}$$

with the former value of the critical index. It can be compared to the simple reference formula of Muthukumar-Nickel (7.60). The two formulas deviate from each other by not more than 0.3%. The formula also reconstructs a_5 with accuracy of 0.7% and a_6 with accuracy of 2.9%.

Critical index for conductivity, elasticity, superconductivity Chapter | 7 **167**

TABLE 7.5 Critical indices for the polymer, α_k, obtained from the optimization conditions $\Delta_{kn}(\alpha_k) = 0$. The starting two sequences demonstrate numerical convergence to the value of $\alpha \approx 0.3439$, leading to $\nu \approx 0.586$, in startling agreement with experimental results [18].

α_k	$\Delta_{k,k+1}(\alpha_k) = 0$	$\Delta_{k6}(\alpha_k) = 0$
α_1	0.2999	0.3277
α_2	0.3594	0.3495
α_3	0.3399	0.3456
α_4	0.3597	0.3503
α_5	0.3439	0.3439

7.11 Concluding remarks

All techniques for calculation of the critical index appear powerful, but there is no such thing as the best method for all problems at once. The very task consists in finding the best method for each concrete physical problem. For instance, in [28] we established that for critical problems based on the coefficients a_1, a_2, the technique of optimized roots has clear advantages. On the other hand, various indices for 2D Ising model, based on a long series from [12], can be better estimated by $DLog$ Padé technique. The examples brought up in [23,28] illustrate a complete breakdown of the $DLog$ Padé technique as documented in Appendix 7.A. In such cases other methods including optimized roots demonstrate very good performance. In the case of graphene analog the two methods complement each other.

Various methods of calculation are applied and results presented for critical indices of several key models of conductivity, superconductivity, elasticity of random media, Ising model in two and three dimensions, sedimentation, polymer coils. The effective conductivity of graphene-type composites with vacancies is considered as well.

Some models appear to be amenable to an almost identical methodology, given similar input. E.g., 3D conductivity, 3D elasticity, 2D Ising model could be treated formally by the same methodology with close results for critical indices.

Accurate calculations with short series are possible in some important cases, because the higher-order coefficients are somewhat redundant and could be estimated from the low-order coefficients and threshold. The redundancy allows us to construct compact formulas for all region, not only calculate the critical index. For instance, by combining the high-temperature expansions with Onsager's form of exact solution we derive novel approximate expressions for the susceptibility of the 2D and 3D Ising model, while the quality of approximations is controlled by exploiting the redundancy [24].

168 Mechanics and Physics of Structured Media

Factor approximants are also amenable to optimization [24], as well as root approximants [19]. But the techniques of accelerating convergence turn out to be different. In the former case we opt for introduction of the unknown additional terms to the series, with the goal to find the unknown coefficient by imposing optimization conditions on the sought critical index. In the latter case the critical index by itself serves as a control parameter to calculate from optimization condition.

Appendix 7.A   Failure of the $DLog$ Padé method

The problems discussed in Appendix 7.A can be also solved by application of a hybrid method of corrected Padé approximants [19,23,27], which combine multiplicatively roots with diagonal Padé approximants. But in the cases discussed above, optimized roots converge much faster.

Consider the correlation function of the Gaussian polymer [32]. The expression is well-known. It is called the Debye-Huckel function. For convenience of presentation we consider inverse of the original expression [32],

$$\Phi(x) = \frac{e^x x^2}{2e^x(x-1)+2}.$$

The expansion for small $x > 0$, is shown below in low-orders,

$$\Phi(x) = 1 + \frac{x}{3} + \frac{x^2}{36} - \frac{x^3}{540} - \frac{x^4}{6480} + O(x^5).$$

For large x the behavior is known as well,

$$\Phi(x) \sim x, \text{ as } x \to \infty.$$

And again, the direct $DLog$ Padé scheme fails. It leads to the sequence of values oscillating between the values of 0 and 2, as shown in Fig. 7.3. The exact value of $\alpha = 1$ for the index is shown as well.

Let us calculate the critical index by applying the same method as in the preceding section 7.10. In the current problem we observed the best, monotonous convergence for the sequence of $\Delta_{k16}(\alpha_k) = 0$, with $k = 1, 2...15$, as shown below,

$$\alpha_1 = 1.26, \ \alpha_2 = 1.52, \ \alpha_3 = 1.522, \ \alpha_4 = 1.28, \ \alpha_5 = 1.05, \ \alpha_6 = 1.04,$$

$$\alpha_7 = 1.036, \ \alpha_8 = 1.03, \ \alpha_9 = 1.025, \ \alpha_{10} = 1.021, \ \alpha_{11} = 1.017, \ \alpha_{12} = 1.014,$$

$$\alpha_{13} = 1.011, \ \alpha_{14} = 1.01, \ \alpha_{15} = 1.006.$$

The final estimate to be deduced from such sequence is $\alpha = 1.006$, brings the error of only 0.59%.

Critical index for conductivity, elasticity, superconductivity Chapter | 7 **169**

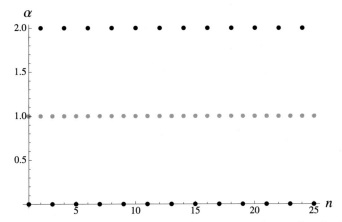

FIGURE 7.3 Critical index for inverse Debye function. The approximate values for the critical index α are shown with varying approximation numbers n. The results are obtained by $DLog$ Padé method. The exact value for the index is shown with gray circles.

The problem of $N = 4$ Super Yang-Mills Circular Wilson Loop [3], can be solved in closed form. For convenience of presentation we consider the inverse of the original expression. After simple variable transformation, $\sqrt{y} = x$, it is expressed in terms of the modified Bessel function of the first kind as follows,

$$\Phi(x) = \frac{x \exp x}{2 I_1(x)}.$$

For small x, one can find the following expansion

$$\Phi(x) = 1 + x + \frac{3x^2}{8} + \frac{x^3}{24} - \frac{x^4}{96} - \frac{x^5}{480} + O(x^6).$$

For large x the following asymptotic expression is available,

$$\Phi(x) \sim x^{\frac{3}{2}}, \text{ as } x \to \infty.$$

The standard $DLog$ Padé method fails to converge, giving meaningless results for the index, as shown in Fig. 7.4.

In the current problem we observed the best, monotonous convergence for the sequence of $\Delta_{k16}(\alpha_k) = 0$, with $k = 1, 2...15$, as shown below,

$\alpha_1 = 2.7, \ \alpha_2 = 2.13, \ \alpha_3 = 2.137, \ \alpha_4 = 2.14, \ \alpha_5 = 1.69, \ \alpha_6 = 1.68,$

$\alpha_7 = 1.66, \ \alpha_8 = 1.65, \ \alpha_9 = 1.64, \ \alpha_{10} = 1.634, \ \alpha_{11} = 1.63, \ \alpha_{12} = 1.62,$

$\alpha_{13} = 1.61, \ \alpha_{14} = 1.606, \ \alpha_{15} = 1.6.$

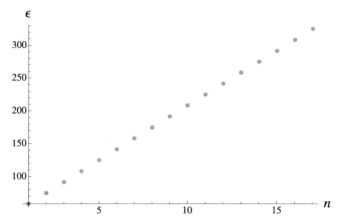

FIGURE 7.4 Critical index for Wilson Loop. The relative percentage error ϵ for the critical index with varying approximation number n, is shown. The results are obtained by $DLog$ Padé method.

The final estimate to be deduced from such sequence is $\alpha = 1.6$, brings reasonable error of 6.7%.

Appendix 7.B Polynomials for the effective conductivity of graphene-type composites with vacancies

The technical details pertining to the polynomials derivation can be found in [20]. The mean values of 35 polynomials for every fixed $p = 1, 2, \ldots, 36$ and up to $O(f^7)$ have the form:

$$1 + 2f + 2f^2 + 2.0431449625677933 f^3 + 4.325376560903891 f^4$$
$$+ 6.653593676968985 f^5 + 9.146502263580448 f^6,$$
$$1 + 2f + 2f^2 + 2.085728259543634 f^3 + 4.503578882551432 f^4$$
$$+ 7.022350274464803 f^5 + 9.889873882233639 f^6,$$
$$1 + 2f + 2f^2 + 2.1309887564966603 f^3 + 4.6945993295106065 f^4$$
$$+ 7.428155309216216 f^5 + 10.732204565996899 f^6,$$
$$1 + 2f + 2f^2 + 2.1757494431530704 f^3 + 4.893752745169728 f^4$$
$$+ 7.863075577145392 f^5 + 11.654501661745625 f^6,$$
$$1 + 2f + 2f^2 + 2.233134965134781 f^3 + 5.1301588079730625 f^4$$
$$+ 8.37660727641368 f^5 + 12.782438827034241 f^6,$$
$$1 + 2f + 2f^2 + 2.2956561991092883 f^3 + 5.3833972235190375 f^4$$
$$+ 8.916132613614705 f^5 + 13.988562234205885 f^6,$$
$$1 + 2f + 2f^2 + 2.3513030719596246 f^3 + 5.633978263563243 f^4$$

$$+\,9.505201514548968\,f^5 + 15.353596276310828\,f^6,$$

$$1 + 2f + 2f^2 + 2.4267342319989553\,f^3 + 5.952924825710871\,f^4$$
$$+\,10.215408230511798\,f^5 + 17.000009114788686\,f^6,$$

$$1 + 2f + 2f^2 + 2.4785087751994466\,f^3 + 6.229980695416432\,f^4$$
$$+\,10.96475719308733\,f^5 + 18.835020913865844\,f^6,$$

$$1 + 2f + 2f^2 + 2.5610221774467856\,f^3 + 6.5830764624482825\,f^4$$
$$+\,11.759638087909211\,f^5 + 20.88296392834553\,f^6,$$

$$1 + 2f + 2f^2 + 2.6354913101367066\,f^3 + 6.937681353800146\,f^4$$
$$+\,12.673008114742498\,f^5 + 23.080703246453126\,f^6,$$

$$1 + 2f + 2f^2 + 2.744204087147164\,f^3 + 7.442853790772685\,f^4$$
$$+\,13.878080891713635\,f^5 + 26.328594239733043\,f^6,$$

$$1 + 2f + 2f^2 + 2.811327787843961\,f^3 + 7.813513309888811\,f^4$$
$$+\,14.991612549672695\,f^5 + 29.360441359782712\,f^6,$$

$$1 + 2f + 2f^2 + 2.926100808445085\,f^3 + 8.314041970931838\,f^4$$
$$+\,16.348821218713763\,f^5 + 33.21865615002286\,f^6,$$

$$1 + 2f + 2f^2 + 3.0620187307125026\,f^3 + 8.919208872271792\,f^4$$
$$+\,17.97510064588165\,f^5 + 38.095270666541516\,f^6,$$

$$1 + 2f + 2f^2 + 3.1356002859493657\,f^3 + 9.517895043281024\,f^4$$
$$+\,19.852478994705884\,f^5 + 43.35304094309354\,f^6,$$

$$1 + 2f + 2f^2 + 3.3176911434096996\,f^3 + 10.335932894481386\,f^4$$
$$+\,22.235452261310755\,f^5 + 51.591319525962845\,f^6,$$

$$1 + 2f + 2f^2 + 3.532155369226726\,f^3 + 11.382628430475668\,f^4$$
$$+\,25.281249256598677\,f^5 + 61.303657278599985\,f^6,$$

$$1 + 2f + 2f^2 + 3.6289248399443865\,f^3 + 12.073909630687467\,f^4$$
$$+\,27.87683164205118\,f^5 + 70.50183701831483\,f^6,$$

$$1 + 2f + 2f^2 + 3.906962446235442\,f^3 + 13.062123397785264\,f^4$$
$$+\,31.444597256959334\,f^5 + 84.17239407537227\,f^6,$$

$$1 + 2f + 2f^2 + 4.07392568436101\,f^3 + 14.624433937163056\,f^4$$
$$+\,36.68984714774904\,f^5 + 103.3074508046892\,f^6,$$

$$1 + 2f + 2f^2 + 4.346979366173176\,f^3 + 15.891963040288054\,f^4$$
$$+\,42.18494510301801\,f^5 + 126.75542971659917\,f^6,$$

172 Mechanics and Physics of Structured Media

$$1 + 2f + 2f^2 + 4.714508397519139 f^3 + 18.210288782077768 f^4$$
$$+ 50.92738175631363 f^5 + 161.3300683927631 f^6,$$
$$1 + 2f + 2f^2 + 4.773454388076464 f^3 + 19.289597340998405 f^4$$
$$+ 56.453972354754285 f^5 + 189.004014684617 f^6,$$
$$1 + 2f + 2f^2 + 5.3481058383603255 f^3 + 23.071729829302065 f^4$$
$$+ 72.11062755615562 f^5 + 262.22042641185834 f^6,$$
$$1 + 2f + 2f^2 + 6.047745255499693 f^3 + 27.09080442453905 f^4$$
$$+ 91.93152604575694 f^5 + 367.2378698733342 f^6,$$
$$1 + 2f + 2f^2 + 6.854376912319701 f^3 + 31.623228426472238 f^4$$
$$+ 118.9158138623377 f^5 + 529.7388747902697 f^6,$$
$$1 + 2f + 2f^2 + 7.091016763118986 f^3 + 34.35334685950311 f^4$$
$$+ 140.87097035997667 f^5 + 678.4080052380075 f^6,$$
$$1 + 2f + 2f^2 + 8.139135304457737 f^3 + 44.31147840441928 f^4$$
$$+ 203.4255268502177 f^5 + 1136.7063834049131 f^6,$$
$$1 + 2f + 2f^2 + 9.333092024635258 f^3 + 57.527298136702825 f^4$$
$$+ 297.57907474057293 f^5 + 1867.0048171916985 f^6,$$
$$1 + 2f + 2f^2 + 11.29975045568634 f^3 + 79.75352725657338 f^4$$
$$+ 477.15363142785037 f^5 + 3559.1422987340734 f^6,$$
$$1 + 2f + 2f^2 + 13.31869770989477 f^3 + 104.83766806979258 f^4$$
$$+ 772.09863789888 f^5 + 7021.786355552324 f^6,$$
$$1 + 2f + 2f^2 + 17.06979338390195 f^3 + 177.0428229057594 f^4$$
$$+ 1704.5210776226313 f^5 + 20306.361175060283 f^6,$$
$$1 + 2f + 2f^2 + 27.918990378770605 f^3 + 364.77734438789344 f^4$$
$$+ 5337.205697831491 f^5 + 94736.23777803621 f^6,$$
$$1 + 2f + 2f^2 + 28.414381382649374 f^3 + 465.78892785428036 f^4$$
$$+ 12464.220525025323 f^5 + 421006.77930792305 f^6. \tag{7.B.1}$$

The set of polynomials was communicated to the Author by Dr. Piotr Drygaś.

References

[1] G.A. Baker Jr., Padé approximant, Scholarpedia 7 (6) (2012) 9756.
[2] G.A. Baker Jr., P. Graves-Moris, Padé Approximants, Cambridge University, Cambridge, UK, 1996.
[3] T. Banks, T.J. Torres, Two point Padé approximants and duality, arXiv:1307.3689v2, 2013.

Critical index for conductivity, elasticity, superconductivity Chapter | 7 **173**

[4] G.K. Batchelor, J.T. Green, The determination of the bulk stress in a suspension of spherical to order c^2, J. Fluid Mech. 56 (1972) 401–427.

[5] T. Bauer, F. Hofling, T. Munk, E. Frey, T. Franosch, The localization transition of the two-dimensional Lorentz model, Eur. Phys. J. Spec. Top. 189 (2010) 103–118.

[6] G.K. Batchelor, Sedimentation in a dilute dispersion of spheres, J. Fluid Mech. 52 (1972) 245–268.

[7] R.J. Baxter, Exactly Solved Models in Statistical Mechanics, Academic Press, 1982.

[8] C.W.J. Beenakker, P. Mazur, Self-diffusion of spheres in a concentrated suspension, Physica A 120 (1984) 388–410.

[9] J.F. Brady, L.J. Durlofsky, The sedimentation rate of disordered suspensions, Phys. Fluids 31 (1988) 717–727.

[10] D.J. Bergman, D. Stroud, Physical properties of macroscopically inhomogeneous media, Solid State Phys. 46 (1992) 148–270.

[11] J.F. Brady, The rheological behavior of concentrated colloidal dispersions, J. Chem. Phys. 99 (1993) 567–581.

[12] P. Butera, M. Comi, A library of extended high-temperature expansions of basic observables for the spin-S Ising models on two- and three-dimensional lattices, J. Stat. Phys. 109 (2002) 311–315.

[13] B. Cichocki, M.L. Ekiel-Jezewska, P. Szymczak, E. Wajnryb, Three-particle contribution to sedimentation and collective diffusion in hard-sphere suspensions, J. Chem. Phys. 117 (2002) 1231–1241.

[14] J.P. Clerc, G. Giraud, J.M. Laugie, J.M. Luck, The electrical conductivity of binary disordered systems, percolation clusters, fractals and related models, Adv. Phys. 39 (1990) 191–309.

[15] N. Clisby, Accurate estimate of the critical exponent for self-avoiding walks via a fast implementation of the pivot algorithm, Phys. Rev. Lett. 104 (2010) 055702.

[16] N. Clisby, B.M. McCoy, Ninth and tenth order virial coefficients for hard spheres in D dimensions, J. Stat. Phys. 122 (2006) 15–57.

[17] C. Cosme, J.M. Viana Parente Lopes, J. Penedones, Conformal symmetry of the critical 3D Ising model inside a sphere, J. High Energy Phys. 08 (2015) 022.

[18] J.P. Cotton, Polymer excluded volume exponent ν: an experimental verification of the n vector model for $n = 0$, J. Phys. Lett. 41 (1980) 231–234.

[19] P. Drygaś, S. Gluzman, V. Mityushev, W. Nawalaniec, Applied Analysis of Composite Media, Woodhead Publishing (Elsevier), Sawston, UK, 2020.

[20] P. Drygaś, L.A. Filshtinski, S. Gluzman, V. Mityushev, Conductivity and elasticity of graphene-type composites, in: R. McPhedran, S. Gluzman, V. Mityushev, N. Rylko (Eds.), 2D and Quasi-2D Composite and Nano Composite Materials, Properties and Photonic Applications, Elsevier, Amsterdam, The Netherlands, 2020, pp. 193–231, Chapter 8.

[21] D. Frenkel, Velocity auto-correlation functions in a 2D lattice Lorentz gas: comparison of theory and computer simulation, Phys. Lett. 121 (1987) 385–389.

[22] S. Gluzman, Nonlinear approximations to critical and relaxation processes, Axioms 9 (2020) 126.

[23] S. Gluzman, Padé and post-Padé approximations for critical phenomena, Symmetry 12 (2020) 1600.

[24] S. Gluzman, Optimized factor approximants and critical index, Symmetry 13 (2021) 903.

[25] S. Gluzman, V. Mityushev, W. Nawalaniec, Computational Analysis of Structured Media, Academic Press (Elsevier), Amsterdam, 2017.

[26] S. Gluzman, V.I. Yukalov, Unified approach to crossover phenomena, Phys. Rev. E 58 (1998) 4197–4209.

[27] S. Gluzman, V.I. Yukalov, Self-similarly corrected Padé approximants for indeterminate problem, Eur. Phys. J. Plus 131 (2016) 340–361.

[28] S. Gluzman, V.I. Yukalov, Critical indices from self-similar root approximants, Eur. Phys. J. Plus 132 (2017) 535.

174 Mechanics and Physics of Structured Media

[29] S. Gluzman, V.I. Yukalov, Extrapolation of perturbation theory expansions by self-similar approximants, Eur. J. Appl. Math. 25 (2014) 595–628.

[30] S. Gluzman, V.I. Yukalov, D. Sornette, Self-similar factor approximants, Phys. Rev. E 67 (2003) 026109.

[31] P. Grassberger, Conductivity exponent and backbone dimension in 2D percolation, Physica A 262 (1999) 251–263.

[32] A.Y. Grosberg, A.R. Khokhlov, Statistical Physics of Macromolecules, AIP Press, Woodbury, NY, USA, 1994.

[33] A.J. Guttman, Validity of hyperscaling for the $d = 3$ Ising model, Phys. Rev. B 33 (1986) 5089–5092.

[34] H. Hayakawa, K. Ichiki, Statistical theory of sedimentation of disordered suspensions, Phys. Rev. E 51 (1995) 3815–3818.

[35] M. Hasenbusch, Finite size scaling study of lattice models in the three-dimensional Ising universality class, Phys. Rev. B 82 (2010) 174433.

[36] F. Hofling, T. Franosch, E. Frey, Localization transition of the three-dimensional Lorentz model and continuum percolation, Phys. Rev. Lett. 96 (2006) 165901.

[37] D.J. Jeffrey, Conduction through a random suspension of spheres, Proc. R. Soc. Lond. A 335 (1973) 355–367.

[38] S. Kirkpatrick, Percolation and conduction, Rev. Mod. Phys. 45 (1973) 574–588.

[39] A.J.C. Ladd, Hydrodynamic transport coefficients of random dispersions of hard spheres, J. Chem. Phys. 93 (1990) 3484–3494.

[40] B. Li, N. Madras, A.D. Sokal, Critical exponents, hyperscaling, and universal amplitude ratios for two- and three-dimensional self-avoiding walks, J. Stat. Phys. 80 (1995) 661–754.

[41] W. Losert, L. Bocquet, T.C. Lubensky, J.P. Gollub, Particle dynamics in sheared granular matter, Phys. Rev. Lett. 85 (2000) 1428–1431.

[42] M.A.G. Maestre, A. Santos, M. Robles, M. Lopez de Haro, On the relation between virial coefficients and the close-packing of hard disks and hard spheres, J. Chem. Phys. 134 (2011) 084502.

[43] A. Mulero, I. Cachadina, J.R. Solana, The equation of state of the hard-disc fluid revisited, Mol. Phys. 107 (2009) 1457–1465.

[44] M. Muthukumar, B.G. Nickel, Expansion of a polymer chain with excluded volume interaction, J. Chem. Phys. 86 (1987) 460–476.

[45] M. Muthukumar, B.G. Nickel, Perturbation theory for a polymer chain with excluded volume interaction, J. Chem. Phys. 80 (1984) 5839–5850.

[46] Th.M. Nieuwenhuizen, P.F.J. van Velthoven, M.H. Ernst, Diffusion and long-time tails in a two-dimensional site-percolation model, Phys. Rev. Lett. 57 (1986) 2477–2480.

[47] A. Pelissetto, E. Vicari, Critical phenomena and renormalization-group theory, Phys. Rep. 368 (2002) 549–727.

[48] A. Santos, M. Lopez de Haro, S. Bravo Yuste, An accurate and simple equation of state for hard disks, J. Chem. Phys. 103 (1995) 4622–4625.

[49] M. Suzuki, On the temperature-dependence of "Effective critical exponents" and Confluent Singularities, Prog. Theor. Phys. 47 (1972) 722–723.

[50] E. Wajnryb, J.S. Dahler, The Newtonian viscosity of a moderately dense suspensions, in: I. Prigogine, S.A. Rice (Eds.), Adv. Chem. Phys., vol. 102, Wiley, New York, 1997, pp. 193–313.

[51] V.I. Yukalov, S. Gluzman, Critical indices as limits of control functions, Phys. Rev. Lett. 79 (1997) 333–336.

[52] V.I. Yukalov, S. Gluzman, Self-similar bootstrap of divergent series, Phys. Rev. E 55 (1997) 6552–6570.

[53] V.I. Yukalov, S. Gluzman, Self-similar exponential approximants, Phys. Rev. E 58 (1998) 1359–1382.

[54] V.I. Yukalov, S. Gluzman, Weighted fixed points in self-similar analysis of time series, Int. J. Mod. Phys. B 13 (1999) 1463–1476.

References

[55] V.I. Yukalov, S. Gluzman, Extrapolation of power series by self-similar factor and root approximants, Int. J. Mod. Phys. B 18 (2004) 3027–3046.

[56] R.M. Ziff, S. Torquato, Percolation of disordered jammed sphere packings, J. Phys. A, Math. Theor. 50 (2017) 085001.

[57] J. Zinn-Justin, Critical phenomena: field theoretical approach, Scholarpedia 5 (5) (2010) 8346, https://doi.org/10.4249/scholarpedia.8346.

Chapter 8

Double periodic bianalytic functions

Piotr Drygaś
University of Rzeszow, Rzeszow, Poland

8.1 Introduction

The theory of elliptic functions has been extensively applying to various two-dimensional stationary problems of mathematical physics governed by Laplace's equation. The key to this theory is the classic Weierstrass functions introduced by means of the absolutely convergent series [1]. In the theory of composites, a homogenization procedure related to periodicity yields conditionally convergent series properly defined by the Eisenstein summation [2]. Use of the Eisenstein functions for the local fields and the effective conductivity is preferable to Weierstrass' formalism in application of the cluster method [3].

The corresponding problem arises in applications of the homogenization approach to the two-dimensional (2D) static elasticity problems described by biharmonic functions. The pioneering paper by Natanzon (Natanson) [4] contains a construction of the double periodic bianalytic function associated to the Weierstrass $\wp$-function. This work [4] was devoted to the local double periodic fields in the elastic plate weakened by the hexagonal regular array of holes. Filshtinsky (Fil'shtinskij) [5] (see also the books [6,7]) extended Natanzon's approach to general double periodic arrays and developed the theory of double periodic bianalytic functions having followed Weierstrass' approach. A numerical algorithm was developed for computation of the effective elastic constants in [6,7]. Extensions to double periodic polyanalytic functions can be found in [7] and [8].

Investigations [9,10] show that the Eisenstein summation method can be extended to derivation of new analytical formulas for the 2D elastic composites. The present chapter is devoted to systematic application of the Eisenstein summation method to derive double periodic bianalytic functions, i.e., to extension of Filshtinsky's theory based on the Weierstrass functions. Besides the formal application of the Eisenstein summation we introduce new low order functions corresponding to the classic Eisenstein functions of first and second orders and their relations to the Natanzon-Filshtinsky functions.

Mechanics and Physics of Structured Media. https://doi.org/10.1016/B978-0-32-390543-5.00013-X
Copyright © 2022 Elsevier Inc. All rights reserved.

178 Mechanics and Physics of Structured Media

8.2 Weierstrass and Natanzon-Filshtinsky functions

We consider a lattice $\Lambda = \Lambda(\omega_1, \omega_2)$ as periodic structure on the complex plane, set by two fundamental translation vectors (generators of Λ) expressed by two complex numbers ω_1, ω_2. The lattice Λ consists the points $\omega = k\omega_1 + l\omega_2$, where $k, l = 0, \pm 1, \pm 2, \ldots$. Without loss of generality we assume that $\omega_1 > 0$ and $\text{Im} \frac{\omega_2}{\omega_1} > 0$, i.e. $\arg \omega_2 = \alpha > 0$. The complex numbers $\pm \frac{1}{2}\omega_1, \pm \frac{1}{2}\omega_2$ define a parallelogram on the complex plane and form the fundamental domain of the translation group generated by ω_1 and ω_2.

Introduce the double summation

$$\sum_{\omega} = \sum_{(l,k)\in\mathbb{Z}^2} \quad \text{and} \quad \sum_{\omega}' = \sum_{(l,k)\in\mathbb{Z}^2\setminus(0,0)} . \tag{8.1}$$

The symbol $\sum'$ means that ω run over all lattice points, except at the point $\omega = 0$. The classic Weierstrass functions are defined by the absolutely convergent series [1]

$$\wp(z) = \frac{1}{z^2} + \sum_{\omega}' \left(\frac{1}{(z-\omega)^2} - \frac{1}{\omega^2} \right) \tag{8.2}$$

and

$$\zeta(z) = \frac{1}{z} + \sum_{\omega}' \left(\frac{1}{z-\omega} + \frac{1}{\omega} + \frac{z}{\omega^2} \right). \tag{8.3}$$

The absolute convergence of these and other similar series follows from the asymptotic equivalence of the general term to $|\omega|^{-3}$.

The Eisenstein summation method is defined as the iterated limits

$$\sum_{\omega e} = \sum_{l=-\infty}^{\infty} \left(\sum_{k=-\infty}^{\infty} \right) = \lim_{M_2\to\infty} \sum_{l=-M_2}^{l=M_2} \left(\lim_{M_1\to\infty} \sum_{k=-M_1}^{k=M_1} \right). \tag{8.4}$$

The summations (8.1) and (8.4) give the same result when they are applied to an absolutely convergent series.

Let n be a natural number ($n \geq 1$). The Eisenstein functions are introduced as follows [2]

$$E_n(z) = \sum_{\omega e} (z-\omega)^{-n}. \tag{8.5}$$

The Eisenstein–Rayleigh lattice sums are introduced analogously

$$S_n = \sum_{\omega e}' \omega^{-n}. \tag{8.6}$$

The series (8.5) and (8.6) are absolutely convergent for $n > 2$ [1] and conditionally convergent for $n = 1, 2$ when $z \neq \omega$. The Eisenstein summation correctly defines the series (8.5) and (8.6) for $n = 1, 2$ [2]. An exact expression for S_2 can be found in [10,11]; $S_n = 0$ for odd n.

The Eisenstein and Weierstrass functions are related by formulas [2]

$$\wp(z) = E_2(z) - S_2, \quad \zeta(z) = E_1(z) + S_2 z$$

and for $n > 2$

$$E_n(z) = \frac{(-1)^n}{(n-1)!} \frac{d^{n-2}}{dz^{n-2}} \wp(z).$$

Properties of the Eisenstein and Weierstrass functions are described in [1,2,12].

We now proceed to discuss polyanalytic functions associated to the Eisenstein functions. Let a number j be fixed ($j = 0, 1, 2 \ldots$). Filshtinsky [7] introduced the functions

$$\wp_j(z) = \sum_{\omega}{}' \left(\frac{\overline{\omega}^j}{(z-\omega)^2} - \sum_{r=0}^{j} (r+1) \frac{\overline{\omega}^j}{\omega^{r+2}} z^r \right). \tag{8.7}$$

The above series is absolutely convergent for any z not belonging to $\Lambda \backslash \{0\}$ and determines a function analytic at $z = 0$. It is worth noting that $\wp_j(z)$ are not periodic for positive j. The function $\wp_1(z)$ was introduced by Natanzon [4]. The function $\wp_0(z)$ coincides with the classic $\wp$-function by Weierstrass (8.2). We call the functions $\wp_j(z)$ of (8.7) by the Natanzon-Filshtinsky functions.

We now consider the series introduced in the paper [13] for $n \geq 1$ by application of the Eisenstein summation

$$E_n^{(j)}(z) = \sum_{\omega}{}_e (\overline{z-\omega})^j (z-\omega)^{-n}. \tag{8.8}$$

The following generalized lattice sums can be also introduced

$$S_n^{(j)} = \sum_{\omega}{}'_e \overline{\omega}^j \omega^{-n}. \tag{8.9}$$

Introduce also new functions (generalized Natanzon-Filshtinsky functions) defined by the absolutely convergent series

$$\wp_{j,n}(z) = \sum_{\omega}{}' \left\{ \frac{\overline{\omega}^j}{(z-\omega)^n} - (-1)^n \sum_{r=0}^{(j+2)-n} \binom{n+r-1}{r} \frac{\overline{\omega}^j}{\omega^n} \left(\frac{z}{\omega} \right)^r \right\} \tag{8.10}$$

for $j = 1, 2, \ldots$ and $n = 1, 2, \ldots$. Here, the sum $\sum_{r=0}^{(j+2)-n)}$ vanishes if $j < n-2$. One can see that $\wp_{j,2}(z)$ coincides with the Natanzon-Filshtinsky function $\wp_j(z)$. The following relations hold between new and the classic functions

$$\wp_{0,1}(z) = \zeta(z) - \frac{1}{z}, \quad \wp_{0,2}(z) = \wp(z) - \frac{1}{z^2}, \quad \wp_{0,k}(z) = E_k(z) - \frac{1}{z^k}$$

for $k = 3, 4, \ldots$.

180 Mechanics and Physics of Structured Media

8.3 Properties of the generalized Natanzon-Filshtinsky functions

First, we note that the functions (8.10) satisfy

$$\wp_{j,n}(-z) = (-1)^{j+n}\wp_{j,n}(z), \tag{8.11}$$

$j, n = 1, 2, \ldots$. We now proceed to derive a relation between the derivatives of the functions (8.10). We have

$$\wp_{j,n}^{(k)}(z) := \frac{\partial^k}{\partial z^k}\wp_{j,n}(z) = \sum_{\omega}{}' \left\{ (-1)^k \frac{(n+k-1)!}{(n-1)!} \frac{\overline{\omega}^j}{(z-\omega)^{n+k}} \right.$$
$$\left. - (-1)^n \sum_{r=k}^{(j+2)-n} \frac{(r+k-1)!}{(r-1)!} \binom{n+r-1}{r} \frac{\overline{\omega}^j}{\omega^{n+r}} z^{r-k} \right\}, \tag{8.12}$$

$j, k = 0, 1, \ldots, n = 1, 2, \ldots$. If $k > \max((j+2)-n, 0)$, then the sum $\sum_{r=k}^{(j+2)-n}$ vanishes. Thus

$$\wp_{j,n}^{(k)}(z) = (-1)^k n(n+1)\ldots(n+k-1)\wp_{j,n+k}(z), \tag{8.13}$$

$j, k = 0, 1, \ldots, n = 1, 2, \ldots$.

Substituting $n = 1$ and replacing k by $n - 1$ in (8.13) we present the function $\wp_{j,n}$ by means of the derivative of the function $\wp_{j,1}$

$$\wp_{j,n}(z) = -\frac{(-1)^n}{(n-1)!}\wp_{j,1}^{(n-1)}(z), \tag{8.14}$$

$j = 0, 1, \ldots, n = 1, 2, \ldots$. Along similar lines we express $\wp_{j,n}(z)$ through the derivative of $\wp_j(z) = \wp_{j,2}(z)$

$$\wp_{j,n}(z) = \frac{(-1)^n}{(n-1)!}\wp_{j,2}^{(n-2)}(z), \tag{8.15}$$

$j = 0, 1, \ldots, n = 2, 3, \ldots$.

From (8.13)

$$\wp_{j,j+1}'(z+\omega_v) - \wp_{j,j+1}'(z) = -(j+1)\left(\wp_{j,j+2}(z+\omega_v) - \wp_{j,j+2}(z)\right) \tag{8.16}$$

and

$$\wp_{j,j+2}(z+\omega_v) - \wp_{j,j+2}(z)$$
$$= \sum_{\omega\neq 0}\left\{\frac{\overline{\omega}^j}{(z+\omega_v-\omega)^{j+2}} - (-1)^j\frac{\overline{\omega}^j}{\omega^{j+2}}\right\} - \sum_{\omega\neq 0}\left\{\frac{\overline{\omega}^j}{(z-\omega)^{j+2}} - (-1)^j\frac{\overline{\omega}^j}{\omega^{j+2}}\right\}$$

$$= \sum_{\omega \neq 0} \left(\sum_{r=1}^{j} \binom{j}{r} \overline{\omega_v}^r \frac{\overline{\omega}^{j-r}}{(z-\omega)^{j+2}} \right) + \frac{\overline{\omega_v}^j}{z^{j+2}}$$

$$= \frac{\overline{\omega_v}^j}{z^{j+2}} + \sum_{r=1}^{j} \binom{j}{r} \overline{\omega_v}^r \wp_{j-r,j+2}(z), \quad (8.17)$$

$j = 0, 1, \ldots, v = 1, 2$. This leads to

$$\wp_{j,j+2}(z+\omega_v) - \wp_{j,j+2}(z)$$

$$= \frac{\overline{\omega_v}^j}{z^{j+2}} - \sum_{r=0}^{j} (-1)^{j-r} \binom{j}{r} \frac{\overline{\omega_v}^j}{(z+\omega_v)^{j+2}} + \sum_{r=1}^{j} \binom{j}{r} \overline{\omega_v}^r \wp_{j-r,j+2}(z), \quad (8.18)$$

$j = 0, 1, \ldots$. The above formula for $j = 0$ becomes the classic jump formula for the Weierstrass $\wp$-function

$$\wp(z+\omega_v) - \wp(z) = 0. \quad (8.19)$$

Using the binomial formula $\sum_{r=0}^{j}(-1)^{j-r}\binom{j}{r} = 0$ in (8.18) we obtain

$$\wp_{j,j+2}(z+\omega_v) - \wp_{j,j+2}(z)$$

$$= (-1)^j \frac{\overline{\omega_v}^j}{(j+1)!} \wp^{(j)}(z) + \sum_{r=1}^{j-1} \binom{j}{r} \overline{\omega_v}^r \wp_{j-r,j+2}(z), \quad (8.20)$$

$j = 0, 1, \ldots$. This formula for $j = 1$ yields Filshtinsky's formula [7]

$$\wp'_{1,2}(z+\omega_v) - \wp'_{1,2}(z) = \overline{\omega_v}\wp'(z). \quad (8.21)$$

Using the derivative formula (8.13) we obtain from (8.20)

$$\wp_{j,j+n}(z+\omega_v) - \wp_{j,j+n}(z)$$

$$= (-1)^{j+n} \frac{\overline{\omega_v}^j}{(j+n-1)!} \wp^{(j+n-2)}(z) + \sum_{r=1}^{j-1} \binom{j}{j-r} \overline{\omega_v}^{j-r} \wp_{r,j+n}(z)', \quad (8.22)$$

$j = 1, 2, \ldots, n = 2, 3, \ldots, v = 1, 2$. Substitution of $z = -\frac{\omega_v}{2}$ into (8.22) yields

$$[1 - (-1)^n] \wp_{j,j+n}\left(\frac{\omega_v}{2}\right) = (-1)^{j+n} \frac{\overline{\omega_v}^j}{(j+n-1)!} \wp^{(j+n-2)}\left(\frac{\omega_v}{2}\right)$$

$$+ \sum_{r=1}^{j-1} (-1)^{n+j-r} \binom{j}{j-r} \overline{\omega_v}^{j-r} \wp_{r,j+n}\left(\frac{\omega_v}{2}\right), \quad (8.23)$$

182 Mechanics and Physics of Structured Media

$j = 1, 2, \ldots, n = 2, 3, \ldots, \nu = 1, 2.$ From (8.22)

$$\wp_{1,n+2}(z + \omega_\nu) - \wp_{1,n+2}(z) = (-1)^n \frac{\overline{\omega_\nu}}{(n+1)!} \wp^{(n)}(z), \qquad (8.24)$$

$n = 1, 2, \ldots, \nu = 1, 2.$

8.4 The function $\wp_{1,2}$

In the present section, we pay attention to the Natanzon-Filshtinsky function $\wp_{1,2}$ denoted in the works [4–7] as $\wp_1$ or Q. Though the main equations were already presented in [7] we simplify them developing Filshtinsky's investigations.

Integrating (8.21) we obtain

$$\wp_{1,2}(z + \omega_\nu) - \wp_{1,2}(z) = \overline{\omega_\nu} \wp(z) + \gamma_\nu, \qquad \nu = 1, 2, \qquad (8.25)$$

where γ_ν are constants. Consider the fundamental domain, the parallelogram with vertices $A = -\frac{\omega_1}{2} - \frac{\omega_2}{2}, B = \frac{\omega_1}{2} - \frac{\omega_2}{2}, C = \frac{\omega_1}{2} + \frac{\omega_2}{2}, D = -\frac{\omega_1}{2} + \frac{\omega_2}{2}.$ The following integral over the curve $ABCD$ vanishes,

$$\int_{ABCD} \wp_{1,2}(z)dz = 0, \qquad (8.26)$$

since the function $\wp_{1,2}(z)$ is analytic in the parallelogram. Calculate the same integral by another method

$$I \equiv \int_{ABCD} \wp_{1,2}(z)dz$$
$$= \int_{AB} \left(\wp_{1,2}(z) - \wp_{1,2}(z + \omega_2)\right) dz + \int_{AD} \left(\wp_{1,2}(z + \omega_1) - \wp_{1,2}(z)\right) dz. \qquad (8.27)$$

Using (8.25) we obtain

$$I = \int_{AB} \left(-\overline{\omega_2} \wp(z) - \gamma_2\right) dz + \int_{AD} \left(\overline{\omega_1} \wp(z) + \gamma_1\right) dz$$
$$= \left(\overline{\omega_2} \zeta(z) + \gamma_2 z\right)\big|_A^B + \left(\overline{\omega_1} \zeta(z) + \gamma_1 z\right)\big|_A^C, \qquad (8.28)$$

where the Weierstrass ζ-function satisfies Legendre's relations [1]

$$\zeta(z + \omega_1) - \zeta(z) = \eta_1, \qquad \zeta(z + \omega_2) - \zeta(z) = \eta_2 \qquad (8.29)$$

with $\eta_1 = 2\zeta\left(\frac{\omega_1}{2}\right), \eta_2 = 2\zeta\left(\frac{\omega_2}{2}\right).$ Then, $I = \overline{\omega_2}\eta_1 - \gamma_2\omega_1 - \overline{\omega_1}\eta_2 + \gamma_1\omega_2$ and because of (8.26)

$$\gamma_2\omega_1 - \gamma_1\omega_2 = \overline{\omega_2}\eta_1 - \overline{\omega_1}\eta_2. \qquad (8.30)$$

Putting in (8.24) $z = -\frac{\omega_\nu}{2}$, $\nu = 1, 2$ we get

$$\wp_{1,n+1}\left(\frac{\omega_\nu}{2}\right) - (-1)^n \wp_{1,n+1}\left(\frac{\omega_\nu}{2}\right) = \frac{\overline{\omega_\nu}}{n!} \wp^{(n-1)}\left(\frac{\omega_\nu}{2}\right), \tag{8.31}$$

$n = 2, 3, 4, \ldots$, $\nu = 1, 2$. Analogously from (8.25)

$$\gamma_{\nu,1} = 2\wp_{1,2}\left(\frac{\omega_\nu}{2}\right) - \overline{\omega_\nu}\wp\left(\frac{\omega_\nu}{2}\right), \tag{8.32}$$

$\nu = 1, 2$.

Natanson and next Filshtinsky proved the identity

$$\alpha\wp'_{1,2}(z) = \frac{1}{3}\wp''(z) + (\zeta(z) - \beta z)\wp'(z) - 2\beta\wp(z) + c, \tag{8.33}$$

where α, β, c, and c_1 are constants. Though these constants were written in [7] we derive below simple expressions for these constants and demonstrate their connections with the lattice sums. Consider the function f of the form

$$f(z) = \alpha\wp'_{1,2}(z) + \beta z\wp'(z) - \zeta(z)\wp'(z). \tag{8.34}$$

Because the function $\wp_{1,2}$ is periodic we consider the periodicity assumptions for the function f. According (8.25)

$$f(z + \omega_\nu) - f(z) = (\alpha\overline{\omega_\nu} + \beta\omega_\nu - \eta_\nu)\wp'(z), \tag{8.35}$$

$\nu = 1, 2$, thus

$$\alpha\overline{\omega_\nu} + \beta\omega_\nu = \eta_\nu. \tag{8.36}$$

This means that

$$\alpha = \frac{2\pi i}{\overline{\omega_1}\omega_2 - \omega_1\overline{\omega_2}}, \quad \beta = \frac{\eta_2\overline{\omega_1} - \eta_1\overline{\omega_2}}{\overline{\omega_1}\omega_2 - \omega_1\overline{\omega_2}}. \tag{8.37}$$

Note that from assumption in Section 8.2 ω_1 is real number. By this (8.37) has the form

$$\alpha = \frac{2\pi i}{\omega_1(\omega_2 - \overline{\omega_2})}, \quad \beta = \frac{\eta_2\omega_1 - \eta_1\overline{\omega_2}}{\omega_1(\omega_2 - \overline{\omega_2})}. \tag{8.38}$$

Note that $\omega_1(\omega_2 - \overline{\omega_2}) = 2i\omega_1\mathrm{Im}\omega_2 = 2i|Q_0|$, where $|Q_0|$ denotes area of the parallelogram defined by complex number ω_1 and ω_2. Then

$$\alpha = \frac{\pi}{|Q_0|}, \quad \beta = \frac{\eta_2\omega_1 - \eta_1\overline{\omega_2}}{2i|Q_0|}. \tag{8.39}$$

From the Legendre's relations we have

$$\beta = \frac{\eta_1\mathrm{Im}\omega_2}{|Q_0|} - \frac{\pi}{|Q_0|}. \tag{8.40}$$

184 Mechanics and Physics of Structured Media

Using the relation (3.4.124) from [12] we have

$$\omega_\nu S_2 = 2\zeta\left(\frac{\omega_\nu}{2}\right) = \eta_\nu.$$

Then we obtain

$$\beta = S_2 - \frac{\pi}{|Q_0|} = S_2 - \alpha. \tag{8.41}$$

Note, that for normalized ($|Q_0| = 1$) hexagonal lattice, $S_2 = \pi$ and $\alpha = \pi$ and $\beta = 0$

The function f is even and has the pole of the form $2z^{-4} - 2\beta z^{-2}$ at the origin. From Liouville's theorem the function f has the form

$$f(z) = \frac{1}{3}\wp''(z) - 2\beta\wp(z) + c, \tag{8.42}$$

where $c = const$. Then holds

$$\alpha\wp'_{1,2}(z) = \frac{1}{3}\wp''(z) + (\zeta(z) - \beta z)\wp'(z) - 2\beta\wp(z) + c. \tag{8.43}$$

Integrating the above equation we obtain

$$\alpha\wp_{1,2}(z) = \frac{1}{3}\wp'(z) + (\zeta(z) - \beta z)\wp(z) + \beta\zeta(z) + \int \wp^2(z)dz + cz + c_1, \tag{8.44}$$

where c and c_1 are complex constants. Using relation

$$2\wp''(z) = 12\wp^2(z) - 60S_4,$$

we obtain from (8.44)

$$\alpha\wp_{1,2}(z) = \frac{1}{2}\wp'(z) + (\zeta(z) - \beta z)\wp(z) + \beta\zeta(z) + (5S_4 + c)z + c_1, \tag{8.45}$$

where c and c_1 are constant. This yields

$$\alpha\left(\wp_{1,2}(z + \omega_\nu) - \wp_{1,2}(z)\right) = (\eta_\nu - \beta\omega_\nu)\wp(z) + \beta\eta_\nu + (5S_4 + c)\omega_\nu, \tag{8.46}$$

$\nu = 1, 2$.

From (8.46) and (8.25)

$$\alpha\gamma_\nu = \beta\eta_\nu - 5\omega_\nu S_4, \tag{8.47}$$

$\nu = 1, 2$. Using equation (8.47) we have

$$\gamma_\nu = \alpha^{-1}\left(S_2^2 - \alpha S_2 - 5S_4\right)\omega_\nu, \tag{8.48}$$

Double periodic bianalytic functions **Chapter** | **8** **185**

and

$$\gamma_v = \frac{|Q_0|}{\pi}\left(S_2^2 - \frac{\pi}{|Q_0|}S_2 - 5S_4\right)\omega_v. \tag{8.49}$$

Note, that for normalized ($|Q_0| = 1$) hexagonal lattice, $S_2 = \pi$ and $S_4 = 0$, thus $\gamma_v = 0$.

We need calculate the undefined constants c and c_1. We consider Taylor series the functions ζ, $\wp$, and $\wp_{1,2}$ of the form

$$\zeta(z) = \frac{1}{z} - \sum_{k=1}^{\infty} S_{2k+2}z^{2k+1}, \tag{8.50}$$

$$\wp(z) = \frac{1}{z^2} + \sum_{k=1}^{\infty}(2k+1)S_{2k+2}z^{2k}, \tag{8.51}$$

and

$$\wp_{1,2} = \sum_{k=2}^{\infty}(k+1)S_{k+2}^{(1)}z^k, \tag{8.52}$$

where $S_{k+2}^{(1)}$ is defined by (8.9). Putting above series (8.50)–(8.52) into (8.45) and taking the coefficients in z^0 and z^1 we obtain $c_1 = 0$ and $c = -10S_4$. The formula (8.43) shows how to find the value of a function $\wp_{1,3}$. An open question is determining the value of the functions $\wp_{n,n+2}$ for $n = 2, 3, \dots$.

8.5 Relation between the generalized Natanzon-Filshtinsky and Eisenstein functions

We can represent each Eisenstein function of the form (8.8) as follows

$$E_n^{(j)} = \sum_{\omega} e^{\frac{(z-\omega)^j}{(z-\omega)^{-n}}} = \sum_{k=0}^{j}(-1)^{j-k}\binom{j}{k}\overline{z}^k\sum_{\omega} e^{\frac{\overline{\omega}^{j-k}}{(z-\omega)^{-n}}}, \tag{8.53}$$

$j = 0, 1, \dots, n = j+2, j+3, \dots.$ We can determine the sum $\sum_{\omega} e^{\frac{\overline{\omega}^{j-k}}{(z-\omega)^{-n}}}$ by Natanzon-Filshtinsky functions $\wp_{l,m}$, $l = 0, 1, \dots, m = 2, 3, \dots.$ For example

$$E_3^{(1)}(z) = -\frac{1}{2}\overline{z}\wp'(z) + \frac{1}{2}\wp'_{1,2}(z) + S_3^1. \tag{8.54}$$

Using relation (8.45), for hexagonal lattice we have

$$E_3^{(1)}(z) = -\frac{1}{2}\overline{z}\wp'(z) + \frac{1}{6\pi}\wp''(z) + \frac{1}{2\pi}\zeta(z)\wp'(z) + \frac{\pi}{2}. \tag{8.55}$$

186 Mechanics and Physics of Structured Media

Analogously

$$E_4^{(2)} = \frac{1}{6}\wp''(z)\bar{z}^2 + \frac{2}{3}\wp'_{1,2}(z)\bar{z} - \frac{1}{3}\left(\wp'_{2,3}(z) - S_4^2\right) \tag{8.56}$$

and so forth.

8.6 Double periodic bianalytic functions via the Eisenstein series

The bianalytic function is a solution of the equation

$$\frac{\partial^2}{\partial \bar{z}^2} F(z) = 0. \tag{8.57}$$

We know that a bianalytic functions can be expressed by formula

$$F(z) = f_1(z) + \bar{z} f_2(z), \tag{8.58}$$

where f_1 and f_2 are analytic functions in considered domain. When we use the methods of the function theory to solve elasticity problems in the plane we consider the biharmonic function $U(x, y)$. We can successfully represent the stress function U as the real part of a certain bianalytic function

$$U(x, y) = \Re F(z),$$

where F is a bianalytic function.

Filshtinsky in [6,7] present double periodic bianalytic function in the form

$$F(z) = \varphi_{0,1}(z) + \varphi_{0,2}(z) + \bar{z}\varphi_{1,2}(z), \tag{8.59}$$

where

$$\varphi_{k,s}(z) = (-1)^{s-k-1}\binom{s-1}{k}\sum_{m=0}^{\infty} A_{m,s}\wp_{s-k-1}^{(s+m-1)}(z), \tag{8.60}$$

$A_{m,s}$ is complex constants.

Using the functions $\wp_{j,n}$ and formula (8.13) we get

$$F(z) = \phi_{0,1}(z) + \phi_{0,2}(z) + \bar{z}\phi_{1,2}(z), \tag{8.61}$$

where

$$\phi_{k,s}(z) = (-1)^k\binom{s-1}{k}\sum_{m=0}^{\infty} B_{m,s}\wp_{s-k-1,s+m+1}(z), \tag{8.62}$$

where $B_{m,s}$ are complex constants satisfying the relation $B_{m,s} = (-1)^m (s + m - 1)! A_{m,s}$. Thus

$$\phi_{0,1}(z) = \sum_{m=0}^{\infty} B_{m,1} \wp_{0,m+2}(z),$$

$$\phi_{0,2}(z) = \sum_{m=0}^{\infty} B_{m,2} \wp_{1,m+3}(z),$$

$$\phi_{1,2}(z) = - \sum_{m=0}^{\infty} B_{m,2} \wp_{0,m+3}(z).$$

Then the bianalytic function is the form

$$F(z) = \sum_{m=0}^{\infty} \left(B_{m,1} \wp_{0,m+2} + B_{m,2} \wp_{1,m+3}(z) - B_{m,2} \bar{z} \wp_{0,m+3}(z) \right), \qquad (8.63)$$

where

$$\wp_{0,m+2}(z) = \begin{cases} \sum_{\omega}{}'_e \frac{1}{(z-\omega)^2} - \sum_{\omega}{}'_e \frac{1}{\omega^2}, & m = 0, \\ \sum_{\omega}{}'_e \frac{1}{(z-\omega)^{m+2}}, & m > 0, \end{cases} \qquad (8.64)$$

and

$$\wp_{1,m+3}(z) = \begin{cases} \sum_{\omega}{}'_e \frac{\bar{\omega}}{(z-\omega)^3} + \sum_{\omega}{}'_e \frac{\bar{\omega}}{\omega^3}, & m = 0, \\ \sum_{\omega}{}'_e \frac{\bar{\omega}}{(z-\omega)^{m+3}}, & m > 0. \end{cases} \qquad (8.65)$$

Using the Eisenstein function we present (8.64) and (8.65) in the form

$$\wp_{0,m+2} = \begin{cases} E_2(z) - \frac{1}{z^2} - S_2, & m = 0, \\ E_{m+2}(z) - \frac{1}{z^{m+2}}, & m > 0, \end{cases}$$

and

$$\wp_{0,m+3}(z)\bar{z} - \wp_{1,m+3}(z) = \begin{cases} E_3^{(1)}(z) - \frac{\bar{z}}{z^3} - S_3^{(1)}, & m = 0, \\ E_{m+3}^{(1)}(z) - \frac{\bar{z}}{z^{m+3}}, & m > 0. \end{cases} \qquad (8.66)$$

188 Mechanics and Physics of Structured Media

From this

$$F(z) = \sum_{m=0}^{\infty} B_{m,1}\left(E_{m+2} - \eta_{m0}S_2 - \frac{1}{z^{m+2}}\right)$$
$$+ \sum_{m=0}^{\infty} B_{m,2}\left(E_{m+3}^{(1)}(z) - \eta_{m0}S_3^{(1)} - \frac{\overline{z}}{z^{m+3}}\right), \qquad (8.67)$$

where η_{m0} denotes Kronecker delta symbol.

It is worth noting that substitution $B_{01} = 1$, $B_{m1} = 0$ $(m = 1, 2, \ldots)$, and $B_{m2} = 0$ $(m = 0, 1, \ldots)$ into (8.67) yields Pokazeev's function represented by the absolutely convergent series [8]

$$\Omega_n(z) = \left(\frac{\overline{z}}{z}\right)^n + {\sum_{\omega}}' \left[\left(\frac{\overline{z-\omega}}{z-\omega}\right)^n - \left(\frac{\overline{\omega}}{\omega}\right)^n - n\left(\frac{\overline{\omega}}{\omega}\right)^n \frac{z}{\omega}\right.$$
$$- \frac{n(n+1)}{2}\left(\frac{\overline{\omega}}{\omega}\right)^n \frac{z^2}{\omega^2} + n\left(\frac{\overline{\omega}}{\omega}\right)^{n-1}\frac{\overline{z}}{\omega} + n^2\left(\frac{\overline{\omega}}{\omega}\right)^{n-1}\frac{\overline{z}z}{\omega^2}$$
$$\left. - \frac{n(n-1)}{2}\left(\frac{\overline{\omega}}{\omega}\right)^{n-2}\frac{\overline{z}^2}{\omega^2}\right]. \qquad (8.68)$$

Using the Eisenstein summation we can rewrite (8.68) in the form

$$\Omega_n(z) = E_n^{(n)}(z) - S_n^{(n)} - nS_{n+1}^{(n)}z - \frac{n(n+1)}{2}S_{n+2}^{(n)}z^2$$
$$+ n^2 S_{n+1}^{(n-1)}\overline{z}z + nS_n^{(n-1)}\overline{z} - \frac{n(n-1)}{2}S_n^{(n-2)}\overline{z}^2. \qquad (8.69)$$

8.7 Conclusion

Natanzon (1935) and Filshtinsky (1964) constructed their double periodic complex potentials having used the classic elliptic Weierstrass functions. They do not use the Eisenstein summation introduced in 1847 and the conditionally convergent lattice sums S_2 introduced by Lord Rayleigh in 1892 and developed in [14,15] including functional equations discussed in [16]. Application of the Eisenstein summation yields a set of new interesting relations in the theory of lattice sums and the corresponding generalizations of the Eisenstein series [17]. In the present chapter, we extend the Eisenstein-Natanzon-Filshtinsky functions to general double periodic polymeromorphic functions. The properties of new functions are systematically investigated. Representations of the generalization Eisenstein-Natanzon-Filshtinsky functions by means of algebraic relations on the Eisenstein type functions are established. These representations can be applied by implementation of the derived formulas to study polydispersed random composites following [18–20].

References

[1] N. Akhiezer, Elements of the Theory of Elliptic Functions, American Mathematical Society, Providence, RI, 1990.

[2] A. Weil, Elliptic Functions According to Eisenstein and Kronecker / Andre Weil, Springer-Verlag, Berlin, New York, 1976.

[3] V. Mityushev, Cluster method in composites and its convergence, Applied Mathematics Letters 77 (2018) 44–48, https://doi.org/10.1016/j.aml.2017.10.001.

[4] V.Y. Natanson, On the stresses in a stretched plate weakened by identical holes located in chessboard arrangement, Matematičeskij Sbornik 42 (5) (1935) 616–636.

[5] L. Fil'shtinskij, Stresses and displacements in an elastic sheet weakened by a doubly periodic set of equal circular holes, Journal of Applied Mathematics and Mechanics 28 (1964) 530–543.

[6] E. Grigolyuk, L. Fil'shtinskij, Perforated Plates and Shells (Perforirovannye Plastiny i Obolochki), Nauka, Moskva, 1970.

[7] E. Grigolyuk, L. Fil'shtinskij, Periodicheskie Kusochno-Odnorodnye Uprugie Struktury, Nauka, Moskva, 1992.

[8] V.V. Pokazeev, Polyanalytic doubly periodic functions, Trudy Seminara po Kraevym Zadacham 18 (1982) 155–167.

[9] P. Drygaś, V. Mityushev, Effective elastic properties of random two-dimensional composites, International Journal of Solids and Structures 97–98 (2016) 543–553, https://doi.org/10.1016/j.ijsolstr.2016.06.034.

[10] S. Yakubovich, P. Drygaś, V. Mityushev, Closed-form evaluation of two-dimensional static lattice sums, Proceedings of the Royal Society A. Mathematical, Physical and Engineering Sciences 472 (2195) (2016), https://doi.org/10.1098/rspa.2016.0510.

[11] L. Rayleigh, On the influence of obstacles arranged in rectangular order upon the properties of a medium, Philosophical Magazine (5) 34 (1892) 481–502.

[12] W.N.S. Gluzman, V. Mityushev, Computational Analysis of Structured Media, Elsevier, Amsterdam, 2017.

[13] P. Drygaś, Generalized Eisenstein functions, Journal of Mathematical Analysis and Applications 444 (2) (2016) 1321–1331, https://doi.org/10.1016/j.jmaa.2016.07.012.

[14] R.C. McPhedran, D.R. McKenzie, The conductivity of lattices of spheres I. The simple cubic lattice, Proceedings of the Royal Society of London. Series A, Mathematical and Physical Sciences 359 (1696) (1978) 45–63.

[15] D.R. Mckenzie, R.C. Mcphedran, G. Derrick, The conductivity of lattices of spheres II. The body centred and face centred cubic lattices, Proceedings of the Royal Society A. Mathematical, Physical and Engineering Sciences 362 (1978) 211–232, https://doi.org/10.1098/rspa.1978.0129.

[16] V. Mityushev, Functional equations in a class of analytic functions and composite materials, Demonstratio Mathematica 30 (1) (1997) 63–70.

[17] P. Chen, M. Smith, R. Mcphedran, Evaluation and regularization of phase-modulated Eisenstein series and application to double Schlömilch-type sums, Journal of Mathematical Physics 59 (2018) 072902, https://doi.org/10.1063/1.5026567.

[18] N. Rylko, Effective anti-plane properties of piezoelectric fibrous composites, Acta Mechanica 224 (11) (2013) 2719–2734.

[19] N. Rylko, Representative volume element in 2D for disks and in 3D for balls, Journal of Mechanics of Materials and Structures 9 (2014) 427–439, https://doi.org/10.2140/jomms.2014.9.427.

[20] P. Kurtyka, N. Rylko, Quantitative analysis of the particles distributions in reinforced composites, Composite Structures 182 (2017) 412–419, https://doi.org/10.1016/j.compstruct.2017.09.048.

Chapter 9

The slowdown of group velocity in periodic waveguides

Yuri A. Godin and Boris Vainberg
Department of Mathematics and Statistics, University of North Carolina at Charlotte, Charlotte, NC, United States

9.1 Introduction

Periodic media offer plenty of possibilities for manipulating wave propagation. The spectrum of propagating frequencies consists of closed intervals (bands) and the propagation of waves is suppressed for the frequencies in complimentary intervals (gaps). While propagating waves do not exist in the frequency gaps, the properties of waves in the bands can be essentially different in terms of their direction, speed, and amplitude of propagation. One example of such a highly dispersive medium is a fluid containing a periodic lattice of scatterers. The propagation of acoustic waves in such media through a lattice of air bubbles or cylinders of air has been studied numerically [1–7], along with some analytical approaches [8–11]. An efficient method for solving some two-dimensional periodic elastic problems was suggested in [12].

The band-gap structures of one-dimensional periodic problems have been studied by the well-known transfer (or propagation) matrix formalism, see [13, Sec. XIII], [14]. These results were applied in numerous publications for acoustic materials [15,16], periodic elastic media [17–20], photonic materials [21,22], etc. The dispersion relations obtained in most of the papers above were studied numerically. Our main goal is to study the dispersion relation analytically and provide asymptotic formulas for the group velocity when the concentration of scatterers is small or the ratio of the impedances is large. The leading order terms in both asymptotic formulas are shown to not depend on the frequency. Thus, a slowdown occurs at any frequency, in any band, and does not have a resonance character. We show that the smallest group velocity is achieved when the scatterers and the host material are taken in equal proportions regardless of their physical properties, provided that the impedance ratio is large. All the asymptotic formulas are verified numerically.

Mechanics and Physics of Structured Media. https://doi.org/10.1016/B978-0-32-390543-5.00014-1
Copyright © 2022 Elsevier Inc. All rights reserved.

9.2 Acoustic waves

We consider the propagation of acoustic waves through a one-dimensional periodic medium with period ℓ consisting of two materials with mass densities $\varrho_\mp$ and wave speed $c_\mp$ when $n\ell < x < n\ell + \delta$ and $n\ell + \delta < x < (n+1)\ell$, $n \in \mathbb{Z}$, respectively (see Fig. 9.1).

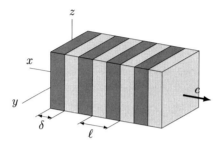

FIGURE 9.1 One-dimensional periodic medium with alternating layers.

Assuming that the excess pressure $p(x,t)$ is time-harmonic $p(x,t) = u(x)e^{-i\omega t}$, the amplitude $u(x)$ of the acoustic wave satisfies the following equation. If $u^\mp$ are the values of u in the inclusions and the medium, respectively, then

$$\frac{d^2 u^-}{dx^2} + k_-^2 u^- = 0, \quad n\ell < x < n\ell + \delta, \tag{9.1}$$

$$\frac{d^2 u^+}{dx^2} + k_+^2 u^+ = 0, \quad n\ell + \delta < x < (n+1)\ell, \tag{9.2}$$

subject to the continuity conditions across all the interfaces $\{n\ell\}$ and $\{n\ell + \delta\}$:

$$u^- = u^+, \quad \frac{1}{\varrho_+} u_x^+ = \frac{1}{\varrho_-} u_x^-, \tag{9.3}$$

where $k_\pm = \dfrac{\omega}{c_\pm}$ is the wavenumber of the corresponding medium. We are looking for the Floquet solutions of (9.1)–(9.3) satisfying the condition

$$e^{i\kappa x} u \text{ is periodic with the period } \ell, \tag{9.4}$$

with some real κ (the wavenumber).

We write the transfer matrices $M_{\alpha,\beta}$ for all intervals (α, β) within the period as well as for the interfaces $x = \delta$ and $x = \ell$:

$$M_{0_+,\delta_-} = \begin{bmatrix} \cos k_- \delta & \dfrac{\sin k_- \delta}{k_-} \\ -k_- \sin k_- \delta & \cos k_- \delta \end{bmatrix}, \tag{9.5}$$

$$M_{\delta_-,\delta_+} = \begin{bmatrix} 1 & 0 \\ 0 & \dfrac{\varrho_+}{\varrho_-} \end{bmatrix}, \tag{9.6}$$

$$M_{\delta_+,\ell_-} = \begin{bmatrix} \cos k_+(\ell - \delta) & \dfrac{\sin k_+(\ell - \delta)}{k_+} \\ -k_+ \sin k_+(\ell - \delta) & \cos k_+(\ell - \delta) \end{bmatrix}, \tag{9.7}$$

$$M_{\ell_-,\ell_+} = \begin{bmatrix} 1 & 0 \\ 0 & \dfrac{\varrho_-}{\varrho_+} \end{bmatrix}. \tag{9.8}$$

Then the transfer matrix M_{0_+,ℓ_+} over the period $(0_+, \ell_+)$ becomes

$$M_{0_+,\ell_+} = M_{\ell_-,\ell_+} M_{\delta_+,\ell_-} M_{\delta_-,\delta_+} M_{0_+,\delta_-}. \tag{9.9}$$

Problem (9.1)–(9.4) has a nontrivial solution if and only if $\lambda = e^{\pm i \kappa \ell}$ is an eigenvalue of M_{0_+,ℓ_+}.

We will assume that $\ell = 1$. The eigenvalues λ of the transfer matrix over the period satisfy the characteristic equation

$$\lambda^2 - 2\lambda F + 1 = 0, \tag{9.10}$$

where

$$F = \cos \frac{(1 - \delta)\,\omega}{c_+} \cos \frac{\omega \delta}{c_-} - \frac{1}{2} \left(\frac{z_+}{z_-} + \frac{z_-}{z_+} \right) \sin \frac{(1 - \delta)\,\omega}{c_+} \sin \frac{\omega \delta}{c_-} \tag{9.11}$$

and

$$z_\pm = c_\pm \varrho_\pm \tag{9.12}$$

are characteristic acoustic impedances of the two media, respectively.

The solution of (9.10) in the form $\lambda = e^{i \kappa}$, with real κ, exists if and only if $|F| \leqslant 1$. Moreover,

$$\cos \kappa = F. \tag{9.13}$$

Thus, the inequality $|F| \leqslant 1$ provides the necessary and sufficient condition for which problem (9.1)–(9.4) has a nontrivial solution. The intervals of ω where this condition holds are called the bands. From (9.13) we can find the group velocity $c_g = \dfrac{\partial \omega}{\partial \kappa}$ in any band:

$$c_g = -\frac{\sin \kappa}{\dfrac{\partial F}{\partial \omega}}. \tag{9.14}$$

194 Mechanics and Physics of Structured Media

Denoting

$$t_+ = \frac{\delta}{c_-} + \frac{1-\delta}{c_+}, \quad t_- = \frac{1-\delta}{c_+} - \frac{\delta}{c_-}, \tag{9.15}$$

$$\eta = \frac{1}{2}\left(\frac{z_+}{z_-} + \frac{z_-}{z_+}\right), \tag{9.16}$$

the group velocity (9.14) takes the form

$$c_g = 2\frac{\sqrt{1 - \frac{1}{4}\left((1+\eta)\cos\omega t_+ + (1-\eta)\cos\omega t_-\right)^2}}{(1+\eta)\,t_+\sin\omega t_+ + (1-\eta)\,t_-\sin\omega t_-}. \tag{9.17}$$

We note that in the bands the square root expression in (9.17) equals $1 - F^2$ and therefore nonnegative, while F is a strictly monotone function of ω [13] and hence the denominator F_ω does not vanish.

9.2.1 Equal impedances

If the characteristic impedances of the two media are equal, $c_+\varrho_+ = c_-\varrho_-$, then $\eta = 1$ and formula (9.17) reduces to $c_g = t_+^{-1}$. This means that the group velocity coincides with the velocity of the wave that propagates through the interval $(0, \delta)$ with the speed c_- and over the interval $(\delta, 1)$ with the speed c_+ without slowing down or acceleration. From (9.13) we have

$$F = \cos\left(\frac{\delta\omega}{c_-} + \frac{(1-\delta)\omega}{c_+}\right) \tag{9.18}$$

and

$$\kappa = \left(\frac{\delta}{c_-} + \frac{1-\delta}{c_+}\right)\omega. \tag{9.19}$$

Then the group velocity c_g, the phase velocity $c_{ph} = \frac{\omega}{\kappa}$, and the average speed of propagation $\langle c \rangle = t_+^{-1}$ coincide and are equal to

$$\langle c \rangle = \left(\frac{\delta}{c_-} + \frac{1-\delta}{c_+}\right)^{-1}. \tag{9.20}$$

Hence, waves propagate without reflection and the speed does not depend on the wave frequency.

9.2.2 Small scatterers

Let us find an approximation of the eigenvalues λ of the transfer matrix when $\delta \ll 1$. We approximate (9.10) in any fixed band using linear in δ approximation of (9.11):

$$\lambda^2 - 2\lambda \left(\cos k_+ - \left(\frac{z_+}{z_-} + \frac{z_-}{z_+} \right) \frac{\omega \delta}{2c_-} \sin k_+ + O\left(\delta^2\right) \right) + 1 = 0. \quad (9.21)$$

Then

$$\lambda = e^{i\kappa} = e^{ik_+} \left(1 + i \left(\frac{z_+}{z_-} + \frac{z_-}{z_+} \right) \frac{\omega \delta}{2c_-} \right) + O\left(\delta^2\right), \quad (9.22)$$

and

$$\kappa = \frac{\omega}{c_+} \left(1 + \left(\frac{z_+}{z_-} + \frac{z_-}{z_+} \right) \frac{c_+}{2c_-} \delta + O\left(\delta^2\right) \right). \quad (9.23)$$

Expression (9.23) allows us to find the group velocity c_g (which coincides with the phase velocity in this approximation)

$$c_g = c_+ \left(1 - \left(\frac{z_+}{z_-} + \frac{z_-}{z_+} \right) \frac{c_+}{2c_-} \delta + O\left(\delta^2\right) \right), \quad \delta \ll 1. \quad (9.24)$$

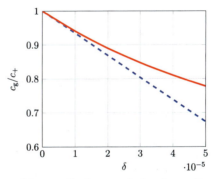

FIGURE 9.2 Dependence of the normalized group velocity of a one-dimensional periodic acoustic waveguide consisting of air layers in water as a function of the volume fraction δ of air when $\delta \ll 1$. The solid red line corresponds to exact velocity (9.17), while the blue dashed line corresponds to the linear approximation (9.24).

For soft scatterers, the coefficient in front of δ in (9.24) can be very large in magnitude which results in a dramatic group velocity reduction, even for a tiny concentration of the scatterers. Such an example is shown in Fig. 9.2, where an acoustic wave with frequency $\omega = 10 \text{ s}^{-1}$ propagates through a periodic waveguide consisting of layers of the air ($\varrho_- = 1.3 \text{ kg/m}^3$, $c_- = 340$ m/s) of width δ followed by a layer of water ($\varrho_+ = 1000 \text{ kg/m}^3$, $c_+ = 1400$ m/s). The solid red line shows the exact value of the group velocity (9.17) while the blue dashed line corresponds to the linear approximation by (9.24), for a very small concentration of the air. A sharp decrease in group velocity of acoustic waves is also valid

9.2.3 Highly mismatched impedances

When the characteristic impedances of the two media differ substantially, i.e. $\eta \gg 1$, and, in addition, $\gamma_\pm = \dfrac{\eta \omega^2}{c_\pm^2} \ll 1$, the group velocity away from the points $\delta = 0$ and $\delta = 1$ is approximated by

$$c_g = \sqrt{\frac{c_+ c_-}{2\delta(1-\delta)\eta}}. \tag{9.25}$$

The latter formula agrees with (9.17) with an accuracy of $O\left(\eta^{-1} + \gamma_+ + \gamma_-\right)$. The minimum of (9.25) is attained at $\delta = \tfrac{1}{2}$ and does not depend on ω. Using (9.16), we have

$$\min c_g = 2\left(\frac{c_+ c_-}{\frac{z_+}{z_-} + \frac{z_-}{z_+}}\right)^{\frac{1}{2}} = 2 c_+ c_- \left(\frac{\varrho_+ + \varrho_-}{c_+^2 \varrho_+^2 + c_-^2 \varrho_-^2}\right)^{\frac{1}{2}}. \tag{9.26}$$

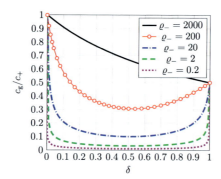

FIGURE 9.3 Dependence of the normalized group velocity of a one-dimensional periodic acoustic waveguide on the volume fraction $\delta \in [0, 1]$ of the scattering medium (from (9.17)). The top solid black line corresponds to equal impedances of the two media. Every subsequent curve below corresponds to a tenfold decrease of the impedance of the scattering medium. The wave frequency for each curve is chosen in such a way to be in the middle of the first band for $\delta = 0.5$.

Graphs of the group velocities for different ratios of the impedances are shown in Fig. 9.3, where the top black solid curve corresponds to a periodic acoustic medium comprised of the scattering acoustic material with $c_- = 1000$ m/s, $\varrho_- = 2000$ kg/m^3 and the volume fraction δ in the host medium with $c_- = 2000$ m/s and $\varrho_- = 1000$ kg/m^3. The speed of propagation for matched impedances is given by (9.20). Every subsequent curve below corresponds to a tenfold decrease of the density ϱ_-, while other parameters are kept the same. To avoid confusion with band-edge effects, the frequencies of wave propagation were chosen in the center of the first band for $\delta = 0.5$ and are equal 1200, 400, 130, and 40 s^{-1}, respectively. On the other hand, similar graphs for a fixed frequency $\omega = 10$ s^{-1} are practically indistinguishable.

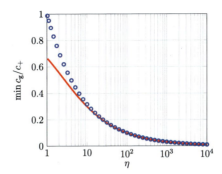

FIGURE 9.4 Dependence of the normalized minimum group velocity min c_g (achieved at $\delta = 0.5$) on the impedance contrast η (9.16) when $\omega = 10$ s^{-1} in a one-dimensional periodic waveguide. The solid red line corresponds to the exact value c_g given by (9.17). The blue circles represent the asymptotic formula (9.26).

We verify formula (9.26) in Fig. 9.4. The solid red line corresponds to the exact value of the minimal group velocity given by (9.17) while the blue circles represent the asymptotic formula (9.26). Below we show that formula (9.17) and its approximation (9.26) are valid for electromagnetic and elastic waves as well.

9.3 Electromagnetic waves

Propagation of electromagnetic waves through a one-dimensional periodic stacked medium with period ℓ consisting of two types of alternating layers is also described by Eqs. (9.1)–(9.2), where

$$k_\pm = \frac{\omega}{c}\sqrt{\varepsilon_\pm \mu_\pm}, \tag{9.27}$$

c is the speed of light, and $\varepsilon_\pm$, $\mu_\pm$ are the relative permittivity and permeability of the layers, respectively. For the magnetic component u of the electromagnetic field continuity conditions have the form

$$u^- = u^+, \quad \frac{1}{\varepsilon_+} u_x^+ = \frac{1}{\varepsilon_-} u_x^-. \tag{9.28}$$

Together with the Floquet condition (9.4), we obtain the characteristic equation (9.10), where

$$z_\pm = \sqrt{\frac{\mu_\pm}{\varepsilon_\pm}} \tag{9.29}$$

198 Mechanics and Physics of Structured Media

is the wave impedance of the corresponding medium. From (9.24) we obtain the group velocity

$$c_g = c_+ \left(1 - \frac{1}{2} \left(\frac{\mu_-}{\mu_+} + \frac{\varepsilon_-}{\varepsilon_+} \right) \delta + O\left(\delta^2\right) \right), \quad \delta \ll 1, \tag{9.30}$$

where $c_+ = \dfrac{c}{\sqrt{\varepsilon_+ \mu_+}}$. Calculation of the minimum group velocity from (9.26) yields

$$\min c_g = \frac{c}{\sqrt{\varepsilon_- \mu_+ + \varepsilon_+ \mu_-}}. \tag{9.31}$$

9.4 Elastic waves

One-dimensional time-harmonic propagation of elastic waves through a periodic in the x-direction stack of layers is governed by the equations [25]

$$\frac{d^2 u_1}{dx^2} + \frac{\omega^2}{c_p^2} u_1 = 0, \tag{9.32}$$

$$\frac{d^2 u_\alpha}{dx^2} + \frac{\omega^2}{c_s^2} u_\alpha = 0, \quad \alpha = 2, 3, \tag{9.33}$$

where the displacement vector $u = (u_1, u_2, u_3)$ depends only on x, and

$$c_p = \sqrt{\frac{\lambda + 2\mu}{\varrho}}, \quad c_s = \sqrt{\frac{\mu}{\varrho}} \tag{9.34}$$

are velocities of the dilatational and shear waves. In this section, μ and λ denote the Lamé parameters and ϱ is the mass density. Also, the displacement vector u and the normal component of the stress tensor σ must be continuous across the interfaces

$$[\![u]\!] = 0, \quad [\![\sigma \cdot n]\!] = 0. \tag{9.35}$$

The latter condition in terms of the displacement components reads

$$(\lambda_- + 2\mu_-) \frac{du_1^-}{dx} = (\lambda_+ + 2\mu_+) \frac{du_1^+}{dx}, \tag{9.36}$$

$$\mu_- \frac{du_\alpha^-}{dx} = \mu_+ \frac{du_\alpha^+}{dx}, \quad \alpha = 2, 3. \tag{9.37}$$

As a result, we obtain the same characteristic equation as in (9.10) with the impedance $z_p^\pm = \varrho_\pm c_p^\pm$ for dilatational waves and $z_s^\pm = \varrho_\pm c_s^\pm$ for shear waves. Formulas (9.17), (9.24)–(9.26) are valid for the two types of waves.

9.5 Discussion

The one-dimensional model considered above agrees with the numerical results obtained in two and three-dimensional cases [2–5,23]. Unlike a slowing down around inflection points of the dispersion curve in [26], in our case, a slowing down happens for a wide frequency range. Since the leading terms of the asymptotic formulas for the group velocity do not depend on the frequency, the slowdown of the group velocity is not related to the resonant phenomenon [2] which occurs only at specific frequencies, but rather to a strong scattering and destructive interference of the waves.

Acknowledgments

The work of BV was supported by the NSF grant DMS-1714402 and the Simons Foundation grant 527180.

References

[1] Anthony A. Ruffa, Acoustic wave propagation through periodic bubbly liquids, J. Accoust. Soc. Jpn. 91 (1) (1992) 1–11.

[2] M. Kafesaki, R.S. Penciu, E.N. Economou, Air bubbles in water: a strongly multiple scattering medium for acoustic waves, Phys. Rev. Lett. 84 (2000) 6050–6053.

[3] A.A. Krokhin, J. Arriaga, L.N. Gumen, Speed of sound in periodic elastic composites, Phys. Rev. Lett. 91 (Dec 2003) 264302.

[4] Daniel Torrent, José Sánchez-Dehesa, Effective parameters of clusters of cylinders embedded in a nonviscous fluid or gas, Phys. Rev. B 74 (Dec 2006) 224305.

[5] Daniel Torrent, José Sánchez-Dehesa, Acoustic metamaterials for new two-dimensional sonic devices, New J. Phys. 9 (9) (Sep 2007) 323.

[6] Valentin Leroy, Alice Bretagne, Mathias Fink, Herve Willaime, Patrick Tabeling, Arnaud Tourin, Design and characterization of bubble phononic crystals, Appl. Phys. Lett. 95 (17) (2009).

[7] Alex Skvortsov, Ian MacGillivray, Gyani Shankar Sharma, Nicole Kessissoglou, Sound scattering by a lattice of resonant inclusions in a soft medium, Phys. Rev. E 99 (6) (2019).

[8] A.Yu. Belyaev, Propagation of waves in a continuum medium with periodically arranged inclusions, Sov. Dokl. 296 (4) (1987) 828–831 (in Russian).

[9] Zhen Ye, Acoustic scattering by periodic array of air-bubbles, Acta Acust. Acust. 89 (2003) 435–444.

[10] Edgar Reyes-Ayona, Daniel Torrent, José Sánchez-Dehesa, Homogenization theory for periodic distributions of elastic cylinders embedded in a viscous fluid, J. Accoust. Soc. Jpn. 132 (4) (2012) 2896–2908.

[11] Claude Boutin, Acoustics of porous media with inner resonators, J. Accoust. Soc. Jpn. 134 (6, 2, SI) (2013) 4717–4729.

[12] E.I. Grigolyuk, L.A. Filshtinsky, Perforated Plates and Shells, Nauka, 1970 (in Russian).

[13] M. Reed, B. Simon, Analysis of Operators, vol. 4, Academic Press, 1978.

[14] P.A. Kuchment, Floquet theory for partial differential equations, Russ. Math. Surv. 37 (4) (1982) 1–60.

[15] Z. Ye, E. Hoskinson, Band gaps and localization in acoustic propagation in water with air cylinders, Appl. Phys. Lett. 77 (26) (2000) 4428–4430.

[16] J.O. Vasseur, P.A. Deymier, B. Chenni, B. Djafari-Rouhani, L. Dobrzynski, D. Prevost, Experimental and theoretical evidence for the existence of absolute acoustic band gaps in two-dimensional solid phononic crystals, Phys. Rev. Lett. 86 (Apr 2001) 3012–3015.

200 Mechanics and Physics of Structured Media

[17] R. Esquivel-Sirvent, G.H. Cocoletzi, Band structure for the propagation of elastic waves in superlattices, J. Accoust. Soc. Jpn. 95 (1) (1994) 86–90.

[18] M.M. Sigalas, C.M. Soukoulis, Elastic-wave propagation through disordered and/or absorptive layered systems, Phys. Rev. B 51 (Feb 1995) 2780–2789.

[19] Mahmoud I. Hussein, Gregory M. Hulbert, Richard A. Scott, Dispersive elastodynamics of 1D banded materials and structures: analysis, J. Sound Vib. 289 (4) (2006) 779–806.

[20] E.H. El Boudouti, B. Djafari-Rouhani, A. Akjouj, L. Dobrzynski, Acoustic waves in solid and fluid layered materials, Surf. Sci. Rep. 64 (11) (2009) 471–594.

[21] Jon M. Bendickson, Jonathan P. Dowling, Michael Scalora, Analytic expressions for the electromagnetic mode density in finite, one-dimensional, photonic band-gap structures, Phys. Rev. E 53 (Apr 1996) 4107–4121.

[22] Pochi Yeh, Optical Waves in Layered Media, Wiley, 2005.

[23] Yuri A. Godin, Boris Vainberg, Dispersion of waves in two and three-dimensional periodic media, Waves Random Complex Media, 1–24, Advanced online publication, https://doi.org/10.1080/17455030.2020.1810822.

[24] Yuri A. Godin, Propagation of longitudinal waves in a random binary rod, Waves Random Complex Media 16 (4) (2006) 409–416.

[25] L.D. Landau, E.M. Lifshits, Theory of Elasticity, 2d English ed., rev. and enl. edition, Course of Theoretical Physics, vol. 7, Pergamon Press, Oxford, 1970.

[26] A. Figotin, I. Vitebskiy, Slow light in photonic crystals, Waves Random Complex Media 16 (3) (Aug 2006) 293–382.

Chapter 10

Some aspects of wave propagation in a fluid-loaded membrane

Julius Kaplunov, Ludmila Prikazchikova, and Sheeru Shamsi
School of Computing and Mathematics, Keele University, Keele, Staffordshire, United Kingdom

10.1 Introduction

Dynamic problems in fluid-structure interaction have been intensively studied for thin elastic plates and shells, inspired by numerous engineering applications, including under water acoustics, e.g. see monographs [1–3], and also papers [4,5] to name a few. Similar considerations for membranes have been given less attention. Among the publications tackling fluid loaded membranes we mention [6], analyzing in a great detail the asymptotic behavior of the Green function, [7–9], investigating time harmonic diffraction, as well as [10] dealing with chaotic motion. A recent paper [11] is an example of fluid-structure interaction analysis in the area of soft robotics.

A growing interest to modeling of soft materials, along with other modern developments, stimulates implementation of the prestress concept within adapted formulations for elastic structures. Asymptotic analysis in [12–14] demonstrates that the leading order low-frequency approximation of the 3D dynamic equations for a prestressed elastic layer corresponds to a membrane, governed by the classical 2D wave equation. In this case, tension in the membrane is expressed through the prestress parameters.

In this chapter, we consider wave motion of an infinite membrane, laying on an infinite compressible fluid. First, we study dispersion of time harmonic waves for various sets of problem parameters. Long wave and short wave approximations are derived along with analysis of the limit of incompressible fluid. The evanescent wave, radiating vibration energy into the fluid, is also investigated.

Next, we proceed to the 2D steady-state problem for a point load moving along the membrane with a uniform speed. Previously, analogous moving load problems have been considered for elastically supported membranes and strings, e.g. see [15,16]. Both subsonic and supersonic regimes are treated, taking into account the relation between the wave speeds in the membrane and fluid as well as other problem parameters. The solution is expressed through the Fourier in-

Mechanics and Physics of Structured Media. https://doi.org/10.1016/B978-0-32-390543-5.00015-3
Copyright © 2022 Elsevier Inc. All rights reserved.

202 Mechanics and Physics of Structured Media

tegral which is calculated using the contour integration technique. The presence of poles at the real axis also necessitates making use of the limiting absorption principle [17,18]. The solutions are expressed in terms of exponential and related integrals. Far and near field asymptotics are obtained.

The explicit model solutions presented in the chapter seem to be of interest for validating more elaborated formulations in fluid-structure interactions. The latter may involve dispersive high order approximations for a membrane, see [12], as well as refined asymptotic theories starting from analysis of the original 3D coupled problem.

10.2 Statement of the problem

Consider a 2D plane problem for an elastic membrane resting on an ideal compressible fluid in Cartesian coordinates $-\infty < x < +\infty$ and $0 < y < +\infty$. The governing equations can be written as:

the equation of membrane motion

$$T\frac{\partial^2 u}{\partial x^2} - \rho_m \frac{\partial^2 u}{\partial t^2} = p\big|_{y=0} + f, \tag{10.1}$$

the equation of fluid motion

$$\Delta p = \frac{1}{c_f^2}\frac{\partial^2 p}{\partial t^2}, \tag{10.2}$$

impenetrability condition

$$\rho_f \frac{\partial^2 u}{\partial t^2} = -\frac{\partial p}{\partial y}\bigg|_{y=0}. \tag{10.3}$$

In the formulas above T is tension of a membrane, $u(x,t)$ – displacement of a membrane, ρ_f and ρ_m are densities of the fluid and membrane, respectively, c_f – sound speed in the fluid, $p(x,y,t)$ – pressure in the fluid, $f(x,t)$ is a vertical force, and t – time.

In the limiting case of an incompressible fluid ($c_f \to \infty$), Eq. (10.2) becomes

$$\Delta p = 0. \tag{10.4}$$

In what follows we consider two setups. The first one corresponds to dispersion of the harmonic waves in the system "membrane-fluid" with the emphasis on long and short-wave limiting behaviors as well as on the effect of compressibility. The second setup is concerned with a steady state moving load problem, including the effect of the ratio of the speed of the load and the sound speed in fluid.

Some aspects of wave propagation in a fluid-loaded membrane Chapter | 10 **203**

10.3 Dispersion relation

First start with Eqs. (10.1)–(10.3) from the previous section, assuming that $f(x,t) = 0$ and

$$p = p_0(y)e^{i(kx-\omega t)},$$
$$u = u_0 e^{i(kx-\omega t)}, \tag{10.5}$$

where k is a wave number, ω is angular frequency, u_0 is a nonzero constant, and p_0 is a function to be found.

On substituting (10.5) into (10.1)–(10.3)

$$\frac{d^2 p_0}{dy^2} + p_0 \left(\frac{\omega^2}{c_f^2} - k^2 \right) = 0,$$
$$u_0(\rho_m \omega^2 - Tk^2) = p_0(0), \tag{10.6}$$
$$\omega^2 \rho_f u_0 = \frac{dp_0}{dy}\bigg|_{y=0},$$

and taking into account the radiation condition at infinity, e.g. see [19] we arrive at the dispersion relation

$$k \frac{\rho_m}{\rho_f} \sqrt{1 - \frac{c^2}{c_f^2}(c^2 - c_m^2)} + c^2 = 0, \tag{10.7}$$

where $c = \omega/k$ is the phase velocity and $c_m = \sqrt{T/\rho_m}$ is the wave speed in a membrane. The analysis of (10.7) requires two considerations: $c < c_f$ and $c > c_f$. In dimensionless form it becomes

$$K \sqrt{1 - c_{mf}^2 C^2} (C^2 - 1) + C^2 = 0, \quad \text{if} \quad c < c_f \tag{10.8}$$

or

$$iK \sqrt{c_{mf}^2 C^2 - 1}(C^2 - 1) - C^2 = 0, \quad \text{if} \quad c > c_f \tag{10.9}$$

where

$$C = \frac{c}{c_m}, \quad K = k\frac{\rho_m}{\rho_f}, \quad c_{mf} = \frac{c_m}{c_f}.$$

The relation given by (10.8) corresponds to exponential decay whereas the relation given by (10.9) corresponds to radiation. Now proceed to the analysis of the derived dispersion relation. In the long-wave limit ($K \ll 1$), we obtain from (10.8) a two-term expansion

$$C^2 = K - K^2 \left(1 + \frac{c_{mf}^2}{2} \right). \tag{10.10}$$

The short wave limit ($K \gg 1$) is more technical and affected by the ratio of the wave speeds in a membrane and fluid. It is separated into three cases, for which two-term asymptotic expansions of (10.8) take the form

(i) $c_m > c_f$

$$C^2 = \frac{1}{c_{mf}^2} - \frac{1}{c_{mf}^2 (c_{mf}^2 - 1)^2} \frac{1}{K^2}, \qquad (10.11)$$

(ii) $c_m = c_f$

$$C^2 = 1 - \frac{1}{K^{\frac{2}{3}}}, \qquad (10.12)$$

(iii) $c_m < c_f$

$$C^2 = 1 - \frac{1}{\sqrt{1 - c_{mf}^2}} \frac{1}{K}. \qquad (10.13)$$

For the dispersion relation (10.9), we only obtain the following two-term expansion in the short wave limit

$$C^2 = 1 - \frac{i}{\sqrt{c_{mf}^2 - 1}} \frac{1}{K} \qquad (10.14)$$

for $c_m > c_f$. It is worth noting that formula (10.14) contains a small imaginary term, corresponding to the dissipation due to radiation of the vibration energy into the fluid. All the other limiting behaviors above do not demonstrate presence of dissipation.

For an incompressible fluid governed by the Laplace equation (10.4) the associated dispersion relation takes a simpler form, i.e.

$$C^2 = \frac{K}{K + 1}. \qquad (10.15)$$

Obviously, it also follows from (10.8) at $c_{mf} \to \infty$. In the long and short-wave limits we have from (10.15), respectively

$$C^2 = K - K^2, \qquad (10.16)$$

and

$$C^2 = 1 - \frac{1}{K}. \qquad (10.17)$$

Numerical results for $c < c_f$ and $c > c_f$ with $c_{mf} = 1.5$ are presented in Figs. 10.1 and 10.2, respectively. In Fig. 10.1 long and short-wave limiting asymptotic relations (10.10) and (10.11) are plotted together with exact dispersion relation (10.8), demonstrating good agreement over the related regions $K \ll 1$ and $K \gg 1$. In Fig. 10.2 the dispersion equation corresponding to radiation into the fluid, see (10.9), is displayed together with short wave limiting behavior (10.14) for real and imaginary parts.

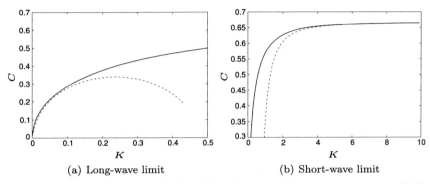

FIGURE 10.1 Dispersion relation (10.8) (solid line) together with asymptotic expansions (10.10) and (10.11) (dashed lines) in (a) and (b), respectively, with $c_{mf} = 1.5$.

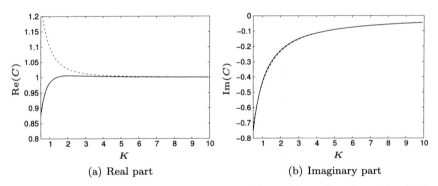

FIGURE 10.2 Real and imaginary parts of wave speed C according to dispersion relation (10.9) (solid line) and their short wave limiting forms (10.14) (dashed line) with $c_{mf} = 1.5$.

10.4 Moving load problem

Consider the steady-state moving problem for which the vertical load in Eq. (10.1) takes the form $f(x,t) = f_0 \delta(x - vt)$, where f_0 and v are given constants. First change the longitudinal variable in Eqs. (10.1) and (10.2) by $\xi = x - vt$. Then these equations become

$$(T - \rho v^2)\frac{\partial^2 u}{\partial \xi^2} = p\big|_{y=0} + f_0 \delta(\xi) \qquad (10.18)$$

206 Mechanics and Physics of Structured Media

and

$$\left(1 - \frac{v^2}{c_f^2}\right)\frac{\partial^2 p}{\partial \xi^2} + \frac{\partial^2 p}{\partial y^2} = 0. \tag{10.19}$$

We also transform (10.3) in terms of ξ

$$\rho_f v^2 \frac{\partial^2 u}{\partial \xi^2} = -\left.\frac{\partial p}{\partial y}\right|_{y=0}. \tag{10.20}$$

Next, applying the Fourier transform in ξ to Eqs. (10.18)–(10.20), we obtain

$$-s^2 U(T - \rho_m v^2) = P + f_0, \tag{10.21}$$

$$s^2 U \rho_f v^2 = \frac{dP}{dy}, \tag{10.22}$$

at $y = 0$ and

$$\frac{d^2 P}{dy^2} - s^2 P \left(1 - \frac{v^2}{c_f^2}\right) = 0, \tag{10.23}$$

where

$$P(s, y) = \int_{-\infty}^{\infty} p(\xi, y)e^{-is\xi}\,d\xi, \qquad U(s) = \int_{-\infty}^{\infty} u(\xi)e^{-is\xi}\,d\xi. \tag{10.24}$$

The solution of the problem (10.21)–(10.23) is given by

$$U = \frac{\alpha\beta|s|f_0}{s^2 v^2 \rho_f(\beta|s| + \alpha)}, \qquad P = -\frac{\alpha f_0}{\beta|s| + \alpha}e^{-\beta|s|y}, \tag{10.25}$$

for the subsonic regime in which $v < c_f$, whereas

$$U = \frac{i\alpha\beta|s|f_0}{s^2 v^2 \rho_f(i\beta|s| - \alpha)}, \qquad P = \frac{\alpha f_0}{i\beta|s| - \alpha}e^{i\beta|s|y}, \tag{10.26}$$

when $v > c_f$, i.e. for the supersonic regime. In the above

$$\alpha = \frac{v^2 \rho_f}{\rho_m v^2 - T}, \qquad \beta = \left|1 - \frac{v^2}{c_f^2}\right|^{1/2}. \tag{10.27}$$

Note that the last formulas are derived using the above mentioned radiation condition. In what follows we concentrate on the fluid pressure at the surface of the half space ($y = 0$). It is given by the inverse Fourier transform

$$p(\xi, 0) = \frac{1}{2\pi}\int_{-\infty}^{\infty} P(s, 0)e^{is\xi}\,ds. \tag{10.28}$$

Some aspects of wave propagation in a fluid-loaded membrane **Chapter | 10 207**

10.5 Subsonic regime

In this case we have from (10.28)

$$p(\xi, 0) = -\frac{\alpha f_0}{2\pi} I, \qquad (10.29)$$

with $I = I_- + I_+$, where

$$I_- = \int_{-\infty}^{0} \frac{e^{is\xi}}{-\beta s + \alpha} \, ds, \qquad I_+ = \int_{0}^{\infty} \frac{e^{is\xi}}{\beta s + \alpha} \, ds. \qquad (10.30)$$

First consider the range of the problem parameters for which $\alpha > 0$, i.e. $c_m < v < c_f$. Then we obtain from (10.30)

$$I = 2 \int_{0}^{\infty} \frac{\cos(s\xi)}{\beta s + \alpha} \, ds = J, \qquad (10.31)$$

with

$$J = -\frac{2}{\beta} \left(\sin\left(\frac{|\alpha\xi|}{\beta}\right) \mathrm{si}\left(\frac{|\alpha\xi|}{\beta}\right) + \cos\left(\frac{|\alpha\xi|}{\beta}\right) \mathrm{Ci}\left(\frac{|\alpha\xi|}{\beta}\right) \right), \qquad (10.32)$$

where

$$\mathrm{si}(z) = \mathrm{Si}(z) - \frac{\pi}{2}, \quad \mathrm{Si}(z) = \int_{0}^{z} \frac{\sin t}{t} \, dt, \quad \mathrm{Ci}(z) = \gamma + \ln z + \int_{0}^{z} \frac{\cos t - 1}{t} \, dt,$$

and γ being an Euler constant, e.g. see [20].

The far ($|\xi| \gg 1$) and near field ($|\xi| \ll 1$) asymptotic behaviors of solution (10.32) are respectively lead to

$$I = \frac{2\beta}{\alpha^2 \xi^2} \qquad (10.33)$$

and

$$I = -\frac{2}{\beta} \left(\gamma + \ln\left(\frac{|\alpha\xi|}{\beta}\right) \right). \qquad (10.34)$$

Here and below the far and near field asymptotics can be derived from original integral (10.31) and similar integrals in what follows. In particular, formula (10.33) is the result of integrating by parts in (10.31).

Over the parameter range in which $\alpha < 0$ ($v < \min[c_f, c_m]$) the poles

$$s_\pm = \mp\frac{\alpha}{\beta} \qquad (10.35)$$

belong to the intervals of integration in (10.30). To this end, insert the small term $-\varepsilon \partial u / \partial t$ into Eq. (10.1) according to the limiting absorption principle [17,18].

208 Mechanics and Physics of Structured Media

A slightly modified integral transform procedure in Section 10.4 validates the integration over the contours, shown in Figs. 10.3 and 10.4 for $\xi < 0$ and $\xi > 0$, respectively.

For $\xi < 0$ (Fig. 10.3) we have

$$I_{\pm} + I_{\pm}^{R} + I_{\pm}^{Im} = 0, \tag{10.36}$$

where the integrals $I_{\pm}^{R}$ correspond to the integration over the quarter circles of the radius R; it is obvious that $I_{\pm}^{R} \to 0$ as $R \to \infty$. Therefore, $I_{\pm} = -I_{\pm}^{Im}$, where

$$I_{-}^{Im} = -\int_{0}^{\infty} \frac{e^{s\xi}}{\beta s - i\alpha} ds, \quad I_{+}^{Im} = -\int_{0}^{\infty} \frac{e^{s\xi}}{\beta s + i\alpha} ds. \tag{10.37}$$

Finally,

$$I = 2\beta \int_{0}^{\infty} \frac{s e^{-s|\xi|}}{\beta^2 s^2 + \alpha^2} ds = J, \tag{10.38}$$

is given by the same expression as (10.31).

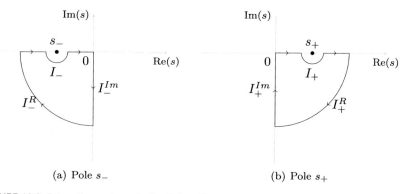

(a) Pole s_{-} (b) Pole s_{+}

FIGURE 10.3 Integration contours for $\xi < 0$ ($\alpha < 0$).

For $\xi > 0$ (Fig. 10.4) we should also account for the contribution of the residues at poles (10.35), having

$$I = J - \frac{4\pi}{\beta} \sin\left(\frac{|\alpha\xi|}{\beta}\right). \tag{10.39}$$

Hence, the far field ($|\xi| \gg 1$) asymptotic expansion is given by

$$I = -\frac{4\pi}{\beta} \sin\left(\frac{|\alpha\xi|}{\beta}\right). \tag{10.40}$$

Some aspects of wave propagation in a fluid-loaded membrane Chapter | 10 **209**

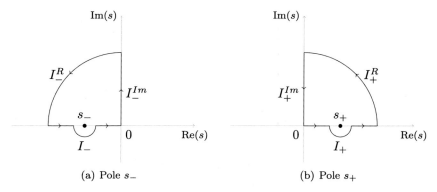

FIGURE 10.4 Integration contours for $\xi > 0$ ($\alpha < 0$).

Numerical examples for the subsonic regime for $\alpha > 0$ (10.38) and $\alpha < 0$ (10.39) are presented in Fig. 10.5 (a) and (b). Both long and short-wave limiting behaviors are compared with the exact solutions for I. Fig. 10.5(a) demonstrates an exponential decay away from the moving load, while in Fig. 10.5(b) a traveling wave is generated in front of it.

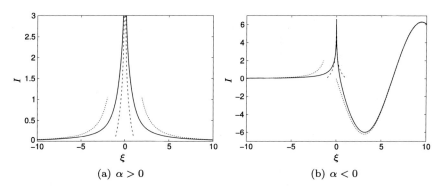

FIGURE 10.5 Numerical results for I (solid lines) corresponding to (10.38) in (a) and (10.39) in (b) with $\beta = 2$, $\alpha = \pm 1$; near field asymptotic expansions (10.34) in both (a) and (b) (dashed lines), far field asymptotic expansions (10.33) in (a) and (10.40) in (b) (dotted lines).

10.6 Supersonic regime

Now we have instead of (10.29) and (10.30)

$$p(\xi, 0) = \frac{\alpha f_0}{2\pi} I \qquad (10.41)$$

and

$$I_- = \int_{-\infty}^{0} \frac{e^{is\xi}}{-i\beta s - \alpha} ds, \quad I_+ = \int_{0}^{\infty} \frac{e^{is\xi}}{i\beta s - \alpha} ds, \qquad (10.42)$$

210 Mechanics and Physics of Structured Media

leading to

$$I = 2 \int_0^\infty \frac{\cos(\xi s)}{i\beta s - \alpha} \, ds = -i2\beta \int_0^\infty \frac{\cos(\xi s)s}{\beta^2 s^2 + \alpha^2} \, ds - 2\alpha \int_0^\infty \frac{\cos(\xi s)}{\beta^2 s^2 + \alpha^2} \, ds.$$

(10.43)

Real and imaginary parts of I can be expressed as

$$\mathrm{Re}(I) = -\frac{\alpha\pi}{|\alpha|\beta} e^{-\frac{|\alpha\xi|}{\beta}}$$

(10.44)

and

$$\mathrm{Im}(I) = \frac{1}{\beta} \left(e^{\frac{|\alpha\xi|}{\beta}} \mathrm{Ei}\left(-\frac{|\alpha\xi|}{\beta}\right) + e^{-\frac{|\alpha\xi|}{\beta}} \mathrm{Ei}\left(\frac{|\alpha\xi|}{\beta}\right) \right),$$

(10.45)

where

$$\mathrm{Ei}(x) = \gamma + \ln|x| + \sum_{n=1}^\infty \frac{x^n}{n!n}, \quad \text{for} \quad x \in \mathbb{R}\backslash\{0\},$$

see [20]. Far and near field expansions of the imaginary part of I are

$$\mathrm{Im}(I) = \frac{2\beta}{\alpha^2\xi^2}$$

(10.46)

and

$$\mathrm{Im}(I) = \frac{2}{\beta} \left(\gamma + \ln\left(\frac{|\alpha\xi|}{\beta}\right) \right),$$

(10.47)

respectively. In Fig. 10.6 (a) and (b) real and imaginary parts of the integral I (10.44) and (10.45) are plotted for $\alpha = 1$, $\beta = 2$. For the imaginary part in Fig. 10.6 (b) far and near field asymptotics (10.46) and (10.47) are also presented.

10.7 Concluding remarks

A set of long-wave and short-wave approximations of the dispersion relation are derived for various parameter setups. Among them, we mention asymptotic formula (10.14), corresponding to an evanescent wave with a small imaginary part at the short-wave limit. All the approximate predictions are compared numerically with their exact counterparts.

The solution of the 2D steady-state moving load problem for a point force is expressed in terms of exponential and related integrals. The related far and near field asymptotic behaviors are obtained and numerically tested. The most technical component of the developed analysis is concerned with contour integration in presence of poles on the real axis, see Eq. (10.39). The latter is tackled using the limiting absorption principle.

Some aspects of wave propagation in a fluid-loaded membrane Chapter | 10   **211**

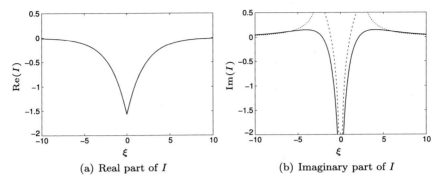

(a) Real part of I (b) Imaginary part of I

FIGURE 10.6 Numerical results for $\alpha = 1$, $\beta = 2$ (a) real part of I (10.44) (solid line); (b) imaginary part of I (10.45) (solid line), near field asymptotic expansions (10.47) (dashed lines), far field asymptotics (10.46) (dotted lines).

Justification of a membrane approximation for modeling of a fluid-loaded prestressed layer apparently needs a further insight. The point is that the asymptotic considerations in [12–14] are oriented to a layer subject to prescribed stresses along its faces without taking into account interaction with the fluid. Ideally, the methodology in the cited papers should be extended to the associated coupled problem.

Finally, we observed that Maple 18.02 gives erroneous results for integrals in (10.43). This drawback of commercial software packages in case of exponential integrals was earlier mentioned in wiki.

Acknowledgment

Support from the Ministry of Science and Education of the Republic of Kazakhstan (Grant No. AP08857255) is gratefully acknowledged.

References

[1] J.D. Kaplunov, L.Y. Kossovich, E.V. Nolde, Dynamics of Thin Walled Elastic Bodies, Academic Press, 1998.
[2] M.C. Junger, D. Feit, Sound, Structures, and Their Interaction, vol. 225, MIT Press, Cambridge, MA, 1986.
[3] N.D. Veksler, Resonance Acoustic Spectroscopy, vol. 11, Springer Science & Business Media, 2012.
[4] A.N. Norris, D.A. Rebinsky, Acoustic coupling to membrane waves on elastic shells, The Journal of the Acoustical Society of America 95 (4) (1994) 1809–1829.
[5] A.V. Belov, J.D. Kaplunov, E.V. Nolde, A refined asymptotic model of fluid-structure interaction in scattering by elastic shells, Flow, Turbulence and Combustion 61 (1) (1998) 255–267.
[6] D.G. Crighton, The Green function of an infinite, fluid loaded membrane, Journal of Sound and Vibration 86 (3) (1983) 411–433.
[7] I.D. Abrahams, A.N. Norris, On the existence of flexural edge waves on submerged elastic plates, Proceedings of the Royal Society of London. Series A: Mathematical, Physical and Engineering Sciences 456 (1999) (2000) 1559–1582.

212 Mechanics and Physics of Structured Media

[8] I.D. Abrahams, J.B. Lawrie, Travelling waves on a membrane: reflection and transmission at a corner of arbitrary angle. I, Proceedings of the Royal Society of London. Series A, Mathematical and Physical Sciences 451 (1943) (1995) 657–683.

[9] J.B. Lawrie, I.D. Abrahams, Travelling waves on a membrane: reflection and transmission at a corner of arbitrary angle. II, Proceedings of the Royal Society of London. Series A: Mathematical, Physical and Engineering Sciences 452 (1950) (1996) 1649–1677.

[10] D. Delande, D. Sornette, Acoustic radiation from membranes at high frequencies: the quantum chaos regime, The Journal of the Acoustical Society of America 101 (4) (1997) 1793–1807.

[11] Z. Lin, A. Hess, Z. Yu, S. Cai, T. Gao, A fluid–structure interaction study of soft robotic swimmer using a fictitious domain/active-strain method, Journal of Computational Physics 376 (2019) 1138–1155.

[12] J.D. Kaplunov, E.V. Nolde, G.A. Rogerson, A low-frequency model for dynamic motion in pre-stressed incompressible elastic structures, Proceedings of the Royal Society of London. Series A: Mathematical, Physical and Engineering Sciences 456 (2003) (2000) 2589–2610.

[13] E.V. Nolde, L.A. Prikazchikova, G.A. Rogerson, Dispersion of small amplitude waves in a pre-stressed, compressible elastic plate, Journal of Elasticity 75 (1) (2004) 1–29.

[14] A.V. Pichugin, G.A. Rogerson, An asymptotic membrane-like theory for long-wave motion in a pre-stressed elastic plate, Proceedings of the Royal Society of London. Series A: Mathematical, Physical and Engineering Sciences 458 (2022) (2002) 1447–1468.

[15] Yu.D. Kaplunov, G.B. Muravskii, Vibrations of an infinite string on a deformable foundation under action of a uniformly accelerating moving load. Passage through critical velocity, Izvestiya Akademii Nauk USSR. Mekhanika Tverdogo Tela 1 (1986) 155–160.

[16] S. Gavrilov, Non-stationary problems in dynamics of a string on an elastic foundation subjected to a moving load, Journal of Sound and Vibration 222 (3) (1999) 345–361.

[17] K.F. Graff, Wave Motion in Elastic Solids, Courier Corporation, 2012.

[18] M. Reed, B. Simon, III: Scattering Theory, vol. 3, Elsevier, 1979.

[19] J.D. Kaplunov, D.G. Markushevich, Plane vibrations and radiation of an elastic layer lying on a liquid half-space, Wave Motion 17 (3) (1993) 199–211.

[20] I.S. Gradshteyn, I.M. Ryzhik, Table of Integrals, Series, and Products, Academic Press, 2014.

Chapter 11

Parametric vibrations of axially compressed functionally graded sandwich plates with a complex plan form

Lidiya Kurpa[a] and Tetyana Shmatko[b]

[a]*Department of Applied Mathematics, National Technical University "Kharkiv Polytechnic Institute", Kharkiv, Ukraine,* [b]*Department of Higher Mathematics, National Technical University "Kharkiv Polytechnic Institute", Kharkiv, Ukraine*

11.1 Introduction

Authors dedicate this work to the blessed memory of the world famous scientist in the field of solid mechanics, Corresponding Member of RAE, Honored Professor of Sumy State University, Doctor of Phys.-Math. Sciences, Prof. L.A. Filshtinsky. Our acquaintance with the scientific results of this great scientist began with his work L.A. Filshtinsky, L.M. Kurshin "Stability of a uniformly compressed polygonal plate" (Izv. Sib. Department of the Academy of Sciences of the USSR. - 1961. - P. 1-8). A reprint of this article in 1981 was kindly presented by L.A. Filshtinsky to the authors, who began to study the application of the theory of R-functions to problems of stability of plates of a complex shape. The results of his research were implemented for comparison, since the exact solution was obtained in his article. Over the years, the R-functions theory and software created on its basis have been applied to solve the stability problems for plates and shallow shells in many publications [22–25]. This work uses and generalizes approaches developed previously and based on the application of the R-functions theory and variational methods to the study of stability and free vibrations of functionally graded sandwich plates with a complex shape and various boundary conditions.

Wide use of functionally graded materials (FGM) for the manufacture of many elements of thin-walled structures is associated with the unique properties of these materials. FGM are heterogeneous composites in which material properties change smoothly and continuously from one surface of the object to another. These properties of the composite significantly reduce the appearance of crack on the surface, lamination of layers, jump tension, and also contribute

214 Mechanics and Physics of Structured Media

to the avoidance of corrosion of the coating and are used as thermal barrier materials. The creation of modern high-strength objects in leading industries, and above all, in the aviation industry, leads to the need to use functionally gradient materials (FGM) with high strength and resistance to temperature and mechanical loads. Despite the FGMs were created in 1984 by Japanese scientists [1], most works devoted to the study of the dynamic and static behavior of FGM plates and membranes have been published in the last two decades [2–5]. A detailed review of these publications is presented in papers [6–8] and others. Laminated functionally graded plates are often used as constituent components of such objects. And if plate is subjected to periodic loads in the middle plate, then it is possible occurrence of parametric resonance or dynamical instability. Such a phenomenon can lead to the destruction of the structure.

This problem is one of the topical problems of modern nonlinear mechanics, in particular case it is the study of stability and parametric vibrations of multi-layered composite plates and shells loaded in the middle plane. Many works [9–15] are devoted to this problem. Many theories and methods have been developed to investigate the stability and parametric oscillations of the FGM plates and shells. Among them – the method of Rayleigh, Rayleigh-Ritz, Galerkin, finite difference method, finite element, differential quadrature differential transformations, boundary integral equations, collocation, etc. [16–21]. An overview of the publications in which the Rayleigh-Ritz method applied was performed [8]. In view of the presented work, we can conclude that the Ritz method was used to study the parametric oscillations of multilayer FGM plates and shallow shells with a rectangular form of the plan [6,16–18]. It should be noted that the works, in which the study of stability and parametric vibrations of FG sandwich-type plates with complex geometric form of the plan at inhomogeneous state are almost absent.

In works [22–24] a method for studying the stability and parametric oscillations of multi-layered orthotropic plates of symmetric structure with a complex form of the plan was proposed. This method is based on the use of the theory of R-functions and variation methods, so it was named as the method of R-functions (RFM).

In the present work, RFM is first developed to study the stability and parametric oscillations of three-layer FG plates loaded in the middle plane with a complex form of plan and different types of fastening. In the paper analytical expressions, obtained early by authors [25] are applied to calculate the effective properties of materials for different arrangement and thicknesses of layers. The proposed approach consists of a few steps. The first step is definition of the prebuckling state as result of the elasticity problem solving. Then critical buckling load and frequencies FGM plate is determined. To study parametric vibrations, unknown functions are presented in a special form due to the initial system of nonlinear differential equations is reduced to a nonlinear ordinary differential equation. The study of the obtained equation is carried out by the Runge-Kutta method. To identify regions of the dynamical instability we lin-

Parametric vibrations **Chapter | 11 215**

earize the obtained nonlinear ordinary differential equations and reduce it to equations by Mathieu. The main region of the dynamical instability is determined by approach of V.V. Bolotin [26]. The developed method and the software are validated on examples of "sandwich" plates with a rectangular form of the plan and applied to investigate stability and parametric vibration of the FGM plates with complex geometric form.

11.2 Mathematical problem

Consider a sandwich plate of an arbitrary plan form under axial compression. The first order shear deformation theory [2] is used in order to analyze stability and parametric vibrations of the plate. The displacements u, v, w at any point in the plate are expressed as functions of the middle surface displacements u_0, v_0, and w in the Ox, Oy, and Oz directions and the independent rotations ψ_x, ψ_y of the transverse normal to middle surface about the Oy and Ox axes, respectively:

$$u(x, y, z, t) = u_0(x, y, t) + z\psi_x(x, y, t),$$

$$v(x, y, z, t) = v_0(x, y, t) + z\psi_y(x, y, t), \qquad (11.1)$$

$$w(x, y, z, t) = w_0(x, y, t).$$

For moderately large deformations the strain components $\{\varepsilon\} = \{\varepsilon_{11}, \varepsilon_{22}, \varepsilon_{12}\}^T$ and $\{\gamma_0\} = \{\varepsilon_{13}, \varepsilon_{23}\}^T$ are expressed as $\{\varepsilon\} = \{\varepsilon_0\} + z\{\chi\}$, where

$$\{\varepsilon_0\} = \left\{u_{0,x} + \frac{1}{2}w_{,x}^2, \quad v_{0,y} + \frac{1}{2}w_{,y}^2 \quad u_{0,y} + v_{0,x} + w_{,x}w_{,y}\right\}^T, \quad (11.2)$$

$$\{\chi\} = \{\phi_{x,x}, \quad \phi_{y,y} \quad \phi_{x,y} + \phi_{y,x}\}^T. \qquad (11.3)$$

Comma defines differentiation of the function with respect to argument that follows for it.

The total force resultants in plane $N = (N_{11}, N_{22}, N_{12})^T$, total moment resultants $M = (M_{11}, M_{22}, M_{12})^T$ and transverse resultants $Q = (Q_x, Q_y)^T$ are calculated as

$$\{N\} = [A]\{\varepsilon\} + [B]\{\chi\}, \quad \{M\} = [B]\{\varepsilon\} + [D]\{\chi\},$$

$$\{Q\} = K_s A_{33}\{\varepsilon_{13}, \varepsilon_{23}\}^T, \qquad (11.4)$$

where K_s^2 denotes the shear correction factor. In this chapter, we take $K_s^2 = 5/6$. Note that elements A_{ij}, B_{ij}, D_{ij} of the matrices $[A]$, $[B]$, and $[D]$ in relations

(11.4) are calculated by formulas:

$$A_{ij} = \sum_{r=1}^{3} \int_{z_r}^{z_{r+1}} Q_{ij}^{(r)} dz, \quad B_{ij} = \sum_{r=1}^{3} \int_{z_r}^{z_{r+1}} Q_{ij}^{(r)} z \, dz,$$

$$D_{ij} = \sum_{r=1}^{3} \int_{z_r}^{z_{r+1}} Q_{ij}^{(r)} z^2 dz, \tag{11.5}$$

where $z_1 = -h/2$, $z_2 = h_1$, $z_3 = h_2$, $z_4 = h/2$. So the general thickness h is a constant value. Values h_1 and h_2 determine position of lower and upper border of the middle layer (core). Values $Q_{ij}^{(r)}$ ($i, j = 1, 2, 6$) are defined by the following expressions

$$Q_{11}^{(r)} = Q_{22}^{(r)} = \frac{E^{(r)}}{1 - \left(v^{(r)}\right)^2}, \quad Q_{12}^{(r)} = \frac{v^{(r)} E^{(r)}}{1 - \left(v^{(r)}\right)^2}, \quad Q_{66}^{(r)} = \frac{E^{(r)}}{2\left(1 + v^{(r)}\right)}. \tag{11.6}$$

Here $E^{(r)}$, $v^{(r)}$ are Young's modulus and Poisson's ratio relatively of the corresponding layer.

In the given study we consider sandwich plates with one or two layers made of functionally graded materials, for example, face layers of the plate are FGM and core is metal or ceramic or converse: core (middle layer) is FGM and the face layers are isotropic (metal or ceramic). The material properties of the plate vary continuously and smoothly in the thickness direction. To calculate the mechanical characteristic of functionally graded layers, we apply approach proposed in [2,20]. According to this approach effective material properties P_{eff} like Young's modulus E, Poisson's ratio v, and mass density ρ for FGM can be estimated by the following Voigt's law:

$$P_{eff} = (P_c - P_m) V_1^{(r)} + P_m, \tag{11.7}$$

where P_c and P_m are corresponding properties of the ceramics and metal respectively; $V_1^{(r)}$ is the volume fraction of material, included in mixture of ceramics and metal. In Fig. 11.1 different schemes of arrangement of plate layers are shown. Value $V_1^{(r)}$ is determined by the following presentation.

Type 1-1

$$\begin{cases} V_1^{(1)} = \left(\frac{z+h/2}{h_1+h/2}\right)^P, \\ V_1^{(2)} = 1, \\ V_1^{(3)} = \left(\frac{z-h/2}{h_2-h/2}\right)^P, \end{cases}$$

Type 2-1

$$\begin{cases} V_1^{(1)} = 1, \\ V_1^{(2)} = \left(\frac{z-h_2}{h_1-h_2}\right)^P, \\ V_1^{(3)} = 0. \end{cases}$$

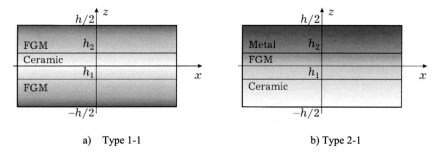

FIGURE 11.1 (a, b). Schemes of arrangement of plate layers.

Note that the value $V_1^{(1)}$ is valid for $z \in [-h/2, h_1]$, $V_1^{(2)}$ is valid for $z \in [h_1, h_2]$, and $V_1^{(3)}$ is valid for $z \in [h_2, h/2]$. Let us note that the value p is the power law exponents (gradient index) of the corresponding layer.

Poisson's ratio of the plate is assumed to be constant in this study, since its influence on the deformation of the plate is essentially less than the effect of Young's modulus [17]. This assumption allows us to obtain the coefficients A_{ij}, B_{ij}, D_{ij} in an analytical form [25].

Let us assume that the plate is subjected to periodic in-plane load $p_N = p_{st} + p_{din} \cos\theta t$, where p_{st} is a static component, p_{din} is the amplitude of the periodic part, and θ is the frequency of the load. Note that all external forces are varied proportionally to the parameter λ. The governing differential equations of equilibrium of the plate subjected to external in-plane loading in terms of displacements and rotations are expressed as follows:

$$A_{11}(L_{11}u + L_{12}v) + B_{11}(L_{14}\psi_x + L_{15}\psi_y) = I_0 \frac{\partial^2 u}{\partial t^2} + I_1 \frac{\partial^2 \psi_x}{\partial t^2} + Nl_1(w), \tag{11.8}$$

$$A_{11}(L_{21}u + L_{22}v) + B_{11}(L_{24}\psi_x + L_{25}\psi_y) = I_0 \frac{\partial^2 v}{\partial t^2} + I_1 \frac{\partial^2 \psi_y}{\partial t^2} + Nl_2(w), \tag{11.9}$$

$$L_{33}w + L_{34}\psi_x + L_{35}\psi_y = I_0 \frac{\partial^2 w}{\partial t^2} + Nl_3(w), \tag{11.10}$$

$$B_{11}(L_{41}u + L_{42}v) + L_{43}w + (D_{11}L_{44} - K_s A_{66})\psi_x + (D_{11}L_{45} - K_s A_{66})\psi_y$$
$$= I_1 \frac{\partial^2 u}{\partial t^2} + I_2 \frac{\partial^2 \psi_x}{\partial t^2} + Nl_4(w), \tag{11.11}$$

$$B_{11}(L_{51}u + L_{52}v) + L_{53}w + D_{11}L_{54}\psi_x + (D_{11}L_{55} - K_s A_{66})\psi_y$$
$$= I_1 \frac{\partial^2 v}{\partial t^2} + I_2 \frac{\partial^2 \psi_y}{\partial t^2} + Nl_5(w), \tag{11.12}$$

where

$$(I_0, I_1, I_2) = \sum_{r=1}^{3} \int_{z_r}^{z_{r+1}} (\rho^{(r)})(1, z, z^2) dz. \tag{11.13}$$

Mass density $\rho^{(r)}$ of the r-th layer is calculated by formula (11.7). Calculation of the values I_0, I_1, I_2 is carried out with help of the analytical expressions that have been obtained in [25]. Linear operators are the following:

$$L_{11} = L_{14} = L_{41} = L_{44} = \frac{1}{1 - v^2} \frac{\partial^2}{\partial x^2} + \frac{1}{2(1 + v)} \frac{\partial^2}{\partial y^2}, \tag{11.14}$$

$$L_{22} = L_{25} = L_{52} = L_{55} = \frac{1}{1 - v^2} \frac{\partial^2}{\partial y^2} + \frac{1}{2(1 + v)} \frac{\partial^2}{\partial x^2}, \tag{11.15}$$

$$L_{12} = L_{21} = L_{24} = L_{42} = L_{45} = L_{54} = L_{15} = L_{51} = \frac{1}{2(1 - v)} \frac{\partial^2}{\partial x \partial y}, \tag{11.16}$$

$$L_{13} = L_{31} = L_{32} = L_{23} = 0, \, L_{33} = K_s A_{66} \left(\frac{\partial^2}{\partial x^2} + \frac{\partial^2}{\partial y^2} \right), \tag{11.17}$$

$$L_{34} = K_s A_{66} \frac{\partial}{\partial x}, \, L_{35} = K_s A_{66} \frac{\partial}{\partial y}, \, L_{43} = -L_{34}, \, L_{53} = -L_{35}. \tag{11.18}$$

Nonlinear expressions $Nl\ell_1(w)$, $Nl_2(w)$, $Nl\ell_3(w)$, $Nl\ell_4(w)$, $Nl\ell_5(w)$ in the right parts of Eqs. (11.8)–(11.12) are defined as:

$$Nl\ell_1(w) = -L_{11}(w) \frac{\partial w}{\partial x} - L_{12}(w) \frac{\partial w}{\partial y}, \tag{11.19}$$

$$Nl\ell_2(w) = -L_{12}(w) \frac{\partial w}{\partial x} - L_{22}(w) \frac{\partial w}{\partial y}, \tag{11.20}$$

$$Nl\ell_3(w) = - \left(N_{11}^0 \frac{\partial^2 w}{\partial x^2} + N_{22}^0 \frac{\partial^2 w}{\partial y^2} + 2N_{12}^0 \frac{\partial^2 w}{\partial x \partial y} \right), \tag{11.21}$$

$$Nl\ell_4(w) = -L_{41}(w) \frac{\partial w}{\partial x} - L_{42}(w) \frac{\partial w}{\partial y}, \tag{11.22}$$

$$Nl\ell_5(w) = -L_{42}(w) \frac{\partial w}{\partial x} - L_{44}(w) \frac{\partial w}{\partial y}. \tag{11.23}$$

Values $\{N^0\} = (N_{11}^0, N_{22}^0, N_{12}^0)^T$ in formula (11.21) denote the force resultant in the prebuckling state.

11.3 Method of solution

Since the prebuckling state can be inhomogeneous, first, we should determine the forces in plane $\{N^0\} = \left(N^0_{11}, N^0_{22}, N^0_{12}\right)^T$.

Take into account that plate remains flat at the prebuckling state, the quantities of w, ψ_x, ψ_y are set to zero. So prebuckling state can be defined by Eqs. (11.8)–(11.9), which in displacement take the form:

$$A_{11}u_{0,xx} + A_{66}u_{0,yy} + (A_{12} + A_{66})\, v_{0,xy} = 0,$$
$$(A_{12} + A_{66})\, u_{0,xy} + A_{66}v_{0,xx} + A_{22}v_{0,yy} = 0. \tag{11.24}$$

System (11.24) is supplemented by the following boundary conditions on the loaded part of the border:

$$N^{(L)}_n (u, v) = -1, \quad T^{(L)}_n (u, v) = 0. \tag{11.25}$$

Operators $N^{(L)}_n, T^{(L)}_n$ are defined by the following formulas:

$$N^{(L)}_n = N^{(L)}_{11}l^2 + N^{(L)}_{22}m^2 + 2N^{(L)}_{12}lm,$$
$$T^{(L)}_n = N^{(L)}_{12}\left(l^2 - m^2\right) + \left(N^{(L)}_{11} - N^{(L)}_{22}\right)lm, \tag{11.26}$$

where $l = \cos(\vec{n}, Ox)$, $m = \cos(\vec{n}, Oy)$, and $\vec{n}$ is a normal vector to a border of the domain. Kind of the boundary conditions on the unloaded part of the border depends on the way fixing the edge.

The problem (11.24)–(11.25) is solved by Ritz's method and the R-functions theory (RFM). Therefore, let us formulate the variational statement of the problem that is reduced to finding the minimum of the following functional:

$$I(u_0, v_0) = \frac{1}{2} \iint_\Omega \left(N^0_{11}\varepsilon^L_{11} + N^0_{22}\varepsilon^L_{22} + N^0_{12}\varepsilon^L_{12}\right) d\Omega$$
$$+ \int_{\partial\Omega_1} p_{st} \left(u_0 \cos\alpha + v_0 \sin\alpha\right) ds, \tag{11.27}$$

where

$$\left\{N^0\right\} = [A]\left\{\varepsilon^L_0\right\}^T, \quad \left\{\varepsilon^L\right\} = \left\{u_{0,x}\, ; v_{0,y}\, ; u_{0,y} + v_{0,x}\right\}. \tag{11.28}$$

Solution of the boundary problem (11.24)–(11.25) or variational problem (11.27) allows to determine the displacements u_0, v_0 and forces $\{N_0\}$ in the middle plane.

To find the buckling load the dynamical approach [28] is applied. Then, the problem is reduced to an equivalent variational problem of the minimization of

220 Mechanics and Physics of Structured Media

the following functional:

$$
\begin{aligned}
I(u, v, w, \psi_x, \psi_y) = \frac{1}{2} \iint_\Omega [& N_{11}^{(L)} \varepsilon_{11}^{(L)} + N_{22}^{(L)} \varepsilon_{22}^{(L)} + N_{12}^{(L)} \varepsilon_{12}^{(L)} + \\
& + M_{11}^{(L)} \chi_{11} + M_{22}^{(L)} \chi_{22} + M_{12}^{(L)} \chi_{12} + Q_x \varepsilon_{13} + Q_y \varepsilon_{23} + \\
& + p_{st} (N_{11}^0 (w_{,x})^2 + N_{22}^0 (w_{,y})^2 + N_{12}^0 w_{,x} w_{,y})] d\Omega - \\
& - \frac{1}{2} \omega_L^2 \iint_\Omega (I_0 (u^2 + v^2 + w^2) + I_1 (\psi_x^2 + \psi_y^2)) d\Omega,
\end{aligned}
$$

$$(11.29)$$

where the terms with superscripts L correspond to the linear terms in formulas (11.4).

The value of the parameter p_{st} increases, when the natural frequency ω_L is a real number. The value of the buckling load N_{cr} is defined by the value of the parameter p_{st} corresponding to the smallest nonnegative value of the frequency. Minimization of the functional (11.29) is performed using Ritz's method. The sequence of coordinate functions has been constructed by the R-functions theory [27].

In order to solve the nonlinear vibration problem we apply the approach proposed in [22,25] and reduce initial original nonlinear movement system to the system of ordinary differential equations. To find instability regions Bolotin's approach is used. Primary instability region of FGM plate under compressed harmonic load can be found from linearized equation

$$
y''(t) + \varepsilon y'(t) + \Omega_L^2 (1 - 2k \cdot \cos(\theta t)) y(t) = 0, \tag{11.30}
$$

where $\Omega_L^2 = \omega_L^2 - p_{st}\alpha$ is the frequency of the plate compressed by the static load p_{st}, ω_L stands for the natural frequency of free vibration, ε is the damping ratio of the plate, and the expression for the coefficient α is obtained in an analytical form, similarly as it has been done in the paper [24] and has the following form:

$$
\alpha = \frac{1}{I_0 \left\| w_1^{(e)} \right\|^2} \iint_\Omega \left(N_{11}^0 w^{(e)}{}_{,xx} + N_{22}^0 w^{(e)}{}_{,yy} + 2 N_{12}^0 w^{(e)}{}_{,xy} \right) w_1^{(e)} dx dy .
$$

$$(11.31)$$

Here w^e is eigenfunction corresponding to the first mode of free linear vibrations of loaded plate. It is known for Eq. (11.30) (Mathieu equation), the primary instability region ($\theta = 2\Omega_L$) is bounded by the following curves [26]:

$$
2\Omega_L \sqrt{1 - \sqrt{k^2 - \left(\frac{\Delta}{\pi}\right)^2}} \leq \theta \leq 2\Omega_L \sqrt{1 + \sqrt{k^2 - \left(\frac{\Delta}{\pi}\right)^2}} \tag{11.32}
$$

where $\Delta = \frac{\pi\varepsilon}{\Omega}$ is the damping decrement. If $\varepsilon = 0$, then we get:

$$\theta_1 = 2\Omega_L\sqrt{1-k}, \qquad \theta_2 = 2\Omega_L\sqrt{1+k}. \tag{11.33}$$

The principal parametric resonance arises in this interval.

11.4 Numerical results

In order to prove the verification of the proposed method the following example of square plate is analyzed. Plan form and acting load are presented in Fig. 11.2.

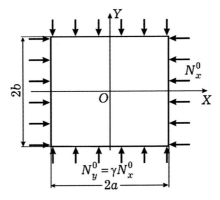

FIGURE 11.2 Plan form of a square plate under acting load.

Three-layer plate is considered, the bottom and top layers are made of a mixture Al/Al_2O_3 and the core is made of metal. Thickness of the layers and gradient index p are taken different. The material properties of the FGM mixture Al/Al_2O_3 are the following [17–21]:

$$Al: \quad E_m = 70 \text{ GPa}, \quad \nu_m = 0.3, \quad \rho_m = 2707 \text{ kg/m}^3,$$
$$Al_2O_3: \quad E_c = 380 \text{ GPa}, \quad \nu_c = 0.3, \quad \rho_c = 3800 \text{ kg/m}^3. \tag{11.34}$$

Assume that the plate is simply supported and compressed evenly along the sides $x = \pm a$ and $y = \pm b$. The total thickness of the plate is equal to $h/(2a) = 0.1$.

A similar problem has been solved in [17,18,21] in the framework of various theories. Table 11.1 demonstrates a comparison of the calculated dimensionless critical load $\widehat{N}_{cr} = \frac{N_{cr}}{100 E_0 h^3}$ (where $E_0 = 1$ GPa, $\rho_0 = 1$ kg/m^3). The values $h^{(1)}, h^{(2)}, h^{(3)}$ define the thickness of layers and calculated as

$$h^{(1)} = h_1 + h/2, \qquad h^{(2)} = h_2 - h_1, \qquad h^{(3)} = h/2 - h_2.$$

Comparison of data in Table 11.1 indicates a good agreement between the results obtained and those known in the literature.

TABLE 11.1 Dimensionless critical buckling load.

p	Method	\multicolumn{6}{c}{Layers thickness ratio $h^{(1)} : h^{(2)} : h^{(3)}$}
		1:0:1	2:1:2	2:1:1	1:1:1	2:2:1	1:2:1
0	RFM	6.5000	6.5000	6.5000	6.5000	6.5000	6.5000
	[21]	6.5030	6.5030	6.5030	6.5030	6.5030	6.5030
	[18]	6.5028	6.5028	6.5028	6.5028	6.5028	6.5028
	[17]	6.4765	6.4765	6.4765	6.4765	6.4765	6.4765
0.5	RFM	3.667	3.958	4.101	4.205	4.392	4.597
	[21]	3.6828	3.9709	4.1127	4.2186	4.4052	4.6083
	[18]	3.6819	3.9702	4.1124	4.2182	4.4051	4.6088
	[17]	3.5809	3.8581	3.9948	4.0964	4.2759	4.4711
1	RFM	2.572	2.906	3.086	3.219	3.462	3.742
	[21]	2.5842	2.9206	3.0973	3.2327	3.4749	3.7531
	[18]	2.5831	2.9197	3.0968	3.2322	3.4748	3.7536
	[17]	2.5306	2.8556	3.0273	3.1575	3.3921	3.6601
5	RFM	1.319	1.512	1.692	1.780	2.048	2.355
	[21]	1.3300	1.5220	1.7022	1.7903	2.0564	2.3674
	[18]	1.3284	1.5207	1.7014	1.7894	2.0558	2.3673
	[17]	1.3183	1.5041	1.6813	1.7651	2.0253	2.3235
10	RFM	1.235	1.363	1.537	1.589	1.844	2.131
	[21]	1.2447	1.3742	1.5672	1.5973	1.5729	2.1909
	[18]	1.2429	1.3725	1.5456	1.5969	1.8534	2.1398
	[17]	1.2360	1.3604	1.5304	1.5789	1.8308	2.1027

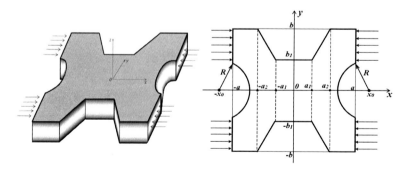

FIGURE 11.3 Geometry and plan form of the sandwich plate.

In order to show the possibilities of the proposed method we study vibration and buckling analysis of the plate shown in Fig. 11.3. It has the following fixed geometrical parameters:

$$b/a = 1, \ a_1/2a = 0.25, \ b_1/2a = 0.35, \ h/2a = 0.1, \ x_0/2a = 0.7.$$

The investigations are conducted for two combinations of metal and ceramic. The first set of materials chosen is aluminium and alumina Al/Al_2O_3 with properties (11.34). Silicon nitride (Si_3N_4) and stainless steel ($SUS304$) were selected for two constituent materials of the substrate FGM layers in the second variant. This material is referred as $Si_3N_4/SUS304$. Material properties of the constituent FGM are [20]:

$$Si_3N_4: \quad E_c = 322.27 \text{ GPa}, \quad v_c = 0.24, \quad \rho_c = 2370 \text{ kg/m}^3,$$

$$SUS304: \quad E_m = 207.787 \text{ GPa}, \quad v_m = 0.317, \quad \rho_m = 8166 \text{ kg/m}^3.$$

$$(11.35)$$

Suppose that harmonic load acts along rectangular part of the boundary, which are parallel to the axis Ox (Fig. 11.3). The plate is movable clamped over the whole border. Then kinematic (main) boundary conditions are:

$$w(x, y) = 0, \quad \psi_x = 0, \quad \psi_y = 0, \quad \forall (x, y) \in \partial\Omega. \quad (11.36)$$

Solution structure [25,27] for plates with complete clamped borders can be taken as:

$$w = \omega\Phi_1, \quad u = \Phi_2, \quad v = \Phi_3, \quad \psi_x = \omega\Phi_4, \quad \psi_y = \omega\Phi_5, \quad (11.37)$$

where Φ_i, $i = \overline{1,5}$, are indefinite components of the structure. They are expended into a series of some complete system (power polynomials, trigonometric polynomials, splines, etc.). The function $\omega(x, y)$ is constructed by the R-functions theory in such a way that the functions w, ψ_x, ψ_y are vanished on whole boundary.

To realize the solution structure (11.37) we should construct the equation of the whole border that is the function $\omega(x, y)$. Using the R-operations [27] we build the equation of the border in the form:

$$\omega(x, y) = ((f_1 \vee_0 f_2) \wedge_0 (f_3 \vee_0 f_4) \vee_0 f_5) \wedge_0 (f_6 \wedge_0 f_7) \wedge_0 (f_8 \wedge_0 f_9). \quad (11.38)$$

Here symbols $\wedge_0, \vee_0$ denote R-operations of the R_0 system, which have the following form:

$x_1 \wedge_0 x_2 \equiv x_1 + x_2 - \sqrt{x_1^2 + x_2^2}$ – R-conjunction describes intersection of the domains,

$x_1 \vee_0 x_2 \equiv x_1 + x_2 + \sqrt{x_1^2 + x_2^2}$ – R-disjunction describes union of the domains. Functions $f_i, i = \overline{1,9}$ in relation (11.38) are defined as:

$$f_1 = k(x - a_1) - y + b_1 \geq 0, \quad f_2 = -k(x + a_1) - y + b_1 \geq 0,$$

$$f_3 = -k(x + a_1) + y + b_1 \geq 0, \quad k = (b - b_1)/(a_2 - a_1),$$

$$f_4 = k(x - a_1) + y + b_1 \geq 0, \quad f_5 = \left(b_1^2 - y^2\right)/2b_1 \geq 0,$$

224 Mechanics and Physics of Structured Media

$$f_6 = \left(b^2 - y^2\right)/2b \geq 0, \quad f_7 = \left(a^2 - x^2\right)/2a \geq 0,$$

$$f_8 = \left((x - x_0)^2 + y^2\right)/2b_1 \geq 0, \quad f_9 = \left((x + x_0)^2 + y^2\right)/2b_1 \geq 0.$$

$$(11.39)$$

The presented below results were obtained by the approximation of indefinite components in structural formulas by power polynomials up to the 14-th and 11-th degrees for the function w and functions u, v, ψ_x, ψ_y, that corresponds to the conservation of 36 and 21 coordinate functions, respectively. To calculate the integrals in the Ritz matrix ten-point Gaussian formulas were used and the integration was performed by a quarter of the domain.

Tables 11.2 and 11.3 show the values of dimensionless natural frequency $\Lambda = \frac{\omega_L (2a)^2}{h}\sqrt{\frac{\rho_c}{E_c}}$ and buckling load $\widehat{N}_{cr} = \frac{N_{cr}}{100 E_0 h^3}$ obtained for a plate made of FGM (Al/Al_2O_3), Type 1-1 and Type 2-1 with scheme of layers 2-1-2 (Fig. 11.3).

Analysis of Tables 11.2 and 11.3 show that while gradient index p is increasing, the critical load and natural frequencies decrease for both types of plates (Types 1-1 and 2-1). Except of this changing of critical load for Type 1-1 is more significant than its changing for Type 2-1. Note that values of natural frequency of plate Type 1-1 exceed the corresponding values of frequency of plate Type 2-1, when gradient index $0 \leq p \leq 1$. If $p > 1$, then we have a converse situation.

The values of critical load and natural frequencies for FG plate ($Si_3N_4/SUS304$) of Type 2-1 with arrangement of layers 2-1-2 are presented in Table 11.4.

Comparison of obtained data from Tables 11.3 and 11.4 shows that values of the critical load for material Al/Al_2O_3 are less than corresponding values for material $Si_3N_4/SUS304$. But values of natural frequencies are bigger for material Al/Al_2O_3. Fig. 11.4 and Fig. 11.5 display dependence on middle layer thickness (1-n-1) and face layers thickness (n-1-1, 1-1-n) for two FG materials (Al/Al_2O_3, $Si_3N_4/SUS304$) and two gradient indices ($p = 0.5$, $p = 5$) of plate Type 1-1.

Fig. 11.6 and Fig. 11.7 present an influence of the gradient index p on values of natural frequencies for different static parameters of critical load and two Types 1-1 and 2-1 of FG plate with layers thickness 2-1-2. Fig. 11.6 depicts the results for material Al/Al_2O_3, Fig. 11.7 depicts results for material $Si_3N_4/SUS304$.

Effect of the power law index p on buckling load is shown in Fig. 11.8.

As can be seen from Fig. 11.8 critical buckling load for plates of Type 2-1 made of Si$_3$N$_4$/SUS304 is greater than critical load for FG plates of Al/Al_2O_3 for all values of the power law index p. This conclusion is valid for plates of Type 1-1 for value $p = 0.3$.

TABLE 11.2 Effect of the power law exponent p on the buckling load and the natural frequency of FG plate (Type 1-1, thickness of layers 2-1-2, Al/Al_2O_3).

p \ p_{st}/N_{cr}	0	0.25	0.5	0.75	0.9	0.95	N_{cr}
0	16.794	16.793	14.404	10.577	6.858	0.277	52.99
0.5	15.268	13.896	11.704	8.562	5.536	0.348	32.13
1	14.278	12.715	10.698	7.814	5.046	0.183	26.05
5	11.055	9.828	8.251	6.009	3.869	0.317	14.21
10	10.602	9.427	7.917	5.772	3.728	0.484	12.78
100	10.312	9.172	7.706	5.618	3.623	0.370	11.83

TABLE 11.3 Effect of the power law exponent p on the buckling load and the natural frequency of FG plate (Type 2-1, thickness of layers 2-1-2, Al/Al_2O_3).

p	p_{st}/N_{cr} = 0	0.25	0.5	0.75	0.9	0.95	N_{cr}
0	13.743	12.220	10.260	7.476	4.815	0.409	24.08
0.5	13.424	11.938	10.027	7.309	4.716	0.531	22.43
1	13.339	11.862	9.964	7.262	4.683	0.479	21.98
5	13.237	11.779	9.899	7.222	4.661	0.507	21.22
10	13.248	11.796	9.916	7.235	4.668	0.473	21.18
100	13.143	11.834	9.951	7.263	4.688	0.470	21.21

TABLE 11.4 Effect of the gradient index p on the buckling load and natural frequency FG plate (Type 2-1, thickness of layers 2-1-2, Si_3N_4/SUS304).

p \ p_{st}/N_{cr}	0	0.25	0.5	0.75	0.9	0.95	N_{cr}
0	11.045	10.900	9.203	6.751	4.373	0.445	36.72
0.5	10.470	10.433	8.813	6.469	4.193	0.446	36.34
1	10.203	10.199	8.636	6.339	4.109	0.421	36.17
5	9.707	9.707	8.314	6.107	3.961	0.428	35.86
10	9.601	9.600	8.247	6.059	3.932	0.449	35.80
100	9.489	9.489	8.176	6.006	3.896	0.403	35.74

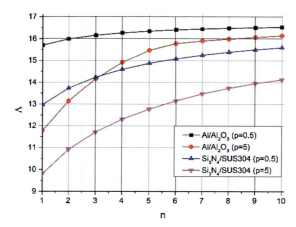

FIGURE 11.4 Effect of middle layer thickness (1-n-1) on fundamental frequencies.

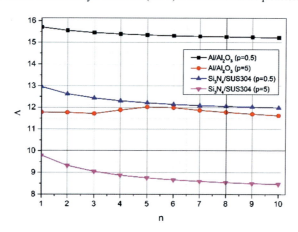

FIGURE 11.5 Effect of face layers thickness (n-1-1, 1-1-n) on fundamental frequencies.

Fig. 11.9 and Fig. 11.10 show the areas of dynamic instability of FG plates (Al/Al_2O_3) Type 1-1 (Fig. 11.9) and Type 2-1 (Fig. 11.10) for different values of static load. The arrangement of layers thickness in both cases is taken as 2-1-2. The value of gradient index is $p = 1$. The dimensionless excitation frequencies $\bar{\theta}$ are determined by the formula:

$$\bar{\theta} = 2\bar{\Lambda}\sqrt{1 \pm k},$$

where $\bar{\Lambda} = \Omega_L(2a^2)\sqrt{\frac{\rho_c}{E_c h^2}}$, Ω_L is a natural frequency, corresponding to the given static load p_{st}.

As we see from Fig. 11.9 and Fig. 11.10 the value of the static component significantly affects on the size and placement of the dynamic instability region. The increase of natural frequencies causes a shift of the instability regions to-

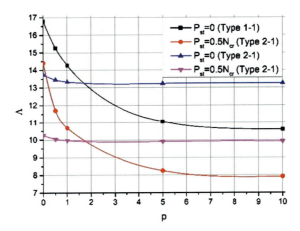

FIGURE 11.6 Effect of gradient index on natural frequencies for FG plate (Al/Al_2O_3).

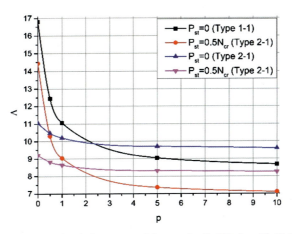

FIGURE 11.7 Effect of gradient index on natural frequencies for FG plate ($Si_3N_4/SUS304$).

wards the smaller values of the exciting frequency. With an increase of the static component the areas of dynamic instability are narrowed.

11.5 Conclusions

In this work for the first time the R-functions method was applied to a new class of problems: parametric vibrations and stability of functionally graded plates of sandwich-type with complex plan form and different boundary conditions. The mathematical formulation of the problem was performed within the framework of a refined theory of the first order type of the theory by Timoshenko. The method takes into account the prebuckling state of the plate, arrangement of layers and their thickness. Effective properties of FGM are calculated according

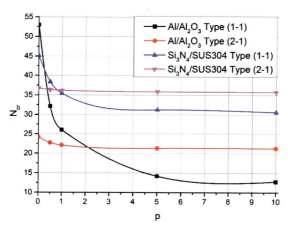

FIGURE 11.8 Effect of power law index p on critical load of FG plates (Fig. 11.3, scheme of layers thickness 2-1-2).

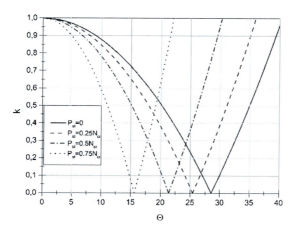

FIGURE 11.9 Regions of instability for FG plate (Al/Al_2O_3, Type 1-1, layers thickness 2-1-2).

to the power law. Verification of the proposed approach and comparison of the results with known are fulfilled on the basis of the developed software. As an illustration of the possibilities of the method, a three-layer plate of complex geometric shape under static and dynamic load is considered. Effect of the power law index on fundamental frequencies and critical loads are studied. Regions of dynamic instability of the FG sandwich plates for different materials and arrangement of layers are drawn. In the future it is planned to develop an approach taking into account the multimode approximation of unknown functions.

Conflict of interest

The authors declare that they have no conflict of interest.

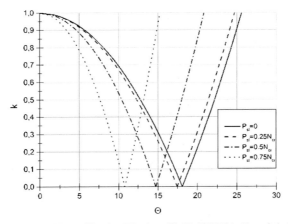

FIGURE 11.10 Regions of instability for FG plate (Si$_3$N$_4$/SUS304, Type 2-1, layers thickness 2-1-2).

References

[1] M. Koizumi, The concept of FGM, Ceramic Transactions 34 (1993) 3–10.
[2] J.N. Reddy, Mechanics of Laminated Composite Plates and Shells, Theory and Analysis, 2nd ed., CRC Press, 2004.
[3] M. Amabili, Nonlinear Vibrations and Stability of Shells and Plates, Cambridge University Press, Cambridge, 2008.
[4] F. Alijani, M. Amabili, K. Karagiozis, F. Bakhtiari-Nejad, Nonlinear vibrations of functionally graded doubly curved shallow shells, Journal of Sound and Vibration 330 (2011) 1432–1454.
[5] M. Strozzi, F. Pellicano, Nonlinear vibrations of functionally graded cylindrical shells, Thin-Walled Structures 67 (2013) 63–77.
[6] H.-T. Thai, S.-E. Kim, A review of theories for the modeling and analysis of functionally graded plates and shells, Composite Structures 128 (2015) 70–86.
[7] K. Swaminathan, D.T. Naveenkumar, A.M. Zenkour, E. Carrera, Stress vibration and buckling analyses of FGV plates – a state-of-art review, Composite Structures 120 (2015) 10–31.
[8] Y. Kumar, The Rayleigh–Ritz method for linear dynamic, static and buckling behavior of beams, shells and plates: a literature review, Journal of Vibration and Control (2017) 1–23, https://doi.org/10.1177/1077546317694724.
[9] A. Argento, R.A. Scott, Dynamic instability of layered anisotropic circular cylindrical shells. II Numerical results, Journal of Sound and Vibration 162 (1993) 323–332.
[10] S. Dash, A.V. Asha, S.K. Sahu, Stability of laminated composite curved panels with cutout using finite element method, in: International Conference on Theoretical, Applied Computation and Experimental Mechanics (ICTACEM 2004), Kharagpur, December 24–31, 2004.
[11] Michael P. Nemeth, Buckling and postbuckling behavior of laminated composite plates with a cutout, NASA technical paper, 1996, 3587.
[12] T.Y. Ng, K.Y. Lam, J.N. Reddy, Dynamic stability of cross-ply laminated composite cylindrical shells, International Journal of Mechanical Sciences 40 (8) (1998) 805–823.
[13] S.K. Sahu, P.K. Datta, Research advances in the dynamic stability behavior of plates and shells: 1987-2005 — part 1: conservative system, Applied Mechanics Reviews 60 (2007) 65–75.
[14] G.J. Simitses, Instability of dynamically loaded structures, Applied Mechanics Reviews (1987) 1403–1408.
[15] M.K. Singha, R. Daripa, Nonlinear vibration and dynamic stability analysis of composite plates, Journal of Sound and Vibration 328 (4) (2009) 541–554.

232 Mechanics and Physics of Structured Media

[16] H. Matsunaga, Free vibration and stability of functionally graded shallow shells according to a 2D higher-order deformation theory, Composite Structures 84 (2008) 132–146.

[17] Q. Li, et al., Three-dimensional vibration analysis of functionally graded material sandwich plates, Journal of Sound and Vibration 311 (2008) 498–515.

[18] N.E. Meiche, et al., A new hyperbolic shear deformation theory for buckling and vibration of functionally graded sandwich plate, International Journal of Mechanical Sciences 53 (4) (2011) 237–247.

[19] A. Neves, et al., Buckling analysis of sandwich plates with functionally graded skins using a new quasi-3D hyperbolic sine shear deformation theory and collocation with radial basis functions, Journal of Applied Mathematics and Mechanics 92 (2012) 749–766.

[20] H.S. Shen, Functionally Graded Materials of Plates and Shells, CRC Press, Florida, 2009.

[21] A.M. Zencour, A comprehensive analysis of functionally graded sandwich plates: part 2–buckling and free vibration, International Journal of Solids and Structures 42 (2005) 5243–5258.

[22] J. Awrejcewicz, L. Kurpa, O. Mazur, Dynamical instability of laminated plates with external cutout, International Journal of Non-Linear Mechanics 81 (2016) 103–114.

[23] Lidiya Kurpa, Olga Mazur, Victoria Tkachenko, Dynamical stability and parametrical vibrations of the laminated plates with complex shape, Latin American Journal of Solids and Structures 10 (2013) 175–188.

[24] L.V. Kurpa, O.S. Mazur, V.V. Tkachenko, Parametric vibrations of multilayer plates of complex shape, Journal of Mathematical Sciences 174 (2) (2014) 101–114.

[25] J. Awrejcewicz, L. Kurpa, T. Shmatko, Analysis of geometrically nonlinear vibrations of functionally graded shallow shells of a complex shape, Latin American Journal of Solids and Structures 14 (2017) 1648–1668.

[26] V.V. Bolotin, Dynamical Stability of Elastic Systems, Gostekhizdat, 1956 (in Russian).

[27] V.L. Rvachev, The R-functions Theory and Its Applications, Nauk Dumka, Kiev, 1982 (in Russian).

[28] S.G. Lekhnitsky, Anisotropic Plates, Gostekhizdat, 1957 (in Russian).

Chapter 12

Application of volume integral equations for numerical calculation of local fields and effective properties of elastic composites

Sergei Kanaun and Anatoly Markov
Tecnologico de Monterrey, School of Engineering and Science, Monterrey, Mexico

12.1 Introduction

Calculation of local fields and effective elastic properties of heterogeneous materials (the homogenization problem) has important applications in geophysics, micro- and nanomechanics, nondestructive evaluation of damage in solids, etc. Microstructures of many important for applications heterogeneous materials can be simulated by random or structured sets of inhomogeneities (cracks, pores, and inclusions) embedded into a homogeneous host medium. The first exact solutions for local fields and effective parameters of periodic heterogeneous structures were obtained in series of works of L.A. Filshtinsky summed in the book [1]. This book anticipates the asymptotic homogenization method for periodic composites developed later in the works of mathematicians [2,3]. The detailed tables of the effective elastic parameters of various periodic structures presented in [1] have been considered as the benchmark database for approximate and numerical solutions of the homogenization problem for decades. Further development of efficient methods of solution of the homogenization problems for periodic structures was performed in series of works of L.A. Filshtinsky and coauthors and summed in the book [4].

For nonperiodic heterogeneous media, a representative volume element (RVE) containing several inhomogeneities is usually introduced. The boundary conditions of first (displacements), second (stresses) kinds or periodic ones are applied on the RVE sides. The effective elastic parameters relate the stress and strain tensors averaged over the RVE. It is obvious that the effective parameters should not depend on the boundary conditions, as well as on the shape of the RVE and on the number of heterogeneities inside. The problem of the optimal

Mechanics and Physics of Structured Media. https://doi.org/10.1016/B978-0-32-390543-5.00017-7
Copyright © 2022 Elsevier Inc. All rights reserved.

234 Mechanics and Physics of Structured Media

choice of shape and size of the RVE has been studied in many works. It was shown that for reliable solution of the homogenization problem by the finite element method, a cubic RVE containing at least 25–30 inhomogeneities must be considered [5,6]. In this case, the boundary conditions don't affect the values of the effective parameters.

An alternative analytic technique for solution of the homogenization problem consists in application of various self-consistent methods. A survey of these methods can be found in [7]. The cluster method [8] that extends the well-known Maxwell methodology to finite or infinite 'clusters' of multiple inhomogeneities relates to this class of methods. After solution of the boundary value problems for clasters of inclusions, the effective elastic parameters are obtained similarly by averaging of local fields. An extensive review of recent developments in area of analytical solutions may be found in the books [9,10]. Also, in the mentioned works, the authors discuss the apparent problems of convergence of classical self-consistent methodologies (like the original Maxwell method) in the case of high inhomogeneity concentrations.

Another approach to the homogenization problem is based on formulation of this problem in terms of volume integral equations. In this method, the RVE of heterogeneous material is embedded into an infinite homogeneous host medium and subjected to a constant external field, thus the interactions of the inhomogeneities are taken into account in a straightforward way. For numerical solution, the integral equations of the problem are discretized using an appropriate class of approximating functions. In this study, Gaussian approximating functions are used for this purpose. The theory of approximation by Gaussians and other similar functions was developed by V. Maz'ya and G. Schmidt [11]. For numerical solution, the RVE is covered by a regular grid of approximating nodes, and the unknown fields are expressed by linear combinations of Gaussian functions centered at the nodes. As a result, the problem is reduced to a system of linear algebraic equations for the coefficients of the approximation [12–15]. An important advantage of the Gaussian functions is that the actions of the integral operators of the problem on such functions can be obtained in explicit analytical forms and numerical integration is excluded from the construction of the matrix of the discretized problem. In addition, for regular node grids, the matrix of the discretized problem has Toeplitz's structure, and fast Fourier transform (FFT) algorithms can be used for the calculation of matrix-vector products in the process of iterative solution of the discretized problem. Note that application of the finite element method requires construction of the numerical solution in the whole RVE volume. Meanwhile, the method of integral equations reduces the problem to the region occupied by inclusions only. It is a substantial advantage in the case of media with cracks [16,17] when instead of 3D-problem for the RVE, the finite system of 2D-integral equations on the crack surfaces can be solved.

The method can be improved by taking into account the heterogeneities outside of the RVE. Presence of the latter can be considered by the additional

Application of volume integral equations **Chapter | 12 235**

(effective) field acting on the RVE. The equation for the effective field is derived by the assumption that the stress field averaged over the inhomogeneities in the infinite composite medium coincides with the average over the RVE region [18,19]. This reduces the problem to a finite system of integral equations in the RVE that are to be solved numerically. In the method, the problem of appropriate boundary conditions on the RVE sides does not emerge.

In this study, the problem of choice of the optimal RVE for the solution of the homogenization problem by the method of volume integral equations is considered. For homogeneous media containing homogeneous in space periodic sets of cracks and inclusions, RVEs of various shapes and sizes are tested, and the appropriate RVEs are indicated.

12.2 Integral equations for elastic fields in heterogeneous media

12.2.1 Heterogeneous inclusions in a homogeneous host medium

The stress field $\sigma_{ij}(x)$ in a heterogeneous region V embedded in an infinite homogeneous elastic host medium (matrix) and subjected to an external stress field $\sigma_{ij}^0(x)$ is presented in the form [7]

$$\sigma_{ij}(x) = \sigma_{ij}^0(x) + \int_V S_{ijkl}\left(x - x'\right) B_{klmn}^1\left(x'\right) \sigma_{mn}\left(x'\right) dx' \qquad (12.2.1)$$

where $B^1(x) = B(x) - B^0$, $x \in V$; $B^1(x) = 0$, $x \notin V$; B^0 and $B(x)$ are the elastic compliance tensors of the matrix and the region V, correspondingly. The kernel $S_{ijkl}(x)$ in Eq. (12.2.1) has the form

$$S_{ijkl}(x) = C_{ijmn}^0 K_{mnpq}(x) C_{pqkl}^0 - C_{ijkl}^0 \delta(x), \qquad (12.2.2)$$

$$K_{ijkl}(x) = -\left[\partial_i \partial_k G_{jl}(x)\right]_{(ij)(kl)}, \qquad (12.2.3)$$

where $\delta(x)$ is the Dirac's delta function; $\mathbf{C}^0 = (\mathbf{B}^0)^{-1}$ is the elastic stiffness tensor of the matrix; indices in parentheses mean symmetrization; $G_{ij}(x)$ is the Green function of the host medium that is the vanishing at infinity solution of the equation

$$\partial_i C_{ijkl}^0 \partial_k G_{lm}(x) = -\delta(x) \delta_{jm}. \qquad (12.2.4)$$

Here δ_{ij} is the Kronecker symbol. The kernel $\mathbf{S}(x)$ of the integral operator in Eq. (12.2.1) formally diverges when $x \to 0$ as $\mathbf{S}(x) \sim |x|^{-3}$. Regularization of this integral for a continuous function $f_{ij}(x)$ is defined by the equation [7]

$$(Sf)_{ij}(x) = D_{ijkl} f_{kl}(x) + pv \int S_{ijkl}\left(x - x'\right) f_{kl}\left(x'\right) dx', \qquad (12.2.5)$$

$$D_{ijkl} = \frac{1}{4\pi} \int_{\Omega^*} S_{ijkl}^*(\mathbf{k}) d\Omega^*. \qquad (12.2.6)$$

236 Mechanics and Physics of Structured Media

Here, pv is Cauchy's principal value of the integral; $S_{ijkl}^*(\mathbf{k})$ is the Fourier transform of the generalized function $S_{ijkl}(x)$; $\mathbf{k}$ is the Fourier transform vector parameter; Ω^* is the surface of a unit sphere in the k-space of Fourier transforms. For an isotropic host medium, the function $S_{ijkl}^*(\mathbf{k})$ has the form

$$S_{ijkl}^*(\mathbf{k}) = \int S_{ijkl}(x) \exp(ik_m x_m)\, dx$$

$$= -2\mu_0 \left[P_{ijkl}^1(\mathbf{m}) + (2\kappa_0 - 1)\, P_{ijkl}^2(\mathbf{m}) \right], \tag{12.2.7}$$

$$\mathbf{m} = \frac{\mathbf{k}}{|\mathbf{k}|}, \quad \kappa_0 = \frac{\lambda_0 + \mu_0}{\lambda_0 + 2\mu_0}, \quad i = \sqrt{-1}, \tag{12.2.8}$$

$$P_{ijkl}^1(\mathbf{m}) = \theta_{i(k}\theta_{l)j}, \quad P_{ijkl}^2(\mathbf{m}) = \theta_{ij}\theta_{kl}, \quad \theta_{ij} = \delta_{ij} - m_i m_j, \tag{12.2.9}$$

where λ_0 and μ_0 are Lame's parameters of the host medium. If $f_{ij}(x)$ is a smooth function which Fourier transform $f_{ij}^*(\mathbf{k})$ tends to zero at infinity faster than $|\mathbf{k}|^{-3}$, the actions of the operator $S_{ijkl}(x)$ on such a function can be calculated as

$$(Sf)_{ij}(x) = \frac{1}{(2\pi)^3} \int S_{ijkl}^*(\mathbf{k})\, f_{kl}^*(\mathbf{k}) \exp(-ik_m x_m)\, d\mathbf{k}. \tag{12.2.10}$$

Because $S_{ijkl}^*(\mathbf{k})$ is a homogeneous function of the order 0, the integral on the right-hand side converges absolutely.

12.2.2 Cracks in homogeneous elastic media

We consider an isolated crack with a smooth surface Ω in an infinite elastic medium subjected to an external stress field $\sigma_{ij}^0(x)$. Let $n_i(x)$ be the unit normal to Ω, and let $b_i(x) = u_i^+(x) - u_i^-(x)$ be the displacement discontinuity vector on Ω (crack opening). The stress field in the medium can be expressed in the form similar to (12.2.1) [14]

$$\sigma_{ij}(x) = \sigma_{ij}^0(x) + \int_\Omega S_{ijkl}\left(x - x'\right) n_k\left(x'\right) b_l\left(x'\right) dx'. \tag{12.2.11}$$

If crack faces are traction-free, the following boundary condition should be satisfied on the crack surface Ω

$$n_i(x)\sigma_{ij}(x)\big|_\Omega = 0. \tag{12.2.12}$$

The integral equation for the crack opening $b_i(x)$ follows from Eqs. (12.2.11) and (12.2.12) in the form

$$\int_\Omega n_k(x) S_{kijl}\left(x - x'\right) n_l\left(x'\right) b_j\left(x'\right) d\Omega' = -n_j(x)\sigma_{ji}^0(x). \tag{12.2.13}$$

The integral in (12.2.13) also formally diverges at $x \to x'$. Regularization of this integral is defined in [14].

12.2.3 Medium with cracks and inclusions

We consider a homogeneous host medium containing sets of M^I inclusions that occupy the volumes $V^{(p)}$, $p = 1, \ldots, M^I$ and M^C cracks with surfaces $\Omega^{(q)}$, $q = 1, \ldots, M^C$. The stress field $\sigma_{ij}(x)$ in such medium can be presented as the sum of the fields in Eqs. (12.2.1) and (12.2.11) [15]

$$\sigma_{ij}(x) = \sigma_{ij}^0(x) + \sum_{p=1}^{M^I} \int_{V^{(p)}} S_{ijkl}(x - x') B_{klmn}^1(x') \sigma_{mn}(x') dx' +$$

$$+ \sum_{q=1}^{M^C} \int_{\Omega^{(q)}} S_{ijkl}(x - x') n_k(x') b_l(x') dx'. \tag{12.2.14}$$

From the boundary conditions (12.2.12) on the crack surfaces $\Omega^{(q)}$, we obtain that

$$-n_i(x)\sigma_{ij}^0(x) = \sum_{p=1}^{M^I} \int_{V^{(p)}} n_i(x) S_{ijkl}(x - x') B_{klmn}^1(x') \sigma_{mn}(x') dx' +$$

$$+ \sum_{q=1}^{M^C} \int_{\Omega^{(q)}} n_i(x) S_{ijkl}(x - x') n_k(x') b_l(x') dx',$$

$$x \in \Omega^{(q)}, \ q = 1, 2, \ldots, M^C. \tag{12.2.15}$$

Eq. (12.2.15) serves on the crack surfaces only; meanwhile Eq. (12.2.14) should be satisfied at the other points of the medium. These two equations compose a closed system for the unknowns of the problem: the stress tensor inside the inclusion volumes $V^{(p)}$ and crack opening vectors on the crack surfaces $\Omega^{(q)}$.

The strain tensor $\varepsilon_{ij}(x)$ in the medium with cracks and inclusions can also be presented in the integral form as follows [15]

$$\varepsilon_{ij}(x) = \varepsilon_{ij}^0 + \int K_{ijkl}(x - x') C_{klmn}^0 B_{mnpq}^1(x) \sigma_{pq}(x') V(x') dx' +$$

$$+ \int K_{ijkl}(x - x') C_{klmn}^0 n_m(x') b_n(x') \Omega(x') dx'. \tag{12.2.16}$$

Here, the kernel $K_{ijkl}(x)$ is defined in Eq. (12.2.3); $V(x) = \sum_{p=1}^{M^I} V^{(p)}(x)$, $\Omega(x) = \sum_{q=1}^{M^C} \Omega^{(q)}(x)$, $V^{(p)}(x)$ is the characteristic function of the region occupied by the pth inclusion ($V^{(p)}(x) = 1$, $x \in V^{(p)}$; $V^{(p)}(x) = 0$, $x \notin V^{(p)}$); $\Omega^{(q)}(x)$ is the delta-function concentrated on the surface of the qth crack.

238 Mechanics and Physics of Structured Media

12.3 The effective field method

12.3.1 The effective external field acting on a representative volume element

We consider an infinite homogeneous medium containing a homogeneous in space set of heterogeneities (cracks and inclusions). The medium is subjected to a constant external field σ_{ij}^0. Let a finite volume W of the medium with a number of inhomogeneities (racks and inclusions) be taken as a representative volume element (RVE) (see Fig. 12.1, left). Eq. (12.2.14) for the stress field in the medium can be written in the form

$$\sigma_{ij}(x) = \sigma_{ij}^*(x) + \int_W S_{ijkl}(x - x') B_{klmn}^1(x') \sigma_{mn}(x') V(x') dx' +$$
$$+ \int_W S_{ijkl}(x - x') n_k(x') b_l(x') \Omega(x') dx'. \tag{12.3.1}$$

$$\sigma_{ij}^*(x) = \sigma_{ij}^0 + \int_{R/W} S_{ijkl}(x - x') B_{klmn}^1(x') \sigma_{mn}(x') V(x') dx' +$$
$$+ \int_{R/W} S_{ijkl}(x - x') n_k(x') b_l(x') \Omega(x') dx'. \tag{12.3.2}$$

Here, R/W is the complement of W to the entire space R. The tensor $\sigma_{ij}^*(x)$ in these equations can be interpreted as the external stress field acting on the region W. This field does not coincide with the external field σ_{ij}^0 applied to the medium and is a sum of this field and the fields induced in W by inclusions and cracks in R/W. Averaging the field $\sigma_{ij}^*(x)$ over the ensemble realizations of the sets inclusions and cracks yields a constant field $\left\langle \sigma_{ij}^*(x) \right\rangle$ that is called the effective external stress field acting on the region W (Fig. 12.1, right).

For homogeneous in space random functions $V(x)$ and $\Omega(x)$, the average of Eq. (12.3.2) yields

$$\left\langle \sigma_{ij}^*(x) \right\rangle = \sigma_{ij}^0 + \int_{R/W} S_{ijkl}(x - x') dx' \left\langle B_{klmn}^1(x) \sigma_{mn}(x) V(x) \right\rangle +$$
$$+ \int_{R/W} S_{ijkl}(x - x') dx' \langle n_k(x) b_l(x) \Omega(x) \rangle. \tag{12.3.3}$$

Here, we take into account that for homogeneous in space sets of inclusions and cracks, the averages $\left\langle B_{klmn}^1(x) \sigma_{mn}(x) V(x) \right\rangle$ and $\langle n_k(x) b_l(x) \Omega(x) \rangle$ are constant. The integral in this equation is calculated as follows

$$\int_{R/W} S_{ijkl}(x - x') dx' = \int_R S_{ijkl}(x - x') dx' - \int_W S_{ijkl}(x - x') dx'$$
$$= -D_{ijkl}(x), \tag{12.3.4}$$

Application of volume integral equations Chapter | 12 **239**

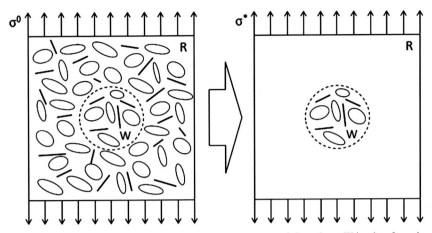

FIGURE 12.1 Schematic representation of the concept of EFM: finite volume W is taken from the infinite medium R and subjected to the effective stress field.

$$D_{ijkl}(x) = \int_W S_{ijkl}(x - x') \, dx', \, x \in W. \quad (12.3.5)$$

Here, we take into account that for a fixed external stress field σ_{ij}^0, the integral operator with the kernel $S_{ijkl}(x)$ vanishes on constants in the entire space R [20]

$$\int_R S_{ijkl}(x - x') \, dx' = 0. \quad (12.3.6)$$

Then, in the spirit of the effective field method [7], we assume that the averages $\langle B^1_{ijkl}(x)\sigma_{kl}(x)V(x)\rangle$ and $\langle n_k(x)b_l(x)\Omega(x)\rangle$ can be calculated not over the entire space but over the RVE W if this volume contains a sufficiently large number of inhomogeneities, i.e.,

$$\langle B^1_{ijkl}(x)\sigma_{kl}(x)V(x)\rangle \approx \langle B^1_{ijkl}(x)\sigma_{kl}(x)V(x)\rangle_W$$
$$= \frac{1}{W}\int_W B^1_{ijkl}(x)\sigma_{kl}(x)V(x)dx, \quad (12.3.7)$$

$$\langle n_k(x)b_l(x)\Omega(x)\rangle \approx \langle n_k(x)b_l(x)\Omega(x)\rangle_W$$
$$= \frac{1}{W}\int_W n_k(x)b_l(x)\Omega(x)dx. \quad (12.3.8)$$

The tensor **D** in Eq. (12.3.4) is constant for the RVEs of ellipsoidal shapes [7]. In the case of a spheroidal RVE with the symmetry axis x_3 and the semiaxes $a_1 = a_2 = a$ and a_3, tensor **D** takes the form

$$\mathbf{D} = d_1\mathbf{P}^2 + d_2\left(\mathbf{P}^1 - \frac{1}{2}\mathbf{P}^2\right) + d_3\left(\mathbf{P}^3 + \mathbf{P}^4\right) + d_5\mathbf{P}^5 + d_6\mathbf{P}^6, \quad (12.3.9)$$

240 Mechanics and Physics of Structured Media

$$d_1 = -\mu_0 \left(4\kappa_0 - 1 - 2\left(3\kappa_0 - 1\right) f_0 - 2f_1\right),$$
$$d_2 = -2\mu_0 \left(1 - \left(2 - \kappa_0\right) f_0 - f_1\right), \tag{12.3.10}$$
$$d_3 = -2\mu_0 \left(\left(2\kappa_0 - 1\right) f_0 + 2f_1\right), \quad d_5 = -4\mu_0 \left(f_0 + 4f_1\right),$$
$$d_6 = -8\mu_0 \left(\kappa_0 f_0 - f_1\right), \tag{12.3.11}$$

$$f_0 = \frac{1-g}{2\left(1-\gamma^2\right)}, \quad f_1 = \frac{\kappa_0}{4\left(1-\gamma^2\right)^2} \left(\left(2+\gamma^2\right) g - 3\gamma^2\right),$$

$$g = \frac{\gamma^2}{\sqrt{\gamma^2 - 1}}, \quad \gamma = \frac{a}{a_3} > 1. \tag{12.3.12}$$

The tensors $\mathbf{P}^1$ and $\mathbf{P}^2$ are defined in Eq. (12.2.9), and the tensors $\mathbf{P}^3, \ldots, \mathbf{P}^6$ have the forms

$$P^3_{ijkl} = \theta_{ij} m_k m_l, \quad P^4_{ijkl} = m_i m_j \theta_{kl}, \quad P^5_{ijkl} = m_i m_k \theta_{lj}|_{(i,j)(k,l)},$$
$$P^6_{ijkl} = m_i m_j m_k m_l. \tag{12.3.13}$$

Here m_i is the unit vector of the x_3-axis. Note that the absolute size of the RVE does not affect the value of tensor $\mathbf{D}$, it depends on the aspect ratios of the ellipsoid W only [7].

Finally, after the substitution of the effective field calculated in (12.3.3) and (12.3.4) into Eq. (12.3.1) instead of $\sigma^*_{ij}(x)$, we obtain:

$$\sigma_{ij}(x) = \sigma^0_{ij} + \int_W \left[S_{ijkl}\left(x - x'\right) - \frac{1}{W} D_{ijkl}\right] B^1_{klmn}(x')\sigma_{mn}(x') V\left(x'\right) dx' +$$
$$+ \int_W \left[S_{ijkl}\left(x - x'\right) - \frac{1}{W} D_{ijkl}\right] n_k\left(x'\right) b_l\left(x'\right) \Omega\left(x'\right) dx', \quad x \in W. \tag{12.3.14}$$

Equation for the crack openings $b_i(x)$ on the crack surfaces follows from (12.3.2) and (12.3.14) in the form

$$-n_i(x)\sigma^0_{ij}$$
$$= \int_W n_i(x) \left[S_{ijkl}\left(x - x'\right) - \frac{1}{W} D_{ijkl}\right] B^1_{klmn}\left(x'\right) \sigma_{mn}\left(x'\right) V\left(x'\right) dx' -$$
$$- \int_W n_i(x) \left[S_{ijkl}\left(x - x'\right) - \frac{1}{W} D_{ijkl}\right] n_k\left(x'\right) b_l\left(x'\right) \Omega\left(x'\right) dx', \quad x \in \Omega. \tag{12.3.15}$$

Eqs. (12.3.14) and (12.3.15) should be solved numerically inside the finite RVE W. In these equations, the terms proportional to the tensor $\mathbf{D}$ take into account presence of heterogeneities outside of W.

Application of volume integral equations **Chapter | 12 241**

12.3.2 The effective compliance tensor of heterogeneous media

For a prescribed constant external stress tensor σ_{ij}^0, the mean values of stress and strain tensors in the inhomogeneous medium are connected by the Hooke's law:

$$\langle \varepsilon_{ij}(x)\rangle = B_{ijkl}^* \langle \sigma_{kl}(x)\rangle, \langle \sigma_{ij}(x)\rangle = \sigma_{ij}^0, \qquad (12.3.16)$$

where B_{ijkl}^* is the effective elastic compliance tensor. The mean value of the strain tensor $\langle \varepsilon_{ij}(x)\rangle$ follows from Eq. (12.2.16) in the form

$$\langle \varepsilon_{ij}(x)\rangle = \varepsilon_{ij}^0 + \int_W K_{ijkl}(x-x')\,dx' C_{klmn}^0 \left\langle B_{mnpq}^1(x)\sigma_{pq}(x)V(x)\right\rangle +$$

$$+ \int_W K_{ijkl}(x-x')\,dx' C_{klmn}^0 \langle n_m(x)b_n(x)\Omega(x)\rangle. \qquad (12.3.17)$$

Here, $\varepsilon_{ij}^0 = B_{ijkl}^0 \sigma_{kl}^0$; and it is taken into account that the averages $\left\langle B_{ijkl}^1(x)\sigma_{kl}(x)V(x)\right\rangle$ and $\langle n_i(x)b_j(x)\Omega(x)\rangle$ are constant. For linearity of the problem, these averages are linear functions of the external field σ_{ij}^0

$$\left\langle B_{ijkl}^1(x)\sigma_{kl}(x)V(x)\right\rangle = M_{ijkl}\sigma_{kl}^0, \qquad (12.3.18)$$

$$\left\langle n_m(x)b_n(x)\Omega^{(q)}(x)\right\rangle = \Lambda_{ijkl}\sigma_{kl}^0, \qquad (12.3.19)$$

where M_{ijkl} and Λ_{ijkl} are constant tensors. For a prescribed external stress field, the integral in Eq. (12.3.17) has the value [7]

$$\int K_{ijkl}(x-x')\,dx' = B_{ijkl}^0. \qquad (12.3.20)$$

As a result, Eq. (12.3.17) takes the form:

$$\langle \varepsilon_{ij}(x)\rangle = \left(B_{ijkl}^0 + M_{ijkl} + \Lambda_{ijkl}\right)\sigma_{kl}^0. \qquad (12.3.21)$$

Comparing Eqs. (12.3.16) and (12.3.21), we obtain the effective compliance tensor of the heterogeneous medium in the form

$$B_{ijkl}^* = B_{ijkl}^0 + M_{ijkl} + \Lambda_{ijkl}. \qquad (12.3.22)$$

12.4 Numerical solution of the integral equations for the RVE

For cracks and inclusions of arbitrary shapes, the system of Eqs. (12.3.14) and (12.3.15) can be solved only numerically. Let a cuboidal RVE W containing a set of inclusions and cracks be covered by a regular node grid (Fig. 12.2).

242 Mechanics and Physics of Structured Media

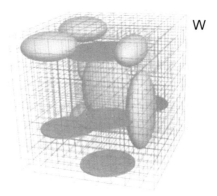

FIGURE 12.2 An example of a volume W of a medium containing cracks and inclusions covered by a regular grid of nodes.

For the discretization of the integral equations given in (12.3.14) and (12.3.15), we approximate the stress tensor in the medium and the crack openings on the crack surfaces by the following equations

$$\sigma_{ij}(x) \approx \sum_{s=1}^{N} \sigma_{ij}^{(s)} \varphi\left(x - x^{(s)}\right), \tag{12.4.1}$$

$$b_i(x) \approx \sum_{s=1}^{N} b_i^{(s)} \bar{\varphi}\left(x - x^{(s)}\right). \tag{12.4.2}$$

Here, $\sigma_{ij}^{(s)}$ and $b_i^{(s)}$ are the coefficients of the approximation, $x^{(s)}$ ($s = 1, 2, \ldots, N$) are the modes of a regular grid covered the RVE W; N is the total number of the nodes, and $b_i^{(s)} = 0$ if $x^{(s)} \notin \Omega$.

In Eq. (12.4.1), $\varphi(x)$ is the 3D-Gaussian function:

$$\varphi(x) = \frac{1}{(\pi H)^{3/2}} \exp\left(-\frac{|x|^2}{Hh^2}\right). \tag{12.4.3}$$

The function $\bar{\varphi}(x)$ is defined in the plane $P^{(s)}$ tangent to the crack surface Ω at the node $x^{(s)} \in \Omega$ by the equation

$$\bar{\varphi}\left(x - x^{(s)}\right) = \frac{1}{\pi H} \exp\left(-\frac{|x - x^{(s)}|^2}{Hh^2}\right), \quad x, x^{(s)} \in P^{(s)}. \tag{12.4.4}$$

Here, h is the grid step; H is a dimensionless parameter of the order of 1. The theory of approximation by Gaussian and other similar functions was developed in [11]. It is shown that for cubic node grids, the coefficient $\sigma_{ij}^{(s)}$ in Eq. (12.4.1) coincide with the values of the approximated function at the nodes

Application of volume integral equations **Chapter | 12** **243**

$\sigma_{ij}^{(s)} = \sigma_{ij}\left(x^{(s)}\right)$. For planar surfaces covered with square node grids, the coefficients in approximation (12.4.2) also coincide with the values of the function $b_i(x)$ at the nodes $b_i^{(s)} = b_i\left(x^{(s)}\right)$. In [8], Eqs. (12.4.1), (12.4.2) are called "approximate approximations" because the approximation errors do not vanish when the node grid step h tends to zero. But this error is small (of order of $\exp\left(-\pi^2 H\right)$) and can be neglected in practical calculations.

After substituting the solutions (12.4.1) and (12.4.2) into the integral equations (12.3.14) and (12.3.15), correspondingly and satisfying the resulting equations at all the nodes (the collocation method), we obtain the following systems of linear algebraic equations for the coefficients of the approximations

$$\sigma_{ij}^{(r)} - \sum_{s=1}^{N} \Gamma_{ijkl}^{(r,s)} B_{klmn}^{1(s)} \sigma_{mn}^{(s)} - \sum_{s=1}^{N} I_{ijk}^{(r,s)} b_k^{(s)} =$$
$$= \sigma_{ij}^{0(r)} - D_{ijkl}\left(\left\langle B_{klmn}^{1}\sigma_{mn} V\right\rangle_W + \left\langle n_i b_j \Omega\right\rangle_W\right), \quad r = 1, \ldots, N, \quad (12.4.5)$$

$$-\sum_{s=1}^{N} n_i^{(r)} \Gamma_{ijkl}^{(r,s)} B_{klmn}^{1(s)} \sigma_{mn}^{(s)} - \sum_{s=1}^{N} n_i^{(r)} I_{ijk}^{(r,s)} b_k^{(s)} =$$
$$= n_i^{(r)} \sigma_{ij}^{0} - n_i^{(r)} D_{ijkl}\left(\left\langle B_{klmn}^{1}\sigma_{mn} V\right\rangle_W + n_i^{(r)} \left\langle n_i b_j \Omega\right\rangle_W\right), \quad x^{(r)} \in \Omega,$$
$$(12.4.6)$$

$$\left\langle B_{klmn}^{1}\sigma_{mn} V\right\rangle_W = \frac{h^3}{W} \sum_{s=1}^{N} B_{klmn}^{1(s)} \sigma_{mn}^{(s)}, \quad (12.4.7)$$

$$\left\langle n_i b_j \Omega\right\rangle_W = \frac{h^2}{W} \sum_{s=1}^{N} n_i^{(s)} b_j^{(s)}, \quad b_i^{(s)} = 0 \quad \text{if } x^{(s)} \notin \Omega. \quad (12.4.8)$$

The terms $\Gamma_{ijkl}^{(r,s)}$ and $I_{ijk}^{(r,s)}$ in these equations can be interpreted as stresses induced at the rth node by sources at the sth node. These tensors depend on the vector of distance between the points $x^{(r)}$ and $x^{(s)}$:

$$\Gamma_{ijkl}^{(r,s)} = \Gamma_{ijkl}\left(x^{(r)} - x^{(s)}\right), \quad I_{ijk}^{(r,s)} = I_{ijk}\left(x^{(r)} - x^{(s)}\right), \quad (12.4.9)$$

$$B_{ijkl}^{1(s)} = B_{ijkl}^{1}\left(x^{(s)}\right), \quad n_i^{(s)} = n_i\left(x^{(s)}\right). \quad (12.4.10)$$

The functions $\mathbf{\Gamma}(x)$ and $\mathbf{I}(x)$ are the following integrals calculated over the entire 3D-space R or 2D-plane P

$$\Gamma_{ijkl}(x) = \int_R S_{ijkl}\left(x - x'\right) \varphi\left(x'\right) dx', \quad (12.4.11)$$

$$I_{ijk}(x) = \int_P S_{ijkl}\left(x - x'\right) \bar{\varphi}\left(x'\right) dP'. \quad (12.4.12)$$

244 Mechanics and Physics of Structured Media

It is taken into account that Gaussian functions are concentrated in small vicinities of the nodes with linear sizes of the order of h. Because h is much smaller than the sizes of inclusions and cracks, integration over the region of inclusions can be changed to the integration over the entire space R (for the function $\varphi(x)$) or the integration over complete plane $P^{(s)}$ tangent to the crack surface at the node $x^{(s)}$ (for the function $\bar{\varphi}(x)$).

The integral $\Gamma(x)$ is calculated explicitly and takes the form:

$$\Gamma_{ijkl}(x) = -2\mu_0 \sum_{q=1}^{6} g^{(q)}\left(\frac{|x|}{h}\right) E_{ijkl}^{(q)}(m), \quad m = \frac{x}{|x|}, \tag{12.4.13}$$

$$E_{ijkl}^1 = \delta_{i(k}\delta_{l)j}, \quad E_{ijkl}^2 = \delta_{ij}\delta_{kl}, \quad E_{ijkl}^3 = \delta_{ij}m_k m_l, \tag{12.4.14}$$

$$E_{ijkl}^4 = m_i m_j \delta_{kl}, \quad E_{ijkl}^5 = m_{(i)}m_{(k}\delta_{l)(j}, \quad E_{ijkl}^6 = m_i m_j m_k m_l. \tag{12.4.15}$$

Scalar functions $g^{(q)}(x)$ in this equation are expressed in terms of three scalar functions $\psi_0(z)$, $\psi_1(z)$, and $\psi_2(z)$:

$$g^1 = (\psi_0 - 2\psi_1) + 4\kappa_0\psi_2, \quad g^2 = (2\kappa_0 - 1)(\psi_0 - 2\psi_1) + 2\kappa_0\psi_2, \tag{12.4.16}$$

$$g^3 = g^4 = (1 - 2\kappa_0)\phi_0 + 2\kappa_0\phi_1, \quad g^5 = -\frac{1}{2}(\phi_0 - 16\kappa_0\phi_1), \quad g^6 = 2\kappa_0\phi_2, \tag{12.4.17}$$

$$\phi_0 = \psi_0 - 3\psi_1, \quad \phi_1 = \psi_0 - 5\psi_1, \quad \phi_2 = \psi_0 - 10\psi_1 + 35\psi_2, \tag{12.4.18}$$

$$\psi_0(z) = \frac{1}{(\pi H)^{3/2}} \exp\left(-\frac{z^2}{H}\right), \tag{12.4.19}$$

$$\psi_1(z) = \frac{1}{4\pi^{3/2}z^3\sqrt{H}}\left[-2z\exp\left(-\frac{z^2}{H}\right) + \sqrt{\pi H}\,erf\left(\frac{z}{\sqrt{H}}\right)\right], \tag{12.4.20}$$

$$\psi_2(z) = \frac{1}{16\pi^2 z^5}\left[6\sqrt{\pi H}z\exp\left(-\frac{z^2}{H}\right) + \pi\left(-3H + 2z^2\right)erf\left(\frac{z}{\sqrt{H}}\right)\right]. \tag{12.4.21}$$

Here $erf(z)$ is the probability integral:

$$erf(z) = \frac{2}{\sqrt{\pi}}\int_0^z \exp\left(-t^2\right)dt. \tag{12.4.22}$$

The integral $\mathbf{I}(x)$ is presented in the form:

$$I_{ijk}(x) = I_{ijk}(r, x_3) = -s_1 q_{(i}\theta_{j)k} + s_2\theta_{ij}q_k + s_3 n_{(i}\theta_{j)k} - s_4\theta_{ij}n_k -$$
$$- s_5 q_{(i}n_{j)}q_k + s_6 q_i q_j n_k + s_7\left(2q_{(i}n_{j)}n_k + n_i n_j q_k\right)$$
$$- s_8 n_i n_j n_k - s_9 q_i q_j q_k, \tag{12.4.23}$$

$$q_{1,2} = \frac{x_{1,2}}{r}, \quad q_3 = 0, \quad r = \sqrt{x_1^2 + x_2^2}, \theta_{ij} = \delta_{ij} - n_i n_j. \tag{12.4.24}$$

Application of volume integral equations **Chapter | 12 245**

Here, $n_i = n_i^{(s)}$ is the normal to Ω at the sth node, scalar coefficients $s_\alpha = s_\alpha(r, x_3)$ are

$$s_1 = 2\mu_0 (g_1 - 4\kappa_0 g_2), \quad s_2 = 2\mu_0 [(1 - 2\kappa_0) g_1 + 2\kappa_0 g_2], \quad (12.4.25)$$

$$s_3 = 2\mu_0 (g_3 - 4\kappa_0 g_4), \quad s_4 = 2\mu_0 [(1 - 2\kappa_0) g_3 + 2\kappa_0 g_4], \quad (12.4.26)$$

$$s_5 = 2\mu_0 (g_5 - 4\kappa_0 g_6), \quad s_6 = 2\mu_0 [(1 - 2\kappa_0) g_5 + 2\kappa_0 g_6], \quad (12.4.27)$$

$$s_7 = 4\mu_0 \kappa_0 g_7, \quad s_8 = 4\mu_0 \kappa_0 g_8, \quad s_9 = 4\mu_0 \kappa_0 g_9, \quad (12.4.28)$$

$$g_1 = \frac{r}{2h^2} \operatorname{sign}(x_3) (F_3 + F_4), \quad g_2 = \frac{x_3}{2hr} F_2, \quad g_3 = (F_2 - F_1), \quad (12.4.29)$$

$$g_4 = \frac{1}{4h} \left(F_1 + F_2 - \frac{|x_3|}{h} (F_3 + F_4) \right), \quad g_5 = \frac{1}{h} F_2,$$

$$g_6 = \frac{1}{2h} \left(F_2 - \frac{|x_3|}{h} F_4 \right), \quad (12.4.30)$$

$$g_7 = \frac{x_3}{2h^2} \left(F_5 - \frac{4h}{h} F_2 \right), \quad g_8 = \frac{1}{2h} \left(F_1 + \frac{|x_3|}{h} F_3 \right), \quad g_9 = \frac{x_3}{2h^2} F_5.$$

$$(12.4.31)$$

Here, $F_m = F_m(\rho, z)$ are 1D-absolutely converging integrals ($\rho = r/h$, $z = x_3/h$):

$$F_1(\rho, z) = \frac{1}{4\pi} \int_0^\infty \exp\left(-k |z| - \frac{k^2 H}{4} \right) J_0(k\rho) k^2 dk, \quad (12.4.32)$$

$$F_2(\rho, z) = \frac{1}{4\pi} \int_0^\infty \exp\left(-k |z| - \frac{k^2 H}{4} \right) J_2(k\rho) k^2 dk, \quad (12.4.33)$$

$$F_3(\rho, z) = \frac{1}{4\pi} \int_0^\infty \exp\left(-k |z| - \frac{k^2 H}{4} \right) J_0(k\rho) k^3 dk, \quad (12.4.34)$$

$$F_4(\rho, z) = \frac{1}{4\pi} \int_0^\infty \exp\left(-k |z| - \frac{k^2 H}{4} \right) J_2(k\rho) k^3 dk, \quad (12.4.35)$$

$$F_5(\rho, z) = \frac{1}{4\pi} \int_0^\infty \exp\left(-k |z| - \frac{k^2 H}{4} \right) J_3(k\rho) k^3 dk. \quad (12.4.36)$$

In these equations, $J_m(\varsigma)$ are Bessel functions of the first kind. These integrals cannot be calculated analytically, however, asymptotics $f_m = f_m(\rho, z)$ of the functions $F_m(\rho, z)$ for large values of the arguments can be presented in closed analytical forms:

$$f_1(\rho, z) = \frac{2z^2 - \rho^2}{4\pi (\rho^2 + z^2)^{5/2}} - \frac{3H (3\rho^4 - 24\rho^2 z^2 + 8z^4)}{16\pi (\rho^2 + z^2)^{9/2}}, \quad (12.4.37)$$

$$f_2(\rho, z) = \frac{3\rho^2}{4\pi (\rho^2 + z^2)^{5/2}} - \frac{15 H \rho^2 (\rho^2 - 6z^2)}{16\pi (\rho^2 + z^2)^{9/2}}, \quad (12.4.38)$$

246 Mechanics and Physics of Structured Media

$$f_3(\rho, z) = \frac{3z\left(2z^2 - 3\rho^2\right)}{4\pi\left(\rho^2 + z^2\right)^{7/2}} - \frac{15Hz\left(15\rho^4 - 40\rho^2z^2 + 8z^4\right)}{16\pi\left(\rho^2 + z^2\right)^{11/2}}, \tag{12.4.39}$$

$$f_4(\rho, z) = \frac{15\rho^2 z}{4\pi\left(\rho^2 + z^2\right)^{7/2}} - \frac{315Hz\rho^2\left(\rho^2 - 2z^2\right)}{16\pi\left(\rho^2 + z^2\right)^{11/2}}, \tag{12.4.40}$$

$$f_5(\rho, z) = \frac{15\rho^3}{4\pi\left(\rho^2 + z^2\right)^{7/2}} - \frac{15H\rho^3\left(\rho^2 - 8z^2\right)}{16\pi\left(\rho^2 + z^2\right)^{11/2}}. \tag{12.4.41}$$

Since the integral $\mathbf{\Gamma}(x)$ is calculated explicitly, and the integral $\mathbf{I}(x)$ is calculated in terms of five 1D-absolutely converging integrals that have explicit asymptotics for large values of arguments and can be tabulated for small arguments, numerical integration is excluded from the construction of the objects $\Gamma_{ijkl}^{(r,s)}$ and $I_{ijk}^{(r,s)}$ in Eqs. (12.4.5) and (12.4.6).

For the numerical solution of the system (12.4.5), (12.4.6), only iterative methods are efficient, and either Biconjugate Gradient Stabilized (BiCGStab) or Generalized Minimal Residual (GMRes) methods can be used for this purpose [14,19]. For regular node grids, the objects $\Gamma_{ijkl}^{(r,s)}$ and $I_{ijk}^{(r,s)}$ have Toeplitz's structure, thus the fast Fourier transform technique can be used for the calculation of sums in Eqs. (12.4.5) and (12.4.6) in the process of iteration [21]. As a result, the computational complexity of the vector-matrix products becomes of order $O\left(N \log N\right)$ instead of $O\left(N^2\right)$.

Once the system of linear equations (12.4.5) and (12.4.6) is solved, the stress tensor at an arbitrary point of the region W is calculated from the equation

$$\sigma_{ij}(x) = \sigma_{ij}^0(x) + \sum_{s=1}^N \Gamma_{ijkl}\left(x - x^{(s)}\right) B_{klmn}^{1(s)} \sigma_{mn}^{(s)} + \sum_{s=1}^N I_{ijk}\left(x - x^{(s)}\right) b_k^{(s)}, \tag{12.4.42}$$

where the functions $\Gamma_{ijkl}(x)$ and $I_{ijk}(x)$ are defined in Eqs. (12.4.13), (12.4.23).

For the numerical solution of the system (12.4.5), (12.4.6), the refinement strategy can be proposed [14]. In this strategy, first, a coarse node grid is used for the global solution of this system. Then, each inclusion and crack is covered by a finer node grid, and the external fields acting on the heterogeneities are calculated from the solution for the coarse grid. Such strategy allows one to increase substantially the accuracy of calculations.

12.5 Numerical examples and optimal choice of the RVE

12.5.1 Periodic system of penny-shaped cracks of the same orientation

We consider a periodic system of penny-shaped cracks of the same radii r and orientations; the crack centers are at the nodes of a regular cuboid lattice with the sides $\alpha_1 = \alpha_2 = \alpha$ and α_3 [19]. The matrix is isotropic with Poisson ratio

$\nu_0 = 0.3$. In the Cartesian coordinate system (x_1, x_2, x_3) with the axes directed along the sides of the cuboid cell and the x_3-axis orthogonal to the crack planes, the tensor Λ in Eq. (12.3.17) has three nonzero components $\Lambda_{1313} = \Lambda_{2323}$, and Λ_{3333}. The effective compliance tensor B^*_{ijkl} in Eq. (12.3.20) takes the form

$$B^*_{ijkl} = B^0_{ijkl} + \Lambda_{1313} P^5_{ijkl}(\mathbf{n}) + \Lambda_{3333} P^6_{ijkl}(\mathbf{n}). \quad (12.5.1)$$

Here the tensors $\mathbf{P}^5(\mathbf{n})$ and $\mathbf{P}^6(\mathbf{n})$ are defined in Eq. (12.3.12), and $\mathbf{n}$ is the unit vector of the x_3-axis. For the calculation of the components Λ_{1313} and Λ_{3333}, the RVE was subjected to two stress fields with nonzero components σ^0_{13} and σ^0_{33}, correspondingly. The crack opening vectors in the RVE were calculated according to the algorithm described in the previous sections. For an ellipsoidal RVE, the tensor $\mathbf{D}$ in Eqs. (12.4.5) and (12.4.6) is defined in Eq. (12.3.8). The ellipsoid aspect ratio is taken equal to the aspect ratio of the elementary cell with the parameters α, α, α_3 (Fig. 12.3). The RVEs containing 1, 8, and 27 cubic cells ($\alpha = \alpha_3$) are considered.

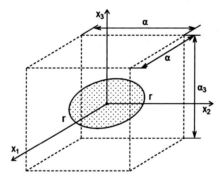

FIGURE 12.3 Elementary cuboidal cell with a penny-shaped crack that is placed into the RVE.

Then, we consider a spherical RVE with 216 cells for a strongly 'oblate' cuboidal periodic crack system ($\alpha/\alpha_3 = 8$). For this case, spheroidal RVEs containing 1, 8, and 27 cuboidal cells with the same aspect ratio as the elementary cell of the crack grid are also considered. All the calculations were performed for the node grid step $h/\alpha = 0.01$ and $H = 1$. The results are presented in Fig. 12.4; the effective elastic parameters E^*_{33} and μ^*_{13} are expressed in terms of the parameters Λ_{1313} and Λ_{3333} as follows

$$E^*_{33} = \left(\frac{1}{E_0} + \Lambda_{3333}\right)^{-1}, \quad \mu^*_{13} = \left(\frac{1}{\mu_0} + \Lambda_{1313}\right)^{-1}, \quad (12.5.2)$$

where E_0 and μ_0 are Young and shear moduli of the host medium.

The bold lines in these figures are the results of [22] that can be considered as exact solutions; the lines with markers are the numerical predictions of the proposed method.

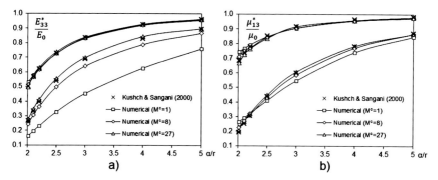

FIGURE 12.4 Dependence of the normalized effective Young's E_{33}^* and shear μ_{13}^* moduli of an isotropic medium with a cuboidal lattice of penny-shaped cracks on the parameter α/r for various aspects of the cuboid α/α_3.

It can be seen from Fig. 12.4 that in the case of a cubic cell, the solutions for the RVE with 1, 8, and 27 cracks and the results of [22] coincide practically. However, in the case of dense crack systems ($\alpha = 8\alpha_3$), the solutions for the RVE with 1 and 8 cracks substantially differ from the results of [22], while there are almost no differences between these results and the numerical solutions for 27 and 216 cracks (thus, the solution for 216 cracks is not shown).

12.5.2 Periodic system of rigid spherical inclusions

If identical spherical inclusions compose a simple cubic lattice in the matrix material, the tensor of the effective elastic stiffness has cubic symmetry. A convenient basis for presentation of such tensors consists of the following three linearly independent rank four tensors

$$H_{ijkl}^1 = \delta_{1i}\delta_{1j}\delta_{1k}\delta_{1l}, \quad H_{ijkl}^2 = E_{ijkl}^2 - H_{ijkl}^1, \quad H_{ijkl}^3 = 2E_{ijkl}^1 - H_{ijkl}^1. \tag{12.5.3}$$

In this basis, the tensors B_{ijkl}^0 and M_{ijkl} in Eq. (12.3.22) for the effective compliance tensor B_{ijkl}^* are presented in the form

$$B_{ijkl}^0 = b_1^0 H_{ijkl}^1 + b_2^0 H_{ijkl}^2 + b_3^0 H_{ijkl}^3, \quad M_{ijkl} = m_1 H_{ijkl}^1 + m_2 H_{ijkl}^2 + m_3 H_{ijkl}^3, \tag{12.5.4}$$

where the scalar coefficients b_1^0, b_2^0, and b_3^0 for an isotropic host medium are

$$b_1^0 = \frac{\mu_0 + 3K_0}{9\mu_0 K_0}, \quad b_2^0 = \frac{2\mu_0 - 3K_0}{18\mu_0 K_0}, \quad b_3^0 = \frac{1}{4\mu_0}. \tag{12.5.5}$$

Here, K_0 and μ_0 are the bulk and shear moduli of the host medium. As the result, the tensor B_{ijkl}^* takes the form:

$$B_{ijkl}^* = b_1^* H_{ijkl}^1 + b_2^* H_{ijkl}^2 + b_3^* H_{ijkl}^3, \tag{12.5.6}$$

Application of volume integral equations **Chapter | 12 249**

$$b_1^* = b_1^0 + m_1, \quad b_2^* = b_2^0 + m_2, \quad b_3^* = b_3^0 + m_3. \tag{12.5.7}$$

The coefficients m_1, m_2, and m_3 are to be calculated from Eq. (12.3.21)

$$\left\langle B_{ijkl}^1 \sigma_{kl} \right\rangle = (m_1 - 2m_3) \sigma_{11}^0 \delta_{1i} \delta_{1j}$$
$$+ m_2 \left[\sigma_{11}^0 \left(\delta_{2i} \delta_{2j} + \delta_{3i} \delta_{3j} \right) + \left(\sigma_{22}^0 + \sigma_{33}^0 \right) \delta_{ij} \right] + 2m_3 \sigma_{ij}^0. \tag{12.5.8}$$

In order to find the values of the coefficients m_1, m_2, and m_3, Eq. (12.4.5) should be solved two times for the external stress fields σ_{ij}^0 in the forms

$$\sigma_{ij}^{0(1)} = \delta_{1i} \delta_{1j}, \quad \sigma_{ij}^{0(2)} = 2\delta_{1(i} \delta_{2j)}. \tag{12.5.9}$$

The products of these tensors with the tensor M_{ijkl} are:

$$M_{ijkl} \sigma_{kl}^{0(1)} = m_1 \delta_{1i} \delta_{1j} + m_2 \left(\delta_{3i} \delta_{3j} + \delta_{3i} \delta_{3j} \right),$$
$$M_{ijkl} \sigma_{kl}^{0(2)} = m_3 \left(\delta_{1i} \delta_{2j} + \delta_{2i} \delta_{1j} \right). \tag{12.5.10}$$

If the solutions of Eq. (12.4.6) for the right-hand sides $\sigma_{ij}^{0(1)}$ and $\sigma_{ij}^{0(2)}$ are $\sigma_{ij}^{(1)}$ and $\sigma_{ij}^{(2)}$ correspondingly, the coefficients m_1, m_2, and m_3 are expressed in terms of the components of the tensors $\left\langle B_{ijkl}^1 \sigma_{kl}^{(1)} \right\rangle$ and $\left\langle B_{ijkl}^1 \sigma_{kl}^{(2)} \right\rangle$ as follows

$$m_1 = \left\langle B_{11kl}^1 \sigma_{kl}^{(1)} \right\rangle, \quad m_2 = \left\langle B_{22kl}^1 \sigma_{kl}^{(1)} \right\rangle, \quad m_3 = \left\langle B_{12kl}^1 \sigma_{kl}^{(2)} \right\rangle. \tag{12.5.11}$$

The effective elastic stiffness tensor $C^* = (B^*)^{-1}$ is presented in the form similar to (12.5.6)

$$C_{ijkl}^* = K^* E_{ijkl}^2 + 2\mu^* \left(E_{ijkl}^1 - H_{ijkl}^2 \right) + 2M^* \left(H_{ijkl}^1 - \frac{1}{3} E_{ijkl}^1 \right), \tag{12.5.12}$$

where the scalar coefficients μ^*, K^*, and M^* are expressed in terms of the coefficients b_1^*, b_2^*, and b_3^* in Eq. (12.5.7)

$$K^* = \frac{1}{3} c_1 + 2c_2, \quad \mu^* = c_3, \quad M^* = \frac{c_1 - c_2}{2}, \tag{12.5.13}$$

$$c_1 = \frac{b_1^* + b_2^*}{\Delta}, \quad c_2 = \frac{-b_1^*}{\Delta}, \quad c_3 = \frac{1}{4b_1^*}, \quad \Delta = \left(b_1^* - b_2^* \right) \left(b_1^* + 2b_2^* \right). \tag{12.5.14}$$

We consider a periodic set of rigid spherical inclusions (the ratio of the Young's moduli of the inclusions E_1 and the medium E_0 is $E_1/E_0 = 1000$) of the same radii r with the centers at the nodes of a cubic lattice with the sides $\alpha_1 = \alpha_2 = \alpha_3 = \alpha$ [18]. The host medium is isotropic; the Poisson's ratios of the matrix

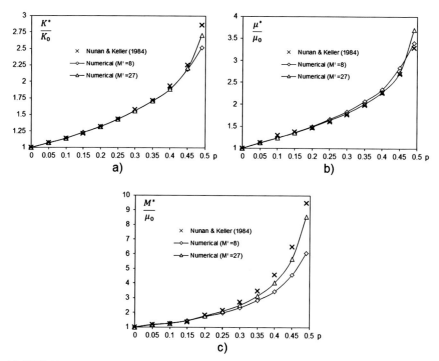

FIGURE 12.5 Dependence of the normalized effective bulk K^*, shear μ^*, and longitudinal M^* moduli of an isotropic medium with a cubic lattice of rigid spherical inclusions on the volume fraction of the inclusions p.

and the inclusions are $\nu_0 = \nu_1 = 0.3$. The stress fields inside the inclusions were calculated according to the algorithm described in the previous sections. The tensor **D** was taken for a spherical RVE. We considered the RVEs with 8 and 27 inclusions inside. All the calculations were performed for the regular node grid with step $h/\alpha = 0.01$ and $H = 1$. The results are given in Fig. 12.5 in terms of volume fraction of inclusions p. The bold lines in these figures are the results of [23] that can be considered as exact ones; lines with markers are the results of the numerical solutions. The numerical solutions are close to the exact ones; some divergence appears only for very high volume fractions of inclusions ($p > 0.4$).

12.6 Conclusions

The homogenization technique developed in this work is a combination of the effective field method and the numerical method based on volume integral equations. The most important features of the proposed technique are the following.

The method reduces the homogenization problem to the problem for a finite RVE of a heterogeneous medium embedded into the homogeneous host medium. Presence of heterogeneities outside of the RVE is taken into account

Application of volume integral equations Chapter | 12 **251**

by the effective stress field applied to the RVE. The problem of boundary conditions on the RVE sides does not emerge in the method.

The effective stress field is found from the condition of self-consistency. This condition is the statement that averages of the elastic fields over the entire medium can be changed with averages over the RVE subjected to the effective external stress field.

Comparison of the results of the method with the solutions of the homogenization problem for periodic sets of inhomogeneities has shown that the RVE should contain about 25–30 inhomogeneities for reliable calculations of the effective elastic stiffness tensor.

The shape of the RVE can be taken ellipsoidal and this choice does not affect the results of calculations. The aspect of the ellipsoid should coincide with the aspect of an elementary cell of heterogeneities in the case of regular composites. For isotropic sets of heterogeneous, the RVE of spherical shape can be taken.

For discretization of the volume integral equations for the RVE problem, Gaussian approximation functions are used. These functions allow avoiding numerical integrations by construction of the matrix of the discretized problem and reducing substantially the amount of calculations. For regular node grid, the matrix of the discretized problem has Toeplitz's properties; and the fast Fourier transform technique can be used for the calculation of matrix-vector products necessary for the iterative solution of the discretized problem, thus accelerating the calculation process. Application of the method to other problems for heterogeneous media is presented in [24–26].

References

[1] E.I. Grigolyuk, L.A. Filshtinsky, Perforated Plates and Shells, Nauka, Moscow, 1970.

[2] G. Bakhvalov, N.S. Panasenko, Homogenization: Averaging Processes in Periodic Media, Springer Netherlands, Dordrecht, 1989.

[3] V.V. Jikov, S.M. Kozlov, O.A. Oleinik, Homogenization of Differential Operators and Integral Functionals, Springer Berlin Heidelberg, Berlin, Heidelberg, 1994.

[4] E.I. Grigolyuk, L.A. Filshtinsky, Regular Piece-Homogeneous Structures With Defects, Fiziko-Matematicheskaya Literatura, Moscow, 1994.

[5] A.A. Gusev, Representative volume element size for elastic composites: a numerical study, J. Mech. Phys. Solids 45 (9) (Sep. 1997) 1449–1459.

[6] J. Segurado, J. Llorca, A numerical approximation to the elastic properties of sphere-reinforced composites, J. Mech. Phys. Solids 50 (10) (Oct. 2002) 2107–2121.

[7] S. Kanaun, V. Levin, Self-Consistent Methods for Composites I, Springer, Dordrecht, 2008.

[8] V. Mityushev, Cluster method in composites and its convergence, Appl. Math. Lett. 77 (Mar. 2018) 44–48.

[9] S. Gluzman, V. Mityushev, W. Nawalaniec, Computational Analysis of Structured Media, Elsevier, 2018.

[10] P. Drygaś, V. Mityushev, S. Gluzman, W. Nawalaniec, Applied Analysis of Composite Media, Elsevier, 2020.

[11] V. Maz'ya, G. Schmidt, Approximate Approximations, AMS Mathematical Surveys and Monographs, Providence, 2007.

[12] S.K. Kanaun, Fast solution of 3D-elasticity problem of a planar crack of arbitrary shape, Int. J. Fract. 148 (4) (Dec. 2007) 435–442.

252 Mechanics and Physics of Structured Media

[13] S. Kanaun, Fast solution of the elasticity problem for a planar crack of arbitrary shape in 3D-anisotropic medium, Int. J. Eng. Sci. 47 (2) (2009) 284–293.

[14] S. Kanaun, A. Markov, S. Babaii, An efficient numerical method for the solution of the second boundary value problem of elasticity for 3D-bodies with cracks, Int. J. Fract. 183 (2) (Oct. 2013) 169–186.

[15] A. Markov, S. Kanaun, Interactions of cracks and inclusions in homogeneous elastic media, Int. J. Fract. 206 (1) (Jul. 2017) 35–48.

[16] V. Grechka, M. Kachanov, Effective elasticity of fractured rocks: a snapshot of the work in progress, Geophysics 71 (6) (Nov. 2006) W45–W58.

[17] V. Grechka, M. Kachanov, Effective elasticity of rocks with closely spaced and intersecting cracks, Geophysics 71 (3) (May 2006) D85–D91.

[18] S. Kanaun, E. Pervago, Combining self-consistent and numerical methods for the calculation of elastic fields and effective properties of 3D-matrix composites with periodic and random microstructures, Int. J. Eng. Sci. 49 (5) (May 2011) 420–442.

[19] A. Markov, S. Kanaun, An efficient homogenization method for elastic media with multiple cracks, Int. J. Eng. Sci. 82 (Sep. 2014) 205–221.

[20] I. Kunin, Elastic Media with Microstructure II, 1st ed., Springer, Berlin, Heidelberg, 1983.

[21] W. Press, B. Flannery, S. Teukolsky, W. Vetterling, Numerical Recipes in FORTRAN: The Art of Scientific Computing, 2nd ed., Cambridge University Press, New York, 1993.

[22] V.I. Kushch, A.S. Sangani, Stress intensity factor and effective stiffness of a solid containing aligned penny-shaped cracks, Int. J. Solids Struct. 37 (44) (Nov. 2000) 6555–6570.

[23] K.C. Nunan, J.B. Keller, Effective elasticity tensor of a periodic composite, J. Mech. Phys. Solids 32 (4) (Jan. 1984) 259–280.

[24] A. Trofimov, A. Markov, S.G. Abaimov, I. Akhatov, I. Sevostianov, Overall elastic properties of a material containing inhomogeneities of concave shape, Int. J. Eng. Sci. 132 (Nov. 2018) 30–44.

[25] A. Markov, A. Trofimov, I. Sevostianov, Effect of non-planar cracks with islands of contact on the elastic properties of materials, Int. J. Solids Struct. 191–192 (May 2020) 307–314.

[26] A. Markov, A. Trofimov, I. Sevostianov, A unified methodology for calculation of compliance and stiffness contribution tensors of inhomogeneities of arbitrary 2D and 3D shapes embedded in isotropic matrix – open access software, Int. J. Eng. Sci. 157 (Dec. 2020) 103390.

Chapter 13

A slipping zone model for a conducting interface crack in a piezoelectric bimaterial

Volodymyr Loboda[a], Alla Sheveleva[b], and Oleksandr Mykhail[a]

[a]*Department of Theoretical and Computational Mechanics, Oles Honchar Dnipro National University, Dnipro, Ukraine,* [b]*Department of Computational Mathematics and Mathematical Cybernetics, Oles Honchar Dnipro National University, Dnipro, Ukraine*

13.1 Introduction

Piezoelectric and piezoelectromagnetic materials are widely used in modern microelectronics. They are also can be found often in combination with other materials to form composites. The cracks, especially microcracks may occur often at the interface of the individual components. Because of small size of piezoelectric components the microcracks can significantly reduce the strength of such structures; therefore, their study is important for practice.

Pioneer results concerning investigation of interface cracks in piezoelectric and piezoelectromagnetic materials were obtained in papers of L.A. Fil'shtinskii and his coauthors [5–7]. Particularly, in-plane and out-of-plane interface crack problems for piezoelectric compounds subjected to piecewise uniform mechanical loading combined with electric loading at infinity, and also a line loading at an arbitrary point were considered in these papers.

It is well-known that physically unrealistic oscillating singularity occurs at interface crack tip in plane case [3,4,16]. For an interface crack in a piezoelectric bimaterial this singularity was revealed by V.Z. Parton, B.A. Kudryavtsev [15]. The way of removing this singularity was suggested in [2] for a crack between two isotropic materials and generalized in [8,9] for an eclectically permeable and electrically insulated crack models, respectively.

Temporary actuators and other electronic devices are often constructed with use of thin film electrodes sandwiched between piezoelectric layers. Such electrodes are usually prepared of a metal powder, conducting polymers, etc., and do not change the mechanical properties of matrices [17]. Delaminating of the mentioned electrodes leads to the appearance of electrically conducting interface cracks. In a plane case conducting conditions were considered in papers [1] and [12] for "open" and contact interface crack models, respectively.

Mechanics and Physics of Structured Media. https://doi.org/10.1016/B978-0-32-390543-5.00018-9
Copyright © 2022 Elsevier Inc. All rights reserved.

Most of interface crack investigations were related to the case of piezoelectric materials polarized in the direction orthogonal to the crack faces. However, very often the crack can be situated parallel to the direction of polarization and it happens that in such cases the mathematical analysis and physical manifestations of the results essentially differ from the first case. The present chapter is just devoted to the development of the way of removing an oscillating singularity for an electrically conducting interface cracks in a piezoelectric bimaterial for the direction of material's polarization parallel to the crack faces.

13.2 Formulation of the problem

Consider an electrically conducting crack $c \leq x_1 \leq a$, $x_3 = 0$ between two semi-infinite piezoelectric half-spaces $x_3 > 0$ and $x_3 < 0$ having both the symmetry class of 6 mm with the poled direction x_1 (Parton and Kudryavtsev [15]). The material properties of the half-spaces are $c_{ijkl}^{(1)}$, $e_{lij}^{(1)}$, $\varepsilon_{ij}^{(1)}$ and $c_{ijkl}^{(2)}$, $e_{lij}^{(2)}$, $\varepsilon_{ij}^{(2)}$, respectively.

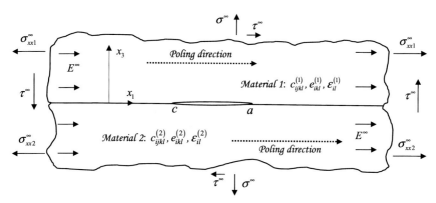

FIGURE 13.1 An electrically conducting interface crack under the action of remote mixed mode mechanical loading σ^∞, τ^∞, and electrical field E^∞.

The loading conditions at infinity are $\sigma_{33}^{(m)} = \sigma^\infty$, $\sigma_{13}^{(m)} = \tau^\infty$, $\sigma_{11}^{(m)} = \sigma_{xxm}^\infty$, $E_1^{(m)} = E^\infty$. Here and in the following, $m = 1$ stands for the upper domain, and $m = 2$ for the lower one. It is assumed that the continuity equations for the strain ε_{11} are satisfied across the interface at infinity. Since the load is independent of the coordinate x_2, the plane strain problem in the (x_1, x_3) plane depicted in Fig. 13.1 can be considered.

The constitutive relations for a linear piezoelectric material in the absence of body forces and free charges can be presented in the form [15]

$$\sigma_{ij} = c_{ijkl}\gamma_{kl} - e_{kij}E_k, \tag{13.1}$$

$$D_i = e_{ikl}\gamma_{kl} + \varepsilon_{ik}E_k, \tag{13.2}$$

A slipping zone model for a conducting interface crack   Chapter | 13   **255**

$$\sigma_{ij,i} = 0, \quad D_{i,i} = 0, \tag{13.3}$$

$$\gamma_{ij} = 0.5\left(u_{i,j} + u_{j,i}\right), \quad E_i = -\varphi_{,i}, \tag{13.4}$$

where u_k, φ, σ_{ij}, γ_{ij}, and D_i are the elastic displacements, electric potential, stresses, strains, and electric displacements, respectively; c_{ijkl}, e_{lij}, and ε_{ij} are the elastic moduli, piezoelectric constants, and dielectric constants, respectively. The subscripts in (13.1)–(13.4) are ranging from 1 to 3 and Einstein's summation convention is used in (13.1)–(13.3).

Substituting Eqs. (13.4) into (13.1), (13.2), and after that into (13.3), one obtains

$$\left(c_{ijkl}u_k + e_{lij}\varphi\right)_{,li} = 0, \quad \left(e_{ikl}u_k - \varepsilon_{il}\varphi\right)_{,li} = 0. \tag{13.5}$$

For the considered case of electro-mechanical fields independent on the co-ordinate x_2, the following presentations

$$\langle \mathbf{L}(x_1)\rangle = \mathbf{W}^+(x_1) - \mathbf{W}^-(x_1), \tag{13.6}$$

$$\mathbf{P}^{(1)}(x_1, 0) = \mathbf{S}\mathbf{W}^+(x_1) - \bar{\mathbf{S}}\mathbf{W}^-(x_1), \tag{13.7}$$

were given in Ref. [12]. In these relations

$$\mathbf{L} = \left[u_1', u_3', D_3\right]^T, \quad \mathbf{P} = \left[\sigma_{31}, \sigma_{33}, E_1\right]^T, \tag{13.8}$$

prime means the differentiation on x_1, $\mathbf{S} = \mathbf{N}^{(1)}\mathbf{D}^{-1}$, $\mathbf{D} = \mathbf{M}^{(1)} - \bar{\mathbf{M}}^{(2)}\left(\bar{\mathbf{N}}^{(2)}\right)^{-1}\mathbf{N}^{(1)}$, $\mathbf{W}(z)$ is a vector-function which is analytic in each semi-infinite plane and $\mathbf{W}^+(x_1) = \mathbf{W}(x_1 + i \cdot 0)$, $\mathbf{W}^-(x_1) = \mathbf{W}(x_1 - i \cdot 0)$. Here and afterwards, the designation $\langle g(x_1)\rangle$ for a function $g(z)$ means the jump of this function across the material interface. Matrices $\mathbf{M}^{(k)}$ and $\mathbf{N}^{(k)}$ have the following structure

$$\mathbf{M}^{(k)} = \begin{bmatrix} a_{1J}^{(k)} \\ a_{3J}^{(k)} \\ b_{4J}^{(k)} \end{bmatrix}_{J=1,3,4}, \quad \mathbf{N}^{(k)} = \begin{bmatrix} b_{1J}^{(k)} \\ b_{3J}^{(k)} \\ -a_{4J}^{(k)} \end{bmatrix}_{J=1,3,4}, \tag{13.9}$$

in which $a_{ij}^{(k)}$, $b_{ij}^{(k)}$ ($k = 1, 2$) are the components of the matrices $\mathbf{A}^{(k)}$ and $\mathbf{B}^{(k)}$ (Suo et al. [19]). It is worth to be mentioned that the presentations (13.6), (13.7) provided the satisfaction of the equation $\mathbf{P}^{(1)}(x_1, 0) = \mathbf{P}^{(2)}(x_1, 0)$ for the whole bimaterial interface $x_1 \in (-\infty, \infty)$.

Similarly to the contracted notations of the anisotropic elasticity (Sokolnikoff [18]), we'll use the following designation for the components c_{ijkl}, e_{lij}, and ε_{ij} related to the (x_1, x_3) plane: $c_{1111} = c_{11}$, $c_{1133} = c_{13}$, $c_{3333} = c_{33}$, $c_{1313} = c_{44}$, $c_{113} = e_{31}$, $e_{333} = e_{33}$, $e_{131} = e_{15}$. Moreover, analysis shows that for the considered class of materials (polarized in the direction x_1) the matrix $\mathbf{S}$

256 Mechanics and Physics of Structured Media

has the following structure:

$$\mathbf{S} = \begin{bmatrix} S_{11} & S_{13} & S_{14} \\ S_{31} & S_{33} & S_{34} \\ S_{41} & S_{43} & S_{44} \end{bmatrix} = \begin{bmatrix} is_{11} & s_{13} & s_{14} \\ s_{31} & is_{33} & is_{34} \\ s_{41} & is_{43} & is_{44} \end{bmatrix}, \tag{13.10}$$

where all s_{ij} are real and $s_{31} = -s_{13}$, $s_{41} = s_{14}$, $s_{43} = -s_{34}$ hold true.

Assume further that the crack faces are traction free and electrically conductive, i.e. the interface conditions are the following

$$\text{for } x_1 \notin [c, a]: \quad \mathbf{P}^{(1)}(x_1, 0) = \mathbf{P}^{(2)}(x_1, 0), \quad \mathbf{L}^{(1)}(x_1, 0) = \mathbf{L}^{(2)}(x_1, 0), \tag{13.11}$$

$$\text{for } x_1 \in (c, a): \quad \sigma_{13}^{\pm}(x_1, 0) = 0, \quad \sigma_{33}^{\pm}(x_1, 0) = 0, \quad E_1^{\pm}(x_1, 0) = 0. \tag{13.12}$$

Applying to Eqs. (13.6), (13.7) the analysis similar to [9], one can arrive to the following presentations

$$ir_{j1}\sigma_{13}^{(1)}(x_1, 0) + \sigma_{33}^{(1)}(x_1, 0) + r_{j4}E_1^{(1)}(x_1, 0) = F_j^+(x_1) + \gamma_j F_j^-(x_1), \tag{13.13}$$

$$t_{j1}\langle u_1'(x_1)\rangle + it_{j3}\langle u_3'(x_1)\rangle + it_{j4}\langle D_3(x_1)\rangle = F_j^+(x_1) - F_j^-(x_1), \tag{13.14}$$

where $\gamma_1 = \frac{1+\delta}{1-\delta}$, $\gamma_3 = 1/\gamma_1$, $\gamma_4 = 1$, $\delta^2 = \frac{s_{13}s_{41}s_{34} + s_{31}s_{43}s_{14} - s_{31}s_{13}s_{44} - s_{14}s_{41}s_{33}}{s_{11}(s_{33}s_{44} - s_{34}s_{43})}$; $r_{11} = (s_{33}s_{44} - s_{34}s_{43})\frac{\delta}{D}$, $r_{14} = \frac{1}{D}(s_{33}s_{14} - s_{34}s_{13})$, $r_{31} = -r_{11}$, $r_{34} = -r_{14}$, $r_{44} = -\frac{s_{31}}{s_{41}}$ $(D = s_{13}s_{44} - s_{14}s_{43})$ are real components of the matrix $\mathbf{R}$ having the following structure:

$$\mathbf{R} = \begin{bmatrix} ir_{11} & 1 & r_{14} \\ ir_{31} & 1 & r_{34} \\ 0 & 1 & r_{44} \end{bmatrix}.$$

The matrix $\mathbf{T}$ with the components T_{ji} $(i, j = 1, 3, 4)$ is defined as

$$\mathbf{T} = \mathbf{RS}. \tag{13.15}$$

For $\delta^2 > 0$ the components of T_{ji} can be presented in the form $T_{j1} = t_{j1}$, $T_{j3} = it_{j3}$, $T_{j4} = it_{j4}$, where all t_{jk} $(j, k = 1, 3, 4)$ are real and $t_{41} = 0$.

The new unknown functions $F_j(z)$ have the same properties as $\mathbf{W}(z)$ and have the following presentation

$$F_j(z) = \mathbf{T}_j\mathbf{W}(z) \tag{13.16}$$

where $\mathbf{T}_j$ are one-line matrices $\mathbf{T}_j = [T_{j1}, T_{j3}, T_{j4}] = \mathbf{R}_j\mathbf{S}$ $(j = 1, 3, 4)$.

A slipping zone model for a conducting interface crack **Chapter | 13 257**

Taking into account that for $x_1 \notin [c, a]$ the relationships $F_j^+(x_1) = F_j^-(x_1) = F_j(x_1)$ hold true, we get from Eq. (13.13) that

$$\left(1 + \gamma_j\right) F_j(x_1) = i r_{j1} \sigma_{13}^{(1)}(x_1, 0) + \sigma_{33}^{(1)}(x_1, 0) + r_{j4} E_1^{(1)}(x_1, 0)$$
$$\text{for } x_1 \to \infty. \tag{13.17}$$

But taking into account that the functions $F_j(z)$ are analytic in the whole plane cut along $x_1 \in (c, a)$ and using the conditions at infinity, one gets from Eq. (13.17)

$$F_j(z)\big|_{z \to \infty} = \tilde{\sigma}_j - i \tilde{\tau}_j, \tag{13.18}$$

where $\tilde{\sigma}_j = \frac{\sigma^\infty + r_{j4} E^\infty}{\vartheta_j}$, $\tilde{\tau}_j = -\frac{r_{j1}\tau^\infty}{\vartheta_j}$, $(j = 1, 3, 4)$, $\vartheta_k = (1 + \gamma_k)$, $(k = 1, 3)$, $\vartheta_4 = 2$.

To solve the formulated problem we use the presentation (13.13) and interface conditions (13.11), (13.12). Satisfaction of (13.11) provides the analyticity of $F_j(z)$ for $x_1 \notin [c, a]$ whilst (13.12) gives the following equations:

$$F_j^+(x_1) + \gamma_j F_j^-(x_1) = 0, \quad (j = 1, 3, 4) \quad \text{for } x_1 \in (c, a). \tag{13.19}$$

According to the results by Muskhelishvili [13] the solution of the problem (13.19) under the conditions at infinity (13.18) and the uniqueness conditions of displacements and the electric field has the form

$$F_j(z) = X_j(z) \left(\tilde{\sigma}_j - i \tilde{\tau}_j\right) \left\{ z - \left[\frac{a + c}{2} + i(a - c)\varepsilon_j\right]\right\}, \tag{13.20}$$

where $X_j(z) = (z - c)^{-1/2 + i\varepsilon_j}(z - a)^{-1/2 - i\varepsilon_j}$, $\varepsilon_j = \frac{\ln \gamma_j}{2\pi}$.

Using (13.13), the combination of stresses and the electric field for $x_1 > a$ can be written in the form

$$i r_{11} \sigma_{13}^{(1)}(x_1, 0) + \sigma_{33}^{(1)}(x_1, 0) + r_{14} E_1^{(1)}(x_1, 0)$$
$$= (1 + \gamma_1)(\tilde{\sigma}_1 - i \tilde{\tau}_1) \left\{ x_1 - \left[\frac{a + c}{2} + i(a - c)\varepsilon_1\right]\right\}$$
$$\times (x_1 - c)^{-1/2 + i\varepsilon_1}(x_1 - a)^{-1/2 - i\varepsilon_1}, \tag{13.21}$$

$$\sigma_{33}^{(1)}(x_1, 0) + r_{44} E_1^{(1)}(x_1, 0) = \frac{2\tilde{\sigma}_4 (x_1 - (c + a)/2)}{\sqrt{(x - c)(x - a)}}. \tag{13.22}$$

The individual components of stresses and the electrical field can easily be found from the system (13.21), (13.22).

It is clearly seen from (13.20)–(13.22) that the obtained solution has an oscillating square root singularity at the crack tips. Despite the fact that this solution gives a good estimate of the stress-strain state at some distance from the crack tip, it demonstrates the physically unrealistic behavior of electromechanical quantities in the immediate vicinity of the crack tips and, particularly, leads to the crack faces overlapping.

13.3 An interface crack with slipping zones at the crack tips

It follows from (13.21), (13.22) the stresses have an oscillation and tend to infinitely large values at the crack tips in the frameworks of the above considered model. To eliminate this physically unrealistic oscillating singularity we assume that the zones of mechanical delamination occur at the crack continuations. We call these regions as slipping zones. As it will be shown later these zones are very small and their influence on each other is negligible. Therefore with high accuracy only longer zone at the right crack tip can be considered for certain class of loadings. For another case of loadings the slipping zones at the left crack tip can be considered by simple transposition of materials. Thus the boundary conditions at the interface can be written in the form

for $x_1 \notin (c, b)$: $\mathbf{P}^{(1)}(x_1, 0) = \mathbf{P}^{(2)}(x_1, 0)$, $\mathbf{L}^{(1)}(x_1, 0) = \mathbf{L}^{(2)}(x_1, 0)$,
(13.23)

for $x_1 \in (c, a)$: $\sigma_{13}^{\pm}(x_1, 0) = 0$, $\sigma_{33}^{\pm}(x_1, 0) = 0$, $E_1^{\pm}(x_1, 0) = 0$,
(13.24)

for $x_1 \in (a, b)$: $\langle u_3'(x_1, 0) \rangle = 0$, $\sigma_{13}^{\pm}(x_1, 0) = 0$, $\langle \sigma_{33}(x_1) \rangle = 0$,
$\langle D_3(x_1, 0) \rangle = 0$, $\langle E_1(x_1, 0) \rangle = 0$.
(13.25)

The interface conditions (13.25) mean that the interface delaminates in (a, b), but it remains close in the normal direction and electrically permeable in this region. This situation is admissible in reality because the interface is usually softer than the surrounding material matrices. The position of the point b is unknown for the time being. (See Fig. 13.2.)

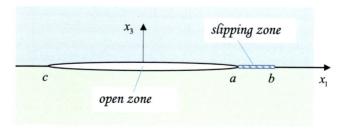

FIGURE 13.2 An electrically conducting interface crack with slipping zone.

The satisfaction of the boundary conditions (13.24) by means of the presentation (13.13) leads to Eqs. (13.19). The boundary conditions (13.25) together

A slipping zone model for a conducting interface crack **Chapter** | 13 **259**

with Eqs. (13.13) and (13.14) give for $x_1 \in (a, b)$ the following relations

$$\text{Im}\left[F_k^+ (x_1) + \gamma_k F_k^- (x_1)\right] = 0, \quad \text{Im}\left[F_k^+ (x_1) - F_k^- (x_1)\right] = 0, \quad (k = 1, 3), \tag{13.26}$$

$$F_4^+ (x_1) - F_4^- (x_1) = 0. \tag{13.27}$$

Relations (13.19) for $(k = 1, 3)$ and (13.26) lead to the following combined Dirichlet-Riemann boundary value problem

$$F_k^+ (x_1) + \gamma_k F_k^- (x_1) = 0, \quad \text{for } x_1 \in (c, a), \tag{13.28}$$

$$\text{Im} \, F_k^{\pm} (x_1) = 0, \quad \text{for } x_1 \in (a, b), \ (k = 1, 3), \tag{13.29}$$

while the relations (13.27) show that $F_4 (z)$ is analytic in the section (a, b).

The boundary conditions (13.24) and presentation (13.13) for $j = 4$ lead to the equation

$$F_4^+ (x_1) + F_4^- (x_1) = 0, \quad \text{for } x_1 \in (c, a). \tag{13.30}$$

Thus, the function $F_4 (z)$ is analytical in the whole plane cut along (c, a).

It follows from a thorough analysis, that the solution of the problem (13.28), (13.29) for $k = 3$ can be easily obtained from the solution of this problem for $k = 1$. Therefore, in the following the solution of the problem (13.28), (13.29) only for $k = 1$ is considered.

A solution of the combined Dirichlet-Riemann boundary value problem (13.28), (13.29) was found and applied to the analysis of a rigid stamp by Nahmein and Nuller [14]. Concerning the problem of an interface crack, this solution has been developed in [11]. Using these results, an exact solution of the problem (13.28), (13.29) for $k = 1$ satisfying the condition at infinity (13.18) can be written in the form

$$F_1 (z) = P (z) X_1 (z) + Q (z) X_2 (z), \tag{13.31}$$

where

$$P (z) = C_1 z + C_2, \quad Q (z) = D_1 z + D_2,$$

$$X_1 (z) = i e^{i\varphi(z)} / \sqrt{(z - c) (z - b)}, \quad X_2 (z) = e^{i\varphi(z)} / \sqrt{(z - c) (z - a)},$$

$$\varphi (z) = 2\varepsilon \ln \frac{\sqrt{(b - a) (z - c)}}{\sqrt{(b - c) (z - a)} + \sqrt{(a - c) (z - b)}}, \quad \varepsilon = \frac{1}{2\pi} \ln \gamma_1,$$

$$C_1 = -\tilde{\tau}_1 \cos \beta - \tilde{\sigma}_1 \sin \beta, \quad D_1 = \tilde{\sigma}_1 \cos \beta - \tilde{\tau}_1 \sin \beta,$$

$$C_2 = -\frac{c + b}{2} C_1 - \beta_1 D_1, \quad D_2 = \beta_1 C_1 - \frac{c + a}{2} D_1,$$

260 Mechanics and Physics of Structured Media

with

$$\beta = \varepsilon \ln \frac{\sqrt{1+\lambda}-1}{\sqrt{1+\lambda}+1}, \quad \beta_1 = \varepsilon \sqrt{(a-c)(b-c)}, \quad \text{and} \quad \lambda = \frac{b-a}{l}, \; l = a-c.$$
(13.32)

Parameter λ introduced by formula (13.32) will play an important role for further analysis, because it defines the relative slipping zone length.

Using solution (13.31) together with the formula (13.13), one gets

$$ir_{11}\sigma_{13}^{(1)}(x_1,0) + \sigma_{33}^{(1)}(x_1,0) + r_{14}E_1^{(1)}(x_1,0)$$
$$= \left[\frac{Q(x_1)}{\sqrt{x_1-a}} + \frac{iP(x_1)}{\sqrt{x_1-b}} \right] \frac{\vartheta_1 \exp[i\varphi(x_1)]}{\sqrt{x_1-c}} \quad \text{for } x_1 > b, \tag{13.33}$$

$$\sigma_{33}^{(1)}(x_1,0) + r_{14}E_1^{(1)}(x_1,0)$$
$$= \frac{P(x_1)}{\sqrt{(x_1-c)(b-x_1)}} \left[\frac{1-\gamma_1}{1+\gamma_1} \cosh\varphi_0(x_1) + \sinh\varphi_0(x_1) \right] +$$
$$+ \frac{Q(x_1)}{\sqrt{(x_1-c)(x_1-a)}} \left[\cosh\varphi_0(x_1) + \frac{1-\gamma_1}{1+\gamma_1} \sinh\varphi_0(x_1) \right] \quad \text{for } x_1 \in (a,b),$$
(13.34)

where $\varphi_0(x_1) = 2\varepsilon \tan^{-1} \sqrt{\frac{(a-c)(b-x_1)}{(b-c)(x_1-a)}}$.

Substituting the solution (13.31) into the presentation (13.14) leads to

$$t_{11}\langle u_1'(x_1)\rangle + it_{13}\langle u_3'(x_1)\rangle + it_{14}\langle D_3(x_1)\rangle = J_1(x_1) \quad \text{for } x_1 \in (c,a),$$
(13.35)

$$t_{11}\langle u_1'(x_1)\rangle = \frac{2}{\sqrt{x_1-c}} \left[\frac{P(x_1)}{\sqrt{b-x_1}} \cosh\varphi_0(x_1) + \frac{Q(x_1)}{\sqrt{x_1-a}} \sinh\varphi_0(x_1) \right]$$
$$\text{for } x_1 \in (a,b), \tag{13.36}$$

where

$$J_1(x_1) = 2\sqrt{\alpha} \left[\frac{P(x_1)}{\sqrt{b-x_1}} - i \frac{Q(x_1)}{\sqrt{a-x_1}} \right] \frac{\exp[i\varphi^*(x_1)]}{\sqrt{x_1-c}},$$
$$\varphi^*(x_1) = 2\varepsilon \ln \frac{\sqrt{(b-a)(x_1-c)}}{\sqrt{(b-a)(a-x_1)} + \sqrt{(a-c)(b-x_1)}}, \quad \alpha = \frac{(\gamma_1+1)^2}{4\gamma_1}.$$

The solution of the Hilbert problem (13.30) can be obtained by using the results of [13] as

$$F_4(z) = \frac{C_{04} + C_{14}z}{\sqrt{(z-c)(z-a)}}. \tag{13.37}$$

To determine the coefficients C_{04}, C_{14}, we use the condition at infinity (13.18) for $j = 4$ and Gaussian theorem concerning the contour that lies on the lower and upper faces of the segment $x_1 \in (c, a)$, which can be presented in the form [10]

$$\int_c^a \left[F_4^+ (x_1) - F_4^- (x_1) \right] dx_1 = 0.$$

These give the following formula

$$F_4 (z) = \frac{h_4}{2} \left(z - \frac{c + a}{2} \right) \frac{1}{\sqrt{(z - c)(z - a)}}, \qquad (13.38)$$

where $h_4 = \sigma^\infty + r_{44} E^\infty$.

It follows from Eq. (13.13) with an account $r_{41} = 0$ and (13.38) that

$$\sigma_{33}^{(1)} (x_1, 0) + r_{44} E_1^{(1)} (x_1, 0)$$
$$= 2 F_4 (x_1) = h_4 \left(x_1 - \frac{c + a}{2} \right) \frac{1}{\sqrt{(x_1 - c)(x_1 - a)}} \quad \text{for } x_1 > a. \quad (13.39)$$

The real part of Eq. (13.33) and Eq. (13.39) provide a system of linear algebraic equations from which the mechanical stress $\sigma_{33}^{(1)} (x_1, 0)$ and the electrical field $E_1^{(1)} (x_1, 0)$ can be easily found for $x_1 > b$. Similarly Eq. (13.34) and Eq. (13.39) provide a system of linear algebraic equations from which the same components can be easily found for the segment (a, b).

Using Eqs. (13.38) and (13.14) with $j = 4$ and taking into account $t_{41} = 0$ we obtain

$$i t_{43} \left\langle u_3' (x_1) \right\rangle + i t_{44} \left\langle D_3 (x_1) \right\rangle = F_4^+ (x_1) - F_4^- (x_1) = J_4 (x_1) \quad \text{for } x_1 \in (c, a) \quad (13.40)$$

where $J_4 (x_1) = h_4 \left(x_1 - \frac{c+a}{2} \right) \frac{1}{i \sqrt{(x_1 - c)(a - x_1)}}$.

Considering further the system composed of imaginary parts of (13.35) and (13.40), one can write

$$\left\langle u_3' (x_1) \right\rangle = \Delta^{-1} \, \text{Im} \left\{ t_{44} J_1 (x_1) - t_{14} J_4 (x_1) \right\}, \qquad (13.41)$$
$$\left\langle D_3 (x_1) \right\rangle = \Delta^{-1} \, \text{Im} \left\{ -t_{43} J_1 (x_1) + t_{13} J_4 (x_1) \right\} \quad \text{for } x_1 \in (c, a) \quad (13.42)$$

where $\Delta = t_{13} t_{44} - t_{14} t_{43}$.

Next we introduce the mechanical stress and electrical field intensity factors

$$k_1 = \lim_{x_1 \to a+0} \sqrt{2\pi (x_1 - a)} \sigma_{33}^{(1)} (x_1, 0),$$
$$k_2 = \lim_{x_1 \to b+0} \sqrt{2\pi (x_1 - b)} \sigma_{13}^{(1)} (x_1, 0), \qquad (13.43)$$
$$k_E = \lim_{x_1 \to a+0} \sqrt{2\pi (x_1 - a)} E_1^{(1)} (x_1, 0).$$

262 Mechanics and Physics of Structured Media

To determine k_2 we multiply the left and right sides of Eq. (13.33) by $\sqrt{2\pi\,(x_1 - b)}$ and consider for $x_1 \to b$. We get the following formula

$$k_2 = \frac{\vartheta_1}{r_{11}}\sqrt{\frac{2\pi}{l}}\,P\,(b)\,. \tag{13.44}$$

To determine k_1 and k_E we multiply the left and right sides of Eqs. (13.34) and (13.39) by $\sqrt{2\pi\,(x_1 - a)}$ and consider for $x_1 \to a$. We arrive to the following system of linear algebraic equations

$$k_1 + r_{14}k_E = \sqrt{\frac{2\pi}{a-c}}\left[\cosh\varphi_0\,(x_1) + \frac{1-\gamma_1}{1+\gamma_1}\sinh\varphi_0\,(x_1)\right]Q\,(a)\,,$$
$$k_1 + r_{44}k_E = h_4\sqrt{\frac{\pi\,(a-c)}{2}}\,. \tag{13.45}$$

Considering that $\varphi_0\,(a) = \ln\sqrt{\gamma_1}$ and $Q\,(a) = \frac{a-c}{2}h_5\,(\beta)$, the system (13.45) takes the form

$$k_1 + r_{14}k_E = \sqrt{\frac{\pi\,(a-c)}{2\alpha}}h_5\,(\beta)\,,$$
$$k_1 + r_{44}k_E = h_4\sqrt{\frac{\pi\,(a-c)}{2}}\,,$$

where $h_5\,(\beta) = \tilde{\sigma}_1\cos\beta - \tilde{\tau}_1\sin\beta - 2\varepsilon_1\sqrt{\lambda+1}\,(\tilde{\tau}_1\cos\beta + \tilde{\sigma}_1\sin\beta)$.
From this system we get

$$k_1 = \frac{\sqrt{\pi\,(a-c)/2}}{r_{44} - r_{14}}\left[r_{44}h_5\,(\beta)\,/\sqrt{\alpha} - r_{14}h_4\right]\,,$$
$$k_E = \frac{\sqrt{\pi\,(a-c)/2}}{r_{44} - r_{14}}\left[h_4 - h_5\,(\beta)\,/\sqrt{\alpha}\right]\,. \tag{13.46}$$

13.4 Slipping zone length

The solution of an interface crack problem, obtained in the previous chapter, will be physically justified if following inequalities

$$\sigma_{33}^{(1)}\,(x_1, 0) \le 0 \quad\text{for } x_1 \in (a, b)\,, \qquad \langle u_3\,(x_1)\rangle \ge 0 \quad\text{for } x_1 \in (c, a) \quad (13.47)$$

are valid. In this case, the crack is open for $x_1 \in (c, a)$ and the slipping zone remains in compressed state for $x_1 \in (a, b)$. A corresponding analysis shows that these inequalities hold true if λ is taken from the segment $[\lambda_1, \lambda_2]$ where λ_1 is the maximum root from the interval $(0, 1)$ of the equation $k_1 = 0$ and λ_2 is the similar root of the equation $\sqrt{a - x_1}\,\langle u_3'(x_1, 0)\rangle\,|_{x_1 \to a-0} = 0$. By using the first Eq. (13.46) and also (13.41) the equations for the determination of λ_1 and λ_2

can be written in the following forms, respectively

$$h_5 \left[\beta \left(\lambda_1 \right) \right] = \sqrt{\alpha} \frac{r_{14}}{r_{44}} h_4, \tag{13.48a}$$

$$h_5 \left[\beta \left(\lambda_2 \right) \right] = \frac{r_1 t_{14} h_4}{2\sqrt{\alpha} t_{44}}. \tag{13.48b}$$

The values of λ_1 and λ_2 can be found numerically. After that the position of the point b can be taken from the segment $[a + \lambda_2(a - c), a + \lambda_1 (a - c)]$.

13.5 The crack faces free from electrodes

Assume now that the electrodes are absent and the crack faces are free from mechanical loading and electrical charge. In this case the electrically permeable or electrically impermeable conditions are used the most often for the modeling of such cracks. Both these types of conditions were considered in [8] and in [9], respectively, for the case of the material polarization orthogonal to the crack faces and the electric flux having the same direction. Because the prefracture zone model has never been considered earlier for the material polarization parallel to the crack faces consider the way of removing the oscillating singularity for this case. Taking into account that the assumption of a permeable crack is more realistic than that of an impermeable crack we will pay attention to electrically permeable crack.

We assume in this case that eclectically permeable crack is located in (c, b) and its part (c, a) is open while the remaining part (a, b) is in frictionless contact. In this case the conditions at the interface can be formulated as follows

for $x_1 \notin (c, b)$: $\quad \mathbf{P}^{(1)} (x_1, 0) = \mathbf{P}^{(2)} (x_1, 0), \quad \mathbf{L}^{(1)} (x_1, 0) = \mathbf{L}^{(2)} (x_1, 0),$
$$\tag{13.49}$$

for $x_1 \in (c, a)$: $\quad \sigma_{13}^{\pm} (x_1, 0) = 0, \quad \sigma_{33}^{\pm} (x_1, 0) = 0,$
$$\langle E_1 (x_1) \rangle = 0, \quad \langle D_3 (x_1) \rangle = 0, \tag{13.50}$$

for $x_1 \in (a, b)$: $\quad \langle u_3' (x_1, 0) \rangle = 0, \quad \langle \sigma_{33} (x_1) \rangle = 0, \quad \sigma_{13}^{\pm} (x_1, 0) = 0,$
$$\langle E_1 (x_1) \rangle = 0, \quad \langle D_3 (x_1) \rangle = 0. \tag{13.51}$$

It follows from Eqs. (13.49)–(13.51) that $\langle D_3 (x_1) \rangle = 0$ for $-\infty < x_1 < \infty$. This relation together with Eq. (13.6) gives $W_4^+ (x_1) - W_4^- (x_1) = 0$ for $-\infty < x_1 < \infty$ and it means that $W_4 (z)$ is an analytic function in the whole plane. Taking into account the constant values of electromechanical quantities for $z \to \infty$ one has $W_4 (z) = W_4^0 = const$. Moreover, by using the appearance (13.10) of the matrix $\mathbf{S}$ the relation (13.7) can be written as

$$\sigma_{13}^{(1)} (x_1, 0) = i s_{11} W_1^+ (x_1) + s_{13} W_3^+ (x_1) + i s_{11} W_1^- (x_1) - s_{13} W_3^- (x_1),$$
$$\sigma_{33}^{(1)} (x_1, 0) = s_{31} W_1^+ (x_1) + i s_{33} W_3^+ (x_1) - s_{31} W_1^- (x_1)$$
$$+ i s_{33} W_3^- (x_1) + 2 i s_{34} W_4^0,$$

264 Mechanics and Physics of Structured Media

$$E_1^{(1)}(x_1, 0) = s_{41} W_1^+(x_1) + i s_{43} W_3^+(x_1) - s_{41} W_1^-(x_1)$$
$$+ i s_{43} W_3^-(x_1) + 2i s_{44} W_4^0. \tag{13.52}$$

Taking into account that for $x_1 \notin (c, b)$ the relationship $W_i^+(x_1) = W_i^-(x_1)$, $(i = 1, 3)$ holds true one gets from Eqs. (13.52)

$$\sigma_{33}^{(1)}(x_1, 0) = 2i s_{33} W_3^+(x_1) + 2i s_{34} W_4^0, \tag{13.53}$$
$$E_1^{(1)}(x_1, 0) = 2i s_{43} W_3^+(x_1) + 2i s_{44} W_4^0 \quad \text{for } x_1 \notin (c, b).$$

Excluding from the last relations $W_3^+(x_1)$ and taking into account $\sigma_{33}^{(1)}(x_1, 0) = \sigma^\infty$, $E_1^{(1)}(x_1, 0) = E^\infty$ for $x_1 \to \infty$, one gets

$$W_4^0 = i \Delta_1^{-1} \left(s_{43} \sigma^\infty - s_{33} E^\infty \right) / 2, \tag{13.54}$$

where $\Delta_1 = s_{33} s_{44} - s_{43} s_{34}$.

By combining the first and second equations of Eqs. (13.52) and taking into account (13.54) one can get

$$\sigma_{33}^{(1)}(x_1, 0) + i m_j \sigma_{13}^{(1)}(x_1, 0) = t_j \left[F_j^+(x_1) + \gamma_j F_j^-(x_1) \right] + 2i s_{34} W_4^0, \tag{13.55}$$

where

$$F_j(z) = W_1(z) + i \rho_j W_3(z), \quad j = 1, 3, \tag{13.56}$$

and

$$\rho_j = \frac{s_{33} + m_j s_{13}}{s_{31} - m_j s_{11}}, \quad \gamma_j = -\left(s_{31} + m_j s_{11} \right) / t_j,$$

$$t_j = s_{31} - m_j s_{11}, \quad m_{1,3} = \mp \sqrt{-\frac{s_{31} s_{33}}{s_{11} s_{13}}}. \tag{13.57}$$

Eqs. (13.6) and (13.56) lead to the following expression for the derivatives of the displacement jumps

$$\langle u_1'(x_1, 0) \rangle + i \rho_j \langle u_3'(x_1, 0) \rangle = F_j^+(x_1) - F_j^-(x_1), \quad j = 1, 3, \tag{13.58}$$

and it is clear from Eqs. (13.49) and (13.58) that the functions $F_j(z)$ are analytic in the whole plane cut along (c, b).

In this case the determination of the contact zone is similar to the paper [8]. The difference is connected with other direction of material polarization and the electric field instead of electric flux prescribed at infinity.

Performing further the analysis similar to [8] one arrives to the following transcendental equation for the determination of the contact zone length

$$\tan \beta = \frac{\sqrt{1 - \lambda} \sigma^\infty + 2\varepsilon_1 m_1 \tau^\infty}{2\varepsilon_1 \sigma^\infty - \sqrt{1 - \lambda} m_1 \tau^\infty}, \tag{13.59}$$

A slipping zone model for a conducting interface crack **Chapter | 13 265**

where $\lambda = \frac{b-a}{l}$, $\beta = \varepsilon_1 \ln \frac{1-\sqrt{1-\lambda}}{1+\sqrt{1-\lambda}}$, $\varepsilon_1 = \frac{1}{2\pi} \ln \gamma_1$.

The SIF $(13.43)_2$ of the shear stress can be found on the following formula

$$k_2 = -\frac{1}{m_1}\sqrt{\frac{\pi l}{2}} \Big[\left(\sigma^\infty \sin \beta - m_1 \tau^\infty \cos \beta \right)$$
$$+ 2\varepsilon_1 \sqrt{1-\lambda} \left(\sigma^\infty \cos \beta + m_1 \tau^\infty \sin \beta \right) \Big].$$

After determination of the contact zone length from Eq. (13.59) the stresses can be found by Eq. (13.55). Further, by use of $(13.53)_2$ the electrical field can be written as

$$E_1^{(1)}(x_1, 0) = \frac{s_{43}}{s_{33}} \Big[\sigma_{33}^{(1)}(x_1, 0) - \sigma^\infty \Big] + E^\infty \quad \text{for } x_1 > b.$$

Detailed analysis of the contact zone model for an electrically impermeable crack was performed by Herrmann et al. [9] and it can be realized here without any difficulties.

13.6 Numerical results and discussion

Consider the influence of the external mechanical and electrical loading on the slipping zone length and the electromechanical intensity factors (IFs). Bimaterial composed of piezoceramics PZT4 and PZT5H are chosen for numerical calculations. For the crack region we take $c = -0.01$ m and $a = 0.01$ m.

The values of λ_1, λ_2, and the mechanical and electrical intensity factors k_2, k_E for $\sigma^\infty = 10^5$ Pa, and different τ^∞ and E^∞ are presented in Table 13.1.

It is clearly seen from Table 13.1 that all monitored parameters essentially depend on the intensity of the electric field. It is worth to mention that the value $\sigma^\infty + r_{44} E^\infty$ plays determinative role in the analysis. Namely, for $E^\infty = -\sigma^\infty/r_{44}$ one has $\lambda_1 = \lambda_2$ (for the Table 13.1 this $E^\infty = 2373.95$ V/m). For $E^\infty < -\sigma^\infty/r_{44}$, particularly for $E^\infty = 0$, we have $\lambda_1 > \lambda_2$ and there are the set of physically admissible positions of the point $b \in [a + \lambda_2 (a - c), a + \lambda_1 (a - c)]$, and for $\sigma^\infty + r_{44} E^\infty > 0$ there are no position of the point b, for which both inequalities (13.47) are satisfied.

The crack opening for $\sigma^\infty = 10^5$ Pa, $E^\infty = -4747.9$ V/m, and $\tau^\infty = -20\sigma^\infty$ (line I) and also $\tau^\infty = -30\sigma^\infty$ (line II) is presented in Fig. 13.3. The correspondent values of the normal stress in the slipping zone are given in Fig. 13.4. These graphs are drawn for $\lambda = \lambda_2$. It is seen that for such λ the crack is closed smoothly and the stresses are substantially compressive.

The behavior of the displacement jump in the vicinity of the right crack tip for $\sigma^\infty = 10^5$ Pa, $E^\infty = -4747.9$ V/m and $\tau^\infty = -50\sigma^\infty$ are shown in Fig. 13.5. Lines I and II are drawn for $\lambda = \lambda_1$ and $\lambda = \lambda_2$, respectively, and the dashed line III is obtained for $\lambda = 0.1$, which does not belong to the segment $[\lambda_1, \lambda_2]$. The correspondent magnitudes of $\sigma_{33}^{(1)}(x_1, 0)$ for the same external

TABLE 13.1 The values of the slipping zone lengths, and the mechanical and electrical intensity factors for $\sigma^\infty = 10^5$ Pa, and different shear stress and electric field.

E^∞ [V/m]	λ_1	λ_2	k_2 [N/m$^{3/2}$]	k_E [V/m$^{1/2}$]
		$\tau^\infty = -2\sigma^\infty$		
2373.95	2.01287×10^{-10}	2.01287×10^{-10}	-39471.7	-18.7281
0	1.40003×10^{-12}	6.4869×10^{-13}	-37160.6	-523.288
-2373.95	4.73359×10^{-15}	3.05196×10^{-16}	-33511.4	-1028.09
		$\tau^\infty = -5\sigma^\infty$		
2373.95	537287×10^{-5}	537287×10^{-5}	-90375.6	-42.8792
0	7.99619×10^{-6}	5.80918×10^{-6}	-89386.3	-548.063
-2373.95	1.13651×10^{-6}	5.93216×10^{-7}	-87937.4	-1053.03
-4747.9	1.51775×10^{-7}	5.59029×10^{-8}	-86002.7	-1557.76
-7121.85	1.86749×10^{-8}	4.72527×10^{-9}	-83547.8	-2062.25
		$\tau^\infty = -20\sigma^\infty$		
2373.95	0.193924	0.193924	-1.05903×10^6	-459.863
0	0.163035	0.158235	-1.03149×10^6	-959.465
-2373.95	0.136767	0.128737	-1.00802×10^6	-1459.88
-4747.9	0.114509	0.104472	-988062.	-1961.02
-7121.85	0.0957095	0.0845961	-961214.	-2518.57

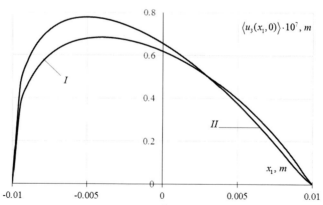

FIGURE 13.3 The crack opening for $\sigma^\infty = 10^5$ Pa, $E^\infty = -4747.9$ V/m and $\tau^\infty = -20\sigma^\infty$ (line I), $\tau^\infty = -30\sigma^\infty$ (line II).

loadings are presented in Fig. 13.6. It is seen that all lines in Fig. 13.6 satisfy the first inequality (13.47). However the displacement jumps in Fig. 13.5 satisfy the second inequality (13.47) for $\lambda = \lambda_1$ and $\lambda = \lambda_2$ and do not satisfy

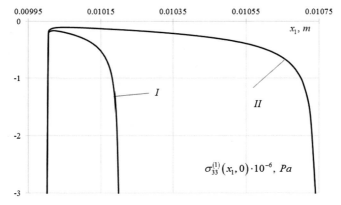

FIGURE 13.4 The normal stress in the slipping zone for the same loadings as in Fig. 13.3.

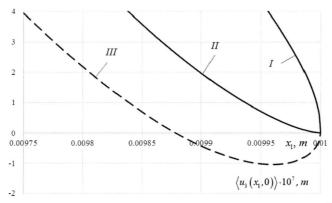

FIGURE 13.5 The displacement jump in the vicinity of the right crack tip for $\sigma^\infty = 10^5$ Pa, $E^\infty = -4747.9$ V/m, and $\tau^\infty = -50\sigma^\infty$.

the mentioned inequality for $\lambda = 0.1$. Therefore, as expected, case 3 cannot be considered as admissible way of the crack opening.

13.7 Conclusion

An electrically conducting interface crack in a piezoelectric bimaterial polarized in the direction parallel to the crack faces and orthogonal to the crack front is considered. Both materials are loaded by remote tension and shear stresses and an electrical field codirected with the material polarization. All fields are assumed to be independent of the coordinate codirected with the crack front, therefore, a 2-D problem is considered. The electromechanical quantities are expressed through the sectionally-analytic vector-function. Using these presentations and satisfying the conditions at the material interface the problem of linear relationship is formulated and solved exactly. The physically unrealis-

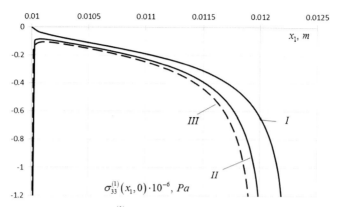

FIGURE 13.6 The magnitudes of $\sigma_{33}^{(1)}(x_1, 0)$ for the same external loadings as in Fig. 13.5.

tic oscillating singularity at the crack tips, which is expressed by the formulas (13.20)–(13.22), is obtained. To remove this singularity a slipping zone of the crack faces is introduced at the crack continuation. In this case a combined Dirichlet-Riemann and Hilbert boundary value problems are formulated and the exact analytical solution of these problems is given by the formulas (13.31) and (13.38). All required mechanical and electrical quantities are presented in a simple analytical form. From the conditions (13.47), which provide the validity of the developed model, the transcendental equations for the determination of the slipping zone length are obtained. The way of removing the oscillating singularity at the electrically conducting crack tip for the above-considered class of polarization is also discussed.

The numerical simulation of the obtained results was performed for the bimaterial PZT4/PZT5H and different mechanical and electric loading. The results for the crack opening, stresses in the slipping zone, mechanical and electrical intensity factors are presented in Table 13.1 and Figs. 13.3–13.6. It is particularly follows from the obtained results that all mentioned quantities substantially depend both on mechanical loading and electric field. Also the presented table and figures confirm the validity of the model, developed in this chapter.

References

[1] H.G. Beom, S.N. Atluri, Conducting cracks in dissimilar piezoelectric media, Int. J. Fract. 118 (2002) 285–301.
[2] M. Comninou, The interface crack, J. Appl. Mech. 44 (1977) 631–636.
[3] A.H. England, A crack between dissimilar media, J. Appl. Mech. 32 (1965) 400–402.
[4] F. Erdogan, Stress distribution in bonded dissimilar materials with cracks, J. Appl. Mech. 32 (1965) 403–410.
[5] L.A. Fil'shtinskii, M.L. Fil'shtinskii, Green's function for a composite piezoceramic plane with a crack between phases, J. Appl. Math. Mech. 58 (1994) 355–362.

A slipping zone model for a conducting interface crack Chapter | 13 **269**

[6] L.A. Fil'shtinskii, M.L. Fil'shtinskii, Anti-plane deformation of a composite piezoceramic space with interphase crack, Int. Appl. Mech. 33 (1997) 655–659.

[7] L.A. Fil'shtinskii, M.L. Fil'shtinskii, Optimal control of physical fields in piezoelectric bodies with defects, Lect. Notes Appl. Comput. Mech. 32 (2006) 465–508 + 517–530.

[8] K.P. Herrmann, V.V. Loboda, Fracture mechanical assessment of electrically permeable interface cracks in piezoelectric bimaterials by consideration of various contact zone models, Arch. Appl. Mech. 70 (2000) 127–143.

[9] K.P. Herrmann, V.V. Loboda, V.B. Govorucha, On contact zone models for an interface crack with electrically insulated crack surfaces in a piezoelectric bimaterial, Int. J. Fract. 111 (2001) 203–227.

[10] P. Knysh, V.V. Loboda, F. Labesse-Jied, Y. Lapusta, An electrically charged crack in a piezoelectric material under remote electromechanical loading, Lett. Fract. Micromech. 175 (1) (2012) 87–94.

[11] V.V. Loboda, The quasi-invariant in the theory of interface cracks, Eng. Fract. Mech. 44 (1993) 573–580.

[12] V.V. Loboda, A. Sheveleva, Y. Lapusta, An electrically conducting interface crack with a contact zone in a piezoelectric bimaterial, Int. J. Solids Struct. 51 (2014) 63–73.

[13] N.I. Muskhelishvili, Some Basic Problems in the Mathematical Theory of Elasticity, Noordhoff, Groningen, 1963.

[14] E.L. Nahmein, B.M. Nuller, Contact of an elastic half plane and a particularly unbonded stamp, Prikl. Mat. Meh. 50 (1986) 663–673 (in Russian).

[15] V.Z. Parton, B.A. Kudryavtsev, Electromagnetoelasticity, Gordon and Breach Science Publishers, New York, 1988.

[16] J.R. Rice, G.C. Sih, Plane problem of cracks in dissimilar media, J. Appl. Mech. 32 (1965) 418–423.

[17] C.Q. Ru, Electrode-ceramic interfacial cracks in piezoelectric multilayer materials, J. Appl. Mech. 67 (2000) 255–261.

[18] I.S. Sokolnikoff, Mathematical Theory of Elasticity, McGraw-Hill, New York, 1956.

[19] Z. Suo, C.-M. Kuo, D.M. Barnett, J.R. Willis, Fracture mechanics for piezoelectric ceramics, J. Mech. Phys. Solids 40 (1992) 739–765.

Chapter 14

Dependence of effective properties upon regular perturbations

Matteo Dalla Riva[a], Paolo Luzzini[b], Paolo Musolino[c], and Roman Pukhtaievych[d]

[a]*Department of Mathematics, The University of Tulsa, Tulsa, OK, United States,* [b]*EPFL, SB Institute of Mathematics, Station 8, Lausanne, Switzerland,* [c]*Dipartimento di Scienze Molecolari e Nanosistemi, Università Ca' Foscari Venezia, Venezia Mestre, Italy,* [d]*Department of Complex Analysis and Potential Theory, Institute of Mathematics of the National Academy of Sciences of Ukraine, Kyiv, Ukraine*

14.1 Introduction

In this chapter, we review some of our recent results on the dependence of effective properties upon perturbations of the geometry and physical parameters. We first consider the case of a Newtonian fluid that flows at low Reynolds numbers around a periodic array of cylinders. By the results of [42,43], we can see that the average of the longitudinal component of the flow velocity depends real analytically on perturbations of the periodicity structure and the cross section of the cylinders. Then we turn our attention to the thermal properties of two-phase composites that are obtained by introducing a periodic set of inclusions in an infinite homogeneous matrix made of a different material. Our aim is to prove that the effective conductivity depends real analytically on perturbations of the shape of the inclusions, the periodicity structure, and the conductivity of each material. First, we present a result of [41] on the case where we have an ideal contact at the interface. Then we show that the result of [41] can be extended to the case of imperfect contact conditions.

The average longitudinal flow and the effective conductivity are defined as specific functionals of the solutions of underlying periodic boundary value problems. In our work on domain perturbations, these problems are set in domains whose shape depends on certain perturbation parameters. Then we adopt a method based on a periodic version of the standard potential theory to transform the boundary value problems into systems of integral equations, which will be defined on the boundary of parameter-dependent domains. Next, with a suitable change of the functional variables, we obtain new systems of integral

272 Mechanics and Physics of Structured Media

equations that depend on the geometry and the parameters under consideration but are defined on the boundary of fixed sets. These last systems can be studied by means of the implicit function theorem for analytic maps in Banach spaces. In particular, we can derive analytic dependence results for the solutions, which eventually yield the desired results for the effective properties. We note that here and throughout the chapter the word 'analytic' always means 'real analytic'. For the definition and properties of analytic operators, we refer to Deimling [17, §15].

It is also worth noting that many existing methods in the literature are applied to periodic structures with specific shapes, *e.g.* two/three-dimensional periodic arrays of circles/spheres or ellipses/ellipsoids (see the references in the next sections). Our method, instead, can be used with arbitrary shapes, provided that they satisfy some reasonable regularity assumption. Moreover, the real analyticity results that we obtain surely imply the differentiability with respect to the parameters. Then one may want to compute the corresponding differentials, with the final goal of characterizing critical configurations. Since our approach is based on periodic potential theory, a preliminary step would be to compute the differentials of the periodic layer potentials. The computation of such differentials can be performed by following the lines of those of classic layer potentials as it is done in [37, Proposition 3.14].

The chapter is organized as follows. In Section 14.2 we introduce the geometric setting of the considered periodic structures. Section 14.3 contains the result on the average longitudinal flow along a periodic array of cylinders. In Section 14.4 we present the result on the effective conductivity of a two-phase periodic composite with ideal contact condition. In Section 14.5 we state a new result on the effective conductivity of a composite with nonideal contact condition. Finally, in Section 14.6 we prove the result of Section 14.5.

14.2 The geometric setting

Throughout the chapter

$$n \in \{2, 3\}$$

plays the role of the space dimension. If $q_{11}, \ldots, q_{nn} \in \,]0, +\infty[$, we use the following notation:

$$q = \begin{pmatrix} q_{11} & 0 \\ 0 & q_{22} \end{pmatrix} \quad \text{if } n = 2, \qquad q = \begin{pmatrix} q_{11} & 0 & 0 \\ 0 & q_{22} & 0 \\ 0 & 0 & q_{33} \end{pmatrix} \quad \text{if } n = 3, \quad (14.1)$$

and

$$Q \equiv \prod_{j=1}^{n} \,]0, q_{jj}[\subseteq \mathbb{R}^n . \tag{14.2}$$

The set Q is the periodicity cell, while q is a diagonal matrix incorporating the information on the periodicity. Clearly, $|Q|_n \equiv \prod_{j=1}^{n} q_{jj}$ is the measure of the cell Q and $q\mathbb{Z}^n \equiv \{qz : z \in \mathbb{Z}^n\}$ is the set of vertices of a periodic subdivision of $\mathbb{R}^n$ corresponding to the cell Q. We denote by q^{-1} the inverse matrix of q. We denote by $\mathbb{D}_n(\mathbb{R})$ the space of $n \times n$ diagonal matrices with real entries and by $\mathbb{D}_n^+(\mathbb{R})$ the set of elements of $\mathbb{D}_n(\mathbb{R})$ with diagonal entries in $]0, +\infty[$. Moreover, we find convenient to set

$$\tilde{Q} \equiv]0, 1[^n .$$

If Ω_Q is a subset of $\mathbb{R}^n$ such that $\overline{\Omega_Q} \subseteq Q$, we define the following two periodic domains:

$$\mathbb{S}_q[\Omega_Q] \equiv \bigcup_{z \in \mathbb{Z}^n} (qz + \Omega_Q), \qquad \mathbb{S}_q[\Omega_Q]^- \equiv \mathbb{R}^n \setminus \overline{\mathbb{S}_q[\Omega_Q]}.$$

The symbol '$\overline{\cdot}$' denotes the closure of a set. If u is a real valued function defined on $\mathbb{S}_q[\Omega_Q]$ or $\mathbb{S}_q[\Omega_Q]^-$, we say that u is q-periodic provided that $u(x + qz) = u(x)$ for all $z \in \mathbb{Z}^n$ and for all x in the domain of definition of u. If $k \in \mathbb{N}$, we set

$$C_b^k(\overline{\mathbb{S}_q[\Omega_Q]^-}) \equiv \left\{ u \in C^k(\overline{\mathbb{S}_q[\Omega_Q]^-}) : D^\gamma u \text{ is bounded } \forall \gamma \in \mathbb{N}^n \text{ s. t. } |\gamma| \le k \right\}.$$

On $C_b^k(\overline{\mathbb{S}_q[\Omega_Q]^-})$ we consider the usual norm

$$\|u\|_{C_b^k(\overline{\mathbb{S}_q[\Omega_Q]^-})} \equiv \sum_{|\gamma| \le k} \sup_{x \in \overline{\mathbb{S}_q[\Omega_Q]^-}} |D^\gamma u(x)| \quad \forall u \in C_b^k(\overline{\mathbb{S}_q[\Omega_Q]^-}),$$

where $|\gamma| \equiv \sum_{i=1}^{n} \gamma_i$ denotes the length of the multiindex $\gamma \equiv (\gamma_1, \dots, \gamma_n) \in \mathbb{N}^n$. Moreover, if $\beta \in]0, 1]$, then we set

$$C_b^{k,\beta}(\overline{\mathbb{S}_q[\Omega_Q]^-})$$
$$\equiv \left\{ u \in C^{k,\beta}(\overline{\mathbb{S}_q[\Omega_Q]^-}) : D^\gamma u \text{ is bounded } \forall \gamma \in \mathbb{N}^n \text{ s. t. } |\gamma| \le k \right\}$$

and on $C_b^{k,\beta}(\overline{\mathbb{S}_q[\Omega_Q]^-})$ we consider the usual norm

$$\|u\|_{C_b^{k,\beta}(\overline{\mathbb{S}_q[\Omega_Q]^-})} \equiv \sum_{|\gamma| \le k} \sup_{x \in \overline{\mathbb{S}_q[\Omega_Q]^-}} |D^\gamma u(x)| + \sum_{|\gamma|=k} |D^\gamma u : \overline{\mathbb{S}_q[\Omega_Q]^-}|_\beta$$
$$\forall u \in C_b^{k,\beta}(\overline{\mathbb{S}_q[\Omega_Q]^-}),$$

where $|D^\gamma u : \overline{\mathbb{S}_q[\Omega_Q]^-}|_\beta$ denotes the β-Hölder constant of $D^\gamma u$ (see, e.g., Gilbarg and Trudinger [24] for the definition of sets and functions of the Schauder class $C^{k,\beta}$). Then $C_q^k(\overline{\mathbb{S}_q[\Omega_Q]^-})$ denotes the Banach subspace of $C_b^k(\overline{\mathbb{S}_q[\Omega_Q]^-})$ defined by

$$C_q^k(\overline{\mathbb{S}_q[\Omega_Q]^-}) \equiv \left\{ u \in C_b^k(\overline{\mathbb{S}_q[\Omega_Q]^-}) : u \text{ is } q\text{-periodic} \right\}$$

274 Mechanics and Physics of Structured Media

and $C_q^{k,\beta}(\overline{\mathbb{S}_q[\Omega_Q]^-})$ denotes the Banach subspace of $C_b^{k,\beta}(\overline{\mathbb{S}_q[\Omega_Q]^-})$ defined by

$$C_q^{k,\beta}(\overline{\mathbb{S}_q[\Omega_Q]^-}) \equiv \left\{ u \in C_b^{k,\beta}(\overline{\mathbb{S}_q[\Omega_Q]^-}) : u \text{ is } q\text{-periodic} \right\}.$$

The spaces $C_b^k(\overline{\mathbb{S}_q[\Omega_Q]})$, $C_b^{k,\beta}(\overline{\mathbb{S}_q[\Omega_Q]})$, $C_q^k(\overline{\mathbb{S}_q[\Omega_Q]})$, and $C_q^{k,\beta}(\overline{\mathbb{S}_q[\Omega_Q]})$ can be defined in a similar way.

We denote by ν_{Ω_Q} the outward unit normal to $\partial\Omega_Q$ and by $d\sigma$ the area element on $\partial\Omega_Q$. We retain the standard notation for the Lebesgue space $L^1(\partial\Omega_Q)$ of Lebesgue integrable functions. We denote by $|\partial\Omega_Q|_{n-1}$ the $(n-1)$-dimensional measure of $\partial\Omega_Q$. To shorten our notation, we denote by $\fint_{\partial\Omega_Q} f\, d\sigma$ the integral mean $\frac{1}{|\partial\Omega_Q|_{n-1}} \int_{\partial\Omega_Q} f\, d\sigma$ for all $f \in L^1(\partial\Omega_Q)$. Also, if $\mathcal{X}$ is a vector subspace of $L^1(\partial\Omega_Q)$ then we set $\mathcal{X}_0 \equiv \left\{ f \in \mathcal{X} : \int_{\partial\Omega_Q} f\, d\sigma = 0 \right\}$.

We now introduce the shape perturbations. In order to consider variable domains, we fix a set and consider a class of diffeomorphisms acting on its boundary. Then a perturbation of the diffeomorphism can be seen as a perturbation of the domain. To this aim, we fix

$$\alpha \in]0, 1[\text{ and a bounded open connected subset } \Omega \text{ of } \mathbb{R}^n \text{ of class } C^{1,\alpha}$$
$$\text{such that } \mathbb{R}^n \setminus \overline{\Omega} \text{ is connected.} \tag{14.3}$$

We denote by $\mathcal{A}_{\partial\Omega}$ the set of functions of class $C^1(\partial\Omega, \mathbb{R}^n)$ which are injective and whose differential is injective at all points of $\partial\Omega$. One can verify that $\mathcal{A}_{\partial\Omega}$ is open in $C^1(\partial\Omega, \mathbb{R}^n)$ (see, *e.g.*, Lanza de Cristoforis and Rossi [38, Lem. 2.2, p. 197] and [37, Lem. 2.5, p. 143]). Then we find it convenient to set

$$\mathcal{A}_{\partial\Omega}^{\widetilde{Q}} \equiv \{\phi \in \mathcal{A}_{\partial\Omega} : \phi(\partial\Omega) \subseteq \widetilde{Q}\},$$

that is the set of diffeomorphisms in $\mathcal{A}_{\partial\Omega}$ whose image is contained in $\widetilde{Q}$ (see Fig. 14.1). If $\phi \in \mathcal{A}_{\partial\Omega}^{\widetilde{Q}}$, the Jordan-Leray separation theorem ensures that $\mathbb{R}^n \setminus \phi(\partial\Omega)$ has exactly two open connected components and we denote by $\mathbb{I}[\phi]$ the bounded one (see, *e.g.*, Deimling [17, Thm. 5.2, p. 26]). Clearly, the set $q\mathbb{I}[\phi] = \{qx : x \in \mathbb{I}[\phi]\}$ is contained in the periodicity cell Q (see Fig. 14.2). Then

$$\mathbb{S}_q[q\mathbb{I}[\phi]] \quad \text{and} \quad \mathbb{S}_q[q\mathbb{I}[\phi]]^-$$

are two unbounded and periodic (q, ϕ)-dependent sets which model the periodic structure of the objects considered in this chapter (see Fig. 14.3). If we modify the entries of q, this will result in a modification of the periodicity of the sets. Instead, perturbing ϕ causes a change in the shape of the periodic inclusions.

Dependence of effective properties upon regular perturbations Chapter | 14 **275**

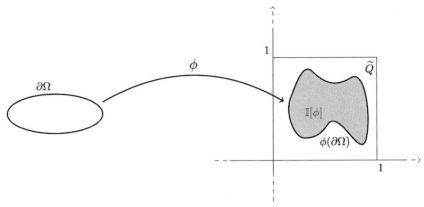

FIGURE 14.1 A diffeomorphism $\phi \in \mathcal{A}_{\partial\Omega}^{\tilde{Q}}$ in $\mathbb{R}^2$.

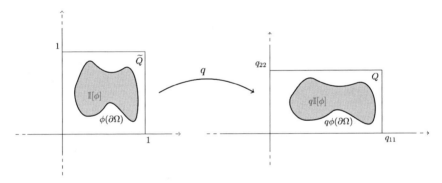

FIGURE 14.2 The transformation induced by q in $\mathbb{R}^2$.

14.3 The average longitudinal flow along a periodic array of cylinders

This section is devoted to the longitudinal flow of a Newtonian fluid flowing at low Reynolds numbers along a periodic array of cylinders. We study the effect of perturbations of the periodicity structure and the shape of the cross section of the cylinders. Since the cylinder's cross section is two-dimensional, in this section we set

$$n = 2.$$

As introduced in the previous section, the shape of the cross section of the cylinders is determined by the image of a fixed domain through a diffeomorphism ϕ and the periodicity cell is a rectangle of sides of length q_{11} and q_{22}, associated

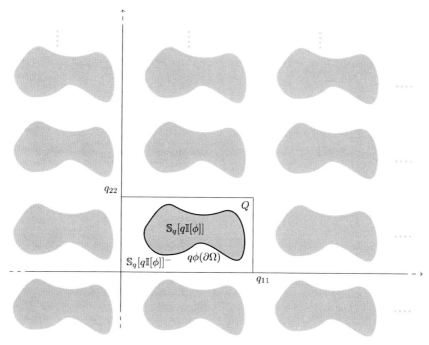

FIGURE 14.3 The sets $\mathbb{S}_q[q\mathbb{I}[\phi]]^-$, $\mathbb{S}_q[q\mathbb{I}[\phi]]$, and $q\phi(\partial\Omega)$ in $\mathbb{R}^2$.

with the matrix

$$q = \begin{pmatrix} q_{11} & 0 \\ 0 & q_{22} \end{pmatrix} \in \mathbb{D}_2^+(\mathbb{R}).$$

We assume that the pressure gradient is parallel to the cylinders. Under these assumptions, the velocity field has only one nonzero component which, by the Stokes equations, satisfies the Poisson equation (see problem (14.4)). Then, by integrating the longitudinal component of the velocity field over the fundamental cell, for each pair (q, ϕ), we define the average of the longitudinal component of the flow velocity $\Sigma[q, \phi]$ (see (14.5)). We note that $\Sigma[q, \phi]$ is a measure of the quantity of fluid flowing through the single periodicity cell and is sometimes referred as the longitudinal permeability of the array of cylinders (or its opposite, see, e.g., Mityushev and Adler [46,47]). Here, we are interested in the dependence of $\Sigma[q, \phi]$ upon the pair (q, ϕ).

The mathematical aspects of fluids in periodic structures have been studied by several authors and with a variety of different methods. With no expectation of being exhaustive, we mention some contributions. Hasimoto [29] has investigated the viscous flow past a cubic array of spheres and he has applied his results to the two-dimensional flow past a square array of circular cylinders. His techniques are based on the construction of a spatially periodic fundamental solution

Dependence of effective properties upon regular perturbations Chapter | 14 **277**

for the Stokes' system and are applied to specific shapes (circular/spherical obstacles and square/cubic arrays). Schmid [59] has investigated the longitudinal laminar flow in an infinite square array of circular cylinders. Sangani and Yao [57,58] have studied the permeability of random arrays of infinitely long cylinders. Mityushev and Adler [46,47] have considered the longitudinal permeability of periodic rectangular arrays of circular cylinders. Finally, the paper [53] with Mityushev deals with the asymptotic behavior of the longitudinal permeability of thin cylinders of arbitrary shape.

Here, instead, we are interested in the dependence of the (average) longitudinal velocity upon the length of the sides of the rectangular array and the shape of the cross section of the cylinders without restricting ourselves to particular shapes, as circles or ellipses.

If $q \in \mathbb{D}_2^+(\mathbb{R})$ and $\phi \in \mathcal{A}_{\partial\Omega}^{\tilde{Q}}$, the set $\overline{\mathbb{S}_q[q\mathbb{I}[\phi]]} \times \mathbb{R}$ represents an infinite array of parallel cylinders. Instead, the set $\mathbb{S}_q[q\mathbb{I}[\phi]]^- \times \mathbb{R}$ is the region where a Newtonian fluid is flowing at low Reynolds numbers. We assume that the driving pressure gradient is constant and parallel to the cylinders. As a consequence, by a standard argument based on the particular geometry of the problem (cf., e.g., Adler [1, Ch. 4], Sangani and Yao [58], and Mityushev and Adler [46,47]), we can transform the Stokes system into a Poisson equation for the nonzero component of the velocity field. Without loss of generality, we may assume that the viscosity of the fluid and the nonzero component of the pressure gradient are both set equal to one. Accordingly, if $q \in \mathbb{D}_2^+(\mathbb{R})$ and $\phi \in \mathcal{A}_{\partial\Omega}^{\tilde{Q}}$, the problem is reduced to the following Dirichlet problem for the Poisson equation:

$$\begin{cases} \Delta u = 1 & \text{in } \mathbb{S}_q[q\mathbb{I}[\phi]]^-, \\ u(x + qe_i) = u(x) & \forall x \in \overline{\mathbb{S}_q[q\mathbb{I}[\phi]]^-}, \ \forall i \in \{1, 2\}, \\ u(x) = 0 & \forall x \in \partial\mathbb{S}_q[q\mathbb{I}[\phi]]^-. \end{cases} \quad (14.4)$$

Here $\{e_1, e_2\}$ is the canonical basis of $\mathbb{R}^2$. We can show that problem (14.4) has a unique solution in the space $C_q^{1,\alpha}(\overline{\mathbb{S}_q[q\mathbb{I}[\phi]]^-})$ of $C^{1,\alpha}$ q-periodic functions on $\overline{\mathbb{S}_q[q\mathbb{I}[\phi]]^-}$, and we denote it by $u[q, \phi]$. From the physical point of view, the function $u[q, \phi]$ represents the nonzero component of the velocity field (see Mityushev and Adler [46, §2]). Then we can define $\Sigma[q, \phi]$ as the integral of the flow velocity $u[q, \phi]$ over the periodicity cell (see Adler [1], Mityushev and Adler [46, §3]), *i.e.*,

$$\Sigma[q, \phi] \equiv \frac{1}{|Q|_2} \int_{Q\backslash q\mathbb{I}[\phi]} u[q, \phi](x) \, dx$$

$$\forall(q, \phi) \in \mathbb{D}_2^+(\mathbb{R}) \times \left(C^{1,\alpha}(\partial\Omega, \mathbb{R}^2) \cap \mathcal{A}_{\partial\Omega}^{\tilde{Q}} \right). \quad (14.5)$$

In [42,43] we have studied the regularity properties of $\Sigma[q, \phi]$ as a function of (q, ϕ). Among other results, we have proven that the map $(q, \phi) \mapsto \Sigma[q, \phi]$ is real analytic, as we state in the following theorem.

278 Mechanics and Physics of Structured Media

Theorem 14.3.1. *Let α, Ω be as in* (14.3). *Then the map from*

$$\mathbb{D}_2^+(\mathbb{R}) \times \left(C^{1,\alpha}(\partial\Omega, \mathbb{R}^2) \cap \mathcal{A}_{\partial\Omega}^{\widetilde{Q}} \right)$$

to $\mathbb{R}$ that takes a pair (q, ϕ) to $\Sigma[q, \phi]$ is real analytic.

14.4 The effective conductivity of a two-phase periodic composite with ideal contact condition

In this section, we recall the results of [41] about the effective conductivity of an n-dimensional periodic two-phase composite ($n \in \{2, 3\}$) with ideal contact at the interface. The composite is obtained by introducing into a homogeneous matrix a periodic set of inclusions of sufficiently smooth shapes. Both the matrix and the set of inclusions are filled with two different homogeneous and isotropic heat conductive materials of conductivity λ^- and λ^+, respectively, with

$$(\lambda^+, \lambda^-) \in [0, +\infty[_*^2 \equiv [0, +\infty[^2 \setminus \{(0, 0)\}.$$

The limit case of a material with zero conductivity corresponds to a thermal insulator. So here we are assuming that the two materials are not both insulators. On the other hand, if the conductivity tends to $+\infty$, the material is a perfect conductor. Similarly to what we have done in the previous section, the inclusions' shape is determined by the image of a fixed domain through a diffeomorphism ϕ, and the periodicity cell is a 'cuboid' of edges of lengths $q_{11}, \ldots, q_{nn}$. As it is known, it is possible to define the composite's effective conductivity matrix $\lambda^{\text{eff,id}}$ by means of the solution of a transmission problem for the Laplace equation (see Definition 14.4.1, cf. Mityushev, Obnosov, Pesetskaya, and Rogosin [49, §5]). The effective conductivity can be thought as the conductivity of a homogeneous material whose global behavior as a conductor is 'equivalent' to the composite. Then we may want to understand the dependence of $\lambda^{\text{eff,id}}$ upon the 'triple' $((q_{11}, \ldots, q_{nn}), \phi, (\lambda^+, \lambda^-))$, *i.e.*, upon perturbations of the periodicity structure of the composite, the inclusions' shape, and the conductivity parameters of each material.

The mathematical literature on the properties of composite materials is too vast to attempt a complete list of references. We confine ourselves to mention some contributions that are more focused on perturbation analysis of the effective properties. For example, in Ammari, Kang, and Touibi [5] the authors have exploited a potential theoretic approach in order to investigate the asymptotic behavior of the effective properties of a periodic dilute composite. Then Ammari, Kang, and Kim [3] and Ammari, Kang, and Lim [4] have studied anisotropic composite materials and elastic composites, respectively. The method of Functional Equations, first proposed in Mityushev [45], has been used to study the dependence on the radius of the inclusions for a wide class of 2D composites. For ideal composites, we mention here, for example, the works of Mityushev, Obnosov, Pesetskaya, and Rogosin [49], Gryshchuk and

Dependence of effective properties upon regular perturbations Chapter | 14 **279**

Rogosin [28], Kapanadze, Mishuris, and Pesetskaya [30]. Berlyand, Golovaty, Movchan, and Phillips [8] have analyzed the transport properties of fluid/solid and solid/solid composites and have investigated how the curvature of the inclusions affects such properties. Berlyand and Mityushev [9] have studied the dependence of the effective conductivity of two-phase composites upon the polydispersity parameter. Gorb and Berlyand [26] considered the asymptotic behavior of the effective properties of composites with close inclusions of optimal shape. For two-dimensional composites, the recent work by Mityushev, Nawalaniec, Nosov, and Pesetskaya [48] studies the effective conductivity of two-phase random composites with nonoverlapping inclusions whose boundaries are $C^{1,\alpha}$ curves. In Lee and Lee [39], the authors have studied how the effective elasticity of dilute periodic elastic composites is affected by its periodic structure. In connection with doubly periodic problems for composite materials, we mention the monograph of Grigolyuk and Fil'shtinskij [27], where the authors have proposed a method of integral equations for planar periodic problems in the frame of elasticity (see also Fil'shtinskij [22] and the more recent work Filshtinsky and Mityushev [23]). Finally, in [55] the fourth named author has explicitly computed the effective conductivity of a periodic dilute composite with perfect contact as a power series in the size of the inclusions (see also [54]).

With the aim of introducing the definition of the effective conductivity, we first have to introduce a family of boundary value problems for the Laplace equation. If $q \in \mathbb{D}_n^+(\mathbb{R})$, $\phi \in C^{1,\alpha}(\partial\Omega, \mathbb{R}^n) \cap \mathcal{A}_{\partial\Omega}^{\tilde{Q}}$, and $(\lambda^+, \lambda^-) \in [0, +\infty[_*^2$, for each $j \in \{1, \ldots, n\}$ we consider the following transmission problem for a pair of functions $(u_j^+, u_j^-) \in C_{loc}^{1,\alpha}(\overline{\mathbb{S}_q[q\mathbb{I}[\phi]]}) \times C_{loc}^{1,\alpha}(\overline{\mathbb{S}_q[q\mathbb{I}[\phi]]^-})$:

$$
\begin{cases}
\Delta u_j^+ = 0 & \text{in } \mathbb{S}_q[q\mathbb{I}[\phi]], \\
\Delta u_j^- = 0 & \text{in } \mathbb{S}_q[q\mathbb{I}[\phi]]^-, \\
u_j^+(x + qe_h) = u_j^+(x) + \delta_{hj}q_{jj} & \forall x \in \overline{\mathbb{S}_q[q\mathbb{I}[\phi]]}, \ \forall h \in \{1, \ldots, n\}, \\
u_j^-(x + qe_h) = u_j^-(x) + \delta_{hj}q_{jj} & \forall x \in \overline{\mathbb{S}_q[q\mathbb{I}[\phi]]^-}, \ \forall h \in \{1, \ldots, n\}, \\
\lambda^+ \dfrac{\partial}{\partial \nu_{q\mathbb{I}[\phi]}} u_j^+ - \lambda^- \dfrac{\partial}{\partial \nu_{q\mathbb{I}[\phi]}} u_j^- = 0 & \text{on } \partial q\mathbb{I}[\phi], \\
u_j^+ - u_j^- = 0 & \text{on } \partial q\mathbb{I}[\phi], \\
\int_{\partial q\mathbb{I}[\phi]} u_j^+ \, d\sigma = 0,
\end{cases}
$$

$$(14.6)$$

where $\nu_{q\mathbb{I}[\phi]}$ is the outward unit normal to $\partial q\mathbb{I}[\phi]$ and $\{e_1, \ldots, e_n\}$ is the canonical basis of $\mathbb{R}^n$. We recall that here above for $(h, j) \in \{1, \ldots, n\}^2$ the symbol δ_{hj} denotes the Kronecker delta symbol, so that $\delta_{hj} = 1$ for $h = j$ and $\delta_{hj} = 0$ otherwise. Problem (14.6) admits a unique solution (u_j^+, u_j^-) in $C_{loc}^{1,\alpha}(\overline{\mathbb{S}_q[q\mathbb{I}[\phi]]}) \times C_{loc}^{1,\alpha}(\overline{\mathbb{S}_q[q\mathbb{I}[\phi]]^-})$, which we denote by $(u_j^+[q, \phi, (\lambda^+, \lambda^-)], u_j^-[q, \phi, (\lambda^+, \lambda^-)])$. This solution is used to define the effective conductivity as follows (cf., e.g., Mityushev, Obnosov, Pesetskaya, and Rogosin [49, §5]).

280 Mechanics and Physics of Structured Media

Definition 14.4.1. Let $q \in \mathbb{D}_n^+(\mathbb{R})$, $\phi \in C^{1,\alpha}(\partial\Omega, \mathbb{R}^n) \cap \mathcal{A}_{\partial\Omega}^{\tilde{Q}}$, and $(\lambda^+, \lambda^-) \in [0, +\infty[_*^2$. Then the effective conductivity

$$\lambda^{\text{eff,id}}[q, \phi, (\lambda^+, \lambda^-)] \equiv (\lambda_{ij}^{\text{eff,id}}[q, \phi, (\lambda^+, \lambda^-)])_{i,j=1,\dots,n}$$

is the $n \times n$ matrix with (i, j)-entry defined by

$$\lambda_{ij}^{\text{eff,id}}[q, \phi, (\lambda^+, \lambda^-)] \equiv \frac{1}{|Q|_n} \left\{ \lambda^+ \int_{q\mathbb{I}[\phi]} \frac{\partial}{\partial x_i} u_j^+[q, \phi, (\lambda^+, \lambda^-)](x) \, dx \right.$$

$$\left. + \lambda^- \int_{Q \setminus \overline{q\mathbb{I}[\phi]}} \frac{\partial}{\partial x_i} u_j^-[q, \phi, (\lambda^+, \lambda^-)](x) \, dx \right\}$$

$$\forall i, j \in \{1, \dots, n\}.$$

As for the average longitudinal flow in Section 14.3, we are interested in the function

$$(q, \phi, (\lambda^+, \lambda^-)) \mapsto \lambda_{ij}^{\text{eff,id}}[q, \phi, (\lambda^+, \lambda^-)].$$

The following result of [41, Thm. 5.1] describes the regularity of the effective conductivity matrix $\lambda^{\text{eff,id}}[q, \phi, (\lambda^+, \lambda^-)]$ of the ideal composite upon the triple 'periodicity-shape-conductivity'. More in details, it shows that the (i, j)-entry $\lambda_{ij}^{\text{eff,id}}[q, \phi, (\lambda^+, \lambda^-)]$ can be expressed in terms of the conductivity λ^- of the matrix, the conductivity λ^+ of the inclusions, and an analytic map of the periodicity q, of the inclusions' shape ϕ, and the ratio $\frac{\lambda^+ - \lambda^-}{\lambda^+ + \lambda^-}$, which is sometimes referred to as the contrast parameter.

Theorem 14.4.2. *Let α, Ω be as in (14.3). Let $i, j \in \{1, \dots, n\}$. Then there exist an open neighborhood $\mathcal{U}$ of $\mathbb{D}_n^+(\mathbb{R}) \times \left(C^{1,\alpha}(\partial\Omega, \mathbb{R}^n) \cap \mathcal{A}_{\partial\Omega}^{\tilde{Q}} \right) \times [-1, 1]$ in the space $\mathbb{D}_n^+(\mathbb{R}) \times \left(C^{1,\alpha}(\partial\Omega, \mathbb{R}^n) \cap \mathcal{A}_{\partial\Omega}^{\tilde{Q}} \right) \times \mathbb{R}$ and a real analytic map Λ_{ij} from $\mathcal{U}$ to $\mathbb{R}$ such that*

$$\lambda_{ij}^{\text{eff,id}}[q, \phi, (\lambda^+, \lambda^-)] \equiv \delta_{ij}\lambda^- + (\lambda^+ + \lambda^-)\Lambda_{ij}\left[q, \phi, \frac{\lambda^+ - \lambda^-}{\lambda^+ + \lambda^-} \right]$$

for all $(q, \phi, (\lambda^+, \lambda^-)) \in \mathbb{D}_n^+(\mathbb{R}) \times \left(C^{1,\alpha}(\partial\Omega, \mathbb{R}^n) \cap \mathcal{A}_{\partial\Omega}^{\tilde{Q}} \right) \times [0, +\infty[_*^2$.

14.5 The effective conductivity of a two-phase periodic composite with nonideal contact condition

We now turn our attention to the study of the effective conductivity of an n-dimensional periodic two-phase composite ($n \in \{2, 3\}$) with imperfect (or non-ideal) contact at the interface.

As in the previous section, the composite consists of a matrix and a periodic set of inclusions. The matrix and the inclusions are filled with two (possibly different) homogeneous and isotropic heat conductive materials. The normal component of the heat flux is assumed to be continuous at the two-phase interface, while we impose that the temperature field displays a jump proportional to the normal heat flux by means of a parameter r. In physics, the appearance of such a discontinuity in the temperature field is a well-known phenomenon and has been largely investigated since 1941, when Kapitza carried out the first systematic study of thermal interface behavior in liquid helium (see, *e.g.*, Swartz and Pohl [61], Lipton [40] and references therein). As in the ideal case, our aim is to study the behavior of the effective conductivity of the nonideal composite upon perturbation of the geometry and the parameters of the problem. The expression defining the effective conductivity of a composite with imperfect contact conditions was introduced by Benveniste and Miloh in [7] by generalizing the dual theory of the effective behavior of composites with perfect contact (see also Benveniste [6] and for a review Drygaś and Mityushev [20]). By the argument of Benveniste and Miloh, in order to evaluate the effective conductivity, one has to study the thermal distribution of the composite when so-called 'homogeneous conditions' are prescribed.

We first introduce the parameters of the problem. Both the matrix and the set of inclusions are filled with two different homogeneous and isotropic heat conductive materials of conductivity λ^- and λ^+, respectively, with

$$(\lambda^+, \lambda^-) \in]0, +\infty[^2.$$

The normal component of the heat flux is assumed to be continuous at the two-phase interface, while we impose that the temperature field displays a jump proportional to the normal heat flux by means of a parameter

$$r \in [0, +\infty[.$$

We find it convenient to set

$$\mathcal{P} \equiv]0, +\infty[^2 \times [0, +\infty[.$$

As in the ideal case, the set $\mathbb{S}_q[q\mathbb{I}[\phi]]^-$ represents the homogeneous matrix made of a material with conductivity λ^- where the periodic set of inclusions $\overline{\mathbb{S}_q[q\mathbb{I}[\phi]]}$ with conductivity λ^+ is inserted. The two-phase composite consists of the union of the matrix and the inclusions.

Let $q \in \mathbb{D}_n^+(\mathbb{R})$, $\phi \in C^{1,\alpha}(\partial\Omega, \mathbb{R}^n) \cap \mathcal{A}_{\partial\Omega}^{\tilde{Q}}$, $(\lambda^+, \lambda^-, r) \in \mathcal{P}$. To define the effective conductivity in the nonideal case we introduce the following boundary

282 Mechanics and Physics of Structured Media

value problem:

$$
\begin{cases}
\Delta u_j^+ = 0 & \text{in } \mathbb{S}_q[q\mathbb{I}[\phi]], \\
\Delta u_j^- = 0 & \text{in } \mathbb{S}_q[q\mathbb{I}[\phi]]^-, \\
u_j^+(x + qe_h) = u_j^+(x) + \delta_{hj}q_{jj} & \forall x \in \overline{\mathbb{S}_q[q\mathbb{I}[\phi]]}, \ \forall h \in \{1,\dots,n\}, \\
u_j^-(x + qe_h) = u_j^-(x) + \delta_{hj}q_{jj} & \forall x \in \overline{\mathbb{S}_q[q\mathbb{I}[\phi]]^-}, \ \forall h \in \{1,\dots,n\}, \\
\lambda^+ \dfrac{\partial}{\partial \nu_{q\mathbb{I}[\phi]}} u_j^+ - \lambda^- \dfrac{\partial}{\partial \nu_{q\mathbb{I}[\phi]}} u_j^- = 0 & \text{on } \partial q\mathbb{I}[\phi], \\
\lambda^+ \dfrac{\partial}{\partial \nu_{q\mathbb{I}[\phi]}} u_j^+ + r\left(u_j^+ - u_j^-\right) = 0 & \text{on } \partial q\mathbb{I}[\phi], \\
\int_{\partial q\mathbb{I}[\phi]} u_j^+ \, d\sigma = 0,
\end{cases}
$$

(14.7)

with $j \in \{1,\dots,n\}$. As we will see, problem (14.7) admits a unique solution (u_j^+, u_j^-) in $C_{\mathrm{loc}}^{1,\alpha}(\overline{\mathbb{S}_q[q\mathbb{I}[\phi]]}) \times C_{\mathrm{loc}}^{1,\alpha}(\overline{\mathbb{S}_q[q\mathbb{I}[\phi]]^-})$, which we denote by

$$
(u_j^+[q, \phi, \lambda^+, \lambda^-, r], u_j^-[q, \phi, \lambda^+, \lambda^-, r]).
$$

Then, with this family of solutions, we can define the effective conductivity as follows. (The reader might note the similarity with Definition 14.4.1 given in the case of an ideal contact.)

Definition 14.5.1. Let $q \in \mathbb{D}_n^+(\mathbb{R})$, $\phi \in C^{1,\alpha}(\partial\Omega, \mathbb{R}^n) \cap \mathcal{A}_{\partial\Omega}^{\widetilde{Q}}$, and $(\lambda^+, \lambda^-, r) \in \mathcal{P}$. Then the effective conductivity

$$
\lambda^{\mathrm{eff,nonid}}[q, \phi, \lambda^+, \lambda^-, r] \equiv (\lambda_{ij}^{\mathrm{eff}}[q, \phi, \lambda^+, \lambda^-, r])_{i,j=1,\dots,n}
$$

is the $n \times n$ matrix with (i, j)-entry defined by

$$
\lambda_{ij}^{\mathrm{eff,nonid}}[q, \phi, \lambda^+, \lambda^-, r] \equiv \frac{1}{|Q|_n} \left\{ \lambda^+ \int_{q\mathbb{I}[\phi]} \frac{\partial}{\partial x_i} u_j^+[q, \phi, \lambda^+, \lambda^-, r](x) \, dx \right.
$$

$$
\left. + \lambda^- \int_{Q \setminus \overline{q\mathbb{I}[\phi]}} \frac{\partial}{\partial x_i} u_j^-[q, \phi, \lambda^+, \lambda^-, r](x) \, dx \right\}
$$

$$
\forall i, j \in \{1,\dots,n\}.
$$

Before describing the main result of this section, we mention that composites with contact conditions different from the ideal ones are studied, for example, in Drygaś and Mityushev [20], in Castro, Kapanadze, and Pesetskaya [10,11] (about nonideal composites), and in Castro and Pesetskaya [12] (about composites with inextensible-membrane-type interface). We also mention that the asymptotic behavior of the effective conductivity of a periodic dilute composite with imperfect contact has been studied in [15,16]. In addition, we note that effective properties of heat conductors with interfacial contact resistance have

Dependence of effective properties upon regular perturbations **Chapter | 14 283**

been studied via homogenization theory (cf. Donato and Monsurrò [18], Faella, Monsurrò, and Perugia [21], Monsurrò [51,52]).

The main goal of the rest of our chapter is to study the regularity of the map

$$(q, \phi, \lambda^+, \lambda^-, r) \mapsto \lambda^{\text{eff,nonid}}[q, \phi, \lambda^+, \lambda^-, r].$$

We will prove the following theorem.

Theorem 14.5.2. *Let α, Ω be as in (14.3). Let i, $j \in \{1, \ldots, n\}$. Then there exist an open neighborhood $\mathcal{V}$ of $\mathbb{D}_n^+(\mathbb{R}) \times \left(C^{1,\alpha}(\partial\Omega, \mathbb{R}^n) \cap \mathcal{A}_{\partial\Omega}^{\tilde{Q}} \right) \times \mathcal{P}$ in the space $\mathbb{D}_n^+(\mathbb{R}) \times \left(C^{1,\alpha}(\partial\Omega, \mathbb{R}^n) \cap \mathcal{A}_{\partial\Omega}^{\tilde{Q}} \right) \times \mathbb{R}^3$ and a real analytic map Λ_{ij} from $\mathcal{V}$ to $\mathbb{R}$ such that*

$$\lambda_{ij}^{\text{eff,nonid}}[q, \phi, \lambda^+, \lambda^-, r] \equiv \delta_{ij}\lambda^- + \Lambda_{ij}\left[q, \phi, \lambda^+, \lambda^-, r\right]$$

for all $(q, \phi, \lambda^+, \lambda^-, r) \in \mathbb{D}_n^+(\mathbb{R}) \times \left(C^{1,\alpha}(\partial\Omega, \mathbb{R}^n) \cap \mathcal{A}_{\partial\Omega}^{\tilde{Q}} \right) \times \mathcal{P}$.

The approach that we use to prove Theorem 14.5.2 was introduced by Lanza de Cristoforis in [33] and then extended to a large variety of singular and regular perturbation problems (cf., *e.g.*, Lanza de Cristoforis [34], Lanza de Cristoforis and the first named author [14], and [13]).

In particular, in the present chapter, we follow the strategy of [43] where we have studied the behavior of the longitudinal flow along a periodic array of cylinders upon perturbations of the shape of the cross section of the cylinders and the periodicity structure (see also Section 14.3), and of [41] where we have considered the effective conductivity of an ideal composite (see also Section 14.4). More precisely, we transform the problem into a set of integral equations defined on a fixed domain and depending on the set of variables $(q, \phi, \lambda^+, \lambda^-, r)$. We study the dependence of the solution of the integral equations upon $(q, \phi, \lambda^+, \lambda^-, r)$ and then we deduce the result on the behavior of $\lambda_{ij}^{\text{eff,nonid}}[q, \phi, \lambda^+, \lambda^-, r]$. In this chapter, the integral equations are derived by a potential theoretic approach. However, integral equations could also be deduced by the generalized alternating method of Schwarz (cf. Gluzman, Mityushev, and Nawalaniec [25] and Drygaś, Gluzman, Mityushev, and Nawalaniec [19]).

Incidentally, we observe that there are several contributions concerning the optimization of effective parameters from many different points of view. For example, one can look for *optimal lattices* without confining to rectangular distributions. In this direction, Kozlov [31] and Mityushev and Rylko [50] have discussed extremal properties of hexagonal lattices of disks. In Rylko [56], the author has studied the influence of perturbations of the shape of the circular inclusion on the macroscopic conductivity properties of 2D dilute composites. For an experimental work concerning the analysis of particle reinforced composites we mention Kurtyka and Rylko [32].

284 Mechanics and Physics of Structured Media

Finally, we note that we do not consider the case where $r \to +\infty$. The asymptotic analysis of such case in a (nonperiodic) transmission problem can be found in Schmidt and Hiptmair [60].

14.6 Proof of Theorem 14.5.2

14.6.1 Preliminaries

Our method is based on a periodic version of the classical potential theory. Periodic layer potentials are constructed by replacing the fundamental solution of the Laplace operator with a q-periodic tempered distribution $S_{q,n}$ such that

$$\Delta S_{q,n} = \sum_{z \in \mathbb{Z}^n} \delta_{qz} - \frac{1}{|Q|_n},$$

where δ_{qz} denotes the Dirac measure with mass at the point $qz \in \mathbb{R}^n$ (see, e.g., Lanza de Cristoforis and the third named author [36, p. 84]). The distribution $S_{q,n}$ is determined up to an additive constant, and we have

$$S_{q,n}(x) = - \sum_{z \in \mathbb{Z}^n \setminus \{0\}} \frac{1}{|Q|_n 4\pi^2 |q^{-1}z|^2} e^{2\pi i(q^{-1}z) \cdot x},$$

where the generalized sum is defined in the sense of distributions in $\mathbb{R}^n$ (see, e.g., Ammari and Kang [2, p. 53], Lanza de Cristoforis and the third named author [36, §3]). It is known that $S_{q,n}$ is real analytic in $\mathbb{R}^n \setminus q\mathbb{Z}^n$ and locally integrable in $\mathbb{R}^n$ (see, e.g., Lanza de Cristoforis and the third named author [36, §3]).

We now introduce periodic layer potentials. Let Ω_Q be a bounded open subset of $\mathbb{R}^n$ of class $C^{1,\alpha}$ for some $\alpha \in {]}0, 1{[}$ such that $\overline{\Omega_Q} \subseteq Q$. We set

$$v_q[\partial\Omega_Q, \mu](x) \equiv \int_{\partial\Omega_Q} S_{q,n}(x - y)\mu(y) \, d\sigma_y \quad \forall x \in \mathbb{R}^n,$$

$$w_{q,*}[\partial\Omega_Q, \mu](x) \equiv \int_{\partial\Omega_Q} \nu_{\Omega_Q}(x) \cdot DS_{q,n}(x - y)\mu(y) \, d\sigma_y \quad \forall x \in \partial\Omega_Q,$$

for all $\mu \in C^0(\partial\Omega_Q)$. Here above, $DS_{q,n}(\xi)$ denotes the gradient of $S_{q,n}$ computed at the point $\xi \in \mathbb{R}^n \setminus q\mathbb{Z}^n$. The function $v_q[\partial\Omega_Q, \mu]$ is called the q-periodic single layer potential, and $w_{q,*}[\partial\Omega_Q, \mu]$ is a function related to the normal derivative of the single layer potential. As is well known, if $\mu \in C^0(\partial\Omega_Q)$, then $v_q[\partial\Omega_Q, \mu]$ is continuous in $\mathbb{R}^n$ and q-periodic. We set

$$v_q^+[\partial\Omega_Q, \mu] \equiv v_q[\partial\Omega_Q, \mu]_{|\overline{\mathbb{S}_q[\Omega_Q]}} \quad \text{and} \quad v_q^-[\partial\Omega_Q, \mu] \equiv v_q[\partial\Omega_Q, \mu]_{|\overline{\mathbb{S}_q[\Omega_Q]^-}}.$$

In the following theorem, we collect some properties of $v_q^{\pm}[\partial\Omega_Q, \cdot]$ and $w_{q,*}[\partial\Omega_Q, \cdot]$ that are the periodic analog of classical regularity results and jump

Dependence of effective properties upon regular perturbations **Chapter | 14** **285**

formulas for the single layer potential. For a proof of statements (i)–(iii) we refer to Lanza de Cristoforis and the third named author [36, Thm. 3.7] and to [15, Lem. 4.2]. For a proof of statement (iv) we refer to [15, Lem. 4.2 (i), (iii)].

Theorem 14.6.1. *Let q, Q be as in (14.1) and (14.2). Let $\alpha \in \,]0, 1[$. Let Ω_Q be a bounded open subset of $\mathbb{R}^n$ of class $C^{1,\alpha}$ such that $\overline{\Omega_Q} \subseteq Q$. Then the following statements hold.*

(i) *The map from $C^{0,\alpha}(\partial\Omega_Q)$ to $C_q^{1,\alpha}(\overline{\mathbb{S}_q[\Omega_Q]})$ that takes μ to $v_q^+[\partial\Omega_Q, \mu]$ is linear and continuous. The map from $C^{0,\alpha}(\partial\Omega_Q)$ to $C_q^{1,\alpha}(\overline{\mathbb{S}_q[\Omega_Q]^-})$ that takes μ to $v_q^-[\partial\Omega_Q, \mu]$ is linear and continuous.*

(ii) *Let $\mu \in C^{0,\alpha}(\partial\Omega_Q)$. Then*

$$\frac{\partial}{\partial\nu_{\Omega_Q}} v_q^\pm[\partial\Omega_Q, \mu] = \mp\frac{1}{2}\mu + w_{q,*}[\partial\Omega_Q, \mu] \quad on \; \partial\Omega_Q.$$

Moreover,

$$\int_{\partial\Omega_Q} w_{q,*}[\partial\Omega_Q, \mu]\, d\sigma = \left(\frac{1}{2} - \frac{|\Omega_Q|_n}{|Q|_n}\right) \int_{\partial\Omega_Q} \mu\, d\sigma.$$

(iii) *Let $\mu \in C^{0,\alpha}(\partial\Omega_Q)_0$. Then*

$$\Delta v_q[\partial\Omega_Q, \mu] = 0 \quad in \; \mathbb{R}^n \setminus \partial\mathbb{S}_q[\Omega_Q].$$

(iv) *The operator $w_{q,*}[\partial\Omega_Q, \cdot]$ is compact in $C^{0,\alpha}(\partial\Omega_Q)$ and in $C^{0,\alpha}(\partial\Omega_Q)_0$.*

Next we turn to problem (14.7). By means of the following proposition, whose proof is of immediate verification, we can transform problem (14.7) into a q-periodic transmission problem for the Laplace equation.

Proposition 14.6.2. *Let q be as in (14.1), Q be as in (14.2), and α, Ω be as in (14.3). Let $(\lambda^+, \lambda^-, r) \in \mathcal{P}$. Let $\phi \in C^{1,\alpha}(\partial\Omega, \mathbb{R}^n) \cap \mathcal{A}_{\partial\Omega}^{\tilde{Q}}$. Let $j \in \{1, \dots, n\}$. A pair*

$$(u_j^+, u_j^-) \in C_{loc}^{1,\alpha}(\overline{\mathbb{S}_q[q\mathbb{I}[\phi]]}) \times C_{loc}^{1,\alpha}(\overline{\mathbb{S}_q[q\mathbb{I}[\phi]]^-})$$

solves problem (14.7) if and only if the pair

$$(\tilde{u}_j^+, \tilde{u}_j^-) \in C_q^{1,\alpha}(\overline{\mathbb{S}_q[q\mathbb{I}[\phi]]}) \times C_q^{1,\alpha}(\overline{\mathbb{S}_q[q\mathbb{I}[\phi]]^-})$$

defined by

$$\tilde{u}_j^+(x) \equiv u_j^+(x) - x_j \quad \forall x \in \overline{\mathbb{S}_q[q\mathbb{I}[\phi]]},$$
$$\tilde{u}_j^-(x) \equiv u_j^-(x) - x_j \quad \forall x \in \overline{\mathbb{S}_q[q\mathbb{I}[\phi]]^-},$$

286 Mechanics and Physics of Structured Media

solves

$$
\begin{cases}
\Delta \tilde{u}_j^+ = 0 & in\ \mathbb{S}_q[q\mathbb{I}[\phi]], \\[4pt]
\Delta \tilde{u}_j^- = 0 & in\ \mathbb{S}_q[q\mathbb{I}[\phi]]^-, \\[4pt]
\tilde{u}_j^+(x + qe_h) = \tilde{u}_j^+(x) & \forall x \in \overline{\mathbb{S}_q[q\mathbb{I}[\phi]]}, \\
& \forall h \in \{1, \dots, n\}, \\[4pt]
\tilde{u}_j^-(x + qe_h) = \tilde{u}_j^-(x) & \forall x \in \overline{\mathbb{S}_q[q\mathbb{I}[\phi]]^-}, \\
& \forall h \in \{1, \dots, n\}, \\[4pt]
\lambda^+ \frac{\partial}{\partial \nu_{q\mathbb{I}[\phi]}} \tilde{u}_j^+ - \lambda^- \frac{\partial}{\partial \nu_{q\mathbb{I}[\phi]}} \tilde{u}_j^- = (\lambda^- - \lambda^+)(\nu_{q\mathbb{I}[\phi]})_j & on\ \partial q\mathbb{I}[\phi], \\[4pt]
\lambda^+ \frac{\partial}{\partial \nu_{q\mathbb{I}[\phi]}} \tilde{u}_j^+ + r\left(\tilde{u}_j^+ - \tilde{u}_j^-\right) = -\lambda^+(\nu_{q\mathbb{I}[\phi]})_j & on\ \partial q\mathbb{I}[\phi], \\[4pt]
\int_{\partial q\mathbb{I}[\phi]} \tilde{u}_j^+ \, d\sigma = -\int_{\partial q\mathbb{I}[\phi]} y_j \, d\sigma_y.
\end{cases}
$$

$$(14.8)$$

By [15, Prop. 5.1, Thm. 5.3], we deduce the validity of the following proposition, stating that (the equivalent) problems (14.7) and (14.8) have unique solution.

Proposition 14.6.3. *Let q be as in* (14.1)*, Q be as in* (14.2)*, and α, Ω be as in* (14.3)*. Let $(\lambda^+, \lambda^-, r) \in \mathcal{P}$. Let $\phi \in C^{1,\alpha}(\partial\Omega, \mathbb{R}^n) \cap \mathcal{A}_{\partial\Omega}^{\tilde{Q}}$. Let $j \in \{1, \dots, n\}$. Then the following statements hold.*

(i) *Problem* (14.7) *has a unique solution (u_j^+, u_j^-) in $C^{1,\alpha}_{loc}(\overline{\mathbb{S}_q[q\mathbb{I}[\phi]]}) \times C^{1,\alpha}_{loc}(\overline{\mathbb{S}_q[q\mathbb{I}[\phi]]^-})$.*

(ii) *Problem* (14.8) *has a unique solution $(\tilde{u}_j^+, \tilde{u}_j^-)$ in $C_q^{1,\alpha}(\overline{\mathbb{S}_q[q\mathbb{I}[\phi]]}) \times C_q^{1,\alpha}(\overline{\mathbb{S}_q[q\mathbb{I}[\phi]]^-})$.*

14.6.2 An integral equation formulation of problem (14.7)

In this section, we convert problem (14.7) into a system of integral equations. As done in [43] for the longitudinal flow and in [41] for the ideal contact, we do so by representing the solution in terms of single layer potentials with densities that solve certain integral equations. We first start with the following proposition on the invertibility of an integral operator that will appear in the integral formulation of problem (14.7).

Proposition 14.6.4. *Let q be as in* (14.1)*, Q be as in* (14.2)*, and α, Ω be as in* (14.3)*. Let $(\phi, \lambda^+, \lambda^-, r) \in (C^{1,\alpha}(\partial\Omega, \mathbb{R}^n) \cap \mathcal{A}_{\partial\Omega}^{\tilde{Q}}) \times \mathcal{P}$. Let $J \equiv (J_1, J_2)$ be the*

Dependence of effective properties upon regular perturbations **Chapter | 14 287**

operator from $(C^{0,\alpha}(\partial q\mathbb{I}[\phi]))^2$ to $(C^{0,\alpha}(\partial q\mathbb{I}[\phi]))^2$ defined by

$$J_1[\mu^+, \mu^-] \equiv \lambda^+ \left(-\frac{1}{2}\mu^+ + w_{q,*}[\partial q\mathbb{I}[\phi], \mu^+]\right)$$
$$- \lambda^- \left(\frac{1}{2}\mu^- + w_{q,*}[\partial q\mathbb{I}[\phi], \mu^-]\right),$$

$$J_2[\mu^+, \mu^-] \equiv \lambda^+ \left(-\frac{1}{2}\mu^+ + w_{q,*}[\partial q\mathbb{I}[\phi], \mu^+]\right)$$
$$+ r \left(v_q^+[\partial q\mathbb{I}[\phi], \mu^+]_{|\partial q\mathbb{I}[\phi]} \right.$$
$$- \frac{1}{|\partial q\mathbb{I}[\phi]|_{n-1}} \int_{\partial q\mathbb{I}[\phi]} v_q^+[\partial q\mathbb{I}[\phi], \mu^+] d\sigma$$
$$- v_q^-[\partial q\mathbb{I}[\phi], \mu^-]_{|\partial q\mathbb{I}[\phi]}$$
$$+ \left. \frac{1}{|\partial q\mathbb{I}[\phi]|_{n-1}} \int_{\partial q\mathbb{I}[\phi]} v_q^-[\partial q\mathbb{I}[\phi], \mu^-] d\sigma \right),$$

for all $(\mu^+, \mu^-) \in (C^{0,\alpha}(\partial q\mathbb{I}[\phi]))^2$, where $|\partial q\mathbb{I}[\phi]|_{n-1}$ denotes the $(n-1)$-dimensional measure of $\partial q\mathbb{I}[\phi]$. Then the following statements hold.

(i) *The operator J restricts to a homeomorphism from $(C^{0,\alpha}(\partial q\mathbb{I}[\phi])_0)^2$ to $(C^{0,\alpha}(\partial q\mathbb{I}[\phi])_0)^2$.*
(ii) *The operator J is a homeomorphism from $(C^{0,\alpha}(\partial q\mathbb{I}[\phi]))^2$ to $(C^{0,\alpha}(\partial q\mathbb{I}[\phi]))^2$.*

Proof. We first notice that the validity of statement (i) follows by [15, Prop. 5.2]. We now consider statement (ii) and we follow the lines of the proof of [15, Prop. 5.2]. Let $\hat{J} \equiv (\hat{J}_1, \hat{J}_2)$ be the linear operator from $(C^{0,\alpha}(\partial q\mathbb{I}[\phi]))^2$ to $(C^{0,\alpha}(\partial q\mathbb{I}[\phi]))^2$ defined by

$$\hat{J}_1[\mu^+, \mu^-] \equiv -(\lambda^-/2)\mu^- - (\lambda^+/2)\mu^+, \quad \hat{J}_2[\mu^+, \mu^-] \equiv -(\lambda^+/2)\mu^+$$

for all $(\mu^+, \mu^-) \in (C^{0,\alpha}(\partial q\mathbb{I}[\phi]))^2$. Clearly, $\hat{J}$ is a linear homeomorphism from $(C^{0,\alpha}(\partial q\mathbb{I}[\phi]))^2$ to $(C^{0,\alpha}(\partial q\mathbb{I}[\phi]))^2$. Then let

$$\tilde{J} \equiv (\tilde{J}_1, \tilde{J}_2)$$

be the operator from $(C^{0,\alpha}(\partial q\mathbb{I}[\phi]))^2$ to $(C^{0,\alpha}(\partial q\mathbb{I}[\phi]))^2$ defined by

$$\tilde{J}_1[\mu^+, \mu^-] \equiv \lambda^+ w_{q,*}[\partial q\mathbb{I}[\phi], \mu^+] - \lambda^- w_{q,*}[\partial q\mathbb{I}[\phi], \mu^-],$$
$$\tilde{J}_2[\mu^+, \mu^-] \equiv \lambda^+ w_{q,*}[\partial q\mathbb{I}[\phi], \mu^+]$$
$$+ r \left(v_q^+[\partial q\mathbb{I}[\phi], \mu^+]_{|\partial q\mathbb{I}[\phi]} \right.$$

$$-\frac{1}{|\partial q \mathbb{I}[\phi]|_{n-1}} \int_{\partial q \mathbb{I}[\phi]} v_q^+[\partial q \mathbb{I}[\phi], \mu^+] \, d\sigma$$

$$- v_q^-[\partial q \mathbb{I}[\phi], \mu^-]_{|\partial q \mathbb{I}[\phi]}$$

$$+ \frac{1}{|\partial q \mathbb{I}[\phi]|_{n-1}} \int_{\partial q \mathbb{I}[\phi]} v_q^-[\partial q \mathbb{I}[\phi], \mu^-] \, d\sigma \Bigg)$$

for all $(\mu^+, \mu^-) \in (C^{0,\alpha}(\partial q \mathbb{I}[\phi]))^2$. Then, by Theorem 14.6.1 (i), the operator from $C^{0,\alpha}(\partial q \mathbb{I}[\phi])$ to $C^{1,\alpha}(\partial q \mathbb{I}[\phi])$ that takes μ to

$$v_q[\partial q \mathbb{I}[\phi], \mu]_{|\partial q \mathbb{I}[\phi]} - \frac{1}{|\partial q \mathbb{I}[\phi]|_{n-1}} \int_{\partial q \mathbb{I}[\phi]} v_q[\partial q \mathbb{I}[\phi], \mu] \, d\sigma \,,$$

is bounded and by Theorem 14.6.1 (iv) the map $w_{q,*}[\partial q \mathbb{I}[\phi], \cdot]$ is compact. Then the compactness of the imbedding of $C^{1,\alpha}(\partial q \mathbb{I}[\phi])$ into $C^{0,\alpha}(\partial q \mathbb{I}[\phi])$ implies that $\tilde{J}$ is a compact operator. Now, since $J = \hat{J} + \tilde{J}$ and since compact perturbations of isomorphisms are Fredholm operators of index 0, we deduce that J is a Fredholm operator of index 0. Thus, to show that J is a linear homeomorphism, it suffices to show that it is injective. So, let $(\mu^+, \mu^-) \in (C^{0,\alpha}(\partial q \mathbb{I}[\phi]))^2$ be such that

$$J[\mu^+, \mu^-] = (0,0) \,. \tag{14.9}$$

Clearly,

$$\int_{\partial q \mathbb{I}[\phi]} r \left(v_q^+[\partial q \mathbb{I}[\phi], \mu^+]_{|\partial q \mathbb{I}[\phi]} - \frac{1}{|\partial q \mathbb{I}[\phi]|_{n-1}} \int_{\partial q \mathbb{I}[\phi]} v_q^+[\partial q \mathbb{I}[\phi], \mu^+] \, d\sigma \right.$$

$$\left. - v_q^-[\partial q \mathbb{I}[\phi], \mu^-]_{|\partial q \mathbb{I}[\phi]} + \frac{1}{|\partial q \mathbb{I}[\phi]|_{n-1}} \int_{\partial q \mathbb{I}[\phi]} v_q^-[\partial q \mathbb{I}[\phi], \mu^-] \, d\sigma \right) d\sigma = 0 \,.$$

Then Theorem 14.6.1 (ii) and the second component of equality (14.9) imply that

$$\int_{\partial q \mathbb{I}[\phi]} \mu^+ \, d\sigma = 0 \,,$$

i.e., $\mu^+ \in C^{0,\alpha}(\partial q \mathbb{I}[\phi])_0$. Then again Theorem 14.6.1 (ii) and the first component of equality (14.9) imply that

$$\int_{\partial q \mathbb{I}[\phi]} \mu^- \, d\sigma = 0 \,,$$

i.e., $\mu^- \in C^{0,\alpha}(\partial q \mathbb{I}[\phi])_0$. In other words, we have shown that if $(\mu^+, \mu^-) \in (C^{0,\alpha}(\partial q \mathbb{I}[\phi]))^2$ is such that $J[\mu^+, \mu^-] = (0,0)$, then we have $(\mu^+, \mu^-) \in (C^{0,\alpha}(\partial q \mathbb{I}[\phi])_0)^2$. As a consequence, statement (i) implies that $(\mu^+, \mu^-) = (0,0)$, and, therefore, the validity of statement (ii). $\quad\square$

Dependence of effective properties upon regular perturbations **Chapter | 14 289**

We are now ready to show that problem (14.7) can be reformulated in terms of a system of integral equations which admits a unique solution.

Theorem 14.6.5. *Let q be as in (14.1), Q be as in (14.2), and α, Ω be as in (14.3). Let $(\lambda^+, \lambda^-, r) \in \mathcal{P}$. Let $\phi \in C^{1,\alpha}(\partial\Omega, \mathbb{R}^n) \cap A_{\partial\Omega}^{\tilde{Q}}$. Let $j \in \{1, \dots, n\}$. Then the unique solution*

$$(u_j^+[q, \phi, \lambda^+, \lambda^-, r], u_j^-[q, \phi, \lambda^+, \lambda^-, r]) \in C_{\text{loc}}^{1,\alpha}(\overline{\mathbb{S}_q[q\mathbb{I}[\phi]]})$$
$$\times C_{\text{loc}}^{1,\alpha}(\overline{\mathbb{S}_q[q\mathbb{I}[\phi]]^-})$$

of problem (14.7) is delivered by

$$u_j^+[q, \phi, \lambda^+, \lambda^-, r](x) = v_q^+[\partial q\mathbb{I}[\phi], \mu_j^+](x) - \int_{\partial q\mathbb{I}[\phi]} v_q^+[\partial q\mathbb{I}[\phi], \mu_j^+](y)\, d\sigma_y$$
$$- \int_{\partial q\mathbb{I}[\phi]} y_j\, d\sigma_y + x_j \quad \forall x \in \overline{\mathbb{S}_q[q\mathbb{I}[\phi]]}, \qquad (14.10)$$

$$u_j^-[q, \phi, \lambda^+, \lambda^-, r](x) = v_q^-[\partial q\mathbb{I}[\phi], \mu_j^-](x) - \int_{\partial q\mathbb{I}[\phi]} v_q^-[\partial q\mathbb{I}[\phi], \mu_j^-](y)\, d\sigma_y$$
$$- \int_{\partial q\mathbb{I}[\phi]} y_j\, d\sigma_y + x_j \quad \forall x \in \overline{\mathbb{S}_q[q\mathbb{I}[\phi]]^-},$$

where (μ_j^+, μ_j^-) is the unique solution in $(C^{0,\alpha}(\partial q\mathbb{I}[\phi]))^2$ of the system of integral equations

$$J[\mu_j^+, \mu_j^-] = \left((\lambda^- - \lambda^+)(v_{q\mathbb{I}[\phi]})_j, -\lambda^+(v_{q\mathbb{I}[\phi]})_j\right), \qquad (14.11)$$

where J is as in Proposition 14.6.4.

Proof. Proposition 14.6.3 (i) implies that problem (14.7) has a unique solution in $C_{\text{loc}}^{1,\alpha}(\overline{\mathbb{S}_q[q\mathbb{I}[\phi]]}) \times C_{\text{loc}}^{1,\alpha}(\overline{\mathbb{S}_q[q\mathbb{I}[\phi]]^-})$. Accordingly, we only need to prove that the pair of functions defined by (14.10) solves problem (14.7). Since

$$(v_{q\mathbb{I}[\phi]})_j \in C^{0,\alpha}(\partial q\mathbb{I}[\phi])_0,$$

Proposition 14.6.4 (i) implies that there exists a unique pair $(\mu_j^+, \mu_j^-) \in (C^{0,\alpha}(\partial q\mathbb{I}[\phi])_0)^2$ that solves the integral equation (14.11). Accordingly, a straightforward computation based on the properties of the single layer potential (see Theorem 14.6.1) together with Proposition 14.6.2 implies that the pair of functions defined by (14.10) solves problem (14.7). $\qquad \square$

290 Mechanics and Physics of Structured Media

Remark 14.6.6. The previous theorem provides an integral equation formulation of problem (14.7) and a representation formula for its solution. Plugging this representation formula into Definition 14.5.1, we can rewrite the effective conductivity in terms of the densities μ_j^+ and μ_j^- solving Eq. (14.11). Let the assumptions of Theorem 14.6.5 hold and let $u_j^+[q, \phi, \lambda^+, \lambda^-, r]$, $u_j^-[q, \phi, \lambda^+, \lambda^-, r]$, and μ_j^+, μ_j^- be as in Theorem 14.6.5. Then the divergence theorem implies that

$$
\int_{q\mathbb{I}[\phi]} \frac{\partial}{\partial x_i} u_j^+[q, \phi, \lambda^+, \lambda^-, r](x)\, dx
$$

$$
= \int_{\partial q\mathbb{I}[\phi]} u_j^+[q, \phi, \lambda^+, \lambda^-, r](y)(\nu_{q\mathbb{I}[\phi]}(y))_i\, d\sigma_y
$$

$$
= \int_{\partial q\mathbb{I}[\phi]} \left(v_q^+[\partial q\mathbb{I}[\phi], \mu_j^+](y) - \fint_{\partial q\mathbb{I}[\phi]} v_q^+[\partial q\mathbb{I}[\phi], \mu_j^+](z)\, d\sigma_z \right.
$$

$$
\left. - \fint_{\partial q\mathbb{I}[\phi]} z_j\, d\sigma_z + y_j \right) (\nu_{q\mathbb{I}[\phi]}(y))_i\, d\sigma_y
$$

$$
= \int_{\partial q\mathbb{I}[\phi]} v_q^+[\partial q\mathbb{I}[\phi], \mu_j^+](y)(\nu_{q\mathbb{I}[\phi]}(y))_i\, d\sigma_y
$$

$$
- \int_{\partial q\mathbb{I}[\phi]} (\nu_{q\mathbb{I}[\phi]}(y))_i\, d\sigma_y \fint_{\partial q\mathbb{I}[\phi]} v_q^+[\partial q\mathbb{I}[\phi], \mu_j^+](z)\, d\sigma_z
$$

$$
- \int_{\partial q\mathbb{I}[\phi]} (\nu_{q\mathbb{I}[\phi]}(y))_i\, d\sigma_y \fint_{\partial q\mathbb{I}[\phi]} z_j\, d\sigma_z + \delta_{ij} |q\mathbb{I}[\phi]|_n.
$$

In the same way,

$$
\int_{Q\backslash \overline{q\mathbb{I}[\phi]}} \frac{\partial}{\partial x_i} u_j^-[q, \phi, \lambda^+, \lambda^-, r](x)\, dx
$$

$$
= \int_{\partial Q} u_j^-[q, \phi, \lambda^+, \lambda^-, r](y)(\nu_Q(y))_i\, d\sigma_y
$$

$$
- \int_{\partial q\mathbb{I}[\phi]} u_j^-[q, \phi, \lambda^+, \lambda^-, r](y)(\nu_{q\mathbb{I}[\phi]}(y))_i\, d\sigma_y
$$

$$
= \delta_{ij} |Q|_n - \int_{\partial q\mathbb{I}[\phi]} v_q^-[\partial q\mathbb{I}[\phi], \mu_j^-](y)(\nu_{q\mathbb{I}[\phi]}(y))_i\, d\sigma_y
$$

$$
+ \int_{\partial q\mathbb{I}[\phi]} (\nu_{q\mathbb{I}[\phi]}(y))_i\, d\sigma_y \fint_{\partial q\mathbb{I}[\phi]} v_q^-[\partial q\mathbb{I}[\phi], \mu_j^-](z)\, d\sigma_z
$$

$$
+ \int_{\partial q\mathbb{I}[\phi]} (\nu_{q\mathbb{I}[\phi]}(y))_i\, d\sigma_y \fint_{\partial q\mathbb{I}[\phi]} z_j\, d\sigma_z - \delta_{ij} |q\mathbb{I}[\phi]|_n.
$$

Indeed,

$$\int_{\partial Q}\left(v_q^-[\partial q\mathbb{I}[\phi],\mu_j^-](y) - \fint_{\partial q\mathbb{I}[\phi]} v_q^-[\partial q\mathbb{I}[\phi],\mu_j^-](z)\,d\sigma_z\right.$$
$$\left.- \fint_{\partial q\mathbb{I}[\phi]} z_j\,d\sigma_z + y_j\right)(v_Q(y))_i\,d\sigma_y$$
$$= \int_{\partial Q} y_j(v_Q(y))_i\,d\sigma_y = \delta_{ij}|Q|_n.$$

Also, by the divergence theorem, we have

$$\int_{\partial q\mathbb{I}[\phi]} (v_{q\mathbb{I}[\phi]}(y))_i\,d\sigma_y = 0 \quad \forall i \in \{1,\dots,n\}.$$

Accordingly, by a straightforward computation we have

$$\lambda_{ij}^{\text{eff,nonid}}[q,\phi,\lambda^+,\lambda^-,r]$$
$$= \frac{1}{|Q|_n}\left\{\lambda^+ \int_{q\mathbb{I}[\phi]} \frac{\partial}{\partial x_i} u_j^+[q,\phi,\lambda^+,\lambda^-,r](x)\,dx\right.$$
$$\left.+ \lambda^- \int_{Q\backslash\overline{q\mathbb{I}[\phi]}} \frac{\partial}{\partial x_i} u_j^-[q,\phi,\lambda^+,\lambda^-,r](x)\,dx\right\}$$
$$= \frac{1}{|Q|_n}\left\{\delta_{ij}\lambda^-|Q|_n + (\lambda^+ - \lambda^-)\delta_{ij}|q\mathbb{I}[\phi]|_n\right.$$
$$+ \lambda^+ \int_{\partial q\mathbb{I}[\phi]} v_q[\partial q\mathbb{I}[\phi],\mu_j^+](y)(v_{q\mathbb{I}[\phi]}(y))_i\,d\sigma_y$$
$$\left.- \lambda^- \int_{\partial q\mathbb{I}[\phi]} v_q[\partial q\mathbb{I}[\phi],\mu_j^-](y)(v_{q\mathbb{I}[\phi]}(y))_i\,d\sigma_y\right\}. \tag{14.12}$$

14.6.3 Analyticity of the solution of the integral equation

Equality (14.12) suggests that the next step in order to study the dependence of the effective conductivity $\lambda_{ij}^{\text{eff,nonid}}[q,\phi,\lambda^+,\lambda^-,r]$ upon the quintuple $(q,\phi,\lambda^+,\lambda^-,r)$ is to analyze the dependence of the solutions μ_j^+,μ_j^- of Eq. (14.11). Before starting with this plan, we note that Eq. (14.11) is defined on the (q,ϕ)-dependent domain $\partial q\mathbb{I}[\phi]$, while a formulation on a fixed domain would be easier to analyze. Thus, we first provide a reformulation on a fixed domain.

292 Mechanics and Physics of Structured Media

Lemma 14.6.7. *Let q be as in (14.1), Q be as in (14.2), and α, Ω be as in (14.3). Let $(\lambda^+, \lambda^-, r) \in \mathcal{P}$. Let $\phi \in C^{1,\alpha}(\partial\Omega, \mathbb{R}^n) \cap \mathcal{A}_{\partial\Omega}^{\widetilde{Q}}$. Let $j \in \{1, \ldots, n\}$. Then the pair $(\theta_j^+, \theta_j^-) \in (C^{0,\alpha}(\partial\Omega))^2$ solves the system of equations*

$$\lambda^+\Bigg(-\frac{1}{2}\theta_j^+(t) + \int_{q\phi(\partial\Omega)} DS_{q,n}(q\phi(t) - s)$$

$$\cdot v_{q\mathbb{I}[\phi]}(q\phi(t))(\theta_j^+ \circ \phi^{(-1)})(q^{-1}s)d\sigma_s\Bigg)$$

$$-\lambda^-\Bigg(\frac{1}{2}\theta_j^-(t) + \int_{q\phi(\partial\Omega)} DS_{q,n}(q\phi(t) - s)$$

$$\cdot v_{q\mathbb{I}[\phi]}(q\phi(t))(\theta_j^- \circ \phi^{(-1)})(q^{-1}s)d\sigma_s\Bigg)$$

$$= (\lambda^- - \lambda^+)(v_{q\mathbb{I}[\phi]}(q\phi(t)))_j \quad \forall t \in \partial\Omega, \tag{14.13}$$

$$\lambda^+\Bigg(-\frac{1}{2}\theta_j^+(t) + \int_{q\phi(\partial\Omega)} DS_{q,n}(q\phi(t) - s)$$

$$\cdot v_{q\mathbb{I}[\phi]}(q\phi(t))(\theta_j^+ \circ \phi^{(-1)})(q^{-1}s)d\sigma_s\Bigg)$$

$$+ r\Bigg(\int_{q\phi(\partial\Omega)} S_{q,n}(q\phi(t) - s)\left(\theta_j^+ \circ \phi^{(-1)}\right)(q^{-1}s)d\sigma_s$$

$$- \frac{1}{|\partial q\mathbb{I}[\phi]|_{n-1}}\int_{q\phi(\partial\Omega)}\int_{q\phi(\partial\Omega)} S_{q,n}(y - s)\left(\theta_j^+ \circ \phi^{(-1)}\right)(q^{-1}s)d\sigma_s\, d\sigma_y$$

$$- \int_{q\phi(\partial\Omega)} S_{q,n}(q\phi(t) - s)\left(\theta_j^- \circ \phi^{(-1)}\right)(q^{-1}s)d\sigma_s$$

$$+ \frac{1}{|\partial q\mathbb{I}[\phi]|_{n-1}}\int_{q\phi(\partial\Omega)}\int_{q\phi(\partial\Omega)} S_{q,n}(y - s)\left(\theta_j^- \circ \phi^{(-1)}\right)(q^{-1}s)d\sigma_s\, d\sigma_y\Bigg)$$

$$= -\lambda^+(v_{q\mathbb{I}[\phi]}(q\phi(t)))_j \quad \forall t \in \partial\Omega, \tag{14.14}$$

if and only if the pair $(\mu_j^+, \mu_j^-) \in (C^{0,\alpha}(\partial q\mathbb{I}[\phi]))^2$ defined by

$$\mu_j^{\pm}(x) \equiv (\theta_j^{\pm} \circ \phi^{(-1)})(q^{-1}x) \quad \forall x \in \partial q\mathbb{I}[\phi] \tag{14.15}$$

solves (14.11). Moreover, system (14.13)–(14.14) has a unique solution in $(C^{0,\alpha}(\partial\Omega))^2$.

Proof. The equivalence of Eqs. (14.13)–(14.14) in the unknown (θ_j^+, θ_j^-) and Eq. (14.11) in the unknown (μ_j^+, μ_j^-), with (μ_j^+, μ_j^-) delivered by (14.15), is a straightforward consequence of a change of variables. Then the existence and uniqueness of a solution of Eqs. (14.13)–(14.14) in $(C^{0,\alpha}(\partial\Omega))^2$ follows from Theorem 14.6.5 and from the equivalence of Eqs. (14.11) and (14.13)–(14.14). $\square$

Dependence of effective properties upon regular perturbations **Chapter | 14 293**

Inspired by Lemma 14.6.7, for all $j \in \{1, \ldots, n\}$ we introduce the map

$$M_j \equiv (M_{j,1}, M_{j,2}) : \mathbb{D}_n^+(\mathbb{R}) \times \left(C^{1,\alpha}(\partial\Omega, \mathbb{R}^n) \cap \mathcal{A}_{\partial\Omega}^{\tilde{Q}} \right) \times \mathbb{R}^3 \times (C^{0,\alpha}(\partial\Omega))^2$$
$$\to (C^{0,\alpha}(\partial\Omega))^2$$

by setting

$$M_{j,1}[q, \phi, \lambda^+, \lambda^-, r, \theta^+, \theta^-](t)$$
$$\equiv \lambda^+ \left(-\frac{1}{2}\theta^+(t) \right.$$
$$\left. + \int_{q\phi(\partial\Omega)} DS_{q,n}(q\phi(t) - s) \cdot v_{q\mathbb{I}[\phi]}(q\phi(t))(\theta^+ \circ \phi^{(-1)})(q^{-1}s)d\sigma_s \right)$$
$$- \lambda^- \left(\frac{1}{2}\theta^-(t) \right.$$
$$\left. + \int_{q\phi(\partial\Omega)} DS_{q,n}(q\phi(t) - s) \cdot v_{q\mathbb{I}[\phi]}(q\phi(t))(\theta^- \circ \phi^{(-1)})(q^{-1}s)d\sigma_s \right)$$
$$- (\lambda^- - \lambda^+)(v_{q\mathbb{I}[\phi]}(q\phi(t)))_j \quad \forall t \in \partial\Omega,$$

$$M_{j,2}[q, \phi, \lambda^+, \lambda^-, r, \theta^+, \theta^-](t)$$
$$\equiv \lambda^+ \left(-\frac{1}{2}\theta^+(t) \right.$$
$$\left. + \int_{q\phi(\partial\Omega)} DS_{q,n}(q\phi(t) - s) \cdot v_{q\mathbb{I}[\phi]}(q\phi(t))(\theta^+ \circ \phi^{(-1)})(q^{-1}s)d\sigma_s \right)$$
$$+ r \left(\int_{q\phi(\partial\Omega)} S_{q,n}(q\phi(t) - s) \left(\theta^+ \circ \phi^{(-1)} \right)(q^{-1}s)d\sigma_s \right.$$
$$- \frac{1}{|\partial q\mathbb{I}[\phi]|_{n-1}} \int_{q\phi(\partial\Omega)} \int_{q\phi(\partial\Omega)} S_{q,n}(y - s) \left(\theta^+ \circ \phi^{(-1)} \right)(q^{-1}s)d\sigma_s \, d\sigma_y$$
$$- \int_{q\phi(\partial\Omega)} S_{q,n}(q\phi(t) - s) \left(\theta^- \circ \phi^{(-1)} \right)(q^{-1}s)d\sigma_s$$
$$\left. + \frac{1}{|\partial q\mathbb{I}[\phi]|_{n-1}} \int_{q\phi(\partial\Omega)} \int_{q\phi(\partial\Omega)} S_{q,n}(y - s) \left(\theta^- \circ \phi^{(-1)} \right)(q^{-1}s)d\sigma_s \, d\sigma_y \right)$$
$$+ \lambda^+ (v_{q\mathbb{I}[\phi]}(q\phi(t)))_j \quad \forall t \in \partial\Omega,$$

for all $(q, \phi, \lambda^+, \lambda^-, r, \theta^+, \theta^-) \in \mathbb{D}_n^+(\mathbb{R}) \times \left(C^{1,\alpha}(\partial\Omega, \mathbb{R}^n) \cap \mathcal{A}_{\partial\Omega}^{\tilde{Q}} \right) \times \mathbb{R}^3 \times$ $(C^{0,\alpha}(\partial\Omega))^2$. As one can readily verify, under the assumptions of Lemma 14.6.7, the system (14.13)–(14.14) can be rewritten as

$$M_j \left[q, \phi, \lambda^+, \lambda^-, r, \theta^+, \theta^- \right] = 0. \tag{14.16}$$

Our aim is to describe the dependence of the pair (θ^+, θ^-) that solves Eq. (14.16) on the periodicity matrix q, the inclusions' shape ϕ, and the parameters λ^+, λ^-, r. To do so, we plan to apply the implicit function theorem

294 Mechanics and Physics of Structured Media

for real analytic maps in Banach spaces to Eq. (14.16). So, as a first step we need to prove that M_j is real analytic. We start with some technical results on the analyticity of certain operators involved in the definition of M_j. The first one concerns integral operators associated with the single layer potential and its normal derivative and shows their analytical dependence upon the periodicity matrix q and the shape ϕ. For a proof we refer to [44].

Lemma 14.6.8. *Let α, Ω be as in (14.3). Then the following statements hold.*

(i) *The map from $\mathbb{D}_n^+(\mathbb{R}) \times \left(C^{1,\alpha}(\partial\Omega, \mathbb{R}^n) \cap \mathcal{A}_{\partial\Omega}^{\widetilde{Q}} \right) \times C^{0,\alpha}(\partial\Omega)$ to $C^{1,\alpha}(\partial\Omega)$ that takes a triple (q, ϕ, θ) to the function $V[q, \phi, \theta]$ defined by*

$$V[q, \phi, \theta](t) \equiv \int_{q\phi(\partial\Omega)} S_{q,n}(q\phi(t) - s)\left(\theta \circ \phi^{(-1)}\right)(q^{-1}s)d\sigma_s \quad \forall t \in \partial\Omega,$$

is real analytic.

(ii) *The map from $\mathbb{D}_n^+(\mathbb{R}) \times \left(C^{1,\alpha}(\partial\Omega, \mathbb{R}^n) \cap \mathcal{A}_{\partial\Omega}^{\widetilde{Q}} \right) \times C^{0,\alpha}(\partial\Omega)$ to $C^{0,\alpha}(\partial\Omega)$ that takes a triple (q, ϕ, θ) to the function $W_*[q, \phi, \theta]$ defined by*

$$W_*[q, \phi, \theta](t)$$
$$\equiv \int_{q\phi(\partial\Omega)} DS_{q,n}(q\phi(t) - s) \cdot v_{q\mathbb{I}[\phi]}(q\phi(t))\left(\theta \circ \phi^{(-1)}\right)\ (q^{-1}s)d\sigma_s$$
$$\forall t \in \partial\Omega,$$

is real analytic.

Next, we need the following lemma about the real analytic dependence of certain maps related to the change of variables in integrals and to the pullback of the outer normal field. For a proof we refer to Lanza de Cristoforis and Rossi [37, p. 166] and to Lanza de Cristoforis [35, Prop. 1].

Lemma 14.6.9. *Let α, Ω be as in (14.3). Then the following statements hold.*

(i) *For each $\psi \in C^{1,\alpha}(\partial\Omega, \mathbb{R}^n) \cap \mathcal{A}_{\partial\Omega}$, there exists a unique $\tilde{\sigma}[\psi] \in C^{0,\alpha}(\partial\Omega)$ such that $\tilde{\sigma}[\psi] > 0$ and*

$$\int_{\psi(\partial\Omega)} w(s)\, d\sigma_s = \int_{\partial\Omega} w \circ \psi(y)\tilde{\sigma}[\psi](y)\, d\sigma_y, \quad \forall w \in L^1(\psi(\partial\Omega)).$$

Moreover, the map $\tilde{\sigma}[\cdot]$ from $C^{1,\alpha}(\partial\Omega, \mathbb{R}^n) \cap \mathcal{A}_{\partial\Omega}$ to $C^{0,\alpha}(\partial\Omega)$ is real analytic.

(ii) *The map from $C^{1,\alpha}(\partial\Omega, \mathbb{R}^n) \cap \mathcal{A}_{\partial\Omega}$ to $C^{0,\alpha}(\partial\Omega, \mathbb{R}^n)$ that takes ψ to $v_{\mathbb{I}[\psi]} \circ \psi$ is real analytic.*

The last technical result that we need is about the analyticity of certain maps related to the measure of sets.

Dependence of effective properties upon regular perturbations **Chapter | 14** **295**

Lemma 14.6.10. *Let α, Ω be as in* (14.3). *Then the following statements hold.*

(i) *The map from $\mathbb{D}_n^+(\mathbb{R}) \times \left(C^{1,\alpha}(\partial\Omega, \mathbb{R}^n) \cap \mathcal{A}_{\partial\Omega}^{\tilde{Q}} \right)$ to $\mathbb{R}$ that takes (q, ϕ) to $|q\mathbb{I}[\phi]|_n$ is real analytic.*

(ii) *The map from $\mathbb{D}_n^+(\mathbb{R}) \times \left(C^{1,\alpha}(\partial\Omega, \mathbb{R}^n) \cap \mathcal{A}_{\partial\Omega}^{\tilde{Q}} \right)$ to $\mathbb{R}$ that takes (q, ϕ) to $|\partial q\mathbb{I}[\phi]|_{n-1}$ is real analytic.*

Proof. The validity of statement (i) follows by the proof of [41, Thm. 5.1]. In order to prove statement (ii) we notice that

$$|\partial q\mathbb{I}[\phi]|_{n-1} = \int_{q\phi(\partial\Omega)} d\sigma = \int_{\partial\Omega} \tilde{\sigma}[q\phi] d\sigma,$$

for all $(q, \phi) \in \mathbb{D}_n^+(\mathbb{R}) \times \left(C^{1,\alpha}(\partial\Omega, \mathbb{R}^n) \cap \mathcal{A}_{\partial\Omega}^{\tilde{Q}} \right)$. By the analyticity of the map from $\mathbb{D}_n^+(\mathbb{R}) \times \left(C^{1,\alpha}(\partial\Omega, \mathbb{R}^n) \cap \mathcal{A}_{\partial\Omega}^{\tilde{Q}} \right)$ to $\left(C^{1,\alpha}(\partial\Omega, \mathbb{R}^n) \cap \mathcal{A}_{\partial\Omega} \right)$ that takes the pair (q, ϕ) to $q\phi$ and by Lemma 14.6.9 (i), we deduce that

$$\int_{q\phi(\partial\Omega)} d\sigma = \int_{\partial\Omega} \tilde{\sigma}[q\phi] d\sigma$$

depends real analytically on $(q, \phi) \in \mathbb{D}_n^+(\mathbb{R}) \times \left(C^{1,\alpha}(\partial\Omega, \mathbb{R}^n) \cap \mathcal{A}_{\partial\Omega}^{\tilde{Q}} \right)$. As a consequence, the validity of statement (ii) follows. $\quad\square$

By Lemmas 14.6.8, 14.6.9, 14.6.10, and by standard calculus in Banach spaces, we immediately deduce the validity of the following.

Proposition 14.6.11. *Let α, Ω be as in* (14.3). *Let $j \in \{1, \dots, n\}$. The map M_j is real analytic from $\mathbb{D}_n^+(\mathbb{R}) \times \left(C^{1,\alpha}(\partial\Omega, \mathbb{R}^n) \cap \mathcal{A}_{\partial\Omega}^{\tilde{Q}} \right) \times \mathbb{R}^3 \times (C^{0,\alpha}(\partial\Omega))^2$ to $(C^{0,\alpha}(\partial\Omega))^2$.*

We are now ready to prove that the solution of (14.16) depends real analytically upon the quintuple $(q, \phi, \lambda^+, \lambda^-, r)$.

Proposition 14.6.12. *Let α, Ω be as in* (14.3). *Let $j \in \{1, \dots, n\}$. Then the following statements hold.*

(i) *There exists an open neighborhood $\mathcal{V}$ of $\mathbb{D}_n^+(\mathbb{R}) \times \left(C^{1,\alpha}(\partial\Omega, \mathbb{R}^n) \cap \mathcal{A}_{\partial\Omega}^{\tilde{Q}} \right) \times \mathcal{P}$ in $\mathbb{D}_n^+(\mathbb{R}) \times \left(C^{1,\alpha}(\partial\Omega, \mathbb{R}^n) \cap \mathcal{A}_{\partial\Omega}^{\tilde{Q}} \right) \times \mathbb{R}^3$ such that for each $(q, \phi, \lambda^+, \lambda^-, r) \in \mathcal{V}$ the operator*

$$M_j[q, \phi, \lambda^+, \lambda^-, r, \cdot, \cdot]$$

is a linear homeomorphism from $(C^{0,\alpha}(\partial\Omega))^2$ onto $(C^{0,\alpha}(\partial\Omega))^2$. In particular, for each $(q, \phi, \lambda^+, \lambda^-, r) \in \mathcal{V}$ there exists a unique pair (θ_j^+, θ_j^-)

in $(C^{0,\alpha}(\partial\Omega))^2$ such that

$$M_j[q,\phi,\lambda^+,\lambda^-,r,\theta_j^+,\theta_j^-]=0.$$

We denote this pair by $(\theta_j^+[q,\phi,\lambda^+,\lambda^-,r],\theta_j^-[q,\phi,\lambda^+,\lambda^-,r])$.

(ii) *The map from $\mathcal{V}$ to $(C^{0,\alpha}(\partial\Omega))^2$ that takes $(q,\phi,\lambda^+,\lambda^-,r)$ to the pair*

$$(\theta_j^+[q,\phi,\lambda^+,\lambda^-,r],\theta_j^-[q,\phi,\lambda^+,\lambda^-,r])$$

is real analytic

Proof. Statement (i) follows by arguing as in the proof of Lemma 14.6.7 and by the fact that the set of linear homeomorphisms is open in the set of linear and continuous operators. To prove statement (ii) we first note that, by Proposition 14.6.11, M_j is a real analytic map from $\mathbb{D}_n^+(\mathbb{R}) \times \left(C^{1,\alpha}(\partial\Omega,\mathbb{R}^n) \cap \mathcal{A}_{\partial\Omega}^Q\right) \times \mathbb{R}^3 \times (C^{0,\alpha}(\partial\Omega))^2$ to $(C^{0,\alpha}(\partial\Omega))^2$. Moreover, for all $(q,\phi,\lambda^+,\lambda^-,r) \in \mathcal{V}$, the partial differential

$$\partial_{(\theta^+,\theta^-)}M_j\left[q,\phi,\lambda^+,\lambda^-,r,\theta^+[q,\phi,\lambda^+,\lambda^-,r],\theta^-[q,\phi,\lambda^+,\lambda^-,r]\right]$$

of M_j at the point

$$(q,\phi,\lambda^+,\lambda^-,r,\theta^+[q,\phi,\lambda^+,\lambda^-,r],\theta^-[q,\phi,\lambda^+,\lambda^-,r])$$

with respect to the variable (θ^+,θ^-) is given by

$$\partial_{(\theta^+,\theta^-)}M_j\left[q,\phi,\lambda^+,\lambda^-,r,\theta^+[q,\phi,\lambda^+,\lambda^-,r],\theta^-[q,\phi,\lambda^+,\lambda^-,r]\right](\psi^+,\psi^-)$$
$$= M_j\left[q,\phi,\lambda^+,\lambda^-,r,\psi^+,\psi^-\right],$$

for all $(\psi^+,\psi^-) \in (C^{0,\alpha}(\partial\Omega))^2$. Accordingly, by statement (i) and by the implicit function theorem for real analytic maps in Banach spaces (see, *e.g.*, Deimling [17, Thm. 15.3]), we deduce the analyticity of the map

$$(q,\phi,\lambda^+,\lambda^-,r) \mapsto (\theta_j^+[q,\phi,\lambda^+,\lambda^-,r],\theta_j^-[q,\phi,\lambda^+,\lambda^-,r])$$

as in the statement. $\qquad\square$

14.6.4 Analyticity of the effective conductivity

We are now ready to prove our main Theorem 14.5.2 for the effective conductivity in the case of nonideal contact conditions. To this aim, we exploit formula (14.12) for $\lambda^{\text{eff,nonid}}$ and the analyticity result of Proposition 14.6.12.

Proof of Theorem 14.5.2. Let (θ_j^+, θ_j^-) and $\mathcal{V}$ be as in Proposition 14.6.12. Then, we set Λ_{ij} to be the map from the $\mathcal{V}$ to $\mathbb{R}$ defined by

$$\Lambda_{ij}[q, \phi, \lambda^+, \lambda^-, r]$$

$$\equiv \frac{1}{|Q|_n} \Bigg\{ \lambda^+ \int_{\partial q\mathbb{I}[\phi]} v_q[\partial q\mathbb{I}[\phi], (\theta_j^+[q, \phi, \lambda^+, \lambda^-, r]$$

$$\circ \phi^{(-1)})(q^{-1}\cdot)](y)(v_{q\mathbb{I}[\phi]}(y))_i \, d\sigma_y$$

$$- \lambda^- \int_{\partial q\mathbb{I}[\phi]} v_q[\partial q\mathbb{I}[\phi], (\theta_j^-[q, \phi, \lambda^+, \lambda^-, r] \circ \phi^{(-1)})(q^{-1}\cdot)](y)(v_{q\mathbb{I}[\phi]}(y))_i \, d\sigma_y$$

$$+ (\lambda^+ - \lambda^-)\delta_{ij} |q\mathbb{I}[\phi]|_n \Bigg\}$$

for all $(q, \phi, \lambda^+, \lambda^-, r) \in \mathcal{V}$. By formula (14.12) for the effective conductivity, by Proposition 14.6.12, by Lemma 14.6.7, and by Theorem 14.6.5, the only thing that remains in order to complete the proof is to show that the map Λ_{ij} is real analytic. Lemma 14.6.9 implies that

$$\Lambda_{ij}[q, \phi, \lambda^+, \lambda^-, r]$$

$$= \frac{1}{|Q|_n} \Bigg\{ \lambda^+ \int_{\partial\Omega} V[q, \phi, \theta_j^+[q, \phi, \lambda^+, \lambda^-, r]](y)(v_{q\mathbb{I}[\phi]}(q\phi(y)))_i \tilde{\sigma}[q\phi](y) \, d\sigma_y$$

$$- \lambda^- \int_{\partial\Omega} V[q, \phi, \theta_j^-[q, \phi, \lambda^+, \lambda^-, r]](y)(v_{q\mathbb{I}[\phi]}(q\phi(y)))_i \tilde{\sigma}[q\phi](y) \, d\sigma_y$$

$$+ (\lambda^+ - \lambda^-)\delta_{ij} |q\mathbb{I}[\phi]|_n \Bigg\}$$

for all $(q, \phi, \lambda^+, \lambda^-, r) \in \mathcal{V}$. Since

$$|Q|_n = \prod_{l=1}^{n} q_{ll} \quad \forall q \in \mathbb{D}_n^+(\mathbb{R}),$$

clearly $|Q|_n$ depends analytically on $q \in \mathbb{D}_n^+(\mathbb{R})$. Lemma 14.6.10 implies that the map from $\mathbb{D}_n^+(\mathbb{R}) \times \left(C^{1,\alpha}(\partial\Omega, \mathbb{R}^n) \cap \mathcal{A}_{\partial\Omega}^{\tilde{Q}} \right)$ to $\mathbb{R}$ that takes (q, ϕ) to $|q\mathbb{I}[\phi]|_n$ is real analytic. Thus, by Proposition 14.6.12, by Lemma 14.6.8 (i), by Lemma 14.6.9, together with the above considerations, we can conclude that the map Λ_{ij} is real analytic from $\mathcal{V}$ to $\mathbb{R}$. Accordingly, the statement of Theorem 14.5.2 holds true. $\square$

14.7 Conclusions

We have presented some of our recent results about the dependence of effective properties upon regular perturbation of the geometric and physical parameters.

We have considered the average flow velocity along a periodic array of cylinders and the effective conductivity of periodic composites, both with ideal and nonideal contact conditions. We have proven that these quantities depend real analytically upon the parameters involved. The method used is based on the so-called *Functional Analytic Approach* proposed by Lanza de Cristoforis for the analysis of regular and singular domain perturbations (cf. [33–35]).

Acknowledgments

M. Dalla Riva, P. Luzzini, and P. Musolino are members of the 'Gruppo Nazionale per l'Analisi Matematica, la Probabilità e le loro Applicazioni' (GNAMPA) of the 'Istituto Nazionale di Alta Matematica' (INdAM). P. Luzzini and P. Musolino acknowledge the support of the Project BIRD191739/19 'Sensitivity analysis of partial differential equations in the mathematical theory of electromagnetism' of the University of Padova. P. Musolino also acknowledges the support of the grant 'Challenges in Asymptotic and Shape Analysis - CASA' of the Ca' Foscari University of Venice. R. Pukhtaievych was supported by the budget program 'Support for the Development of Priority Areas of Scientific Research' (КПКВК 6541230) of the National Academy of Sciences of Ukraine.

References

[1] P.M. Adler, Porous Media: Geometry and Transports, Butterworth/Heinemann, 1992.
[2] H. Ammari, H. Kang, Polarization and Moment Tensors. With Applications to Inverse Problems and Effective Medium Theory, Applied Mathematical Sciences, vol. 162, Springer, New York, 2007.
[3] H. Ammari, H. Kang, K. Kim, Polarization tensors and effective properties of anisotropic composite materials, J. Differ. Equ. 215 (2) (2005) 401–428.
[4] H. Ammari, H. Kang, M. Lim, Effective parameters of elastic composites, Indiana Univ. Math. J. 55 (3) (2006) 903–922.
[5] H. Ammari, H. Kang, K. Touibi, Boundary layer techniques for deriving the effective properties of composite materials, Asympt. Anal. 41 (2) (2005) 119–140.
[6] Y. Benveniste, Effective thermal conductivity of composites with a thermal contact resistance between the constituents: nondilute case, J. Appl. Phys. 61 (1987) 2840–2844.
[7] Y. Benveniste, T. Miloh, The effective conductivity of composites with imperfect thermal contact at constituent interfaces, Int. J. Eng. Sci. 24 (1986) 1537–1552.
[8] L. Berlyand, D. Golovaty, A. Movchan, J. Phillips, Transport properties of densely packed composites. Effect of shapes and spacings of inclusions, Q. J. Mech. Appl. Math. 57 (4) (2004) 495–528.
[9] L. Berlyand, V. Mityushev, Increase and decrease of the effective conductivity of two phase composites due to polydispersity, J. Stat. Phys. 118 (3–4) (2005) 481–509.
[10] L.P. Castro, D. Kapanadze, E. Pesetskaya, A heat conduction problem of 2D unbounded composites with imperfect contact conditions, Z. Angew. Math. Mech. 95 (9) (2015) 952–965.
[11] L.P. Castro, D. Kapanadze, E. Pesetskaya, Effective conductivity of a composite material with stiff imperfect contact conditions, Math. Methods Appl. Sci. 38 (18) (2015) 4638–4649.
[12] L.P. Castro, E. Pesetskaya, A composite material with inextensible-membrane-type interface, Math. Mech. Solids 24 (2) (2019) 499–510.
[13] M. Dalla Riva, Stokes flow in a singularly perturbed exterior domain, Complex Var. Elliptic Equ. 58 (2) (2013) 231–257.
[14] M. Dalla Riva, M. Lanza de Cristoforis, Weakly singular and microscopically hypersingular load perturbation for a nonlinear traction boundary value problem: a functional analytic approach, Complex Anal. Oper. Theory 5 (3) (2011) 811–833.

Dependence of effective properties upon regular perturbations Chapter | 14 **299**

[15] M. Dalla Riva, P. Musolino, A singularly perturbed nonideal transmission problem and application to the effective conductivity of a periodic composite, SIAM J. Appl. Math. 73 (1) (2013) 24–46.

[16] M. Dalla Riva, P. Musolino, R. Pukhtaievych, Series expansion for the effective conductivity of a periodic dilute composite with thermal resistance at the two-phase interface, Asymptot. Anal. 111 (3–4) (2019) 217–250.

[17] K. Deimling, Nonlinear Functional Analysis, Springer-Verlag, Berlin, 1985.

[18] P. Donato, S. Monsurrò, Homogenization of two heat conductors with an interfacial contact resistance, Anal. Appl. (Singap.) 2 (2004) 247–273.

[19] P. Drygaś, S. Gluzman, V. Mityushev, W. Nawalaniec, Applied Analysis of Composite Media. Analytical and Computational Results for Materials Scientists and Engineers, Woodhead Publishing Series in Composites Science and Engineering, Woodhead Publishing, 2020.

[20] P. Drygaś, V. Mityushev, Effective conductivity of unidirectional cylinders with interfacial resistance, Q. J. Mech. Appl. Math. 62 (3) (2009) 235–262.

[21] L. Faella, S. Monsurrò, C. Perugia, Homogenization of imperfect transmission problems: the case of weakly converging data, Differ. Integral Equ. 31 (2018) 595–620.

[22] L. Fil'shtinskii, Stresses and displacements in an elastic sheet weakened by a doubly-periodic set of equal circular holes, J. Appl. Math. Mech. 28 (1964) 530–543.

[23] L. Filshtinsky, V. Mityushev, Mathematical models of elastic and piezoelectric fields in two-dimensional composites, in: Mathematics Without Boundaries, Springer, New York, 2014, pp. 217–262.

[24] D. Gilbarg, N.S. Trudinger, Elliptic Partial Differential Equations of Second Order, 2nd edition, Grundlehren der Mathematischen Wissenschaften (Fundamental Principles of Mathematical Sciences), vol. 224, Springer-Verlag, Berlin, 1983.

[25] S. Gluzman, V. Mityushev, W. Nawalaniec, Computational Analysis of Structured Media. Mathematical Analysis and Its Applications, Academic Press, London, 2018.

[26] Y. Gorb, L. Berlyand, Asymptotics of the effective conductivity of composites with closely spaced inclusions of optimal shape, Q. J. Mech. Appl. Math. 58 (1) (2005) 84–106.

[27] E. Grigolyuk, L. Fil'shtinskij, Periodic Piecewise Homogeneous Elastic Structures, Nauka, Moskva, 1992 (in Russian).

[28] S. Gryshchuk, S. Rogosin, Effective conductivity of 2D disk-ring composite material, Math. Model. Anal. 18 (3) (2013) 386–394.

[29] H. Hasimoto, On the periodic fundamental solutions of the Stokes' equations and their application to viscous flow past a cubic array of spheres, J. Fluid Mech. 5 (1959) 317–328.

[30] D. Kapanadze, G. Mishuris, E. Pesetskaya, Improved algorithm for analytical solution of the heat conduction problem in doubly periodic 2D composite materials, Complex Var. Elliptic Equ. 60 (1) (2015) 1–23.

[31] S.M. Kozlov, Geometric aspects of averaging, Usp. Mat. Nauk 44 (2) (1989) 79–120; (in Russian); translation in Russ. Math. Surv. 44 (2) (1989) 91–144.

[32] P. Kurtyka, N. Rylko, Quantitative analysis of the particles distributions in reinforced composites, Compos. Struct. 182 (2017) 412–419.

[33] M. Lanza de Cristoforis, Asymptotic behavior of the conformal representation of a Jordan domain with a small hole in Schauder spaces, Comput. Methods Funct. Theory 2 (1) (2002) 1–27.

[34] M. Lanza de Cristoforis, A domain perturbation problem for the Poisson equation, Complex Var. Theory Appl. 50 (7–11) (2005) 851–867.

[35] M. Lanza de Cristoforis, Perturbation problems in potential theory, a functional analytic approach, J. Appl. Funct. Anal. 2 (3) (2007) 197–222.

[36] M. Lanza de Cristoforis, P. Musolino, A perturbation result for periodic layer potentials of general second order differential operators with constant coefficients, Far East J. Math. Sci.: FJMS 52 (1) (2011) 75–120.

[37] M. Lanza de Cristoforis, L. Rossi, Real analytic dependence of simple and double layer potentials upon perturbation of the support and of the density, J. Integral Equ. Appl. 16 (2) (2004) 137–174.

300 Mechanics and Physics of Structured Media

[38] M. Lanza de Cristoforis, L. Rossi, Real analytic dependence of simple and double layer potentials for the Helmholtz equation upon perturbation of the support and of the density, in: Analytic Methods of Analysis and Differential Equations: AMADE 2006, Camb. Sci. Publ., Cambridge, 2008, pp. 193–220.

[39] H. Lee, J. Lee, Array dependence of effective parameters of dilute periodic elastic composite, in: Imaging, Multi-Scale and High Contrast Partial Differential Equations, in: Contemp. Math., vol. 660, Amer. Math. Soc., Providence, RI, 2016, pp. 59–71.

[40] R. Lipton, Heat conduction in fine scale mixtures with interfacial contact resistance, SIAM J. Appl. Math. 58 (1998) 55–72.

[41] P. Luzzini, P. Musolino, Perturbation analysis of the effective conductivity of a periodic composite, Netw. Heterog. Media 15 (4) (2020) 581–603.

[42] P. Luzzini, P. Musolino, Domain perturbation for the solution of a periodic Dirichlet problem, in: Proceedings of the 12th ISAAC congress (Aveiro, 2019), Research Perspectives, Birkhäuser, in press.

[43] P. Luzzini, P. Musolino, R. Pukhtaievych, Shape analysis of the longitudinal flow along a periodic array of cylinders, J. Math. Anal. Appl. 477 (2) (2019) 1369–1395.

[44] P. Luzzini, P. Musolino, R. Pukhtaievych, Real analyticity of periodic layer potentials upon perturbation of the periodicity parameters and of the support, in: Proceedings of the 12th ISAAC congress (Aveiro, 2019), Research Perspectives, Birkhäuser, in press.

[45] V. Mityushev, Functional equations in a class of analytic functions and composite materials, Demonstr. Math. 30 (1) (1997) 63–70.

[46] V. Mityushev, P.M. Adler, Longitudinal permeability of spatially periodic rectangular arrays of circular cylinders. I. A single cylinder in the unit cell, Z. Angew. Math. Mech. 82 (5) (2002) 335–345.

[47] V. Mityushev, P.M. Adler, Longitudinal permeability of spatially periodic rectangular arrays of circular cylinders. II. An arbitrary distribution of cylinders inside the unit cell, Z. Angew. Math. Phys. 53 (3) (2002) 486–517.

[48] V. Mityushev, W. Nawalaniec, D. Nosov, E. Pesetskaya, Schwarz's alternating method in a matrix form and its applications to composites, Appl. Math. Comput. 356 (2019) 144–156.

[49] V. Mityushev, Yu. Obnosov, E. Pesetskaya, S. Rogosin, Analytical methods for heat conduction in composites, Math. Model. Anal. 13 (1) (2008) 67–78.

[50] V. Mityushev, N. Rylko, Optimal distribution of the nonoverlapping conducting disks, Multiscale Model. Simul. 10 (1) (2012) 180–190.

[51] S. Monsurrò, Homogenization of a two-component composite with interfacial thermal barrier, Adv. Math. Sci. Appl. 13 (2003) 43–63.

[52] S. Monsurrò, Erratum for the paper: "Homogenization of a two-component composite with interfacial thermal barrier" [Adv. Math. Sci. Appl. 13 (2003) 43–63], Adv. Math. Sci. Appl. 14 (2004) 375–377.

[53] P. Musolino, V. Mityushev, Asymptotic behavior of the longitudinal permeability of a periodic array of thin cylinders, Electron. J. Differ. Equ. (290) (2015) 20.

[54] R. Pukhtaievych, Asymptotic behavior of the solution of singularly perturbed transmission problems in a periodic domain, Math. Methods Appl. Sci. 41 (9) (2018) 3392–3413.

[55] R. Pukhtaievych, Effective conductivity of a periodic dilute composite with perfect contact and its series expansion, Z. Angew. Math. Phys. 69 (3) (2018) 83, 22 pp.

[56] N. Rylko, Dipole matrix for the 2D inclusions close to circular, Z. Angew. Math. Mech. 88 (12) (2008) 993–999.

[57] S.A. Sangani, C. Yao, Transport processes in random arrays of cylinders. I. Thermal conduction, Phys. Fluids 31 (1988) 2426–2434.

[58] S.A. Sangan, C. Yao, Transport properties in random arrays of cylinders. II. Viscous flow, Phys. Fluids 31 (1988) 2435–2444.

[59] J. Schmid, Longitudinal laminar flow in an array of circular cylinders, Int. J. Heat Mass Transf. 9 (9) (1966) 925–937.

[60] K. Schmidt, R. Hiptmair, Asymptotic expansion techniques for singularly perturbed boundary integral equations, Numer. Math. 137 (2) (2017) 397–415.

[61] E.T. Swartz, R.O. Pohl, Thermal boundary resistance, Rev. Mod. Phys. 61 (1989) 605–668.

Chapter 15

Riemann-Hilbert problems with coefficients in compact Lie groups

Gia Giorgadze[a] and Giorgi Khimshiashvili[b]

[a] *Faculty of Exact and Natural Sciences, Tbilisi State University, Tbilisi, Georgia,*
[b] *Institute of Fundamental and Interdisciplinary Mathematical Research, Ilia State University, Tbilisi, Georgia*

15.1 Introduction

We discuss several generalizations of two classical problems traditionally referred to as Riemann-Hilbert problems (RHPs). The first one, the so-called Hilbert's 21st problem, is concerned with the monodromy of Fuchsian systems of ordinary differential equations on the Riemann sphere [4], while the second one, usually called Hilbert's boundary value problem, is concerned with piecewise holomorphic functions satisfying the so-called transmission (or linear conjugation) condition on a contour in the Riemann sphere [32]. In the context of our presentation it is convenient to refer to these two classical problems as Riemann-Hilbert monodromy problem (RHMP) and Riemann-Hilbert transmission problem (RHTP).

The generalizations discussed in the sequel are formulated in terms of representations of compact Lie groups and loop groups [36]. The key ingredients of our approach are a recent generalization of Birkhoff factorization in terms of loop groups of compact Lie groups [36] and interpretation of the monodromy problem in terms of principal bundles on Riemann surfaces [38], [14]. We also describe a natural connection between these two problems in group-theoretic setting analogous to Plemelj's classical reduction of monodromy problem to a boundary value problem for piecewise holomorphic vector-functions [35]. Thus Birkhoff factorization in loop groups becomes a crucial tool for solving both these problems on the Riemann sphere and suggests various generalizations of RHTP with coefficients in compact Lie groups. This line of development is described in the next two sections of the chapter. We also present two applications of generalized Birkhoff factorization in mathematical physics given in [6] and [29].

Mechanics and Physics of Structured Media. https://doi.org/10.1016/B978-0-32-390543-5.00020-7
Copyright © 2022 Elsevier Inc. All rights reserved.

304 Mechanics and Physics of Structured Media

The rest of the chapter is devoted to RHMP for differential G-systems on Riemann surfaces. An adequate language for such developments is provided by meromorphic connections in principal bundles. We recall necessary algebraic and topological tools in Section 15.4. In Section 15.5 we present a number of results on solvability of RHMP on Riemann surfaces of genus at least two.

The generalizations of RHPs discussed below have many applications in geometry and mathematical physics, see, e.g. [7], [8], [16], [19], [25], [29], [36]. It should also be mentioned that in last two decades there appeared further important generalizations of Birkhoff factorization, in particular, for loop groups of certain noncompact Lie groups [1], [9], double loop groups [28], and in terms of Hopf algebra approach to renormalization of quantum field theory developed by A. Connes and D. Kreimer [13]. These topics go far beyond the aims and scope of the present chapter and an interested reader should consult the sources listed above.

As was already mentioned, a general paradigm for our considerations is provided by an approach to Hilbert's 21st problem developed in the classical treatise of J. Plemelj [35]. To clarify the concepts and ideas playing a central role in the sequel we begin with necessary recollections from [35] and related classical papers [32], [20], [40].

15.2 Recollections on classical Riemann-Hilbert problems

To clarify the context and paradigm accepted below, let us introduce some notation and recall a few necessary concepts and results concerned with the classical Riemann-Hilbert problems.

Recall the original formulation of Hilbert's 21st problem in modern terms. One considers a system of linear ordinary differential equations (ODE) for vector functions in the complex plane, having the form

$$\frac{dY}{dz} = A(z)Y(z), \qquad (15.1)$$

where $Y(z)$ is a vector-function (n-column) and $A(z) \in \mathbf{C}^{n \times n}$ is a matrix-function in $\mathbf{C}$. Denote by $\mathbf{CP}^1 = \overline{\mathbf{C}}$ the Riemann sphere and by $X_m = \mathbf{CP}^1 \setminus \{s_1, \dots, s_m\}$ the punctured Riemann sphere, where $s_1, \dots, s_m$ are m distinct points in $\mathbf{CP}^1$ and $S = \{s_1, \dots, s_m\}$. Assume that matrix function $A(z)$ is holomorphic in X_m. Then by the Picard-Lindelöf theorem any fundamental (matrix) solution $F(z)$ of (15.1) is holomorphic in simple connected domain $U \subset X_m$ containing $z_0 \in U$, $z_0 \bar{\in} X_m$, and analytic continuation along a simple loop encircling any point s_j yields an invertible monodromy matrix M_j defined up to conjugation in $GL(n, \mathbf{C})$ which only depends on the homotopy class of loop in the fundamental group $\pi_1(X_m, z_0)$ [4]. Thus for any point $z_0 \in X_m$, the group generated by monodromy matrices M_j, $j = 1, \dots, m$ in $GL(n, \mathbf{C})$ is a homomorphic image of the fundamental group $\pi_1(X_m, z_0)$. These observations suggest to

Riemann-Hilbert problems with coefficients in compact Lie groups Chapter | 15 **305**

define the monodromy group of Eq. (15.1) as an element of the space

$$\mathcal{M} = Hom(\pi_1(S, z_0), GL(n, \mathbf{C}))/GL(n, \mathbf{C}) \tag{15.2}$$

of conjugacy classes of representations of $\pi_1(X_m, z_0)$ (see, e.g., [4]).

Recall that system (15.1) is called Fuchsian if all of its singular points $S = \{s_1, \ldots, s_m\}$ are first order poles of $A(z)$. Without loss of generality the matrix of such a system can be written in the form

$$A(z) = \sum_{i=1}^{m} \frac{R_i}{z - s_i}, \quad \sum_{i=1}^{m} R_i = 0, \tag{15.3}$$

where $R_i \in \mathbf{C}^{n \times n}$ are residues of $A(z)$ at s_i and $s_j \neq \infty$, $j = 1, ..., m$.

The Hilbert's 21st problem in classical formulation is a question whether for any element M in the space $\mathcal{M}$ there exists a Fuchsian system with fixed singularities at $\{s_1, \ldots, s_m\}$ such that its monodromy coincides with M. The same question can be asked for the class of systems having the so-called regular singular points at s_i, i.e. such that their solutions have polynomial growth at each singular point s_i [4]. Both versions of this problem will be referred to as RHMP (Fuchsian or regular). A more general and detailed discussion of these concepts is given in Section 15.5, where they are considered in the context of differential G-systems.

Fundamental results on RHMP have been obtained by J. Plemelj [35]. In particular, J. Plemelj obtained a positive solution to RHMP in the class of regular systems and presented an argument which was supposed to yield a positive answer to RHMP in the class of Fuchsian systems. However the latter argument contained a serious flaw which was only noticed much later and which appeared irreparable. Namely, A. Bolibruch has constructed relevant counter-examples and gave a criterion of solvability of RHMP in the class of Fuchsian systems on $\mathbf{CP}^1$ [12], [4].

The approach and results developed in [35] appeared seminal and influential. In particular, Plemelj established important connection of RHMP with transmission problems for piecewise holomorphic functions which paved the way for many achievements in complex analysis, topology, and mathematical physics (see, e.g., [29]). We now recall the main ingredients of this connection since one of our aims is to place it in the context of Lie groups.

The main idea of Plemelj's solution was to reduce RHMP to a certain boundary value problem for holomorphic functions. To this end let us choose a simple closed oriented contour Γ which contains all given points s_i in the given order. Next, define a piecewise constant matrix-valued invertible function $g(t)$ as follows. For $i = 1, \ldots, m, t \in [s_i, s_{i+1})$, put

$$g(t) = (M_i M_{i-1} \cdot \ldots \cdot M_1)^{-1}, \tag{15.4}$$

306 Mechanics and Physics of Structured Media

where M_i are monodromy matrices of the system (15.1) and it is assumed that $s_{m+1} = s_1$. Notice that from the cocycle property of monodromy matrices follows that $g(t) = Id$ for $t \in [s_m, s_1)$.

Denote by U_+ the inner domain of Γ and by U_- the complement of closure $\overline{U_+}$ in $\mathbf{CP}^1$. In this setting, Plemelj introduced and investigated several boundary value problems for piecewise holomorphic functions of which the most relevant for our purposes is the following problem called in [35] the accompanying Hilbert boundary value problem.

This problem requires to find all vector-valued (n-column) functions $\Phi(z)$ such that $\Phi_\pm(z)$ is holomorphic in $U_\pm$, extends Hölder continuously to Γ from both sides, its boundary values satisfy the boundary condition

$$\Phi_+(t) = \Phi_-(t)(g(t))^{-1}, \tag{15.5}$$

for any $t \in \Gamma$, and Φ_- vanishes at infinity. Later on, the boundary condition (15.5) was called the *linear conjugation condition* [32], or *transmission condition* [29], and acquired many generalizations going under those titles. For this reason the "accompanying Hilbert boundary value problem" is nowadays called the Riemann-Hilbert linear conjugation problem (RHLCP) or the Riemann-Hilbert transmission problem (RHTP). Specifically, the second term is often used for various nonlinear generalizations [29]. In our context it is also natural and convenient to use the second term, RHTP.

Omitting further analytic details of Plemelj's reduction we wish to emphasize two important circumstances. Firstly, a solution to the above RHTP can be obtained using the Birkhoff factorization theorem [32], [36]. Secondly, RHTP naturally leads to construction of a holomorphic vector bundle on the Riemann sphere with the transition function defined by an extension of matrix-function $g(t)$ to a neighborhood of contour Γ [4]. One can also consider a principal $GL(n, \mathbf{C})$-bundle associated with the transmission coefficient $g(t)$. The latter interpretation of RHTP appeared extremely useful in many algebraic and topological problems. In particular, applying the Birkhoff factorization theorem to arising holomorphic bundle A. Grothendieck [27] obtained a complete classification of holomorphic bundles on the Riemann sphere. An analogous classification appeared possible for principal $G_\mathbf{C}$-bundles where $G_\mathbf{C}$ is the complexification of a compact Lie group G [15]. Moreover, Plemelj's reduction of RHMP to RHTP suggested that the language of principal bundles is relevant to some aspects of monodromy problem and eventually led to wide generalizations of monodromy problem for differential G-systems on Riemann surfaces [38], [14]. The aforementioned aspects of RHMP and RHTP play a central role in our presentation and will be considered in more detail in the sequel.

Finally, we wish to note that by virtue of Riemann mapping theorem in many topics concerned with holomorphic vector bundles and principal bundles it is sufficient to consider the case where contour Γ is the unit circle $\{|z| = 1\}$ in the complex plane. In other words, it is often sufficient to work with invertible matrix functions on the unit circle, i.e. with loops in the complex linear group

$GL(n, \mathbf{C})$. Obviously, one may also work with loops in a complex Lie group G and arising principal G-bundles on the Riemann sphere. This idea was developed in various aspects in a fundamental monograph [36] on results of which we heavily rely on in the sequel. For further reference we now introduce the standard decomposition of the Riemann sphere

$$\mathbf{CP}^1 = U_+ \cup \mathbf{T} \cup U_-, \tag{15.6}$$

where U_+ is the unit disc, $\mathbf{T}$ is the unit circle, and U_- is the exterior domain, which contains the infinite point ∞ which will be sometimes denoted by N ("north pole").

15.3 Generalized Riemann-Hilbert transmission problem

We proceed by presenting the first generalization of RHTP along the lines described above. The main innovation here is to permit more general coefficients in the transmission equation (15.5). It is natural to take a function on the circle with values in a Lie group G as a coefficient in (15.5) and search for piecewise holomorphic mappings with values in a given complex representation space of G. It turns out that in the case of a compact Lie group G one can develop a reasonable theory analogous to the classical one [40] which relies on the recent generalization by A. Pressley and G. Segal [36] of the well-known factorization theorem due to G. Birkhoff [4], [32]. The precise statement of the problem is given below, and the rest of the section is devoted to its investigation.

Generalized RHTP for linear representation of compact Lie group

Let G be a connected compact Lie group of rank p with the Lie algebra $\mathbf{g}$. As is well known [36], each of such groups has a complexification $G_{\mathbf{C}}$ with the Lie algebra $\mathbf{g}_{\mathbf{C}} = \mathbf{g} \otimes \mathbf{C}$. This fact is very important as it provides complex structures on loop groups and this is the main reason why our discussion is restricted to compact groups.

Let LG denote the group of continuous based (i.e., sending the number 1 to the unit of G) loops on G endowed with the point-wise multiplication and usual topology [36]. We need some regularity conditions on loops and for the sake of simplicity let us first assume that all loops under consideration are (at least once) continuously differentiable.

For an open set U in $\mathbf{CP}^1$ let $\mathcal{A}(U, \mathbf{C}^n)$ denote the subset of $C(\overline{U}, \mathbf{C}^n)$ formed by those vector-functions which are holomorphic in U, where $\overline{U}$ denotes the closure of U. Assume also that we are given a fixed linear representation r of the group G in a vector space V. For our purposes it is natural to assume that V is a complex vector space. Notice that, for a compact group, one has a complete description of all of its complex linear representations [2].

We are now in a position to formulate the desired generalization. Let us again consider the standard decomposition (15.6) of the Riemann sphere $\mathbf{CP}^1$.

308 Mechanics and Physics of Structured Media

Given a loop $f \in LG$, the (homogeneous) generalized Riemann-Hilbert transmission problem (GRHTP) with coefficient f is formulated as a question about the existence and structure of pairs $(\Phi_+, \Phi_-) \in \mathcal{A}(U_+, V) \times \mathcal{A}(U_-, V)$ with $\Phi_-(N) = 0$ satisfying the following transmission condition on $\mathbf{T}$

$$\Phi_+(z) = r(f(z)) \cdot \Phi_-(z). \tag{15.7}$$

For any loop h in V we also obtain a nonhomogeneous problem (with the right-hand side h) by replacing the transition equation (15.7) by the condition

$$\Phi_+(z) - r(f(z)) \cdot \Phi_-(z) = h(z). \tag{15.8}$$

In other words, we are interested in the kernel and cokernel of the natural linear operator T_f expressed by the left-hand side of the formula (15.8) and acting from the space of piecewise holomorphic vector-functions on $\mathbf{CP}^1$ with values in V into the loop space LV. To avoid annoying repetition when dealing with the inhomogeneous GRHP, it will always be assumed that the loop h is Hölder-continuous, which is a usual assumption in the classical theory [32].

Remark 15.1. In the particular case where $G = U(n)$ is the unitary group we get that $G_{\mathbf{C}} = GL(n, \mathbf{C})$ is the general linear group. If we take r to be the standard representation on $\mathbf{C}^n$, then Eqs. (15.7) and (15.8) coincide and we obtain the classical Riemann-Hilbert transmission problem. Note that even in this classical case one obtains a plenty of such problems taking various representations of $U(n)$, and the results below can be best illustrated in this situation.

One can choose various natural regularity classes for a coefficient such that the problem is described by a Fredholm operator in corresponding functional spaces, which is a direct generalization of the corresponding classical result [32]. A natural framework for our discussion is provided by the generalized Birkhoff factorization theorem and Birkhoff stratification of a loop group so we have to present first some auxiliary concepts and results from [36].

Birkhoff factorization in loop groups

Needless to say, the same problem can be formulated for all groups but as a matter of fact only irreducible representations of simple groups are essential. Moreover, the exceptional groups of Cartan's list will also be excluded and the remaining groups will be termed as "classical compact groups".

It would not be appropriate to reproduce and discuss here all necessary concepts and constructions from the theory of Lie groups. All necessary results on Lie groups, in a form suitable for our purposes, are contained in a book of J. Adams [2] and we repeatedly refer to this book in the sequel.

Let f be a loop in G. We would like to associate with f some numerical invariant analogous to the classical *partial indices* [32]. To this end let us choose a maximal torus $\mathbf{T}^p$ in G and a system of positive roots. Then following [36] one

Riemann-Hilbert problems with coefficients in compact Lie groups Chapter | 15 **309**

can define the nilpotent subgroups $N_0^{\pm}$ of $G_{\mathbf{C}}$ whose Lie algebras are spanned by the root vectors of $\mathbf{g}_{\mathbf{C}}$ corresponding to the positive (respectively negative) roots. We also introduce subgroups $L^{\pm}$ of $LG_{\mathbf{C}}$ formed by the loops which are the boundary values of holomorphic mappings of the domain B_+ (respectively B_-) into the group $G_{\mathbf{C}}$, and the subgroups $N^{\pm}$ consisting of the loops from L^+ (respectively L_-) such that $f(0)$ belongs to N_0^+ (respectively $f(N)$ belongs to N_0^-).

The following fundamental result was proved in [36].

Decomposition Theorem. *Let G be a classical simple compact Lie group, and $H = L^2(\mathbf{T}, \mathbf{g}_{\mathbf{C}})$ be the polarized Hilbert space with $H = H_+ \oplus H_-$, where H_+ is the usual Hardy space of boundary values of holomorphic loops on $\mathbf{g}_{\mathbf{C}}$. Then we have the following decomposition of the groups of based loops LG:*

(i) *LG is the union of subsets B_K indexed by the lattice of homomorphisms of $\mathbf{T}$ into the maximal torus $\mathbf{T}^p$.*

(ii) *B_K is the orbit of $K \cdot H_+$ under N^- where the action is defined by the usual adjoint representation of G. Every B_K is a locally closed contractible complex submanifold of finite codimension d_K in LG, and it is diffeomorphic to the intersection L_K^+ of N^- with $K \cdot L_1^- \cdot K^{-1}$, where L_1^- consists of loops equal to the unit at the infinite point N.*

(iii) *The orbit of $K \cdot H_+$ under N^+ is a complex cell C_K of dimension d_K. It is diffeomorphic to the intersection L_K^+ of N^+ with $K \cdot L_1^- \cdot K^{-1}$, and meets B_K transversally at the single point $K \cdot H_+$.*

(iv) *The orbit of $K \cdot H_+$ under $K \cdot L_1^- \cdot K^{-1}$ is an open subset U_K of LG, and the multiplication of loops gives a diffeomorphism from $B_K \times C_K$ into U_K.*

Recall that in the classical case this result reduces essentially to the Birkhoff factorization theorem for matrix loops [40].

Let us introduce the corresponding concept in the above setting. Namely, for a loop f on G the (left) Birkhoff factorization will be called its representation in the form

$$f = f_+ \cdot D_{\texttt{diag}} \cdot f_-, \tag{15.9}$$

where f_+ belongs to the corresponding group $L^{\pm}G$ and d is some homomorphism of $\mathbf{T}$ into $\mathbf{T}^p$.

Now it is evident that the points (ii) and (iv) of the Decomposition Theorem imply the following existence result.

Proposition 15.1. *Every Hölder class loop in a classical simple compact group has a (left) Birkhoff factorization.*

Note that we could also introduce the right factorization with the reversed order of f_+ and f_- and the result would also be valid. Our choice of the factorization type is consistent with the problem under consideration.

310 Mechanics and Physics of Structured Media

Taking into account that any homomorphism d from (15.9) is determined by a sequence of p integer numbers $(k_1, \ldots, k_p)$, we get that this sequence can be associated with any loop f. These integers are called (left) *G-exponents* (or *partial G-indices*) of f. Their collection will be denoted $K(f)$.

It is easy to prove that $K(f)$ (up to the order) does not actually depend neither on the factors from the representation (15.9) nor on the choice of the maximal torus. For a given maximal torus the proof of this fact can be obtained as in the classical case, while the independence on the choice of a maximal torus follows from the well-known fact that any two maximal tori are conjugate [2].

The exponents provide basic analytical invariants of loops and have some topological meaning.

Proposition 15.2. *Two loops lie in the same connected component of LG if and only if they have the same sum of exponents.*

This follows easily from the contractibility of subgroups $L^{\pm}$ and the point (ii) of the theorem.

Remark 15.2. In the classical case where $G = U(n)$ we obtain the usual partial indices, and Proposition 15.2 reduces to the evident observation that the connected components of LU_n are classified by the sum of partial indices which is known to coincide with the increment of the determinant argument of an invertible matrix function along the unit circle [32].

Having at hand exponents of loops we may identify each subset B_K with the collection of loops having a given collection of G-exponents equal to K (up to the order) and use the corresponding decomposition of LG in the topological study of GRHTP. Note that in the classical theory of RHP the geometry of B_K (so-called Birkhoff strata) was the subject of an intensive investigation [10], [26]. Later B. Bojarski proposed an approach in the spirit of global analysis (see [11]) which can also be treated from the viewpoint of the theory of Fredholm structures [31]. The following result is a direct consequence of the generalized Birkhoff factorization theorem.

Proposition 15.3. *([31]) For each Hölder loop $f \in LG$, the operator defined by the left hand side of (15.8) in the space of square integrable V-valued functions on $\mathbf{T}$, is Fredholm.*

One can also obtain other versions of the index formula for a GRHP as above and express the dimension of its kernel and cokernel in terms of G-exponents [31]. We would also like to mention one aspect of the loop group theory which is closely related to the above results. Namely, a given loop, generally speaking, can be attributed to various ambient loop groups by considering some natural embeddings of the groups under consideration (e.g., $U(n) \subset O(2n)$). Simple examples show that the exponents of a given loop may change under such an operation and there arises a natural problem of describing possible changes of exponents under such embeddings of coefficient groups [23].

Riemann-Hilbert problems with coefficients in compact Lie groups **Chapter | 15** **311**

This problem is not of a merely theoretical interest because it is closely related with the problem of effective computation of the exponents and factorization of a given loop. A natural way is to realize the group in question as a matrix group by considering the matrix realization of the representation r involved in the definition of GRHP, and then compute the partial indices of the corresponding matrix function using the aforementioned results of [18] and [3]. It should be noted that for wide classes of matrix functions there exist effective algorithms for such computations which are easy to implement on computer [3]. In this way one can investigate behavior of G-exponents under embeddings of coefficient groups. Interesting general results which shed some light on this problem have been obtained by G. Schatz [39]. These results of G. Schatz will be discussed in more detail in Section 15.4.

15.4 Lie groups and principal bundles

It is well known that an appropriate context for studying topological and algebraic properties of the classical RHTP is provided by holomorphic vector bundles over $\mathbf{CP}^1$ [36]. In the case of arbitrary compact Lie group G there also exists a natural connection between GRHTP and principal $G_{\mathbf{C}}$-bundles over the Riemann sphere [36]. This connection "works in both directions". In particular, the results on the structure of solutions to GRHP enable one to get some information on deformations of $G_{\mathbf{C}}$-bundles. In this section we present necessary concepts and auxiliary results from the theory of principal bundles.

Let G be a connected complex Lie group, M – a complex manifold and P – a holomorphic principal G-bundle on M. Then there is an exact sequence of vector bundles on M

$$0 \to \mathrm{ad}\, P \to Q(P) \to TM, \qquad (15.10)$$

where TM is the tangent bundle of M, $\mathrm{ad}\, P$ is the vector bundle associated to P and $Q(P)$ is the bundle of G-invariant tangent vector fields on P. Here and in the sequel P also denotes the total space of the bundle.

One may consider an exact sequence equivalent to (15.10) in the following way. Let $\mathrm{der}\, M$ and $\mathrm{der}\, P$ be the sets of all derivations of $C^\infty(M)$ and $C^\infty(P)$ respectively. By definition $D \in \mathrm{der}(P)$ if one has

$$D(af) = d(a) \cdot f + aDf$$

for any $a \in C^\infty(M)$ and $f \in C^\infty(P)$, where $d \in \mathrm{der}\, M$ is uniquely determined by $D \in \mathrm{der}\, P$. Thus D determines a homomorphism of $C^\infty(M)$-modules

$$\sigma : \mathrm{der}\, P \to \mathrm{der}\, M$$

defined by

$$\sigma(D) = d.$$

312 Mechanics and Physics of Structured Media

One can check that $\ker \sigma = \operatorname{ad} P$, so that there is a short exact sequence of vector bundles

$$0 \to \operatorname{ad} P \to \operatorname{der} P \to \operatorname{der} M \to 0. \tag{15.11}$$

$\operatorname{der} P$ and $\operatorname{der} M$ are Lie algebras, and σ is a homomorphism of Lie algebras. A holomorphic (resp. smooth) *connection* on a principal bundle $P \to M$ is a splitting of the exact sequence (15.10) (resp., (15.11)).

Let

$$\nabla : \operatorname{der} M \to \operatorname{der} P$$

be a splitting of the sequence (15.10). Then ∇ need not be a Lie algebra homomorphism, i.e. it can happen that $[\nabla_{\tau_1}, \nabla_{\tau_2}] \neq \nabla_{[\tau_1, \tau_2]}$, where τ and ∇_τ denote vector fields on M and P respectively. Deviation of ∇ from being a Lie algebra homomorphism is measured by the curvature F. Another tool for describing this deviation is the Atiyah class defined below.

The exact sequence (15.10) determines an element $a(P)$ of the cohomology group

$$H^1(M; \operatorname{Hom}(TM, \operatorname{ad} P)) \cong H^1(M; \operatorname{ad} P \otimes \Lambda^1)$$

called the *Atiyah class* of the bundle P.

If the sequence (15.10) splits, then $a(P)$ is the trivial element of $H^1(M; \operatorname{ad} P \otimes \Lambda^1)$. Thus the bundle P admits a holomorphic connection if and only if $a(P) = 0$.

Locally the sequence (15.10) always splits. Let $\theta_i \in \Gamma(U_i, \operatorname{ad} P \otimes \Lambda^1|_{U_i})$ and $\rho_{ij} \in \Gamma(U_i \cap U_j, \operatorname{ad} P \cong \Lambda^1_{U_i \cap U_j})$ be induced by the transition functions $g_{ij} : U_i \cap U_j \to G$, then on the intersections $U_i \cap U_j$ the sections θ_i and θ_j are related by the relation

$$\theta_i = \operatorname{ad}(g_{ij})\theta_j - \rho_{ij}. \tag{15.12}$$

Remark 15.3. If $G = GL_n(\mathbf{C})$, then (15.12) has the form

$$\theta_i = g_{ij}\theta_j g_{ij}^{-1} - dg_{ij} \cdot g_{ij}^{-1}.$$

Let the principal bundle P have a σ-connection and let the collection of 1-forms $\{\theta_j\}$ satisfy the condition (15.12). Suppose that the forms θ_j are of type $(1, 0)$. Then the curvatures $F(\theta_i) = d\theta_i$ are of the type $(1, 1)$ and thus can be considered as elements of the Dolbeault cohomology group $H^{1,1}(M; \operatorname{ad} P)$. Let

$$\tau : H^{1,1}(M; \operatorname{ad} P) \to H^1(M; \operatorname{ad} P \otimes \Lambda^1)$$

be the Dolbeault isomorphism, then $\tau(F(\theta_i)) = a(P)$.

If $G = GL_n(\mathbf{C})$ then it is known that there is a one-to-one correspondence between principal G-bundles P and associated vector bundles E. Moreover, the

Riemann-Hilbert problems with coefficients in compact Lie groups **Chapter | 15** **313**

vector bundles $\text{End}\,E$ and $\text{ad}\,P$ are isomorphic, and thus there is an isomorphism of cohomology groups

$$H^1(M; \text{ad}\,P \otimes \Lambda^1) \cong H^1(M; \text{End}\,E \otimes \Lambda^1). \tag{15.13}$$

Let $b(E)$ be the element of $H^1(M; \text{End}\,E \otimes \Lambda^1)$ corresponding to the exact sequence

$$0 \to \text{End}\,E \to Q(E) \to TM \to 0.$$

The cohomology class $b(E)$ is called the *Atiyah class* of the vector bundle E.

It can be proved that $\mu(a(P)) = -b(E)$, where μ denotes the isomorphism (15.13), and that for a line bundle $L \to M$ one has $b(L) = -2\pi i c_1(L)$, where $c_1(L)$ is the Chern class of L.

Let $\Omega(P)$ be the sheaf of germs of sections of the principal bundle P. The group G acts on the sheaf of sets $\Omega(P)$. If $U \subset M$ is an open set and $s : U \times G \to P|_U$ is an isomorphism given by the formula

$$U \times G \ni (x, g) \mapsto s(x) \cdot g \in P,$$

then s and $s(x)g$ give the same splitting of the exact sequence

$$0 \to \text{ad}\,P|_U \to Q(P)|_U \to TU \to 0.$$

Consequently, global G-invariant sections of the sheaf $\Omega(P)$ determine splitting of the exact sequence (15.10). A holomorphic connection on a principal bundle $P \to M$ is called *integrable*, if the splitting of (15.10) is G-invariant.

The following fundamental result serves as a paradigm in the sequel.

Proposition 15.4. *A holomorphic principal bundle $P \to M$ with structure group G possesses an integrable connection if and only if it is induced by a representation of the fundamental group $\rho : \pi_1(M) \to G$.*

Let $G = GL_n(\mathbf{C})$ and let $E \to M$ be a vector bundle. If E is induced by a representation $\rho : \pi_1(M) \to G$, then there is a system of differential equations with holomorphic coefficients $df = \omega f$ monodromy of which coincides with the given representation. Moreover, ω will be a connection in this bundle, and its holomorphy implies its complete integrability.

Proposition 15.4 implies that a holomorphic vector bundle $E \to \mathbf{CP}^1$ possesses a holomorphic connection if and only if the type of the splitting has the form $K = (0, ..., 0)$, i.e. if and only if the bundle is holomorphically trivial.

Let G be a compact connected Lie group of rank r and let $G_\mathbf{C}$ be the complexification of G, so that $G_\mathbf{C}$ is a reductive group. Let us, as above, denote by $\mathbf{g}$ and $\mathbf{g}_\mathbf{C}$ the Lie algebras of the groups G and $G_\mathbf{C}$ respectively. If G is a Lie group and $\mathbf{g}$ its Lie algebra, then under the complexification of G is understood a complex Lie group $G_\mathbf{C}$ whose Lie algebra is $\mathbf{g}_\mathbf{C}$. Such a complexification need not exist in general. If G is isomorphic to a subgroup of a unitary group $U(n)$

314 Mechanics and Physics of Structured Media

for sufficiently large n, then $G_{\mathbb{C}}$ can be considered as a subgroup of the complexification of the unitary group $U(n)_{\mathbb{C}} = GL_n(\mathbb{C})$. Thus for compact groups there always exists a complexification, unique up to isomorphism.

In the case of an arbitrary compact group G there exists a natural connection between GRHTP and principal $G_{\mathbb{C}}$-bundles over the Riemann sphere [36]. In particular, the results on the structure of solutions to GRHTP enable one to get some information on deformations of $G_{\mathbb{C}}$-bundles.

Theorem 15.1. *The base of the versal deformation of a holomorphic principal $G_{\mathbb{C}}$-bundle corresponding to a loop f has dimension d_K, where $K = K(f)$ is the collection of G-exponents of f. In other words, this dimension is given by the formula*

$$d_K = \sum_{k_i > k_j} (k_i - k_j - 1). \tag{15.14}$$

This can be derived from the geometric description of the Birkhoff stratification provided by the above Decomposition Theorem. Indeed, each stratum corresponds to a fixed isomorphism class of the bundles under consideration [36]. In fact, the point (iii) of the Decomposition Theorem shows that such strata possess natural transversals which, due to the smoothness of strata, yield the germs of base of versal deformation. At the same time, the dimension d_K may be computed by the general technique of the deformation theory in terms of the first cohomology group of $\mathbb{CP}^1$ with coefficients in the adjoint representation of G. The formula (15.14) then follows from (15.9) and Serre duality.

Corollary 15.1. *A holomorphic principal $G_{\mathbb{C}}$-bundle is holomorphically trivial if and only if all exponents of the corresponding loop are equal to zero.*

Corollary 15.2. *A holomorphic principal $G_{\mathbb{C}}$-bundle is stable if and only if all pairwise differences of its exponents do not exceed 1.*

In Corollary 15.2 stability is understood in the sense of Gohberg-Krein-Bojarski [26], [10] which is weaker than stability in the sense of Mumford. There are no bundles on the Riemann sphere which are stable in the sense of Mumford, whereas the semistable ones are only those with holomorphic type $(k, k, \ldots, k)$.

One can also use the exponents for investigating which principal bundles can be realized as subbundles of a given $G_{\mathbb{C}}$-bundle. Notice that for $U(n)$-bundles a complete solution of this problem was obtained in [39].

Embeddings of principal bundles

In [39] S. Shatz, in terms of partial indices of the corresponding matrix function, has given a necessary condition for a holomorphic vector bundle to be a subbundle of the given vector bundle. We present an analogous result using the splitting type of holomorphic principal bundle on the Riemann sphere, which,

Riemann-Hilbert problems with coefficients in compact Lie groups Chapter | 15 **315**

in particular, provides a partial answer to the problem formulated at the end of the preceding section.

Let

$$E = \mathcal{O}(k_1) \oplus \mathcal{O}(k_2) \oplus \ldots \oplus \mathcal{O}(k_n), \quad F = \mathcal{O}(\kappa_1) \oplus \mathcal{O}(\kappa_2) \oplus \ldots \oplus \mathcal{O}(\kappa_m)$$

be two vector bundles on the Riemann sphere and $m > n$. As it is known, in order for E to be a holomorphic subbundle of F it suffices that the following inequalities be satisfied [39]:

$$k_i \leq \kappa_i, \quad i = 1, \ldots, n. \tag{15.15}$$

Let P_1 and P_2 be principal bundles on the Riemann sphere with structure groups G_1 and G_2. Let $\varphi : P_1 \to P_2$ be an injective morphism. We will then say that P_1 is a principal holomorphic subbundle of P_2, and P_2 is obtained from P_1 by extension of the structure group from G_1 to G_2. Suppose moreover that holomorphic types of P_1 and P_2 are respectively $(k_1, k_2, \ldots, k_r)$ and $(\kappa_1, \kappa_2, \ldots, \kappa_s)$. Let us introduce notation $k_{ij} = k_i - k_j, \kappa_{ij} = \kappa_i - \kappa_j, i < j$.

Theorem 15.2. *If P_1 is a principal holomorphic subbundle of P_2. Then for each pair (i, j), $i < j$, there exists a pair (p, q), $p < q$, such that the inequality $k_{ij} \leq \kappa_{pq}$ holds.*

Proof. Consider vector bundles $\mathrm{ad}\, P_1$ and $\mathrm{ad}\, P_2$ on $\mathbf{CP}^1$. For each $x \in \mathbf{CP}^1$ one has $\mathrm{ad}_x P_1 \cong \mathbf{g}_1 \otimes \mathbf{g}_1^*$, $\mathrm{ad}_x P_2 \cong \mathbf{g}_2 \otimes \mathbf{g}_2^*$. By the Birkhoff-Grothendieck theorem each holomorphic vector bundle on the Riemann sphere decomposes into the sum of line bundles, hence for $\mathrm{ad}\, P_1$ and $\mathrm{ad}\, P_2$ one has

$$\mathcal{O}(k_{12}) \oplus \mathcal{O}(k_{13}) \oplus \ldots \oplus \mathcal{O}(k_{r-1,r}) \oplus \mathcal{O} \oplus \mathcal{O} \oplus \ldots \to \mathbf{CP}^1,$$
$$\mathcal{O}(\kappa_{12}) \oplus \mathcal{O}(\kappa_{13}) \oplus \ldots \oplus \mathcal{O}(\kappa_{s-1,s}) \oplus \mathcal{O} \oplus \mathcal{O} \oplus \ldots \to \mathbf{CP}^1.$$

The above sums contain trivial bundles of dimensions not smaller than the ranks of the corresponding structure groups. Now referring to (15.15) one completes the proof of the theorem. $\square$

Converse statement of this theorem is not true. From the fact that $G_1 \to G_2$ is a holomorphic embedding and the above condition holds for the holomorphic splitting types, it does not follow that P_1 is a holomorphic principal subbundle of P_2. The following example illustrates this phenomenon. Suppose $G_1 = U(2)$ and the holomorphic type of the corresponding bundle is $(2, -7)$, $G_2 = U(3)$ and the bundle has holomorphic type $(8, 3, 0)$, then

$$\mathrm{ad}\, P_1 \cong \mathcal{O}(9) \oplus \mathcal{O} \oplus \mathcal{O} \oplus \mathcal{O}(-9),$$
$$\mathrm{ad}\, P_2 \cong \mathcal{O}(8) \oplus \mathcal{O}(5) \oplus \mathcal{O}(3) \oplus \mathcal{O} \oplus \mathcal{O} \oplus \mathcal{O} \oplus \mathcal{O}(-3) \oplus \mathcal{O}(-5) \oplus \mathcal{O}(-8),$$

and $\mathrm{ad}\, P_1$ cannot be embedded into $\mathrm{ad}\, P_2$ as a holomorphic vector subbundle.

316 Mechanics and Physics of Structured Media

Corollary 15.3. *If P_1 is a holomorphic subbundle of a principal bundle P_2 and P_2 is stable, then P_1 is also stable.*

Here the stability is understood in the sense of Gohberg-Krein-Bojarski as above.

Theorem 15.3. *[21] If $G = U(n), SO(n), Sp(n)$, then after identifying $\pi_1(G)$ as*

$$\pi_1(G) \cong \mathbf{Z}, \quad if \ G = U(1); \ SO(2),$$
$$\pi_1(G) \cong \mathbf{Z}_2, \quad if \ G = SO(n), n \geq 3,$$
$$\pi_1(G) \cong 0, \quad if \ G = SU(n), Sp(n)$$

characteristic classes of principal bundles are expressed by their holomorphic types as follows: $\chi(P) = k_1 + k_2 + \ldots + k_r$.

To prove this theorem it suffices to map $\mathbf{T}^r$ isomorphically onto the maximal torus of $U(r)$ via the mapping $\exp_{\mathbf{T}^r}(k_1\tau, k_2\tau, \ldots, k_r\tau) \mapsto \mathrm{diag}(k_1, k_2, \ldots, k_r)$. It remains to use functoriality of this isomorphism and the fact that a similar equality holds for the full Chern number of the induced bundle [22].

15.5 Riemann-Hilbert monodromy problem for a compact Lie group

As was mentioned in the previous section, for any vector bundle, there exists connection with regular singularities at the given points (Plemelj's theorem). This result can be generalized for holomorphic principal G-bundles. To describe this generalization we consider a system of differential equations of the form $Df = \alpha f$, where α is a $\mathbf{g}$-valued 1-form defined on Riemann surface X, and $f : X \to G$ is a G-valued unknown function.

Let $r_g : G \to G$ be right shift on the group G, let $C(X, G)$ be the group of all smooth functions $f : X \to G$ and let $\Lambda^p(X, G)$, $p = 0, 1, 2$, be the space of all $\mathbf{g}$-valued p-forms on X.

Let us define the operator

$$D : \Lambda^0(X, \mathbf{g}) \to \Lambda^1(X, \mathbf{g}) \tag{15.16}$$

by the formula $D_x(f)(u) = dr_{f(x)}^{-1}(df)_x(u)$.

An expression of the form

$$Df = \omega, \tag{15.17}$$

where α is a $\mathbf{g}_{\mathbf{C}}$-valued 1-form on X and $f : X \to G_{\mathbf{C}}$ is an unknown smooth function, is called a *G-system of differential equations*.

For a G-system, it is possible to formulate *Riemann-Hilbert monodromy problem* as follows: whether, for a given discrete set $S = \{s_1, \ldots, s_m\} \subset X$ and for a given homomorphism $\rho : \pi_1(X \setminus S, z_0) \to G_{\mathbf{C}}$, there exists a G-system of

Riemann-Hilbert problems with coefficients in compact Lie groups Chapter | 15 **317**

the type (15.17) with a 1-form ω which is holomorphic in $X \setminus S$ and monodromy of which coincides with ρ.

It is known that solution of above problem depends on group G.

Example 15.1. Let X be a compact Riemann surface of genus $g = 1$, $G = SO(3)$, γ_1 and γ_2 are generators of the fundamental group $\pi_1(X)$. Let $\rho : \pi_1(x) \to SO(3)$ be the representation:

$$\rho(\gamma_1) = \begin{pmatrix} 1 & 0 & 0 \\ 0 & -1 & 0 \\ 0 & 0 & -1 \end{pmatrix}, \quad \rho(\gamma_2) = \begin{pmatrix} -1 & 0 & 0 \\ 0 & -1 & 0 \\ 0 & 0 & 1 \end{pmatrix}.$$

This representation cannot be a monodromy homomorphism of any G-system $Df = \omega$ [34].

If $G = U(n)$, then $Df = df \cdot f^{-1}$ and α is a matrix of 1-forms on X, so that one obtains a usual system of the form (15.1) or (15.3). If $n = 1$, then $G_{\mathbf{C}} = \mathbf{C}^*$ and $Df = d \log f$, the logarithmic derivative of the function f.

Let

$$* : \Lambda^1(X; \mathbf{g}) \to \Lambda^1(X; \mathbf{g})$$

be the Hodge operator, then the complexification of de Rham complex $\Lambda_{\mathbf{C}}^p(X; \mathbf{g})$, $p = 0, 1, 2$, decomposes into the direct sum

$$\Lambda_{\mathbf{C}}^1(X; \mathbf{g}) = \Lambda^{1,0}(X; \mathbf{g}) \oplus \Lambda^{0,1}(X; \mathbf{g})$$

by the requirement that $* = -i$ on $\Lambda^{1,0}(X; \mathbf{g})$ and $* = i$ on $\Lambda^{0,1}(X; \mathbf{g})$. The operator D decomposes into the direct sum $D = D' \oplus D''$, where

$$D' : \Lambda^0(X; \mathbf{g}) \to \Lambda^{1,0}(X; \mathbf{g}), \quad D'' : \Lambda^0(X; \mathbf{g}) \to \Lambda^{0,1}(X; \mathbf{g}),$$

are determined by the formulas

$$D'_x(f)(u) = d'_{r^{-1}_{f(x)}}(d'f)_x(u), \quad D''_x(f)(u) = d'' r^{-1}_{f(x)}(d''f)_x(u).$$

A $G_{\mathbf{C}}$-valued function $f : X \to G_{\mathbf{C}}$ is called holomorphic (resp. antiholomorphic) if $D'' f = 0$ (resp. $D' f = 0$).

The operator D has the following properties:

1) it is a crossed homomorphism, i.e.

$$D(f \cdot g) = (Df)_x + (\mathrm{ad} f(x)) \circ (Dg)_x$$

for any $f, g \in C(X, G)$. Note that the operator D'' is also a crossed homomorphism,

2) the kernel $\ker D$ consists of constant functions.

318 Mechanics and Physics of Structured Media

We will say that system (15.17) is *integrable* if, for any $x_0 \in X$ and $g_0 \in G$, there exists a solution f of this system in a neighborhood of x_0 satisfying $f(x_0) = g_0$. A point x_0 is called an *isolated singular point* of a map $f : U \to G_{\mathbf{C}}$ if there is a punctured neighborhood U_{x_0} such that the map f is analytic in U_{x_0}. We also will say that a $G_{\mathbf{C}}$-valued function $f \in \Omega(U_\epsilon(x_0))$ is of *polynomial growth* in x_0 if for each sector

$$S = \{ z | \theta_0 \le \arg z \le \theta_1, 0 \le |z| < \epsilon \},$$

where z denotes a local coordinate system on X, for sufficiently small ϵ, there exist an integer $k > 0$ and a constant $c > 0$ such that the inequality

$$d(f(z), \mathbf{1}) < c|z|^{-k}$$

holds, where $d(\_, \mathbf{1})$ denotes the distance to the unit of group $G_{\mathbf{C}}$.

The properties 1), 2) of the operator D imply that if f_0 is some solution of system (15.17), then $f = f_0 h$ is also a solution for any $h \in \ker D$, i.e. the solution is uniquely determined up to multiplication by a constant. G-system (15.17) is called *regular*, if all solutions of this system at all singular points have at most polynomial growth [4].

In the sequel integration of $G_{\mathbf{C}}$-valued functions is understood as a multiplicative integral for Lie groups and algebras. Let $\gamma \subset U$ be a smooth arc, with a parameterized map $z : [a, b] \to U$. Multiplicative integral along the arc γ is by definition $\int_\gamma (1 + f(z)) dz := \int_a^b (1 + f(z)) z'(t) dt$, where 1 denotes the unit element of $G_{\mathbf{C}}$. If γ is a closed arc, then $M_f(\gamma) = \oint_\gamma (1 + f(z)) dz$ is an invertible element of $\mathbf{g}$ called holonomy of the map f with respect to γ.

G-systems on the Riemann sphere

Consider a G-system (15.17) on $\mathbf{CP}^1$. Let f_0 be a solution of (15.17) in a neighborhood $U \subset \mathbf{CP}^1$ of the point z_0, of polynomial growth at the points from the set $S = \{s_1, ..., s_m\}$. After continuation of f_0 along a path $\gamma_i \in \pi_1(\mathbf{CP}^1 \setminus S, z_0)$ starting and ending in z_0 and circling once around a singular point s_i, the solution f_0 transforms into another solution f_1. As noted before, $\gamma_i^* f_0 = g_i f_1$ for some $g_i \in G$. Thus f_0 determines a representation

$$\rho : \pi_1(\mathbf{CP}^1 \setminus S, z_0) \to G_{\mathbf{C}}. \tag{15.18}$$

The subgroup $\mathrm{Im}\rho \subset G_{\mathbf{C}}$ is called the *monodromy group* of G-system (15.17) and the representation (15.18) induces a principal $G_{\mathbf{C}}$-bundle $P'_\rho \to \mathbf{CP}^1 \setminus S$, the form ω being a holomorphic connection for this bundle.

Consider an extension of the bundle $P'_\rho \to \mathbf{CP}^1 \setminus S$ a holomorphic principal bundle $P_\rho \to \mathbf{CP}^1$ [38], [14]. To extend the bundle $P'_\rho \to \mathbf{CP}^1 \setminus S$ to some point $s_i \in S$, consider a simple covering $\{U_j\}$ of $X_m = CP^1 \setminus S$, such that every intersection $U_{\alpha_1} \cap U_{\alpha_2} \cap ... \cap U_{\alpha_k}$ is simply connected. For each U_α, we choose

Riemann-Hilbert problems with coefficients in compact Lie groups **Chapter | 15 319**

a point $z_\alpha \in U_\alpha$ and join z_0 and s_α by a simple path γ_α starting at z_0 and ending at s_α. For a point $z \in U_\alpha \cap U_\beta$, we choose a path $\tau_\alpha \subset U_\alpha$ which starts at s_α and ends at z. Consider

$$g_{\alpha\beta}(z) = \rho \left(\gamma_\alpha \tau_\alpha(z) \tau_\beta^{-1}(z) \gamma_\beta^{-1} \right). \tag{15.19}$$

We see that $g_{\alpha\beta}(z) = g_{\beta\alpha}^{-1}(z)$ on $U_\alpha \cap U_\beta$ and $g_{\alpha\beta} g_{\beta\gamma}(z) = g_{\alpha\gamma}(z)$ on $U_\alpha \cap U_\beta \cap U_\gamma$.

The cocycle $\{g_{\alpha\beta}(z)\}$ is constant [4]. Hence from this cocycle we obtain a flat principal bundle, which is denoted by P'_ρ. Let $\{t_\alpha(z)\}$ be a trivialization of our bundle, i.e.

$$t_\alpha : p^{-1}(U_\alpha) \to G_{\mathbf{C}}$$

is a holomorphic mapping. Consider the **g** valued 1-form $\omega_\alpha = -t_\alpha^{-1} dt_\alpha$.

The cocycle $\{g_{\alpha\beta}(z)\}$ is constant on the intersection $U_\alpha \cap U_\beta$ and $g_{\alpha\beta}(z) t_\beta(z) = t_\alpha(z)$, so the identity $\omega_\alpha = \omega_\beta$ holds on $U_\alpha \cap U_\beta$. Indeed, replacing t_β by $t_\beta^{-1} g_{\alpha\beta}$ in the expression $\omega_\beta = -t_\beta^{-1} dt_\beta$, we obtain

$$\omega_\beta = -t_\alpha^{-1} g_{\alpha\beta}(z) dt_\alpha g_{\alpha\beta}^{-1}(z) = -t_\alpha^{-1} dt_\alpha.$$

So, $\omega = \{\omega_\alpha\}$ is a holomorphic 1-form on X_m and therefore it defines a connection 1-form of the bundle $P'_\rho \to X_m$. The corresponding connection is denoted by ∇'. We will extend the pair $\left(P'_\rho, \nabla'\right)$ to X. As the required construction is of local character, we shall extend $P'_\rho \to X_m$ to the bundle $P''_\rho \to X_m \cup \{s_i\}$, where $s_i \in S$.

Let a neighborhood V_i of the point s_i intersect each of the open sets $U_{\alpha_1}, U_{\alpha_2}, ..., U_{\alpha_k}$ having s_i in its closure. As we noted when constructing the bundle from transition functions (15.19), only one of them is different from identity. Let us denote it by g_{1k}, then $g_{1k} = M_i$, where M_i is the monodromy which corresponds to the singular point s_i and is obtained from representation (15.18). Mark a branch of the multi-valued function $(\tilde{z} - s_i)^{A_i}$ containing the point $\tilde{s}_i \in \tilde{U}_i$ (where $2\pi i \exp(A_i) = M_i$). Thus the marked branch defines a function

$$g_{01} := \exp(A_j \ln(z - s_j)), \tag{15.20}$$

Denote by g_{02} the extension of g_{01} along the path which goes around s_i counterclockwise, and similarly for other points. Hence on $U_i \cap U_{\alpha_k} \cap U_{\alpha_1}$ we shall have:

$$g_{0k}(z) = g_{01}(z) M_i = g_{01}(z) g_{0k}(z).$$

The function $g_{0k} : V_i \to G_{\mathbf{C}}$ is defined at the point s_i and takes there the value coinciding with the monodromy. In a neighborhood of s_i one will have

$$\omega_i = dg_{0k} g_{0k}^{-1}.$$

320 Mechanics and Physics of Structured Media

If we use the above construction of extension for all points from S we obtain a holomorphic principal bundle $P_\rho \to \mathbf{CP}^1$ on the Riemann sphere $\mathbf{CP}^1$. The holomorphic sections of P_ρ are solutions of the equation

$$\nabla f = 0 \iff Df = \omega, \tag{15.21}$$

where ω is the meromorphic 1-form of connection ∇. It means that $P_\rho \to \mathbf{CP}^1$ is induced by the system of the form (15.17) and the Atiyah class $a(P_\rho)$ is nontrivial [24]. Therefore $P_\rho \to \mathbf{CP}^1$ does not admit holomorphic connections and hence the system (15.17) must necessary have singular points. Here and in the sequel under singular points will be meant critical singular points, i.e. ramification points of the solution.

Remark 15.4. The construction of extension of the bundle at the singular points of equation described above has local character. From this follows that this construction may be applied for any Riemann surface of higher genus.

The Birkhoff stratum Ω_κ consists of the loops from $L_p G_\mathbf{C}$ with fixed partial indices $K = (k_1, ..., k_r)$. Topology of Ω_K is investigated in [30], [31]. Existence of a one-to-one correspondence between the Birkhoff strata Ω_K and holomorphic equivalence classes of principal bundles on $\mathbf{CP}^1$ is a generalization of Birkhoff-Grothendieck theorem for holomorphic vector bundles on Riemann sphere. More precisely, the following theorem holds.

Theorem 15.4. *[36] Each loop $f \in \Omega G$ determines a pair (P, ξ), where P is a holomorphic principal $G_\mathbf{C}$-bundle on $\mathbf{CP}^1$ and ξ is a smooth section of the bundle $P|_{\bar{X}_\infty}$ holomorphic in X_∞, and if (P', ξ') and (P, ξ) are holomorphically equivalent bundles, then f' and f lie in the same Birkhoff stratum.*

The theorem implies that to each principal bundle with a fixed trivialization corresponds a tuple of integers $(k_1, ..., k_r)$ which completely determine holomorphic type of the principal bundle and hence if a holomorphic principal G-bundle is induced by a system of the form (15.17) without singular points, then this bundle is trivial.

Theorem 15.5. *If $\mathrm{Im}\rho$ is connected then the Riemann-Hilbert monodromy problem is solvable for any m points $s_1, ..., s_m$.*

Proof. Let $\gamma_1, ..., \gamma_m$ be generators of $\pi_1(X_m, z_0)$. Let $\rho_1 = \rho(\gamma_1), ..., \rho_m = \rho(\gamma_m)$. If $\mathrm{Im}\rho$ is a connected subgroup then there exists a continuous path $\rho_j(t)$, such that $\rho_i(0) = 1$ and $\rho_j(1) = \rho_j$. From this follows that there exists homomorphism $\chi_t : \pi_1(X_m, z_0) \to G$, such that $\chi_t(\gamma_j) = \rho_j(t)$. To finish the proof we use the following general result from homological algebra: if G_1 is some connected group, then the homomorphism $h : \pi_1(X_m, z_0) \to G_1$ is the monodromy homomorphism of a G-system if and only if it is possible to connect h to 1 by continuous path in group of cochains $Z^1(\pi_1(M), G)$. The theorem is proved. $\qquad\square$

Riemann-Hilbert problems with coefficients in compact Lie groups Chapter | 15 **321**

In general, ω may have a pole at infinity whose order is more that 1. For example, see [5].

Proposition 15.5. *Suppose* $\rho(\gamma_j) \in \mathbf{T}$ *for some* j. *Then* $\rho : \pi_1(X \setminus S) \to G$ *is the monodromy of a regular G-system.*

Proof. The proposition is proved using the Plemelj's scheme. The properties of $\rho(\gamma_j) \in \mathbf{T}$ guarantee that the gauge transformation reduces the regular system to a system with singularities of first order at all singular points. From this follows the proof of proposition. $\qquad\qquad\square$

G-system on the Riemann surfaces of genus $g \geq 2$

As above suppose that G is a connected compact Lie group and $G_\mathbf{C}$ is its complexification; $\mathbf{g}$ and $\mathbf{g}_\mathbf{C}$ are the Lie algebras of the group G and $G_\mathbf{C}$, respectively; Z is the center of the group $G_\mathbf{C}$, and Z_0 is the connected component of the unit; X is a compact connected Riemann surface of genus $g \geq 2$. If $\tilde{X} \to X$ is a universal covering and $\rho : \pi_1(X) \to G_\mathbf{C}$ is a representation, then the corresponding principal bundle will be denoted P_ρ.

Let $x_0 \in X$ be a fixed point and $p : \tilde{X} \to X \setminus \{x_0\}$ be a universal cover, then the triple $(\tilde{X}, p, X \setminus \{x_0\})$ is a principal bundle whose structure group Γ is a free group on $2g$ generators, and if γ is a loop circling around x_0 then $\gamma = \prod_{i=1}^{g}[a_i, b_i]$, where a_i, b_i are generators of $\Gamma \cong \pi_1(X \setminus \{x_0\})$, and $[\_,\_]$ denotes the commutator.

Let $P'_\rho \to X \setminus \{x_0\}$ be the principal bundle corresponding to the representation $\rho : \pi_1(X \setminus \{x_0\}) \to G_\mathbf{C}$. Since by Theorem 15.4 each loop $f : S_X^1 \to G$ determines a holomorphic principal $G_\mathbf{C}$-bundle, using f one can extend the bundle $P'_\rho \to X \setminus \{X_0\}$ to X in the following way: let U_{x_0} be a neighborhood of x_0 homeomorphic to a unit disc and consider the trivial bundles $U_{x_0} \times G_\mathbf{C} \to U_{x_0}$ and $P'_\rho \to X \setminus \{x_0\}$. Let us glue these bundles over the intersection $(X \setminus \{x_0\}) \cap U_{x_0} = U_{x_0} \setminus \{x_0\}$ using the loop f. We thus obtain an extended bundle $P_\rho \to X$.

Consider the homomorphism of fundamental groups

$$f_* : \pi_1(S_X^1) \to \pi_1(G_\mathbf{C})$$

induced by f and suppose that γ is a generator of $\pi_1(S_X^1)$ mapped to $+1$ under the isomorphism $\pi_1(S_X^1) \cong \mathbf{Z}$. If $f' : S_X^1 \to G_\mathbf{C}$ is homotopic to f, then $f'_* = f_*$, and f and f' correspond to topologically equivalent $G_\mathbf{C}$-bundles on X. Conversely, for any element $c \in \pi_1(G_\mathbf{C})$ there exists $f_* : \pi_1(S_X^1) \to \pi_1(G_\mathbf{C})$ with $f_*(\gamma) = c$.

Let $P \to X$ be a principal bundle and f the corresponding loop. The element

$$\chi(P) := f_*(\gamma) \in \pi_1(G_\mathbf{C})$$

of the fundamental group is called the *characteristic class* of the bundle P.

322 Mechanics and Physics of Structured Media

It is easy to see that the map

$$\chi : H^1(X; C^\infty(G_{\mathbf{C}})) \to \pi_1(G_{\mathbf{C}})$$

determined by the formula $\chi(P) = c$ for each $P \in H^1(X; C^\infty(G_{\mathbf{C}}))$ is surjective. Here $C^\infty(G_{\mathbf{C}})$ denotes the sheaf of germs of continuous maps $X \to G_{\mathbf{C}}$.

Let $p : \tilde{G}_{\mathbf{C}} \to G_{\mathbf{C}}$ be the universal cover of the group, with fiber $\pi_1(G_{\mathbf{C}}) \cong \ker p$. The exact sequence of groups

$$1 \to \pi_1(G_{\mathbf{C}}) \to \tilde{G}_{\mathbf{C}} \to G_{\mathbf{C}} \to 1 \tag{15.22}$$

induces the exact sequence of sheaves

$$1 \to \pi_1(G_{\mathbf{C}}) \to C^\infty(\tilde{G}_{\mathbf{C}}) \to C^\infty(G_{\mathbf{C}}) \to 1. \tag{15.23}$$

Since $\pi_1(G_{\mathbf{C}})$ is contained in the center of the group $\tilde{G}$, the sequences (15.22) and (15.23) yield the following commutative diagram

$$
\begin{array}{ccc}
\delta : H^1(\pi_1(X); G_{\mathbf{C}}) & \to & H^2(\pi_1(X); \pi_1(G_{\mathbf{C}})) \\
\downarrow \mu & & \downarrow \nu \\
\delta : H^1(X; G_{\mathbf{C}}) & \to & H^2(X; \pi_1(G_{\mathbf{C}})) \\
\downarrow i^* & & \downarrow \text{id} \\
\delta : H^1(X; C^\infty(G_{\mathbf{C}})) & \to & H^2(X; \pi_1(G_{\mathbf{C}}))
\end{array}
\tag{15.24}
$$

where μ, ν are isomorphisms and i^* is induced by the embedding $i : G_{\mathbf{C}} \hookrightarrow C^\infty(G_{\mathbf{C}})$. The coboundary operator

$$\delta : H^1(X; C^\infty(G_{\mathbf{C}})) \to H^2(X; \pi_1(G_{\mathbf{C}})) \cong \pi_1(G_{\mathbf{C}})$$

from the last row of the diagram (15.24) equals χ (see [37]).

Let $\rho : \pi_1(X \setminus x_0) \to G_{\mathbf{C}}$ be a representation such that $\rho(S_X^1) = c \in Z_0$. If $\tilde{Z}_0$ is the Lie algebra of group Z_0, then $\exp : \tilde{Z}_0 \to Z_0$ is a universal covering. Let us choose an element $\alpha \in \tilde{Z}_0$ such that $\exp \alpha = c$. Extend the bundle $P'_\rho \to X \setminus x_0$ to X using the loop $f : S_X^1 \to G$, with

$$f(z) = \exp(\alpha \ln(z - x_0))$$

on S_X^1. Denote the obtained principal bundle by $P_{\rho,\alpha} \to X$.

The space $H \subset G$ is called *irreducible* if

$$\{Y \in \mathbf{g} \mid \forall h \in H \ \mathrm{ad}h(Y) = Y\} = \text{center } \mathbf{g}.$$

The representation $\rho : \Gamma \to G_{\mathbf{C}}$ is called *unitary* if $\rho(\Gamma) \subset G$, and $\rho : \Gamma \to G$ is called *irreducible*, if $\rho(\Gamma)$ is irreducible. The following theorem gives a useful criterion of holomorphic equivalence of G-bundles.

Riemann-Hilbert problems with coefficients in compact Lie groups **Chapter | 15** **323**

Theorem 15.6. *[37] Let ρ and ρ' be unitary representations of the group $\Gamma \cong \pi_1(X \setminus \{x_0\})$ in G. The bundles $P_{\rho,\beta}$ and $P_{\rho',\beta'}$ are holomorphically equivalent if and only if ρ and ρ' are equivalent in a maximal compact subgroup of $G_{\mathbf{C}}$ and $\beta = \beta'$.*

Let M be any connected smooth manifold (compact or not) and let $\rho : \pi_1(M) \to G_{\mathbf{C}}$ be any homomorphism. The following theorem from [34] is important for our considerations.

Theorem 15.7. *[34] 1) If $\pi_1(M)$ is a free group and $G_{\mathbf{C}}$ is connected, then ρ is a monodromy homomorphism for a G-system (15.17).*

2) If $\pi_1(M)$ is a free abelian group and G is a connected compact Lie group with torsion free cohomology, and if $Im\rho \subset G$, then ρ is a monodromy homomorphism for some G-system of the type (15.17).

Theorem 15.7 yields a solution of the Riemann-Hilbert monodromy problem for holomorphic systems of type (15.17). In particular 1) implies that if $M = X \setminus \{x_0\}$, then for any representation $\rho : \pi_1(X \setminus \{x_0\}) \to G_{\mathbf{C}}$ there exists a G-system with the monodromy homomorphism ρ.

To proceed we need some concepts and constructions used in [34].

Lemma 15.1. *If there is a lifting of ρ to $\tilde{\rho} : \pi_1(M) \to \tilde{G}_{\mathbf{C}}$, then ρ is the monodromy homomorphism of a G-system (15.17).*

Proof. Let $[\rho] \in H^1(\pi_1(M); G_{\mathbf{C}})$ be the class of a given representation ρ. Commutativity of the diagram (15.24) implies that if

$$i^*\mu([\rho]) = 1 \in H^1(M; C^\infty(G_{\mathbf{C}})),$$

then $\nu\delta([\rho]) = 1 \in H^2(M; \pi_1(G_{\mathbf{C}}))$. Since ν is an isomorphism, this implies $\delta([\rho]) = 1$, i.e. there is a $\tilde{\rho} \in H^1(\pi_1(M); \tilde{G}_{\mathbf{C}})$ such that $[\rho] = p^*[\tilde{\rho}]$, where p^* is the homomorphism induced from p. Hence there is a $\tilde{\rho} : \pi_1(M) \to \tilde{G}_{\mathbf{C}}$ satisfying $p\tilde{\rho} = \rho$. The lemma is proved. $\qquad\square$

A holomorphic principal $G_{\mathbf{C}}$-bundle $P \to X$ is called *stable* (resp. *semistable*) if for any reduction $\sigma : X \to P/B$ the degree of the vector bundle $T_{G/B}$ is positive (resp. nonnegative), where B is a maximal parabolic subgroup of G and $T_{G/B}$ is the tangent bundle along the fibers of the bundle $P/B \to X$.

The following theorem is an analog of the Narasimhan-Seshadri-Donaldson theorem [33], [17] and provides a criterion for stability of holomorphic principal bundles on X.

Theorem 15.8. *A holomorphic $G_{\mathbf{C}}$-bundle $P \to X$ is stable if and only if it is of the form $P_{\rho,\alpha}$ for some irreducible unitary representation $\rho : \pi_1(X \setminus \{x_0\}) \to G$ such that $\rho(\gamma) = c \in Z_0$, $\alpha \in \tilde{Z}_0$, and $\exp \alpha = c$.*

324 Mechanics and Physics of Structured Media

For a given Riemann surface X, consider a G-system of differential equations (15.17) which has a regular singularity at the point x_0 and the monodromy homomorphism of the system (15.17) is such that $\rho(\gamma) = c \in Z_0$. Let P_ρ be a principal G-bundle over the noncompact Riemann surface $X \setminus \{x_0\}$. Let us use the above extension technique and extend this bundle to the whole X in the following way: suppose $\alpha \in \tilde{Z}_0$ be an element with $\exp \alpha = c$, and let $\tilde{\rho}(\gamma) = \beta$, where $\tilde{\rho} : \pi_1(X \setminus x_0) \to \tilde{G}_{\mathbf{C}}$ is a lifting of ρ to the covering of $G_{\mathbf{C}}$. As the transition function take the $G_{\mathbf{C}}$-valued function $g_{12}(z) = \exp(-z\beta)$. After gluing trivial $G_{\mathbf{C}}$-bundles over U and $X \setminus x_0$ using the function $g_{12}(z)$ one obtains a $G_{\mathbf{C}}$-bundle $P_{\rho,\alpha} \to X$ which is an extension of $P_\rho \to X \setminus x_0$. From Theorem 15.8 follows that $P_{\alpha,\beta}$ is stable.

The converse is also true.

Theorem 15.9. *A stable holomorphic principal $G_{\mathbf{C}}$-bundle has a connection θ with a single regular singularity at the given point x_0.*

Proof. Indeed, let $H = \{z \in \mathbf{C} \mid \Im z > 0\}$ be the upper half-plane and $H \to X$ be a covering with the single ramification point $x_0 \in X$ with ramification index m. Then the Fuchsian group Γ realizing X as a quotient $X = H/\Gamma$ is generated by elements $\alpha_1, \beta_1, ..., \alpha_g, \beta_g, \gamma$ with the relations

$$\left(\prod_{i=1}^{g} \alpha_i \beta_i \alpha_i^{-1} \beta_i^{-1} \right) \gamma = 1, \quad \gamma^m = 1. \tag{15.25}$$

It is clear that $\Gamma \cong \pi_1(X \setminus x_0)$ and by Theorem 15.8 the bundle $P \to X$ has the form $P_{\rho,\alpha}$. By Lemma 15.1 for the representation $\rho : \pi_1(X \setminus x_0) \to G_{\mathbf{C}}$ there exists a G-system $Df = \theta$ with a singularity at the point x_0, whose monodromy representation coincides with ρ. The form θ is Γ-invariant and thus is a connection for the bundle P. The second identity of relations (15.25) implies that x_0 is a regular singular point of the equation $Df = \theta$. The theorem is proved. $\square$

Proposition 15.6. *Let the monodromy representation of a G-system $Df = \omega$ with one regular singular point x_0 be unitary, irreducible and $\rho(\gamma) = c \in Z_0$, where γ is a loop circling around x_0. Then the characteristic class of the principal bundle $P_{\rho,\alpha}$ corresponding to this G-system equals $\beta - \alpha$.*

Proof. We will apply the following fact well known in algebraic geometry. If G is a reductive group with connected center $Z(G)$, then $G_1 = [G, G]$ is a semisimple group and the homomorphism $Z(G) \times G_1 \to G$ has finite kernel. Let $\tilde{G}_1$ be the universal cover of G_1 and $\tilde{Z}_0$ be the universal cover of $Z(G)$, then $\tilde{G} = \tilde{Z}_0 \times \tilde{G}_1 \to G$ is a universal cover of G.

The embedding $Z_0 \subset G$ induces canonically an embedding $\tilde{Z}_0 \subset \tilde{Z}_0 \times \tilde{G}_1$, and $\tilde{Z}_0$ can be identified with its image. Since $S_X^1 = \prod_{i=1}^{g}[a_i, b_i]$ and $\pi_1(X \setminus x_0)$ is a free group, there exists a lifting of the homomorphism $\rho : \pi_1(X \setminus x_0) \to G$, i.e. the diagram

$$
\begin{array}{ccc}
 & & \tilde{G} \\
 & \overset{\tilde{\rho}}{\nearrow} & \\
\pi_1(X \setminus x_0) & & \downarrow{\tilde{\rho}} \\
 & \overset{\rho}{\searrow} & \\
 & & G
\end{array}
$$

commutes. Since $\rho(S_X^1)$ lies in the center of the group G and $\pi\tilde{\rho} = \rho$, it follows that $\tilde{\rho}(\gamma) = \beta$ lies in the kernel of p, i.e. in $\pi_1(G)$. The element β does not depend on the lifting of the homomorphism ρ and by the definition of the characteristic class one obtains

$$
\chi(P_{\rho,\alpha}) = \beta - \alpha.
$$

The theorem is proved. $\qquad\square$

References

[1] Alase Abhijeet, Boundary Physics and Bulk-Boundary Correspondence in Topological Phases of Matter, Springer, 2019.

[2] J. Adams, Lectures on the Lie Groups, University of Chicago Press, 1969.

[3] V. Adukov, Wiener-Hopf factorization, Algebra Anal. 4 (1) (1992) 54–74.

[4] D.V. Anosov, A.A. Bolibruch, The Riemann-Hilbert Problem, Aspects of Mathematics, Vieweg, Braunschweig, Wiesbaden, 1994.

[5] V.I. Arnold, Yu.S. Il'yashenko, Ordinary Differential Equations, Dynamical Systems I, Encyclopaedia of Mathematical Sciences, vol. 1, Springer, 1988.

[6] M.F. Atiyah, Vector bundles over an elliptic curve, Proc. Lond. Math. Soc. 3 (7) (1957) 414–452.

[7] M.F. Atiyah, R. Bott, The Yang-Mills equation on the Riemann surface, Philos. Trans. R. Soc. Lond. A 308 (1982) 523–615.

[8] M.F. Atiyah, Instantons in two and four dimensions, Commun. Math. Phys. 93 (4) (1984) 437–451.

[9] V. Balan, J. Dorfmeister, Birkhoff decompositions and Iwasawa decompositions for loop groups, Tohoku Math. J. 53 (4) (2001) 593–615.

[10] B. Bojarski, Connection between complex and global analysis: analytical and geometrical aspects of the Riemann-Hilbert transition problem, in: Complex Analysis, Methods, Application, Berlin, A.V., 1983.

[11] B. Bojarski, G. Giorgadze, Some analytical and geometrical aspects of stable partial indices, Proc. I. Vekua Inst. Appl. Math. 61–62 (2010) 14–32.

[12] A.A. Bolibruch, The Riemann-Hilbert problem, Russ. Math. Surv. 45 (2) (1990) 1–47.

[13] A. Connes, D. Kreimer, Renormalization in quantum field theory and the Riemann-Hilbert problem. I. The Hopf algebra structure of graphs and the main theorem, Commun. Math. Phys. 210 (1) (2000) 249–273.

[14] P. Deligne, Equations différentiales á points singuliers réguliers, Lect. Notes Math., vol. 163, 1970.

[15] Sh. Disney, The exponents of loops on complex general linear group, Topology 12 (1973) 297–315.

[16] H. Doi, Nonlinear equations on a Lie group, Hiroshima Math. J. 17 (3) (1987) 535–560.

326 Mechanics and Physics of Structured Media

[17] S.K. Donaldson, A new proof of a theorem of Narasimhan and Seshadri, J. Differ. Geom. 18 (1983) 269–277.

[18] T. Ehrhardt, I. Spitkovsky, Factorization of piecewise constant matrix functions and systems of linear differential equations, Algebra Anal. 13 (6) (2001) 56–123.

[19] A.S. Fokas, A.R. Its, A.A. Kapaev, V.Yu. Novokshenov, Painlevé Transcendents: The Riemann-Hilbert Approach, AMS, 2006.

[20] F. Gakhov, Boundary Value Problems, Elsevier, 1966.

[21] G. Giorgadze, G-systems and holomorphic principal bundles on Riemann surfaces, J. Dyn. Control Syst. 8 (2) (2002) 245–291.

[22] G. Giorgadze, G. Khimshiashvili, Factorization of loops in loop groups, Bull. Georgian Natl. Acad. Sci. 5 (3) (2011) 35–38.

[23] G. Giorgadze, G. Khimshiashvili, The Riemann-Hilbert problem in loop spaces, Dokl. Math. 73 (2) (2006) 258–260.

[24] G. Giorgadze, On holomorphic principal bundles on a Riemann surface, Bull. Georgian Acad. Sci. 166 (1) (2002) 27–31.

[25] S. Gluzman, V. Mityushev, W. Nawalaniec, Computational Analysis of Structured Media, Academic Press, 2017.

[26] I. Gohberg, M.G. Krein, Systems of integral equations on semiaxis with kernel depending on the arguments difference, Usp. Mat. Nauk 13 (1958) 3–72.

[27] A. Grothendieck. , Sur le mémoire de Weil: Généralisation des fonctions abéllianes, Sém. Bourbaki 141 (1956).

[28] U. Helmke, P. Slodowy, Loop groups, elliptic singularities and principal bundles over elliptic curves, Banach Cent. Publ. 62 (2003) 87–99.

[29] A. Its, The Riemann-Hilbert problem and integrable systems, Not. Am. Math. Soc. 50 (11) (2003) 1389–1400.

[30] G.N. Khimshiashvili, Lie groups and transmission problems on Riemann surface, Contemp. Math. 131 (1992) 164–178.

[31] G.N. Khimshiashvili, On a Riemann-Hilbert problem for a compact Lie group, Dokl. Akad. Nauk SSSR 310 (5) (1990) 1055–1058.

[32] N.I. Muskhelishvili, Singular Integral Equations, Noordhoff, Groningen, 1953.

[33] M.S. Narasimhan, T.R. Seshadri, Stable and unitary vector bundles on a compact Riemann surface, Ann. Math. 82 (1965) 540–564.

[34] A. Onishchik, Some concepts and applications of non-abelian cohomology theory, Trans. Mosc. Math. Soc. 17 (1967) 49–97.

[35] J. Plemelj, Problems in the Sense of Riemann and Klein, Intersience Publishers. A division of J. Wiley & Sons Inc., New York, London, Sidney, 1964.

[36] A. Pressley, G. Segal, Loop Groups, Clarendon Press, Oxford, 1984.

[37] A. Ramanathan, Stable principal bundles on a compact Riemann surface, Math. Ann. 213 (1975) 129–152.

[38] H. Röhrl, Holomorphic vector bundles over Riemann surfaces, Bull. Am. Math. Soc. 68 (3) (1962) 125–160.

[39] S.S. Shatz, On subbundles of vector bundles over $\mathbf{CP}^1$, J. Pure Appl. Algebra 10 (1977) 315–322.

[40] N.P. Vekua, Systems of Singular Integral Equations and Certain Boundary Value Problems, Nauka, Moscow, 1970.

Chapter 16

When risks and uncertainties collide: quantum mechanical formulation of mathematical finance for arbitrage markets

Simone Farinelli[a] and Hideyuki Takada[b]

[a]*Core Dynamics GmbH, Zurich, Switzerland,* [b]*Department of Information Science, Narashino Campus, Toho University, Funabashi-Shi, Chiba, Japan*

16.1 Introduction

This chapter further develops a conceptual structure – called geometric arbitrage theory – to link arbitrage modeling in generic markets with quantum mechanics.

Geometric arbitrage theory rephrases classical stochastic finance in stochastic differential geometric terms in order to characterize arbitrage. The main idea of the geometric arbitrage theory approach consists of modeling markets made of basic financial instruments together with their term structures as principal fiber bundles. Financial features of this market – like no arbitrage and equilibrium – are then characterized in terms of standard differential geometric constructions – like curvature – associated to a natural connection in this fiber bundle. Principal fiber bundle theory has been heavily exploited in theoretical physics as the language in which laws of nature can be best formulated by providing an invariant framework to describe physical systems and their dynamics. These ideas can be carried over to mathematical finance and economics. A market is a financial-economic system that can be described by an appropriate principle fiber bundle. A principle like the invariance of market laws under change of numéraire can be seen then as gauge invariance. Concepts like No-Free-Lunch-with-Vanishing-Risk (NFLVR) and No-Unbounded-Profit-with-Bounded-Risk (NUPBR) have a geometric characterization, which have the Capital Asset Pricing Model (CAPM) as a consequence.

The fact that gauge theories are the natural language to describe economics was first proposed by Malaney and Weinstein in the context of the economic index problem ([23], [30]). Ilinski (see [19] and [20]) and Young ([32]) proposed to view arbitrage as the curvature of a gauge connection, in analogy to some physical theories. Independently, Cliff and Speed ([27]) further devel-

Mechanics and Physics of Structured Media. https://doi.org/10.1016/B978-0-32-390543-5.00021-9
Copyright © 2022 Elsevier Inc. All rights reserved.

oped Flesaker and Hughston seminal work ([12]) and utilized techniques from differential geometry (indirectly mentioned by allusive wording) to reduce the complexity of asset models before stochastic modeling.

The contributions of Cliff and Speed are independent from those of Ilinski, Malaney, and Young. The former represent a base financial asset as an ordered couple of a deflator and a term structure, where the deflator is the asset value expressed in term of some numéraire and the term structure models the price structure of futures, i.e., linear derivatives of the base assets. Gauge transforms correspond to the linear portfolio construction and are utilized to represent a financial market by means of a minimal set of gauges on which stochastic modeling is to be applied. But Cliff and Speed do not use the formalism of differential stochastic geometry. Ilinski, Malaney, and Young, on the contrary, after having introduced a different principal bundle structure corresponding to dilations, utilize a connection to define the parallel transport. Ilinski, moreover, is the first one to consider the concept of curvature applied to financial markets.

This chapter is structured as follows. Section 16.2 reviews classical stochastic finance and geometric arbitrage theory. Arbitrage is seen as curvature of a principal fiber bundle representing the market which defines the quantity of arbitrage associated to it. Proofs are omitted and can be found in [9], [10], and in [11], where geometric arbitrage theory has been given a rigorous mathematical foundation utilizing the background of stochastic differential geometry as in Schwartz ([26]), Elworthy ([6]), Eméry ([7]), Hackenbroch and Thalmaier ([14]), Stroock ([28]), and Hsu ([17]).

Section 16.3 describes the intertwined dynamics of assets, term structures, and market portfolio as constrained Lagrange system deriving it from a stochastic variational principle whose Lagrange function measures the arbitrage quantity allowed by the market. This constrained Lagrange system and its stochastic Euler-Lagrange equation is equivalent to a constrained Hamilton system, obtained by Legendre transform, with its stochastic Hamilton equations. These stochastic Hamilton system is, on its turn, equivalent to a quantum mechanical system, obtained by quantizing the deterministic version of the Hamilton system. This is shown in Section 16.4 where we reformulate mathematical finance in terms of quantum mechanics. The Schrödinger equation describes then both the asset and market portfolio dynamics, which can be explicitly computed once the spectrum of the Hamilton operator is known. Without knowledge of the spectrum it is still possible by means of Ehrenfest's theorem to determine stochastic properties of future asset values and market portfolio nominals. Their expected values are identical to those computed with the stochastic Euler Lagrange equation, demonstrating the consistency of the quantum mechanical approach. Moreover, we prove that for a closed market, the returns on market portfolio nominals, asset values, and term structures are centered and serially uncorrelated. Hence, the justification of econometrics with its autocorrelated models or those with stochastic volatilities lies in the fact that markets are not closed: there is always an asset category which has not been modeled, to which

When risks and uncertainties collide **Chapter | 16 329**

wealth can escape, destroying the uncorrelated identical distributional behavior of the remaining asset categories. By applying Heisenberg's uncertainty relation to the quantum mechanical model of the market we obtain one more econometric result: the volatilities of asset values on one hand and their weights in the market portfolio are mutually exclusive, meaning by this that if the former increases, the latter decreases, and vice versa.

In Section 16.5 we solve the Schrödinger equation representing the arbitrage market dynamics by using Feynman's path integrals. Appendix 16.A reviews and generalizes Nelson's stochastic derivatives. Section 16.6 concludes.

16.2 Geometric arbitrage theory background

In this section we explain the main concepts of geometric arbitrage theory introduced in [9], to which we refer for proofs and examples.

16.2.1 The classical market model

In this subsection we will summarize the classical set up, which will be rephrased in Section 16.2.4 in differential geometric terms. We basically follow [18] and [3].

We assume continuous time trading and that the set of trading dates is $[0, +\infty[$. This assumption is general enough to embed the cases of finite and infinite discrete times as well as the one with a finite horizon in continuous time. This motivates the technical effort of continuous time stochastic finance.

The outcome of chance is modeled by a filtered probability space $(\Omega, \mathcal{A}, \mathbb{P})$, where $\mathbb{P}$ is the statistical (physical) probability measure, $\mathcal{A} = \{\mathcal{A}_t\}_{t \in [0,+\infty[}$ an increasing family of sub-σ-algebras of $\mathcal{A}_\infty$, and $(\Omega, \mathcal{A}_\infty, \mathbb{P})$ is a probability space. The filtration $\mathcal{A}$ is assumed to satisfy the usual conditions: $\mathcal{A}_t = \bigcap_{s>t} \mathcal{A}_s$ for all $t \in [0, +\infty[$ (right continuity), and $\mathcal{A}_0$ contains all null sets of $\mathcal{A}_\infty$.

The market consists of finitely many **assets** indexed by $j = 1, \ldots, N$, whose **nominal prices** are given by the vector valued semimartingale $S : [0, +\infty[\times \Omega \to \mathbb{R}^N$ denoted by $(S_t)_{t \in [0,+\infty[}$ adapted to the filtration $\mathcal{A}$. The stochastic process $(S_t^j)_{t \in [0,+\infty[}$ describes the price at time t of the jth asset in terms of unit of cash *at time* $t = 0$. More precisely, we assume the existence of a 0th asset, the **cash**, a strictly positive semimartingale, which evolves according to $S_t^0 = \exp(\int_0^t du\, r_u^0)$, where the integrable semimartingale $(r_t^0)_{t \in [0,+\infty[}$ represents the continuous interest rate provided by the cash account: one always knows in advance what the interest rate on the own bank account is, but this can change from time to time. The cash account is therefore considered the locally risk less asset in contrast to the other assets, the risky ones. In the following we will mainly utilize **discounted prices**, defined as $\hat{S}_t^j := S_t^j / S_t^0$, representing the asset prices in terms of *current* unit of cash.

We remark that there is no need to assume that asset prices are positive. But, there must be at least one strictly positive asset, in our case the cash. If

330 Mechanics and Physics of Structured Media

we want to renormalize the prices by choosing another asset instead of the cash as reference, i.e., by making it to our **numéraire**, then this asset must have a strictly positive price process. More precisely, a generic numéraire is an asset, whose nominal price is represented by a strictly positive stochastic process $(B_t)_{t \in [0,+\infty[}$, and which is a portfolio of the original assets $j = 0, 1, 2, \dots, N$. The discounted prices of the original assets are then represented in terms of the numéraire by the semimartingales $\hat{S}_t^j := S_t^j / B_t$.

We assume that there are no transaction costs and that short sales are allowed. Remark that the absence of transaction costs can be a serious limitation for a realistic model. The filtration $\mathcal{A}$ is not necessarily generated by the price process $(S_t)_{t \in [0,+\infty[}$: other sources of information than prices are allowed. All agents have access to the same information structure, that is to the filtration $\mathcal{A}$.

Let $v \geq 0$. A v-admissible **strategy** $x = (x_t)_{t \in [0,+\infty[}$ is a predictable semimartingale for which the Itô integral $\int_0^t x \cdot dS \geq -v$ for all $t \geq 0$. A strategy is admissible if it is v-admissible for some $v \geq 0$. An admissible strategy x is said to be **self-financing** if and only if $V_t := x_t \cdot S_t$, the portfolio value at time t, is given by

$$V_t = V_0 + \int_0^t x_u \cdot dS_u. \tag{16.1}$$

Definition 16.1 (Arbitrage). Let the process $(S_t)_{[0,+\infty[}$ be a semimartingale and $(x_t)_{t \in [0,+\infty[}$ be admissible self-financing strategy. Let us consider trading up to time $T \leq \infty$. The portfolio wealth at time t is given by $V_t(x) := V_0 + \int_0^t x_u \cdot dS_u$, and we denote by K_0 the subset of $L^0(\Omega, \mathcal{A}_T, P)$ containing all such $V_T(x)$, where x is any admissible self-financing strategy. We define:

1. $C_0 := K_0 - L_+^0(\Omega, \mathcal{A}_T, P)$.
2. $C := C_0 \cap L_+^\infty(\Omega, \mathcal{A}_T, P)$.
3. $\bar{C}$: the closure of C in L^∞ with respect to the norm topology.
4. $\mathcal{V}^{V_0} := \{(V_t)_{t \in [0,+\infty[} \mid V_t = V_t(x), \text{ where } x \text{ is } V_0\text{-admissible}\}$.
5. $\mathcal{V}_T^{V_0} := \{V_T \mid (V_t)_{t \in [0,+\infty[} \in \mathcal{V}^{V_0}\}$: terminal wealth for V_0-admissible self-financing strategies.

We say that S satisfies

1. **(NA), no arbitrage**, if and only if $C \cap L^\infty(\Omega, \mathcal{A}_T, P) = \{0\}$.
2. **(NFLVR), no-free-lunch-with-vanishing-risk**, if and only if $\bar{C} \cap L^\infty(\Omega, \mathcal{A}_T, P) = \{0\}$.
3. **(NUPBR), no-unbounded-profit-with-bounded-risk**, if and only if $\mathcal{V}_T^{V_0}$ is bounded in L^0 for some $V_0 > 0$.

The relationship between these three different types of arbitrage has been elucidated in [2] and in [21] with the proof of the following result.

Theorem 16.2.

$$(NFLVR) \Leftrightarrow (NA) + (NUPBR). \tag{16.2}$$

When risks and uncertainties collide Chapter | 16 **331**

Remark 16.3. We recall that, as shown in [2,21], (NUPBR) is equivalent to (NAA_1), i.e., no asymptotic arbitrage of the 1st kind, and equivalent to (NA_1), i.e., no arbitrage of the 1st kind.

Theorem 16.4 (First fundamental theorem of asset pricing). *The market* $(S, \mathcal{A})$ *satisfies the (NFLVR) condition if and only if there exists an equivalent local martingale measure* P^*.

Remark 16.5. In the first fundamental theorem of asset pricing we just assumed that the price process S is locally bounded. If S is bounded, then (NFLVR) is equivalent to the existence of a martingale measure. But without this additional assumption (NFLVR) only implies the existence of a *local* martingale measure, i.e., a local martingale which is *not* a martingale. This distinction is important, because the difference between a security price process being a strict local martingale versus a martingale under a probability P^* relates to the existence of asset price bubbles.

16.2.2 Geometric reformulation of the market model: primitives

We are going to introduce a more general representation of the market model introduced in Subsection 16.2.1, which better suits to the arbitrage modeling task.

Definition 16.6. A **gauge** is an ordered pair of two $\mathcal{A}$-adapted real valued semi-martingales (D, P), where $D = (D_t)_{t \geq 0} : [0, +\infty[\times \Omega \to \mathbb{R}$ is called **deflator** and $P = (P_{t,s})_{t,s} : \mathcal{T} \times \Omega \to \mathbb{R}$, which is called **term structure**, is considered as a stochastic process with respect to the time t, termed **valuation date**, and $\mathcal{T} := \{(t, s) \in [0, +\infty[^2 \, | \, s \geq t\}$. The parameter $s \geq t$ is referred as **maturity date**. The following properties must be satisfied a.s. for all t, s such that $s \geq t \geq 0$:

  (i) $P_{t,s} > 0$,
  (ii) $P_{t,t} = 1$.

Remark 16.7. Deflators and term structures can be considered *outside the context of fixed income*. An arbitrary financial instrument is mapped to a gauge (D, P) with the following economic interpretation:

- Deflator: D_t is the value of the financial instrument at time t expressed in terms of some numéraire. If we choose the cash account, the 0-th asset as numéraire, then we can set $D_t^j := \hat{S}_t^j = \frac{S_t^j}{S_t^0}$ $(j = 1, \ldots N)$.
- Term structure: $P_{t,s}$ is the value at time t (expressed in units of deflator at time t) of a synthetic zero coupon bond with maturity s delivering one unit of financial instrument at time s. It represents a term structure of forward prices with respect to the chosen numéraire.

We point out that there is no unique choice for deflators and term structures describing an asset model. For example, if a set of deflators qualifies, then we can

332 Mechanics and Physics of Structured Media

multiply every deflator by the same positive semimartingale to obtain another suitable set of deflators. Of course term structures have to be modified accordingly. The term "deflator" is clearly inspired by actuarial mathematics and was first introduced by Smith and Speed in [27]. In the present context it refers to an asset value up division by a strictly positive semimartingale (which can be the state price deflator if this exists and it is made to the numéraire). There is no need to assume that a deflator is a positive process. However, if we want to make an asset to our numéraire, then we have to make sure that the corresponding deflator is a strictly positive stochastic process.

16.2.3 Geometric reformulation of the market model: portfolios

We want now to introduce transforms of deflators and term structures in order to group gauges containing the same (or less) stochastic information. That for, we will consider *deterministic* linear combinations of assets modeled by the same gauge (e.g. zero bonds of the same credit quality with different maturities).

Definition 16.8. Let $\pi : [0, +\infty[\longrightarrow \mathbb{R}$ be a deterministic cashflow intensity (possibly generalized) function. It induces a **gauge transform** $(D, P) \mapsto \pi(D, P) := (D, P)^\pi := (D^\pi, P^\pi)$ by the formulas

$$D_t^\pi := D_t \int_0^{+\infty} dh\, \pi_h P_{t,t+h}, \qquad P_{t,s}^\pi := \frac{\int_0^{+\infty} dh\, \pi_h P_{t,s+h}}{\int_0^{+\infty} dh\, \pi_h P_{t,t+h}}. \qquad (16.3)$$

Remark 16.9. The cashflow intensity π specifies the bond cashflow structure. The bond value at time t expressed in terms of the market model numéraire is given by D_t^π. The term structure of forward prices for the bond future expressed in terms of the bond current value is given by $P_{t,s}^\pi$.

Proposition 16.10. *Gauge transforms induced by cashflow vectors have the following property:*

$$((D, P)^\pi)^\nu = ((D, P)^\nu)^\pi = (D, P)^{\pi * \nu}, \qquad (16.4)$$

where $*$ denotes the convolution product of two cashflow vectors or intensities respectively:

$$(\pi * \nu)_t := \int_0^t dh\, \pi_h \nu_{t-h}. \qquad (16.5)$$

The convolution of two noninvertible gauge transform is noninvertible. The convolution of a noninvertible with an invertible gauge transform is noninvertible.

Definition 16.11. If the term structure is differentiable with respect to the maturity date, it can be written as a functional of the **instantaneous forward rate**

f defined as

$$f_{t,s} := -\frac{\partial}{\partial s} \log P_{t,s}, \quad P_{t,s} = \exp\left(-\int_t^s dh f_{t,h}\right), \tag{16.6}$$

and

$$r_t := \lim_{s \to t^+} f_{t,s} \tag{16.7}$$

is termed **short rate**.

Remark 16.12. The special choice of vanishing interest rate $r \equiv 0$ or flat term structure $P \equiv 1$ for all assets corresponds to the classical model, where only asset prices and their dynamics are relevant.

16.2.4 Arbitrage theory in a differential geometric framework

Now we are in the position to rephrase the asset model presented in Subsection 16.2.1 in terms of a natural geometric language. Given N base assets we want to construct a portfolio theory and study arbitrage and thus we cannot a priori assume the existence of a risk neutral measure or of a state price deflator. The market model is seen as a principal fiber bundle of the (deflator, term structure) pairs, discounting and foreign exchange as a parallel transport, numéraire as global section of the gauge bundle, arbitrage as curvature. The no-free-lunch-with-vanishing-risk condition is proved to be equivalent to a zero curvature condition.

16.2.4.1 Market model as principal fiber bundle

Let us consider – in continuous time – a market with N assets and a numéraire. A general portfolio at time t is described by the vector of nominals $x \in \mathfrak{X}$, for an open set $\mathfrak{X} \subset \mathbb{R}^N$. Following Definition 16.6, the asset model consisting in N synthetic zero bonds is described by means of the gauges

$$(D^j, P^j) = ((D_t^j)_{t \in [0, +\infty[}, (P_{t,s}^j)_{s \geq t}), \tag{16.8}$$

where D^j denotes the deflator and P^j the term structure for $j = 1, \ldots, N$. This can be written as

$$P_{t,s}^j = \exp\left(-\int_t^s f_{t,u}^j du\right), \tag{16.9}$$

where f^j is the instantaneous forward rate process for the j-th asset and the corresponding short rate is given by $r_t^j := \lim_{u \to 0+} f_{t,u}^j$. For a portfolio with

334 Mechanics and Physics of Structured Media

nominals $x \in \mathfrak{X} \subset \mathbb{R}^N$ we define

$$D_t^x := \sum_{j=1}^N x_j D_t^j, \qquad f_{t,u}^x := \sum_{j=1}^N \frac{x_j D_t^j}{\sum_{j=1}^N x_j D_t^j} f_{t,u}^j,$$

$$P_{t,s}^x := \exp\left(-\int_t^s f_{t,u}^x du\right). \tag{16.10}$$

The short rate writes

$$r_t^x := \lim_{u \to 0^+} f_{t,u}^x = \sum_{j=1}^N \frac{x_j D_t^j}{\sum_{j=1}^N x_j D_t^j} r_t^j. \tag{16.11}$$

The image space of all possible strategies reads

$$M := \{(t, x) \in [0, +\infty[\times \mathfrak{X}\}. \tag{16.12}$$

In Subsection 16.2.3 cashflow intensities and the corresponding gauge transforms were introduced. They have the structure of an Abelian semigroup

$$H := \mathcal{E}'([0, +\infty[, \mathbb{R}) = \{F \in \mathcal{D}'([0, +\infty[) \mid \mathrm{supp}(F) \subset [0, +\infty[\text{ is compact}\}, \tag{16.13}$$

where the semigroup operation on distributions with compact support is the convolution (see [16], Chapter IV), which extends the convolution of regular functions as defined by formula (16.5).

Definition 16.13. The **Market Fiber Bundle** is defined as the fiber bundle of gauges

$$\mathcal{B} := \{(D_t^x, P_{t,\cdot}^x)^\pi \mid (t, x) \in M, \pi \in G\}. \tag{16.14}$$

The cashflow intensities defining invertible transforms constitute an Abelian group

$$G := \{\pi \in H \mid \text{it exists } \nu \in H \text{ such that } \pi * \nu = \delta\} \subset \mathcal{E}'([0, +\infty[, \mathbb{R}). \tag{16.15}$$

From Proposition 16.10 we obtain

Theorem 16.14. *The market fiber bundle $\mathcal{B}$ has the structure of a G-principal fiber bundle ([1]) given by the action*

$$\mathcal{B} \times G \longrightarrow \mathcal{B},$$
$$((D, P), \pi) \mapsto (D, P)^\pi = (D^\pi, P^\pi). \tag{16.16}$$

The group G acts freely and differentiably on $\mathcal{B}$ to the right.

16.2.4.2 Stochastic parallel transport

Let us consider the projection of $\mathcal{B}$ onto M

$$
\begin{aligned}
p : \mathcal{B} &\cong M \times G \longrightarrow M, \\
(t, x, g) &\mapsto (t, x),
\end{aligned}
\tag{16.17}
$$

and its differential map at $(t, x, g) \in \mathcal{B}$ denoted by $T_{(t,x,g)}p$, see for example, Definition 0.2.5 in [1]

$$
T_{(t,x,g)}p : \quad \underbrace{T_{(t,x,g)}\mathcal{B}}_{\cong \mathbb{R}^N \times \mathbb{R} \times \mathbb{R}^{[0,+\infty[}} \longrightarrow \underbrace{T_{(t,x)}M}_{\cong \mathbb{R}^N \times \mathbb{R}} .
\tag{16.18}
$$

The vertical directions are

$$
\mathcal{V}_{(t,x,g)}\mathcal{B} := \ker\left(T_{(t,x,g)}p\right) \cong \mathbb{R}^{[0,+\infty[},
\tag{16.19}
$$

and the horizontal ones are

$$
\mathcal{H}_{(t,x,g)}\mathcal{B} \cong \mathbb{R}^{N+1}.
\tag{16.20}
$$

An Ehresmann connection on $\mathcal{B}$ is a projection $T\mathcal{B} \to \mathcal{V}\mathcal{B}$. More precisely, the vertical projection must have the form

$$
\begin{aligned}
\Pi^v_{(t,x,g)} : T_{(t,x,g)}\mathcal{B} &\longrightarrow \mathcal{V}_{(t,x,g)}\mathcal{B}, \\
(\delta x, \delta t, \delta g) &\mapsto (0, 0, \delta g + \Gamma(t, x, g).(\delta x, \delta t)),
\end{aligned}
\tag{16.21}
$$

and the horizontal one must read

$$
\begin{aligned}
\Pi^h_{(t,x,g)} : T_{(t,x,g)}\mathcal{B} &\longrightarrow \mathcal{H}_{(t,x,g)}\mathcal{B}, \\
(\delta x, \delta t, \delta g) &\mapsto (\delta x, \delta t, -\Gamma(t, x, g).(\delta x, \delta t)),
\end{aligned}
\tag{16.22}
$$

such that

$$
\Pi^v + \Pi^h = \mathbb{1}_{\mathcal{B}}.
\tag{16.23}
$$

Stochastic parallel transport on a principal fiber bundle along a semimartingale is a well defined construction (cf. [14], Chapter 7.4 and [17] Chapter 2.3 for the frame bundle case) in terms of Stratonovich integral. Existence and uniqueness can be proved analogously to the deterministic case by formally substituting the deterministic time derivative $\frac{d}{dt}$ with the stochastic one $\mathcal{D}$ corresponding to the Stratonovich integral.

Following Ilinski's idea ([20]), we motivate the choice of a particular connection by the fact that it allows to encode foreign exchange and discounting as parallel transport.

336 Mechanics and Physics of Structured Media

Theorem 16.15. *With the choice of connection*

$$\chi(t, x, g).(\delta x, \delta t) := \left(\frac{D_t^{\delta x}}{D_t^x} - r_t^x \delta t \right) g, \tag{16.24}$$

the parallel transport in $\mathcal{B}$ has the following financial interpretations:

- *Parallel transport along the nominal directions (x-lines) corresponds to a multiplication by an exchange rate.*
- *Parallel transport along the time direction (t-line) corresponds to a division by a stochastic discount factor.*

Recall that time derivatives needed to define the parallel transport along the time lines have to be understood in Stratonovich's sense. We see that the bundle is trivial, because it has a global trivialization, but the connection is not trivial.

16.2.4.3 Nelson $\mathcal{D}$ weak differentiable market model

We continue to reformulate the classic asset model introduced in Subsection 16.2.1 in terms of stochastic differential geometry.

Definition 16.16. A **Nelson $\mathcal{D}$ weak differentiable market model** for N assets is described by N gauges which are Nelson $\mathcal{D}$ weak differentiable with respect to the time variable. More exactly, for all $t \in [0, +\infty[$ and $s \geq t$ there is an open time interval $I \ni t$ such that for the deflators $D_t := [D_t^1, \ldots, D_t^N]^\dagger$ and the term structures $P_{t,s} := [P_{t,s}^1, \ldots, P_{t,s}^N]^\dagger$, the latter seen as processes in t and parameter s, there exists a $\mathcal{D}$ weak t-derivative (see Appendix 16.A). The short rates are defined by $r_t := \lim_{s \to t^-} \frac{\partial}{\partial s} \log P_{ts}$.

A strategy is a curve $\gamma : I \to X$ in the portfolio space parameterized by the time. This means that the allocation at time t is given by the vector of nominals $x_t := \gamma(t)$. We denote by $\bar{\gamma}$ the lift of γ to M, that is $\bar{\gamma}(t) := (\gamma(t), t)$. A strategy is said to be **closed** if it represented by a closed curve. A **weak $\mathcal{D}$-admissible strategy** is predictable and $\mathcal{D}$-weak differentiable.

Remark 16.17. We require weak $\mathcal{D}$-differentiability and not strong $\mathcal{D}$-differentiability because imposing a priori regularity properties on the trading strategies corresponds to restricting the class of admissible strategies with respect to the classical notion of Delbaen and Schachermayer. Every (no-)arbitrage consideration depends crucially on the chosen definition of admissibility. Therefore, restricting the class of admissible strategies may lead to the automatic exclusion of potential arbitrage opportunities, leading to vacuous statements of FTAP-like results. An admissible strategy in the classic sense (see Section 16.2) is weak $\mathcal{D}$-differentiable.

In general the allocation can depend on the state of the nature, i.e., $x_t = x_t(\omega)$ for $\omega \in \Omega$.

When risks and uncertainties collide   Chapter | 16   **337**

Proposition 16.18. *A weak $\mathcal{D}$-admissible strategy is self-financing if and only if*

$$\mathcal{D}(x_t \cdot D_t) = x_t \cdot \mathcal{D}D_t - \frac{1}{2}\mathfrak{D}_* \langle x, D \rangle_t \quad or \quad \mathcal{D}x_t \cdot D_t = -\frac{1}{2}\mathfrak{D}_* \langle x, D \rangle_t$$
$$or \quad \mathfrak{D}x_t \cdot D_t = 0, \tag{16.25}$$

almost surely. The bracket $\langle \cdot, \cdot \rangle$ denotes the continuous part of the quadratic covariation.

For the remainder of this chapter unless otherwise stated we will deal only with $\mathcal{D}$ differentiable market models, $\mathcal{D}$ differentiable strategies, and, when necessary, with $\mathcal{D}$ differentiable state price deflators. All Itô processes are $\mathcal{D}$ differentiable, so that the class of considered admissible strategies is very large.

16.2.4.4 Arbitrage as curvature

The Lie algebra of G is

$$\mathfrak{g} = \mathbb{R}^{[0, +\infty[} \tag{16.26}$$

and therefore commutative. The $\mathfrak{g}$-valued curvature 2-form is defined by means the $\mathfrak{g}$-valued connection 1-form as

$$R := d\chi + [\chi, \chi], \tag{16.27}$$

meaning by this, that for all $(t, x, g) \in \mathcal{B}$ and for all $\xi, \eta \in T_{(t,x)}M$

$$R(t, x, g)(\xi, \eta) := d\chi(t, x, g)(\xi, \eta) + [\chi(t, x, g)(\xi), \chi(t, x, g)(\eta)]$$
$$= d\chi(t, x, g)(\xi, \eta). \tag{16.28}$$

Remark that, being the Lie algebra commutative, the Lie bracket $[\cdot, \cdot]$ vanishes.

Proposition 16.19 (Curvature formula). *Let R be the curvature. Then, the following quality holds:*

$$R(t, x, g) = g \, dt \wedge d_x \left[\mathcal{D} \log(D_t^x) + r_t^x \right]. \tag{16.29}$$

The following result characterizes arbitrage as curvature.

Theorem 16.20 (No arbitrage). *The following assertions are equivalent:*

(i) *The market model (with base assets and futures with discounted prices D and P) satisfies the no-free-lunch-with-vanishing-risk condition.*

(ii) *There exists a positive local martingale $\beta = (\beta_t)_{t \geq 0}$ such that deflators and short rates satisfy for all portfolio nominals and all times the condition*

$$r_t^x = -\mathcal{D} \log(\beta_t D_t^x). \tag{16.30}$$

338 Mechanics and Physics of Structured Media

(iii) *There exists a positive local martingale* $\beta = (\beta_t)_{t \geq 0}$ *such that deflators and term structures satisfy for all portfolio nominals and all times the condition*

$$P_{t,s}^x = \frac{\mathbb{E}_t[\beta_s D_s^x]}{\beta_t D_t^x}. \tag{16.31}$$

This motivates the following definition.

Definition 16.21. The market model satisfies the **zero curvature (ZC)** if and only if the curvature vanishes a.s.

Therefore, we have following implications connecting two different definitions of no-arbitrage:

Corollary 16.22.

$$(NFLVR) \Rightarrow (ZC). \tag{16.32}$$

Remark 16.23. The positive local martingale $\beta = (\beta_t)_{t \geq 0}$ in Theorem 16.20 is termed pricing kernel or state price deflator throughout the literature.

16.3 Asset and market portfolio dynamics as a constrained Lagrangian system

In [9] and [10] the minimal arbitrage principle, stating that asset dynamics and market portfolio choose the path guaranteeing the minimization of arbitrage, was encoded as the Hamilton principle under constraints for a Lagrangian measuring the arbitrage. Then, the SDE describing asset deflators, term structures, and market portfolio were derived by means of a stochasticization procedure of the Euler-Lagrange equations following a technique developed by Cresson and Darses ([4] who follow previous works of Yasue ([31]) and Nelson ([25]). Since we need this set up to proceed with its quantization, we briefly summarize it here below.

Definition 16.24. Let γ be the market $\mathcal{D}$-admissible strategy, and $\delta\gamma$, δD, δr be perturbations of the market strategy, deflators', and short rates' dynamics. The **variation** of (γ, D, r) with respect to the given perturbations is the following one parameter family:

$$\epsilon \longmapsto (\gamma^\epsilon, D^\epsilon, r^\epsilon) := (\gamma, D, r) + \epsilon(\delta\gamma, \delta D, \delta r). \tag{16.33}$$

Thereby, the parameter ϵ belongs to some open neighborhood of $0 \in \mathbb{R}$. The **arbitrage action** with respect to a positive local martingale β can be consistently defined by

$$A^\beta(\gamma; D, r) := \int_\gamma dt \left\{ \mathcal{D} \log(\beta_t D_t^{x_t}) + r_t^{x_t} \right\} =$$
$$= \int_0^T dt \, \frac{x_t \cdot (\mathcal{D}D_t + r_t D_t)}{x_t \cdot D_t} + \log \frac{\beta_1}{\beta_0}, \tag{16.34}$$

When risks and uncertainties collide **Chapter | 16 339**

where $x = x_t$ is an admissible self-financing strategy taking values on the curve γ, and the first variation of the arbitrage action as

$$\delta A^\beta(\gamma; D, r) := \frac{d}{d\epsilon} A^\beta(\gamma^\epsilon; D^\epsilon, r^\epsilon)|_{\epsilon:=0}. \tag{16.35}$$

This leads to the following

Definition 16.25. Let us introduce the notation $q := (x, D, r)$ and $q' := (x', D', r')$ for two vectors in $\mathbb{R}^{3N}$. The **Lagrangian (or Lagrange function)** is defined as

$$L(q, q') := L(x, D, r, x', D', r') := \frac{x \cdot (D' + rD)}{x \cdot D}. \tag{16.36}$$

The self-financing **constraint** is defined as

$$C(q, q') := x' \cdot D. \tag{16.37}$$

Lemma 16.26. *The arbitrage action for a self-financing strategy γ is the integral of the Lagrange function along the D-admissible strategy:*

$$
\begin{aligned}
A^\beta(\gamma; D, r) &= \int_\gamma dt\, L(q_t, q'_t) + \log \frac{\beta_1}{\beta_0} \\
&= \int_\gamma dt\, L(x_t, D_t, r_t, x'_t, D'_t, r'_t) + \log \frac{\beta_1}{\beta_0}.
\end{aligned} \tag{16.38}
$$

A fundamental result of classical mechanics allows to compute the extrema of the arbitrage action in the *deterministic* case as the solution of a system of ordinary differential equations.

Theorem 16.27 (Hamilton principle). *Let us denote the derivative with respect to time as $\frac{d}{dt} =: '$ and assume that all quantities observed are deterministic. The local extrema of the arbitrage action satisfy the Lagrange equations under the self-financing constraints*

$$
\begin{cases}
\delta A^\beta(\gamma; D, r) = 0 \text{ for all } (\delta\gamma, \delta D, \delta r) \\
\text{such that } x_t'^\epsilon \cdot D_t^\epsilon = 0 \text{ for all } \epsilon
\end{cases}
\Longleftrightarrow
\begin{cases}
\frac{d}{dt}\frac{\partial L_\lambda}{\partial q'} - \frac{\partial L_\lambda}{\partial q} = 0, \\
C(q, q') := x' \cdot D = 0,
\end{cases} \tag{16.39}
$$

where $\lambda \in \mathbb{R}$ denotes the self-financing constraint Lagrange multiplier and $L_\lambda := L - \lambda C$.

Let $L = L(q, q')$ be the Lagrange function of a deterministic Lagrangian system with the nonholonomic constraint $C(q, q') = 0$. Setting $L_\lambda := L - \lambda C$ for the constraint Lagrange multipliers the dynamics is given by the extended Euler-Lagrange equations

$$
\text{(EL)} \quad
\begin{cases}
\frac{d}{dt}\frac{\partial L_\lambda}{\partial q'}(q, q') - \frac{\partial L_\lambda}{\partial q}(q, q') = 0, \\
C(q, q') = 0
\end{cases} \tag{16.40}
$$

340 Mechanics and Physics of Structured Media

meaning by this that the deterministic solution $q = q_t$ and $\lambda \in \mathbb{R}$ satisfy the constraint and

$$\frac{d}{dt}\frac{\partial L_\lambda}{\partial q'}\left(q_t, \frac{dq_t}{dt}\right) - \frac{\partial L_\lambda}{\partial q}\left(q_t, \frac{dq_t}{dt}\right) = 0. \tag{16.41}$$

Definition 16.28. The formal **stochastic embedding of the Euler-Lagrange equations** is obtained by the formal substitution

$$S : \frac{d}{dt} \longmapsto \mathcal{D}, \tag{16.42}$$

and allowing the coordinates of the tangent bundle to be stochastic

$$(\text{SEL}) \quad \begin{cases} \mathcal{D}\frac{\partial L_\lambda}{\partial q'}(q, q') - \frac{\partial L_\lambda}{\partial q}(q, q') = 0, \\ C(q, q') = 0 \end{cases} \tag{16.43}$$

meaning by this that the stochastic solution $Q = Q_t$ and the random variable λ satisfy the constraint and

$$\begin{cases} \mathcal{D}\frac{\partial L_\lambda}{\partial q'}(Q_t, \mathcal{D}Q_t) - \frac{\partial L_\lambda}{\partial q}(Q_t, \mathcal{D}Q_t) = 0, \\ C(Q_t, \mathcal{D}Q_t) = 0. \end{cases} \tag{16.44}$$

Let $L = L(q, q')$ be the Lagrange function of a deterministic Lagrangian system on a time interval I with constraint $C = 0$. Set

$$\Xi := \left\{ Q \in C^1(I) \mid \mathbb{E}\left[\int_I |L_\lambda(Q_t, \mathcal{D}Q_t)|dt\right] < +\infty \right\}. \tag{16.45}$$

Definition 16.29. The action functional associated to L_λ defined by

$$\begin{aligned} F : \quad &\Xi \longrightarrow \mathbb{R}, \\ &Q \longmapsto \mathbb{E}\left[\int_I L_\lambda(Q_t, \mathcal{D}Q_t)dt\right] \end{aligned} \tag{16.46}$$

is called **stochastic analogue of the classic action** under the constraint $C = 0$.

For a sufficiently smooth extended Lagrangian L_λ a necessary and sufficient condition for a stochastic process to be a critical point of the action functional F is the fulfillment of the stochastic Euler-Lagrange equations (SEL), as it can be seen in Theorem 7.1 page 54 in [4]. Moreover we have the following

Lemma 16.30 (Coherence). *The following diagram commutes*

$$L_\lambda(q_t, q'_t) \xrightarrow{\ S\ } L_\lambda(Q_t, \mathcal{D}Q_t) \tag{16.47}$$

$$\left\downarrow{\scriptstyle Critical\ Action\ Principle}\qquad\qquad \left\downarrow{\scriptstyle Stochastic\ Critical\ Action\ Principle}$$

$$(EL) \xrightarrow[\ S\ ]{} (SEL)$$

16.4 Asset and market portfolio dynamics as solution of the Schrödinger equation: the quantization of the deterministic constrained Hamiltonian system

There are two ways of obtaining a stochastic theory starting from a deterministic one. The first way is the one adopted by Cresson and Darses [4], which analyzes a deterministic Lagrangian system or an equivalent deterministic Hamilton system and studies what happens if we make all variables stochastic variables. The biggest difficult is the time t-dependence, which cannot be defined pathwise and is related to the definitions of Itô's integrals or, equivalently, to Stratonovich's integrals. Since for the Stratonovich's time derivative corresponding to the Stratonovich's integral we have the same chain rule for the composition of two functions, Itô's Lemma for the Stratonovich derivative has the same form as the chain rule for deterministic functions. This is the main ingredient for the Cresson-Darses stochasticization construction. Their conclusion is that, in order to solve the stochastic Euler-Lagrange equations or their equivalent, the stochastic Hamilton equations, it suffices to solve their deterministic counterpart and add a stochastic perturbation with zero expectation, satisfying several constraints. This is the way summarized in Section 16.3 and has been carried out in [10].

The second way to obtain a stochastic theory from a deterministic one, relies on quantization of a deterministic system. In classical physics one describes the time evolution of the system by either Newton's second law of the dynamics, or the equivalent deterministic Euler-Lagrange equations, or the equivalent Hamilton equations. In the Euler-Lagrange equations the Lagrange function describing the system appears, while in the Hamilton equations, the Hamilton function appears. The Hamilton function is obtained by Legendre transform of the Lagrange function. The quantization step consists in introducing an Hilbert Space of L^2 complex valued functions over the coordinate space for the deterministic (real) variables (all but the time), in our case x, D, and r: this coordinate space is a manifold, and the Hilbert space is an L^2 space over this manifold. We consider the unit ball of this Hilbert space whose elements are defined as the states of the quantum mechanical system: the interpretation of the square of the absolute value of those complex valued functions with L^2 norm one is that of a probability density. To obtain the time dynamics we have to construct the Hamilton operator derived by the Hamilton function. This procedure is called quantization of the Hamilton function and, according to the type of the Hamilton function, it is not always a well defined procedure. In our case we are lucky. By replacing the coordinate functions of the manifold by the corresponding multiplication operators with those coordinates and by replacing the tangential plane coordinates by the partial derivatives with respect to the corresponding coordinates in the manifold, we obtain an operator, which can be symmetrized, and later made to a selfadjoint operator, by an appropriate choice of its domain of definition: the Hamilton operator. Given the Hamilton operator we can solve the

342 Mechanics and Physics of Structured Media

Schrödinger equation, which gives the time evolution of the quantum mechanical initial state as a function of time.

Finally we have two approaches who should be equivalent, that is leading to the same solution. The asset dynamics and the market portfolio dynamics. With the first approach we obtain the stochastic processes (x_t, D_t, r_t) as a solution of the stochastic Euler-Lagrange equations. With the second approach we obtain an L^2 function $\psi_t(x, D, r)$ such that $|\psi_t(x, D, r)|^2$ is the probability density of the random vector (x_t, D_t, r_t).

We proceed now by introducing an equivalent quantum mechanical representation of the asset and market portfolio dynamics. As a general background to the mathematics of quantum mechanics we refer to [29] and [15].

Proposition 16.31. *The Hamilton function H defined as Legendre transform of the Lagrangian L is*

$$H(p, q) := \left(p \cdot q' - L_\lambda(q, q'))\right)\big|_{p:=\frac{\partial L}{\partial q'}} = -\frac{x \cdot (rD)}{x \cdot D}, \qquad (16.48)$$

where $q = (x, D, r, \lambda)$ and $p := \frac{\partial L}{\partial q'} = (p_x, p_D, p_r, p_\lambda)$. Thereby, following [22] we have elevated the Lagrange multiplier corresponding to the self-financing constraint C to an additional dynamic variable λ with its conjugate momentum p_λ. The Hessian matrix of L_λ is singular, which translates into the open first class additional constraints $p_r = 0$ and $p_\lambda = 0$.

Proof. It follows directly by inserting

$$
\begin{aligned}
p_x &= \frac{\partial L_\lambda}{\partial x'} = -\lambda D, \\
p_D &= \frac{\partial L_\lambda}{\partial D'} = \frac{x}{x \cdot D}, \\
p_r &= \frac{\partial L_\lambda}{\partial r'} = 0, \\
p_\lambda &= \frac{\partial L_\lambda}{\partial \lambda'} = 0
\end{aligned}
\qquad (16.49)
$$

into Eqs. (16.36) and (16.39).

The last two equations are constraints on the conjugate momenta. They are first class because we have the following equalities among Poisson brackets for all i, j

$$
\begin{cases}
\{p_r^i, p_r^j\} = 0, \\
\{p_\lambda, p_\lambda\} = 0, \\
\{p_r^i, p_\lambda\} = 0, \\
\{p_r^i, H\} = \frac{x^i D^i}{x \cdot D}, \\
\{p_\lambda^i, H\} = 0. \qquad \square
\end{cases}
\qquad (16.50)
$$

When risks and uncertainties collide Chapter | 16 **343**

Proposition 16.32. *The selfadjoint Hamilton operator obtained by the standard quantization procedure*

$$q \longrightarrow q \quad (multiplication\ operator),$$
$$p \longrightarrow \frac{1}{\iota}\frac{\partial}{\partial q} \quad (differential\ operator) \tag{16.51}$$

is

$$H := -\frac{x \cdot (rD)}{x \cdot D} \tag{16.52}$$

with domain of definition

$$\mathrm{dom}(H)$$
$$:= \left\{ \varphi \in L^2(\mathfrak{X} \times \mathbb{R}^{2N+1}, \mathbb{C}, d^{3N}q d\lambda) \;\middle|\; H\varphi \in L^2(\mathfrak{X} \times \mathbb{R}^{2N+1}, \mathbb{C}, d^{3N}q d\lambda) \right.$$
$$\left. \frac{\partial\varphi}{\partial r^i} = 0\ for\ all\ i,\ \frac{\partial\varphi}{\partial\lambda} = 0 \right\}. \tag{16.53}$$

Proof. The quantization procedure of first class constrained Hamiltonian systems, explained in [5], [22], and [8], is directly applied here. $\square$

Remark 16.33. The Hamilton operator H is a multiplication operator with a real function, and is selfadjoint. By Dirac's theory of first constrained quantized systems, if the constraints are satisfied at time $t = 0$, then they are automatically satisfied for all times for the solution of Schrödinger's equation. Since H does not explicitly depend on the Lagrange multiplier λ as well as any φ in its domain of definition, we can drop any reference to λ and write

$$\mathrm{dom}(H) := \left\{ \varphi \in L^2(\mathfrak{X} \times \mathbb{R}^{2N}, \mathbb{C}, d^{3N}q) \;\middle|\; H\varphi \in L^2(\mathfrak{X} \times \mathbb{R}^{2N}, \mathbb{C}, d^{3N}q) \right.$$
$$\left. \frac{\partial\varphi}{\partial r^i} = 0\ for\ all\ i \right\}. \tag{16.54}$$

Theorem 16.34. *The asset and market portfolio dynamics is given by the solution of the Schrödinger equation*

$$\begin{cases} \iota\frac{d}{dt}\psi(q,t) = H\psi(q,t), \\ \psi(q,0) = \psi_0(q), \end{cases} \tag{16.55}$$

where ψ_0 is the initial state satisfying $C\psi_0 = 0$ and $\int_{\mathfrak{X} \times \mathbb{R}^{2N}} dq^{3N} |\psi_0(q)|^2 = 1$. The solution is given by

$$\psi(q,t) = e^{-\iota Ht}\psi_0, \tag{16.56}$$

344 Mechanics and Physics of Structured Media

where $\{e^{-\iota Ht}\}_{t \geq 0}$ is the strong continuous, unitary one parameter group associated to the selfadjoint H by Stone's theorem.

Remark 16.35. This is the quantum mechanical formulation of the constrained stochastic Lagrangian system described by the SDE (16.44). The interpretation of $|\psi(q,t)|^2$ is the probability density at time t for the coordinates q:

$$P[q_t \in \mathcal{Q}] = \int_{\mathcal{Q}} dq^{3N} |\psi(q,t)|^2. \tag{16.57}$$

Therefore, if we have a random variable $a_t = a(p,q,t)$, by mean of its quantization

$$A := a\left(\frac{1}{\iota}\frac{\partial}{\partial q}, q, t\right), \tag{16.58}$$

we can compute its expectation by means of both the **Schrödinger and Heisenberg representation** as

$$\mathbb{E}_0[a_t] = (A\psi, \psi) = \int_{\mathcal{X} \times \mathbb{R}^{2N}} dq^{3N} A\psi(q,t)\bar{\psi}(q,t)$$

$$= \int_{\mathcal{X} \times \mathbb{R}^{2N}} dq^{3N} A_t \psi(q,0)\bar{\psi}(q,0), \tag{16.59}$$

where the time dependent operator A_t, the **Heisenberg representation** of the operator A is defined as

$$A_t := e^{iHt} A e^{-iHt}. \tag{16.60}$$

Higher moments of random variables a_t, like any measurable functions $f(a_t)$ of them can be computed by means of this technique as

$$\mathbb{E}_0[f(a_t)] = (f(A)\psi, \psi) = \int_{\mathcal{X} \times \mathbb{R}^{2N}} dq^{3N} f(A_t)\psi(q,0)\bar{\psi}(q,0), \tag{16.61}$$

transforming the problem in one of operator calculus.

Theorem 16.36 (Ehrenfest). *The time derivative of the expectation of a selfadjoint operator A is given by*

$$\frac{d}{dt}(A\psi, \psi) = \frac{1}{\iota}([A, H]\psi, \psi) + \left(\frac{\partial A}{\partial t}\psi, \psi\right). \tag{16.62}$$

A direct consequence of Ehrenfest's result is the "energy conservation" theorem.

Corollary 16.37.

$$\frac{d}{dt}(H\psi, \psi) \equiv 0. \tag{16.63}$$

When risks and uncertainties collide Chapter | 16 **345**

Corollary 16.38. *The dynamics of the expected values of market portfolio, asset values and term structures is given by*

$$\mathbb{E}_0[x_t] \equiv const,$$
$$\mathbb{E}_0[D_t] \equiv const, \qquad (16.64)$$
$$\mathbb{E}_0[r_t] \equiv const.$$

Proof. It follows direct from Ehrenfest's Theorem 16.36, because the multiplication operators x, D, and r commute with the Hamilton operator H. $\qquad\square$

Remark 16.39. Note that formulas (16.64) coincide with those in ([10]), where the stochastic Lagrange equations (16.44) have been explicitly solved. These demonstrate the consistency and compatibility of the quantum mechanical reformulation to mathematical finance.

We can now derive the consequences of Ehrenfest's theorem in the standard QM representation for the stochastic Hamiltonian/Lagrangian representation of geometric arbitrage theory.

Corollary 16.40. *The stochastic processes $(x_t)_t$, $(D_t)_t$, and $(r_t)_t$ for market portfolio nominals, asset values, and term structures are identically distributed along time.*

Proof. It suffices to apply Ehrenfest's theorem to any nonnegative power of the operators x, D, and r which do not depend on time and commute with the Hamilton operator. Hence, the claim must hold true for the distribution functions of x_t, D_t, and r_t for any time t. $\qquad\square$

Corollary 16.41. *The stochastic processes $(\mathcal{D}x_t)_t$, $(\mathcal{D}D_t)_t$, and $(\mathcal{D}r_t)_t$ for the returns market portfolio nominals, asset values, and term structures are centered and serially uncorrelated.*

Proof. It suffices to prove it for the nominals, because for the asset values and term structures the proofs are formally the same. By taking the time derivative of the first equality in (16.64) we obtain

$$\mathbb{E}_0[\mathcal{D}x_t] = 0. \qquad (16.65)$$

Let $t_1 \neq t_2$. By applying Ehrenfest's Theorem 16.36 we obtain

$$\mathbb{E}_0\left[x_{t_1} x_{t_2}\right] \equiv const, \qquad (16.66)$$

where componentwise multiplication of the vector components is meant. Hence, together with the first equality in (16.64)

$$\mathrm{Cov}_0\left(x_{t_1}, x_{t_2}\right) \equiv const, \qquad (16.67)$$

346 Mechanics and Physics of Structured Media

and by differentiating with respect to time t_1 and time t_2

$$\mathrm{Cov}_0 \left(\mathcal{D}_{t_1} x_{t_1}, \mathcal{D}_{t_2} x_{t_2} \right) \equiv 0. \tag{16.68}$$

Since $t_1 \neq t_2$ are arbitrary we conclude that all autocovariances for any nonzero lag of $(\mathcal{D}x_t)_t$ vanish, meaning that the process is serially uncorrelated. $\qquad\square$

Theorem 16.42 (Heisenberg's uncertainty relation). *Let A and B two self-adjoint operators on $\mathcal{H}$. The variance of the corresponding observables in the state $\varphi \in \mathrm{dom}(A) \cap \mathrm{dom}(B)$ is*

$$\sigma_\varphi^2(A) := \| A\phi - \|A\phi\|^2 \|^2, \qquad \sigma_\varphi^2(B) := \| B\phi - \|B\phi\|^2 \|^2, \tag{16.69}$$

where $\| \cdot \|$ and $(\cdot, \cdot)$ are the norm and the scalar product in $\mathcal{H} = L^2(\mathfrak{X} \times \mathbb{R}^{2N}, \mathbb{C}, d^{3N}q)$. Then,

$$\sigma_\varphi^2(A)\sigma_\varphi^2(B) \geq \frac{1}{4} \| [A, B]\varphi \|^2. \tag{16.70}$$

The proof of Theorem 16.42 can be found f.i. in [29] or in [15].

By applying Heisenberg's uncertainty relation to the quantum mechanical representation of our market model we obtain the following

Proposition 16.43. *The dynamics of the volatilities of market portfolio and asset values satisfies the inequalities*

$$\mathrm{Var}_0 \left(D_t^j \right) \mathrm{Var}_0 \left(\frac{x_t^j}{x_t \cdot D_t} \right) \geq \frac{1}{4}, \tag{16.71}$$

for all indices $j = 1, \ldots, N$.

Proof. For $q = (x, D)$ we choose $A := q^i$ and $B := \frac{1}{\iota}\frac{\partial}{\partial q^j}$, obtaining by Theorem 16.42, since $[A, B] = \iota\delta^{i,j}$, $\varphi_t = e^{-\iota t H}\varphi_0$, and $\|\varphi_t\|^2 = 1$,

$$\sigma_{\varphi_t}^2(q^j)\sigma_{\varphi_t}^2 \left(\frac{1}{\iota}\frac{\partial}{\partial q^j} \right) \geq \frac{1}{4}. \tag{16.72}$$

Now we can identify

$$\sigma_{\varphi_t}^2(q^j) = \mathrm{Var}_0(q_t^j),$$
$$\sigma_{\varphi_t}^2 \left(\frac{1}{\iota}\frac{\partial}{\partial q^j} \right) = \mathrm{Var}_0(p_t^j), \tag{16.73}$$

and (16.71) follows after inserting the second equations of (16.49)

$$p_D = \frac{\partial L}{\partial D} = \frac{x}{x \cdot D}. \tag{16.74}$$

The proof is completed. $\qquad\square$

When risks and uncertainties collide Chapter | 16 **347**

Heisenberg's uncertainty relation has an interesting econometric consequence. In Proposition 16.43, where we assume that the market forces minimize arbitrage, the dynamics of the market portfolio (x_t, D_t, r_t) is such that

$$\text{Var}_0\left(D_t^j\right)\text{Var}_0\left(\frac{w_t^j}{D_t^j}\right) \geq \frac{1}{4}, \tag{16.75}$$

where $w_t^j := \frac{x_t^j D_t^j}{x_t \cdot D_t}$ is the market weight of the j-th asset. This means that the less volatile the j-th asset value is, the more volatile its market weight must be, so that (16.75) is fulfilled.

Remark 16.44. If we try to obtain the analogon of Proposition 16.43 for the conjugate variables r and p_r, we face a technical problem. As a matter of fact we can apply Theorem 16.42 to $q = r$, but, since $p_r = 0$ as computed in the third equation (16.49), we cannot give the interpretation of $\text{Var}(p_{r_t}^j) \equiv 0$ to $\sigma_{\varphi_t}^2\left(\frac{1}{i}\frac{\partial}{\partial r^j}\right) \neq 0$, similarly to (16.73). The same difficulty occurs for $q = D$.

Example 16.45. Let us consider an asset with no futures, where the asset discounted prices and the market portfolio nominals are Brownian processes

$$\begin{aligned}\hat{S}_t &= \hat{S}_0 + \alpha_t + \sigma_t W_t, \\ x_t &= x_0 + a_t + s_t W_t,\end{aligned} \tag{16.76}$$

where $(W_t)_{t\in[0,+\infty[}$ is a standard P-Brownian motion in $\mathbb{R}^K$, for some $K \in \mathbb{N}$, $(\sigma_t)_{t\in[0,+\infty[}$ and $(s_t)_{t\in[0,+\infty[}$ are $\mathbb{R}^{N\times K}$-valued differentiable functions, and $(\alpha_t)_{t\in[0,+\infty[}$ and $(a_t)_{t\in[0,+\infty[}$ are $\mathbb{R}^N$-valued differentiable functions. The coefficients σ, s, α, a must satisfy certain conditions implied by Corollaries 16.40 and 16.41. Since

$$\begin{aligned}\mathbb{E}_0[\hat{S}_t] &= \hat{S}_0 + \alpha_t, \quad \text{Var}_0(\hat{S}_t) = \text{diag}(\sigma_t\sigma_t^\dagger)t, \\ \mathbb{E}_0[x_t] &= x_0 + \alpha_t, \quad \text{Var}_0(x_t) = \text{diag}(s_t s_t^\dagger)t,\end{aligned} \tag{16.77}$$

from Corollary 16.40 we derive the conditions

$$\alpha_t \equiv 0, \quad a_t \equiv 0, \quad \text{diag}(\sigma_t\sigma_t^\dagger) = \frac{\text{diag}(\sigma_0\sigma_0^\dagger)}{t}, \quad \text{diag}(s_t s_t^\dagger) = \frac{\text{diag}(s_0 s_0^\dagger)}{t}, \tag{16.78}$$

and hence,

$$\hat{S}_t \sim \mathcal{N}(\hat{S}_0, \sigma_0\sigma_0^\dagger), \quad x_t \sim \mathcal{N}(x_0, s_0 s_0^\dagger) \tag{16.79}$$

under which Corollary 16.41 is automatically satisfied.

348 Mechanics and Physics of Structured Media

16.5 The (numerical) solution of the Schrödinger equation via Feynman integrals

16.5.1 From the stochastic Euler-Lagrangian equations to Schrödinger's equation: Nelson's method

Following Chapter 14 of [24] we consider diffusions on N-dimensional Riemannian manifold satisfying the SDE

$$d\xi_t = b(t, \xi_t)dt + \sigma(\xi_t)dW_t, \tag{16.80}$$

where $(W_t)_{t\geq0}$ is a K-dimensional Brownian motion, and

$$b : [0, +\infty[\times\mathbb{R}^N \to \mathbb{R}^N \text{ and } \sigma : \mathbb{R}^N \to \mathbb{R}^{N\times K} \tag{16.81}$$

are vector and matrix valued functions with appropriate regularity. We assume that

$$\sigma^2(q)(q', q') := q'\sigma(q)\sigma^\dagger(q)q' = \sum_{j=1}^{N} q'^j q'_j \tag{16.82}$$

defines a Riemannian metric, and introduce the notation $v^j := \sum_{i=1}^{N}(\sigma\sigma^\dagger)_{j,i} v_i$.

We consider a Lagrangian on M given as

$$L(q, q', t) := \sum_{j=1}^{N} \left[\frac{1}{2}q'^j q'_j - \Phi(q) + A_j(q)q'^j\right], \tag{16.83}$$

for given potentials Φ and A. For the diffusion (16.80) the Guerra-Morato Lagrangian writes

$$L_+(\zeta, t) := \sum_{j=1}^{N} \left[\frac{1}{2}b^j(t, \zeta)b_j(t, \zeta) + \frac{1}{2}\nabla_j b^j(t, \zeta) - \Phi(\zeta)\right.$$

$$\left. + A_j(\zeta)b^j(t, \zeta) + \frac{1}{2}\nabla_j A^j(\zeta)\right]. \tag{16.84}$$

We define

$$R(t, q) := \frac{1}{2}\log\rho(t, q), \tag{16.85}$$

where ρ is the density of the process $(\xi_t)_{t\geq0}$, and

$$S(t, q) := \mathbb{E}\left[.\int_0^t L_+(\xi_s, s)ds \,\middle|\, \xi_t = q\right]. \tag{16.86}$$

Hamilton's principle for the Guerra-Morato Lagrangian implies that

$$\left(\frac{\partial}{\partial t} + \sum_{j=1}^{N} b^j \nabla_j + \frac{1}{2}\Delta\right) S = \sum_{j=1}^{N}\left[\frac{1}{2}b^j b_j + \frac{1}{2}\nabla_j b^j - \Phi + A_j b^j + \frac{1}{2}\nabla_j b^j\right],$$

(16.87)

which, since

$$b^j = \nabla^j S - A^j + \nabla^j R,$$

(16.88)

becomes the Hamilton-Jacobi equation

$$\frac{\partial S}{\partial t} + \frac{1}{2}\sum_{j=1}^{N}(\nabla^j S - A^j)(\nabla_j S - A_j) + \Phi - \frac{1}{2}\sum_{j=1}^{N}\nabla^j R \nabla_j R - \frac{1}{2}\Delta R = 0.$$

(16.89)

The continuity equation

$$\frac{\partial \rho}{\partial t} = -\sum_{j=1}^{N}\nabla_j(v^j \rho),$$

(16.90)

where

$$v^j = \frac{1}{2}(b^j + b^{j*}), \qquad b^{j*} = b^j - \nabla^j \log \rho,$$

(16.91)

becomes

$$\frac{\partial R}{\partial t} + \sum_{j=1}^{N}(\nabla_j R)(\nabla^j S - A^j) + \frac{1}{2}\Delta S - \frac{1}{2}\nabla_j A^j = 0.$$

(16.92)

The nonlinear Hamilton-Jacobi and continuity PDE lead to the linear Schrödinger equation

$$i\frac{\partial \psi}{\partial t} = \underbrace{\left[\frac{1}{2}\sum_{j=1}^{N}\left(\frac{1}{i}\nabla^j - A^j\right)\left(\frac{1}{i}\nabla_j - A_j\right) + \Phi\right]}_{=:H}\psi,$$

(16.93)

for the Schrödinger operator H, if we define the probability amplitude

$$\psi(q,t) := e^{R(q,t)+iS(q,t)}.$$

(16.94)

Note that

$$\rho(q,t) = |\psi(q,t)|^2.$$

(16.95)

350 Mechanics and Physics of Structured Media

16.5.2 Solution to Schrödinger's equation via Feynman's path integral

The Hamilton function is the Legendre transformation of the Lagrangian:

$$H(p, q, t) := \left(\sum_{j=1}^{N} p^j q'_j - L(q, q', t) \right) \Bigg|_{p := \frac{\partial L}{\partial q'} = q' + A}$$

$$= \frac{1}{2} \sum_{j=1}^{N} \left(p^j - A^j \right) \left(p_j - A_j \right) + \Phi, \qquad (16.96)$$

and the Schrödinger operator is obtained by the quantization

$$q \to q \quad \text{(Multiplication operator)}, \qquad p \to \frac{1}{i} \nabla \quad \text{(Differential operator)}. \qquad (16.97)$$

The solution of the Schrödinger initial value problem

$$\begin{cases} i \frac{\partial \psi}{\partial t} = H \psi, \\ \psi(q, 0) = \psi_0(q), \end{cases} \qquad (16.98)$$

can be obtained as the convolution of the initial condition with Feynman's path integral:

$$\psi(y, t) = \int \psi_o(q) \left(\int_{q(0)=q}^{q(t)=y} \exp \left(i \int_0^t L(u(s), u'(s), s) ds \right) Du \right) dq. \qquad (16.99)$$

An approximation of Feynman's path integral can be obtained by averaging over a number of possible paths. If the original Lagrangian problem has to fulfill some constraints, these can be enforced in the choice of the paths to be averaged over in the integral.

16.5.3 Application to geometric arbitrage theory

The geometric arbitrage theory Lagrangian reads

$$L(q, q', t) := \frac{x \cdot (D' + r D)}{x \cdot D}, \qquad (16.100)$$

for $q := (x, D, r) \in \mathbb{R}^{3N}$, where x, D, and r represent portfolio nominals, deflators, and short rates. The portfolios under consideration have to satisfy the self-financing condition

$$x' \cdot D = 0. \qquad (16.101)$$

Let us assume that the diffusions can be written separately as

$$\begin{cases} dx_t = b^x(t, x_t)dt + \sigma^x(x_t)dW_t, \\ dD_t = b^D(t, D_t)dt + \sigma^D(D_t)dW_t, \\ dr_t = b(t, r_t)dt + \sigma^r(r_t)dW_t, \end{cases} \qquad (16.102)$$

where

$$\begin{aligned} b^x, b^D, b^r &: [0, +\infty[\times\mathbb{R}^N \to \mathbb{R}^N, \\ \sigma^x, \sigma^D, \sigma^r &: \mathbb{R}^N \to \mathbb{R}^{N \times K} \end{aligned} \qquad (16.103)$$

are vector and matrix valued functions with appropriate regularity.

The geometric arbitrage theory Lagrangian can be written in the form (16.83)

$$L(q, q', t) = \left(\frac{1}{2}\sum_{j=1}^{3N} q'^j q'_j - \frac{1}{2}\right) + \Phi(q) + \sum_{j=1}^{3N} A_j(q)q'^j, \qquad (16.104)$$

if we set

$$\Phi(q) := -\frac{x \cdot (rD)}{x \cdot D} - \frac{1}{2}, \qquad A_j^D(q) := -\frac{\sigma^{D^{-2}}_{j,i} x_i}{x \cdot D}, \qquad (16.105)$$
$$A_j^x(q) := 0, \qquad\qquad\qquad A_j^r(q) := 0.$$

Therefore, the solution of Schrödinger's initial value problem (16.98) reads

$$\psi(y, t) = \int \psi_o(q)\left(\int_{q(0)=q}^{q(t)=y} \exp\left(i\int_0^t \frac{x_s \cdot (D'_s + r_s D_s)}{x_s \cdot D_s}ds\right)Du\right)dq, \qquad (16.106)$$

where the Feynman integration is over all paths satisfying the constraint

$$x' \cdot D \equiv 0. \qquad (16.107)$$

Formula (16.106) can be approximately computed by Monte Carlo methods simulating system trajectories satisfying the constraints (16.107), leading to numerical efficient computations.

16.6 Conclusion

By introducing an appropriate stochastic differential geometric formalism, the classical theory of stochastic finance can be embedded into a conceptual frame-

352 Mechanics and Physics of Structured Media

work called geometric arbitrage theory, where the market is modeled with a principal fiber bundle with a connection whose curvature corresponds to the instantaneous arbitrage capability. The market and its dynamic can be seen as a stochastic Lagrangian system, or, equivalently, as a stochastic Hamiltonian system. Instead of trying to compute direct a solution of the stochastic Hamilton equations, we quantize a deterministic version of the Hamiltonian system. The asset and market dynamics have then a quantum mechanical formulation in terms of Schrödinger equation which can be solved numerically by means of Feynman's integrals. Ehrenfest's theorem and Heisenberg's uncertainty relation lead to new econometric results: in the equilibrium minimal arbitrage returns on asset values and nominal values are centered and serially uncorrelated; the variances of an asset value and its corresponding weight in the market portfolio cannot be both arbitrarily small: if one increases, the other decreases, and vice versa.

Appendix 16.A   Generalized derivatives of stochastic processes

In stochastic differential geometry one would like to lift the constructions of stochastic analysis from open subsets of $\mathbb{R}^N$ to N dimensional differentiable manifolds. To that aim, chart invariant definitions are needed and hence a stochastic calculus satisfying the usual chain rule, and not Itô's Lemma is required, (cf. [14], Chapter 7, and the remark in Chapter 4 at the beginning of page 200). That is why the papers about geometric arbitrage theory are mainly concerned in by stochastic integrals and derivatives meant in *Stratonovich*'s sense and not in *Itô*'s. Of course, at the end of the computation, Stratonovich integrals can be transformed into Itô's. Note that a fundamental portfolio equation, the self-financing condition cannot be directly formally expressed with Stratonovich integrals, but first with Itô's and then transformed into Stratonovich's, because it is a nonanticipative condition.

Definition 16.A.1. Let I be a real interval and $Q = (Q_t)_{t \in I}$ be a $\mathbb{R}^N$-valued stochastic process on the probability space $(\Omega, \mathcal{A}, P)$. The process Q determines three families of σ-subalgebras of the σ-algebra $\mathcal{A}$:

  (i) "Past" $\mathcal{P}_t$, generated by the preimages of Borel sets in $\mathbb{R}^N$ by all mappings $Q_s : \Omega \to \mathbf{R}^N$ for $0 < s < t$.
 (ii) "Future" $\mathcal{F}_t$, generated by the preimages of Borel sets in $\mathbb{R}^N$ by all mappings $Q_s : \Omega \to \mathbf{R}^N$ for $0 < t < s$.
(iii) "Present" $\mathcal{N}_t$, generated by the preimages of Borel sets in $\mathbb{R}^N$ by the mapping $Q_s : \Omega \to \mathbf{R}^N$.

When risks and uncertainties collide Chapter | 16 **353**

Let $Q = (Q_t)_{t \in I}$ be continuous. Assuming that the following limits exist, **Nelson's stochastic derivatives** are defined as

$$
\begin{aligned}
\mathfrak{D} Q_t &:= \lim_{h \to 0^+} \mathbb{E}\left[\frac{Q_{t+h} - Q_t}{h} \middle| \mathcal{P}_t \right] : \text{forward derivative}, \\
\mathfrak{D}_* Q_t &:= \lim_{h \to 0^+} \mathbb{E}\left[\frac{Q_t - Q_{t-h}}{h} \middle| \mathcal{F}_t \right] : \text{backward derivative}, \\
\mathcal{D} Q_t &:= \frac{\mathfrak{D} Q_t + \mathfrak{D}_* Q_t}{2} : \text{mean derivative}.
\end{aligned}
\tag{16.A.1}
$$

Let $\mathcal{S}^1(I)$ the set of all processes Q such that $t \mapsto Q_t$, $t \mapsto \mathfrak{D} Q_t$, and $t \mapsto \mathfrak{D}_* Q_t$ are continuous mappings from I to $L^2(\Omega, \mathcal{A})$. Let $\mathcal{C}^1(I)$ the completion of $\mathcal{S}^1(I)$ with respect to the norm

$$
\|Q\| := \sup_{t \in I} \left(\|Q_t\|_{L^2(\Omega, \mathcal{A})} + \|\mathfrak{D} Q_t\|_{L^2(\Omega, \mathcal{A})} + \|\mathfrak{D}_* Q_t\|_{L^2(\Omega, \mathcal{A})} \right). \tag{16.A.2}
$$

Remark 16.A.2. The stochastic derivatives $\mathfrak{D}$, $\mathfrak{D}_*$, and $\mathcal{D}$ correspond to Itô's, to the anticipative, and, respectively, to Stratonovich's integral (cf. [13]). The process space $\mathcal{C}^1(I)$ contains all Itô processes. If Q is a Markov process, then the sigma algebras $\mathcal{P}_t$ ("past") and $\mathcal{F}_t$ ("future") in the definitions of forward and backward derivatives can be substituted by the sigma algebra $\mathcal{N}_t$ ("present"), see Chapters 6.1 and 8.1 in [13].

Stochastic derivatives can be defined pointwise in $\omega \in \Omega$ outside the class $\mathcal{C}^1$ in terms of generalized functions.

Definition 16.A.3. Let $Q : I \times \Omega \to \mathbb{R}^N$ be a continuous linear functional in the test processes $\varphi : I \times \Omega \to \mathbb{R}^N$ for $\varphi(\cdot, \omega) \in C_c^\infty(I, \mathbb{R}^N)$. We mean by this that for a fixed $\omega \in \Omega$ the functional $Q(\cdot, \omega) \in \mathcal{D}(I, \mathbb{R}^N)$, the topological vector space of continuous distributions. We can then define **Nelson's generalized stochastic derivatives:**

$$
\begin{aligned}
\mathfrak{D} Q(\varphi_t) &:= -Q(\mathfrak{D} \varphi_t) : \text{forward generalized derivative}, \\
\mathfrak{D}_* Q(\varphi_t) &:= -Q(\mathfrak{D}_* \varphi_t) : \text{backward generalized derivative}, \\
\mathcal{D}(\varphi_t) &:= -Q(\mathcal{D} \varphi_t) : \text{mean generalized derivative}.
\end{aligned}
\tag{16.A.3}
$$

If the generalized derivative is regular, then the process has a derivative in the classic sense. This construction is nothing else than a straightforward pathwise lift of the theory of generalized functions to a wider class stochastic processes which do not a priori allow for Nelson's derivatives in the strong sense. We will utilize this feature in the treatment of credit risk, where many processes with jumps occur.

References

[1] D. Bleecker, Gauge Theory and Variational Principles, Addison-Wesley Publishing, 1981 (republished by Dover 2005).

[2] F. Delbaen, W. Schachermayer, A general version of the fundamental theorem of asset pricing, Mathematische Annalen 300 (1994) 463–520.

[3] F. Delbaen, W. Schachermayer, The Mathematics of Arbitrage, Springer, 2008.

[4] J. Cresson, S. Darses, Stochastic embedding of dynamical systems, Journal of Mathematical Physics 48 (2007).

[5] P.A.M. Dirac, Lectures on Quantum Mechanics, Belfer Graduate School of Science, Yeshiva University Press, 1964.

[6] K.D. Elworthy, Stochastic Differential Equations on Manifolds, London Mathematical Society Lecture Notes Series, 1982.

[7] M. Eméry, Stochastic Calculus on Manifolds-With an Appendix by P.A. Meyer, Springer, 1989.

[8] L. Faddeev, R. Jackiw, Hamiltonian reduction of unconstrained and constrained systems, Physical Review Letters 60 (1988) 1692–1694.

[9] S. Farinelli, Geometric arbitrage theory and market dynamics, Journal of Geometric Mechanics 7 (4) (2015) 431–471.

[10] S. Farinelli, Geometric arbitrage theory and market dynamics reloaded, Preprint Arxiv, reviewed version of [9], 2020.

[11] S. Farinelli, H. Takada, The Black-Scholes equation in presence of arbitrage, Preprint Arxiv, 2020.

[12] B. Flesaker, L. Hughston, Positive interest, Risk 9 (1) (1996) 46–49.

[13] Y.E. Gliklikh, Global and Stochastic Analysis with Applications to Mathematical Physics, Theoretical and Mathematical Physics, Springer, 2010.

[14] W. Hackenbroch, A. Thalmaier, Stochastische Analysis. Eine Einführung in die Theorie der stetigen Semimartingale, Teubner Verlag, 1994.

[15] B.C. Hall, Quantum Theory for Mathematicians, Springer Graduate Texts in Mathematics, 2013.

[16] L. Hörmander, The Analysis of Linear Partial Differential Operators I: Distribution Theory and Fourier Analysis, Springer, 2003.

[17] E.P. Hsu, Stochastic Analysis on Manifolds, Graduate Studies in Mathematics, vol. 38, AMS, 2002.

[18] P.J. Hunt, J.E. Kennedy, Financial Derivatives in Theory and Practice, Wiley Series in Probability and Statistics, 2004.

[19] K. Ilinski, Gauge geometry of financial markets, Journal of Physics. A, Mathematical and General 33 (2000) L5–L14.

[20] K. Ilinski, Physics of Finance: Gauge Modelling in Non-Equilibrium Pricing, Wiley, 2001.

[21] Y.M. Kabanov, On the FTAP of Kreps-Delbaen-Schachermayer, in: Statistics and Control of Stochastic Processes, World Scientific Publishing Company, Moscow, 1997, pp. 191–203.

[22] J.R. Klauder, Quantization of Constrained Systems, part of Lecture Notes in Physics book series, LNP, vol. 572, 2001, pp. 143–182.

[23] P.N. Malaney, The Index Number Problem: A Differential Geometric Approach, PhD Thesis, Harvard University Economics Department, 1996.

[24] E. Nelson, Quantum Fluctuations, Princeton University Press, 1985.

[25] E. Nelson, Dynamical Theories of Brownian Motion, second edition, Princeton University Press, 2001.

[26] L. Schwartz, Semi-martingales sur des variétés et martingales conformes sur des variétés analytiques complexes, Lecture Notes in Mathematics, Springer, 1980.

[27] A. Smith, C. Speed, Gauge transforms in stochastic investment, in: Proceedings of the 1998 AFIR Colloquium, Cambridge, England, 1998.

[28] D.W. Stroock, An Introduction to the Analysis of Paths on a Riemannian Manifold, Mathematical Surveys and Monographs, vol. 74, AMS, 2000.

[29] L.A. Takhtajan, Quantum Mechanics for Mathematicians, Graduate Studies in Mathematics, vol. 95, AMS, 2008.

[30] E. Weinstein, Gauge Theory and Inflation: Enlarging the Wu-Yang Dictionary to a unifying Rosetta Stone for Geometry in Application, Talk given at Perimeter Institute, 2006. Available online at: http://pirsa.org/06050010/, 2006.

[31] K. Yasue, Stochastic calculus of variations, Journal of Functional Analysis 41 (1981) 327–340.

[32] K. Young, Foreign exchange market as a lattice gauge theory, American Journal of Physics 67 (1999).

Chapter 17

Thermodynamics and stability of metallic nano-ensembles

Michael Vigdorowitsch[a,b,c]

[a]*Angara GmbH, Düsseldorf, Germany,* [b]*All-Russian Scientific Research Institute for the Use of Machinery and Oil Products in Agriculture, Tambov, Russia,* [c]*Tambov State Technical University, Tambov, Russia*

17.1 Introduction

Ultra-dispersed structures appeared to be much spoken about [1,2] after famous studies by H. Gleiter with coauthors [3,4] had been presented to the scientific and industrial communities. The term *nano-materials* originates from so-called nano-crystallines of the size up to 10 nm introduced there. Nevertheless, the idea to treat substance as the matter with size-dependent properties wasn't virginally new, and we trace back its progress here briefly.

17.1.1 Nano-substance: inception

Such notions as, e.g., *parent phase* (macroscopic crystal of the same nature as a nano-sized particle under consideration) [5] and *thermodynamics of small objects* [6–9] had been introduced in the middle of the twentieth century and served as the basis for functional size-dependencies of surface tension for small particles like the equation (δ is a distance between equimolar separation surface and tension surface, subscript ∞ stays for the parent phase hereinafter)

$$\sigma(r) = \sigma_\infty \left(1 - \frac{2\delta}{r} \right).$$

Study [10] by P.E. Strebejko is likely to have been the first topical dissertation devoted to small objects. The understanding that particle size affects the melting temperature arose first with the works by P. Pawlow [11–13]. Starting with equality condition for chemical potentials of small liquid and spherical crystalline particles (both with size r and equal masses) in equilibrium with their own vapor at the triple point, he obtained the equation

$$\frac{T_r}{T_s} = 1 - \frac{3}{\lambda r} \left[\sigma_s - \sigma_l \left(\frac{\rho_l}{\rho_s} \right)^{1/3} \right],$$

358 Mechanics and Physics of Structured Media

where T_r and T_s are melting temperatures of a particle with size r and of the massive crystal respectively, λ is melting heat, and ρ_l and ρ_s stand for density of crystalline and liquid phases respectively.

Factual genesis, indeed, has to be associated with the nineteenth century as the studies by R. Helmholtz [14] and L. Houllevigue [12] affirmed the equation for reduction of evaporation heat of small droplets with radius r, when compared to macroscopic liquids (a^2 stands for the capillary constant, α is the temperature coefficient):

$$\Delta\lambda = \frac{a^2(1 - \alpha T)}{r}$$

having been derived by B. Sreznevsky [16] and first improved just in the sixtieth [7].

Nevertheless, works [5–16] did not invoke any public resonance what the author is inclined to assume a number of reasons for. Thus, high-tech ideas reasonably pretend to become successful with their applications typically upon having gained a comprehensive theoretical knowledge which enables their promoters to choose optimal configurations and/or operating modes to demonstrate impressive outcomes. Modern development of computing technique has had the research community to rethink the role and place of analytical mathematical methods whose further progress rate naturally remained behind that of computers. Should any complex problem be rigorously formulated, it is highly likely to be quickly solved numerically. This is especially typical for materials sciences whose structures and their properties admit clear computer modeling. Another reason, in the last decades, finds itself in the nature of such a PR- and market economy product as hype. Thus, according to public releases of the eighties, the tokamak technologies showed promise for the fast breakthrough in power generation but today they still have a status of much promising. Despite really avalanche-like accretion of theoretical and experimental studies in the nano-branch, applications other than "car-washing nano-facilities" are still coming soon. And one more, sacramental, reason is well-known as that any brilliant idea has to wait for the proper time. We believe that a huge number of nano-related researches are to bring a qualitative leap before long.

To provide now a deeper insight into the origin of somehow quaint properties of the nano-substance, we consider further its thermodynamic and kinetic basics [17,18].

17.1.2 Nano-substance: thermodynamics basics

The principles of Gibbs's thermodynamics make one to realize that any interaction of nano-structures, no matter whether that is of physical or chemical nature, has to have common thermodynamic reasons. The most obvious thing is that a nano-particle radius (or effective size) r pretends to be present among thermodynamic parameters of the whole system and to condition the nano-substance properties as well as (or even more than) the temperature does. Other

Thermodynamics and stability of metallic nano-ensembles Chapter | 17 **359**

than nano-, classic thermodynamics whose objects are massive particles with $r > 100$ nm considers chemical potential μ_∞ in a conventional state as a constant. For $r < 100$ nm, one has to introduce its dependence on r [18] in order to consistently interpret dimensional phenomena:

$$\mu_i = \mu_{i,\infty} + f_i(r), \quad i = \overline{1 \ldots m}, \tag{17.1}$$

where subscript i enumerates particles being distinguished through size. Therefore, in the case of substance transformation, one has now to write down for the TPs as Helmholtz free energy F, enthalpy H, Gibbs free energy G, and the system's internal energy U respectively:

$$dF = -pdV - SdT + \sum_{i=1}^{m} \left[\mu_{i,\infty} + f_i(r)\right] dN_i, \tag{17.2}$$

$$dH = TdS + Vdp + \sum_{i=1}^{m} \left[\mu_{i,\infty} + f_i(r)\right] dN_i, \tag{17.3}$$

$$dG = Vdp - SdT + \sum_{i=1}^{m} \left[\mu_{i,\infty} + f_i(r)\right] dN_i, \tag{17.4}$$

$$dU = TdS - pdV + \sum_{i=1}^{m} \left[\mu_{i,\infty} + f_i(r)\right] dN_i. \tag{17.5}$$

In particular, from Eq. (17.5) follows the change of internal energy because of effective size of any of interacting participants. To such interactions belong not only processes like

$$aA + bB \rightarrow cC + dD,$$

but also adsorption phenomena, mutual solubility of solids and further related processes. Thus, low-nuclear clusters are known to play often a decisive role in processes with low-dimensional particles. Eq. (17.1) can take the following form for such clusters Cu_m, Ag_m, Au_m [19] (m stands for the number of atoms in a low-atomic cluster):

$$\mu_i = \mu_{i,\infty} + \hat{f}_i(m).$$

A functional form of dimensional factor $f(r)$ in the equations above can be uncovered in the framework of approach in [20] to description of low-dimensional systems. The surface energy depends on particles dispersity and concentration C_r of vacancies (defects). How the latter depends on the particle size, was derived in [20,21] while interpreting the difference in melting and polymorphic transformation temperatures for massive samples and small particles:

$$C_r = C_\infty \exp\left(\frac{\Delta v}{r} \frac{B}{kT}\right), \tag{17.6}$$

360 Mechanics and Physics of Structured Media

where Δv is change in volume due to vacancy nucleation, B is the substance characteristics directly proportional to surface tension. The vacancy concentration appears to contribute essentially for $r \leq 10$ nm. The change in the Gibbs energy at dispersing occurs, in general, on account of the two factors. One of them is a low dimension of the vacancy-related dispersed phase whereas another one arises because of surface tension:

$$\delta G = -kT \left(C_r - C_\infty\right) + \frac{BA}{\rho N_A} \frac{1}{r}, \tag{17.7}$$

where ρ is density of the dispersed phase, A is atomic weight, and N_A the Avogadro number. It was concluded in [20] that dispersing a material is capable of initiating the chemical transformations with energy threshold 10–100 kJ/mole.

The functional form of $f(r)$ follows from Eqs. (17.6) and (17.7) and appears to be

$$f(r) = -kT C_\infty \left[\exp\left(\frac{B\Delta v}{kT} \cdot \frac{1}{r} \right) - 1 \right] + \frac{BA}{\rho N_A} \frac{1}{r}. \tag{17.8}$$

It represents a correction to the classic chemical potential due to low-dimensional nature of the system, should the number of particles be changed with no work to be done.

17.1.3 Nano-substance: kinetics basics

Kinetics of processes with the substance in nano-state bears its imprint [17]. A nano-particle appears to exist in an intermediate state between (i) monoatomic or -molecular form with a negligible energy of cooperative interactions and (ii) a condensed phase with particle size >100 nm. To assume this intermediate state is stable, one has to produce some prerequisites. At the best, this could be related with a local energy minimum. With such a stipulation, one has to consider a manifold increased nonlocal reactivity of the nano-structure. As an example, a fresh meta-stable surface of a metal can be pointed out to, whose juvenile active (adsorption) centers are to become quickly occupied by particles to be adsorbed out of the coterminous medium [22].

Instead of process kinetic indicators such as direct and reverse reaction rate constants, diffusion coefficients, composition of intermediate active complexes *et cetera* under an impact of a particle effective size, one can address the integral kinetic factor, i.e., activation energy E_a according to Arrhenius equation [23]:

$$E_a = -RT \ln k_r / k_0, \tag{17.9}$$

where k_0 and k_r are dimensional and rate constants respectively, R is the gas constant. For low-dimensional structures, an explicit dependence of E_a on particle size is being looked for but still absent in Eq. (17.9). It is reasonable to assume, at least, the variety of dependencies $E_a = E_a(r)$ presented in Fig. 17.1.

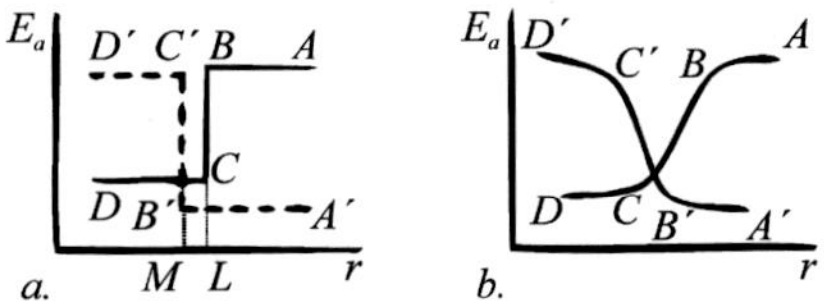

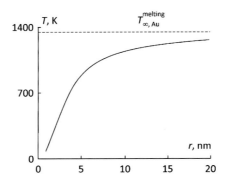

FIGURE 17.1 Particular graphical forms of dependencies $E_a = E_a(r)$ for low-dimensional systems.

FIGURE 17.2 Dependence of the gold melting temperature on particle size.

Their forms follow from both experimental results (Fig. 17.2 [24]) and theoretical representations [25] as well as, more widely, from [23].

Consider $E_a = E_a(r)$ in more detail. The solid lines in Fig. 17.1 show reduction of E_a with increase of r. The reverse situation is produced there with the dashed lines. Fig. 17.1a proposes step-like dependencies whereas Fig. 17.1b smooth ones. For the values of r within AB in Fig. 17.1a, the micro- and nano-structures ($r < 100$ nm) interact but do not produce any nano-effect in the sense of dimensionally conditioned energetic characteristics of the process. Where $r = L$ or $r = M$ (Fig. 17.1a), the kinetic parameter (E_a in the case under consideration) abruptly changes. Note that under L and M no fundamental constants independent of system nature and interaction conditions are implied.

As r decreases ($r < L$ or $r < M$ respectively, Fig. 17.1a), the system enters in the nano-dimensional area and finds itself in the apt state potentially subjected to termination because of, e.g., adsorption of micro-pollutants out of the air in the corresponding case. Should r decrease further, the system migrates to the atomic or molecular phase state (gases) where the nano-state doesn't exist anymore.

According to Fig. 17.1, evolution of E_a can lead to both increase and reduction of the process activation energy depending on its nature. Technologically, both routes represent interest. Route $ABCD$ (Fig. 17.1a) is related to a reaction rate increase and, therefore, considered as valuable, should the process produce a desired product. Route $A'B'C'D'$ (Fig. 17.1a) is unwanted as it results in production of by-products and inhibits the yield of desired product.

362 Mechanics and Physics of Structured Media

The routes in Fig. 17.1a are simplified models. Dependencies in Fig. 17.1b are to be perceived as more reasonable (compare Fig. 17.2).

Thus, the nano-substance [in]stability appears to be determined by thermodynamics but mediately through kinetics. Although the instability origins can widely vary, the common thing is the changes arising in TPs of the ultra-dispersed substance, compared to those of relevant macroscopic crystals.

The present work proposes a novel approach to determination of stability conditions for metallic nano-ensembles. This is based on a model of two effects oppositely affecting the ensemble stability, namely the surface tension and bulk- (vacancy-) related effects in Eq. (17.7). The forerunner [26] proposed a rough estimate of the respective quantities for systems of this kind.

17.2 Vacancy-related reduction of the metallic nano-ensemble's TPs

A vacancy-related effect of TPs reduction (in particular, for Gibbs energy) is sure to contribute to the system stability and resistance to the external inputs mentioned above. We perform here the corresponding modeling with various distributions of particles on their size (radii).

17.2.1 Solution in quadrature of the problem of vacancy-related reduction of TPs

We discuss Eq. (17.1) first which improves the classic chemical potential of the system of mono-sized particles of the i-th sort, with the correction represented by the first term in Eq. (17.8).

Neither substance parameter B nor arising vacancy's volume Δv depends on particle size. Let subscript i enumerate groups of particles according to the particle size starting from $r_{\min}$ ($i = 1$) up to $r_{\max}$ ($i = m$). The value of $f_i(r) \, dN_i$ represents the change in the TPs due to vacancy-related effect of the i-th sort of particles, namely (hereinafter the notation δE is used for all the TPs (17.2)–(17.5), i.e., $\delta F = \delta H = \delta G = \delta U \equiv \delta E$)

$$\delta E_i = -kTC_\infty \left(e^{\frac{B\Delta v}{kT} \cdot \frac{1}{r_i}} - 1 \right) \delta N_i$$

or, introducing the number of particles in a volume unit

$$n_i(r_i) = N_i(r_i)/V, \tag{17.10}$$

now in an integral form:

$$\delta E = -kTC_\infty V \sum_{i=1}^{m} \left(e^{\frac{B\Delta v}{kT} \cdot \frac{1}{r_i}} - 1 \right) \delta n_i(r_i) = \cdots .$$

Thermodynamics and stability of metallic nano-ensembles Chapter | 17 **363**

Physically, concentration of particles n_i cannot possess peculiarities like singularities or discontinuities and is sure to belong, at least, to a class of continuously differentiable functions. The first term in brackets is the monotonously decreasing down to zero function of r_i in the whole domain. For r_{max} big enough, it is nearly zero. These considerations substantiate the following transition from the sum to an integral:

$$\cdots = -kTC_\infty V \int_{r_{min}}^{r_{max}+\delta r_{max}} \left(e^{\frac{B\Delta v}{kT} \cdot \frac{1}{r}} - 1 \right) n(r) \, dr, \tag{17.11}$$

where $\delta r_{max} > 0$, and the continuous function $n(r)$ takes the sense of distribution of particles on their radii. An exact determination of the integration upper limit is not a simple task, but it does not necessarily need to be solved here because the expression in the brackets tends exponentially to zero in the area of r_{max} what makes the contribution into the integral within the segment $[r_{max}, r_{max} + \delta r_{max}]$ exponentially small. Neglecting then δr_{max}, we obtain:

$$\delta E \approx -kTC_\infty V \int_{r_{min}}^{r_{max}} \left(e^{\frac{B\Delta v}{kT} \cdot \frac{1}{r}} - 1 \right) n(r) \, dr. \tag{17.12}$$

Eq. (17.12) represents the solution of the problem of vacancy-related reduction of TPs, compared to those for the massive crystal, in quadrature. Integration in Eq. (17.12) is to be performed from the lowest up to the highest radius of nano-structures, i.e., in general case from 1 to 100 nm. This reduction of TPs disappears with a transition to the parent phase inasmuch as $r_{min} \to r_{max} \to \infty$, and because of an obvious continuity of the integrand, the integral tends to zero in the sense of Riemann.

17.2.2 Particle distributions on their radii

We consider further consequences from Eq. (17.12) for different distributions $n(r)$ with the normalization condition related to an overall number n_t of particles of all sizes in a volume unit:

$$\int_{r_{min}}^{r_{max}} n(r) \, dr = n_t. \tag{17.13}$$

As those, we employ the even distribution (i.e., n does not depend on r)

$$n(r) = \frac{n_t}{r_{max} - r_{min}}; \tag{17.14}$$

linear distributions incl. an increasing

$$n(r) = n_t \left[n_{min} + g(r - r_{min}) \right] \tag{17.15}$$

364 Mechanics and Physics of Structured Media

and decreasing one

$$n(r) = n_t [n_{max} - g(r - r_{min})] \qquad (17.16)$$

with the "concentration gradient"

$$g = \frac{n_{max} - n_{min}}{r_{max} - r_{min}} = \frac{2}{r_{max} - r_{min}} \left[\frac{1}{r_{max} - r_{min}} - n_{min} \right],$$

which produce a quasiholonomic constraint because of normalization condition (17.13);

$$n_{max} + n_{min} = \frac{2}{r_{max} - r_{min}}; \qquad (17.17)$$

exponentially increasing distribution

$$n(r) = n_t \frac{1}{\gamma} e^{r/\gamma}, \qquad (17.18)$$

where $\gamma > 0$ is usually called a distribution modulus, with the quasiholonomic constraint according to Eq. (17.13)

$$e^{r_{max}/\gamma} - e^{r_{min}/\gamma} = 1; \qquad (17.19)$$

and the normal (asymmetrically truncated in a general case) distribution

$$n(r) = n_t \frac{1}{\sigma} \exp \left[-\frac{(r - r_0)^2}{2\sigma^2} \right], \qquad (17.20)$$

where $r_{min} < r_0 < r_{max}$, with the quasiholonomic constraint

$$\Phi \left(\frac{r_{max} - r_0}{\sqrt{2}\sigma} \right) - \Phi \left(\frac{r_{min} - r_0}{\sqrt{2}\sigma} \right) = \sqrt{\frac{2}{\pi}} \qquad (17.21)$$

involving dispersion σ^2, lower- and uppermost particle sizes and the most probable size r_0, where $\Phi(x) = \frac{2}{\sqrt{\pi}} \int_0^x e^{-t^2} dt$ is the error function integral [27].

Quasiholonomic constraint (17.17) in the case of distributions (17.15) and (17.16) links lower- and uppermost radii of the particles altogether with their concentrations. Thus, for example, concentration of particles with the uppermost radius n_{max} in the case of linearly increasing distribution should meet the rule

$$n_{max} = \frac{2}{r_{max} - r_{min}} - n_{min} > n_{min},$$

from where we have

$$r_{max} < \frac{1}{n_{min}} + r_{min}.$$

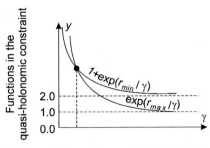

FIGURE 17.3 Functions in the quasiholonomic constraint (17.22) for the exponentially increasing distribution.

And *vice versa*, a constraint may be imposed on concentration of particles with the lowermost radius:

$$n_{min} < \frac{1}{r_{max} - r_{min}}.$$

Then for the lower- and uppermost radii considered in this chapter, $n_{min} < 0.01$ nm^{-1} takes place.

In the case of exponentially increasing distribution (17.18), the quasiholonomic constraint (17.19) enables one to calculate modulus γ. Let us consider the constraint (17.19) as (Fig. 17.3)

$$e^{r_{max}/\gamma} = 1 + e^{r_{min}/\gamma}. \tag{17.22}$$

Solution of Eq. (17.22) for γ can be sought by using the method of successive approximations. For the zeroth approximation, one can consider the limit $\gamma \to \infty$ and represent the exponents by their McLaurin expansions in powers of $1/\gamma$ taking into account the first two expansion terms:

$$\gamma^{(0)} = r_{max} - r_{min}. \tag{17.23}$$

The successive, k-th, approximation is to be found using the recurrency due to Eq. (17.22):

$$e^{r_{max}/\gamma^{(k+1)}} = 1 + e^{r_{min}/\gamma^{(k)}}, \quad k = 0, 1, \ldots.$$

This enables one to build the sequence $\{\gamma^{(k)}\}$ approximating the true solution:

$$\gamma^{(k)} = r_{max} \ln^{-1}\left[1 + \left[1 + \ldots \left(1 + e^{\frac{r_{min}}{r_{max}-r_{min}}}\right)^{\frac{r_{min}}{r_{max}}}\right]^{\frac{r_{min}}{r_{max}}}\right]. \tag{17.24}$$

$\{(k-1) \text{ times}\}$

366 Mechanics and Physics of Structured Media

Thus, for $\gamma^{(3)}$ we have

$$\gamma^{(3)} = \frac{r_{max}}{\ln\left[1 + \left[1 + \left(1 + e^{\frac{r_{min}}{r_{max}-r_{min}}}\right)^{\frac{r_{min}}{r_{max}}}\right]^{\frac{r_{min}}{r_{max}}}\right]}.$$

We obtain a sequence of solutions for various combinations of r_{min} and r_{max} (Table 17.1).

TABLE 17.1 The first 5 terms of sequence $\gamma^{(k)}$ (in units of r) for the solution of Eq. (17.22) due to Eqs. (17.23) and (17.24) for different combinations r_{min} and r_{max}.

r_{min}	r_{max}	$\gamma^{(0)}$	$\gamma^{(1)}$	$\gamma^{(2)}$	$\gamma^{(3)}$	$\gamma^{(4)}$
1	15	14	20.56	21.57	20.93	20.91
	35	34	49.44	50.46	49.78	49.77
	75	74	107.15	108.19	107.48	107.48
	100	99	143.22	144.26	143.55	143.55
15	100	85	127.34	141.76	133.79	133.20
35	100	65	100.18	129.64	119.44	117.60
75	100	25	32.80	74.49	75.86	76.63

As it follows from the data of Table 17.1, the broader the range of particles radii $[r_{min}, r_{max}]$, the faster the sequence of solutions converges. For a range broad enough, one does not need to use 5 or more iterations.

The quasiholonomic constraint (17.21) in the case of normal distribution (17.20) enables one to approximately find the square root of dispersion σ^2 by using the following method. We'll look for the solution of Eq. (17.21) in the form

$$\sigma = \sigma^{(0)} + \sigma^{(1)} + \sigma^{(2)} \tag{17.25}$$

assuming $\left|\frac{\sigma^{(2)}}{\sigma^{(0)}}\right| < \left|\frac{\sigma^{(1)}}{\sigma^{(0)}}\right| < 1$ (easy to check later). Taking into account the asymptotic approximation of the error function integral $\Phi(x)$ [27] and substituting Eq. (17.25) into Eq. (17.21), we have for small arguments

$$\Phi(x) \approx \frac{2}{\sqrt{\pi}}\left(x - \frac{x^3}{3} + \frac{x^5}{10}\right)$$

and come then to the equation

$$\frac{r_{max} - r_{min}}{\sigma^{(0)}\left(1 + \frac{\sigma^{(1)}}{\sigma^{(0)}} + \frac{\sigma^{(6)}}{\sigma^{(0)}}\right)} + \frac{(r_{min} - r_0)^3 - (r_{max} - r_0)^3}{6\sigma^{(0)3}\left(1 + \frac{\sigma^{(1)}}{\sigma^{(0)}} + \frac{\sigma^{(6)}}{\sigma^{(0)}}\right)^3}$$

$$- \frac{(r_{min} - r_0)^5 - (r_{max} - r_0)^5}{40\sigma^{(0)5}\left(1 + \frac{\sigma^{(1)}}{\sigma^{(0)}} + \frac{\sigma^{(2)}}{\sigma^{(0)}}\right)^5} = 1.$$

Transformation of trinomials in the denominators due to McLaurin expansion leads to the following equation:

$$\frac{r_{max} - r_{min}}{\sigma^{(0)}}\left(1 - \frac{\sigma^{(1)}}{\sigma^{(0)}} - \frac{\sigma^{(2)}}{\sigma^{(0)}} + \frac{\sigma^{(1)2}}{\sigma^{(0)2}} + \cdots\right)$$

$$+ \frac{(r_{min} - r_0)^3 - (r_{max} - r_0)^3}{6\sigma^{(0)3}}\left(1 - 3\frac{\sigma^{(1)}}{\sigma^{(0)}} - 3\frac{\sigma^{(2)}}{\sigma^{(0)}} + 6\frac{\sigma^{(1)2}}{\sigma^{(0)2}} + \cdots\right)$$

$$- \frac{(r_{min} - r_0)^5 - (r_{max} - r_0)^5}{40\sigma^{(0)5}}\left(1 - 5\frac{\sigma^{(1)}}{\sigma^{(0)}} - 5\frac{\sigma^{(2)}}{\sigma^{(0)}} + 15\frac{\sigma^{(1)2}}{\sigma^{(0)2}} + \cdots\right) = 1.$$

Treating it as the series expansion in powers of $\frac{\sigma^{(i)}}{\sigma^{(0)}}$, we derive equations

(a) for $\sigma^{(0)}$ taking into account linear terms for $i = 0$

$$\frac{r_{max} - r_{min}}{\sigma^{(0)}} = 1$$

from where it follows

$$\sigma^{(0)} = r_{max} - r_{min}, \tag{17.26}$$

(b) for $\sigma^{(1)}$ with linear terms for $i = 1$

$$\frac{(r_{min} - r_0)^3 - (r_{max} - r_0)^3}{6\sigma^{(0)3}} = \frac{\sigma^{(1)}}{\sigma^{(0)}}\left[\frac{(r_{min} - r_0)^3 - (r_{max} - r_0)^3}{2\sigma^{(0)3}} + 1\right]$$

from where simple transformations give the expression

$$\sigma^{(1)} = \frac{\sigma^{(0)}}{3}\frac{(r_{min} - r_0)^3 - (r_{max} - r_0)^3}{(r_{min} - r_0)^3 - (r_{max} - r_0)^3 + 2\sigma^{(0)3}}, \tag{17.27}$$

and (c) for $\sigma^{(2)}$ with linear terms for $i = 2$ and up to quadratic terms for $i = 1$:

$$\frac{\sigma^{(2)}}{\sigma^{(0)}}\left[-1 - \frac{1}{2}\frac{(r_{min} - r_0)^3 - (r_{max} - r_0)^3}{\sigma^{(0)3}} + \frac{1}{8}\frac{(r_{min} - r_0)^5 - (r_{max} - r_0)^5}{\sigma^{(0)5}}\right]$$

$$+\frac{\sigma^{(1)^2}}{\sigma^{(0)^2}}+\frac{(r_{min}-r_0)^3-(r_{max}-r_0)^3}{\sigma^{(0)^3}}\frac{\sigma^{(1)^2}}{\sigma^{(0)^2}}$$

$$-\frac{1}{40}\frac{(r_{min}-r_0)^5-(r_{max}-r_0)^5}{\sigma^{(0)^5}}\left(1-5\frac{\sigma^{(1)}}{\sigma^{(0)}}+15\frac{\sigma^{(1)^2}}{\sigma^{(0)^2}}\right)=0,$$

which finally results in the expression

$$\sigma^{(2)}=\sigma^{(0)}$$

$$\times\frac{\frac{1}{40}\frac{(r_{min}-r_0)^5-(r_{max}-r_0)^5}{\sigma^{(0)^5}}\left(1-5\frac{\sigma^{(1)}}{\sigma^{(0)}}+15\frac{\sigma^{(1)^2}}{\sigma^{(0)^2}}\right)-\frac{\sigma^{(1)^2}}{\sigma^{(0)^2}}-\frac{(r_{min}-r_0)^3-(r_{max}-r_0)^3}{\sigma^{(0)^3}}\frac{\sigma^{(1)^2}}{\sigma^{(0)^2}}}{-1-\frac{1}{2}\frac{(r_{min}-r_0)^3-(r_{max}-r_0)^3}{\sigma^{(0)^3}}+\frac{1}{8}\frac{(r_{min}-r_0)^5-(r_{max}-r_0)^5}{\sigma^{(0)^5}}}.$$

$$(17.28)$$

Numerical calculations of σ^2 enable us to judge if the approximate solution (17.25)–(17.28) is good. Table 17.2 presents such a comparison for different r_{min}, r_{max} and the Gaussian's vertex abscissa r_0.

As it follows from the data of Table 17.2, Eqs. (17.25)–(17.28) give a pretty good approximation in the case of symmetrically truncated Gaussian, which deteriorates as the asymmetry develops. Furthermore, for small differences $r_{max}-r_{min}$ (within 1 nm, absent in Table 17.2) the analytical approximation error grows but such differences do not have much practical meaning. We also note that in many configurations one can afford to omit $\sigma^{(2)}$. The relative error differs then from that in calculations taking $\sigma^{(2)}$ into account just negligibly.

In the case of exponentially decreasing distribution

$$n(r)=n_t\frac{1}{\gamma}e^{-r/\gamma},$$

the quasiholonomic constraint

$$e^{-r_{min}/\gamma}-e^{-r_{max}/\gamma}=1$$

does not enable one to calculate γ on $\mathbb{R}$ since there is no solution there at all (Fig. 17.4).

Indeed, for any γ, inequalities $1<1+e^{-r_{max}/\gamma}<2$ and $e^{-r_{min}/\gamma}<1$ take place. Solutions can be found on set $\mathbb{C}$ of complex numbers only but this debunks $n(r)$ as a distribution function. That an exponentially decreasing distribution of nano-particles on their radii is inconsistent, is not related to a passage from the sum to integral in Eq. (17.11) as this quasiholonomic constraint does not have a solution on $\mathbb{R}$ without replacement $r_{max}+\delta r_{max}\to r_{max}$ either. Thus, inconsistence of an exponentially decreasing distribution follows instantly from the nature of substance dispersity in the nano-sized area. Since this outcome is directly related to a thermodynamic description of the nano-ensemble, it is limited to equilibrium systems only. In other words, should such a distribution be

Thermodynamics and stability of metallic nano-ensembles Chapter | 17 **369**

TABLE 17.2 Numerical and approximate analytical solutions (17.25)–(17.28) of Eq. (17.21) for different lower-, uppermost and most probable radii of particles.

r_{min}, nm	r_{max}, nm	r_0, nm	σ, nm (numerically)	σ, nm (analytical approximation)	Relative error
1	100	50.50	94.668	94.286	0.404%
		34.00	93.209	92.400	0.867%
		67.00	93.209	92.400	0.867%
		1.00	77.643	66.001	14.995%
		100.00	77.643	66.001	14.995%
	75	38.00	70.749	70.476	0.385%
		25.67	69.671	69.067	0.867%
		50.33	69.671	69.067	0.867%
		1.00	58.036	49.334	14.994%
		75.00	58.036	49.334	14.994%
	50	25.50	46.858	46.667	0.409%
		17.33	46.120	45.733	0.839%
		33.67	46.120	45.733	0.839%
		1.00	38.414	32.668	14.956%
		50.00	38.414	32.668	14.956%
	25	13.00	22.946	22.857	0.385%
		9.00	22.596	22.400	0.866%
		17.00	22.596	22.400	0.866%
		1.00	18.823	16.004	14.977%
		25.00	18.823	16.004	14.977%
25	100	62.50	71.721	71.429	0.408%
		50.00	70.613	70.000	0.867%
		75.00	70.613	70.000	0.867%
		25.00	58.821	50.001	14.994%
		100.00	58.821	50.001	14.994%
50	100	75.00	47.804	47.619	0.387%
		66.67	47.062	46.667	0.840%
		83.33	47.062	46.667	0.840%
		50.00	39.197	33.335	14.956%
		100.00	39.197	33.335	14.956%
75	100	87.50	23.907	23.810	0.408%
		83.33	23.538	23.334	0.866%
		91.67	23.538	23.334	0.866%
		75.00	19.607	16.670	14.978%
		100.00	19.607	16.670	14.978%

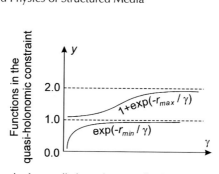

FIGURE 17.4 Functions in the quasiholonomic constraint for exponentially increasing distribution.

arranged artificially and find then itself free of any external inputs, this, as thermodynamically inconsistent, would migrate towards some of equilibrium-stable distributions.

17.2.3 Derivation of equations for TPs reduction

Hereinafter we'll use parameter $\alpha \equiv \frac{B\Delta v}{kT}$ which has the dimension of length. It depends on surface- (B) and bulk-related (Δv) properties of the material.

17.2.3.1 Even distribution of particles on their radii

Here we consider the reduction of TPs in the case of even distribution (17.14). Upon substituting Eq. (17.14) into Eq. (17.12), we obtain (note the change of integration variable, $t \equiv 1/r$)

$$\frac{\delta E_e}{kTC_\infty V n_t} = -\frac{1}{r_{max} - r_{min}} \int_{r_{min}}^{r_{max}} \left(e^{\frac{\alpha}{r}} - 1\right) dr$$

$$= -\frac{1}{r_{max} - r_{min}} \left[\int_{\frac{1}{r_{max}}}^{\frac{1}{r_{min}}} e^{\alpha t} t^{-2} dt - (r_{max} - r_{min})\right].$$

The integral in the right part can be taken with the use of the special function of mathematical physics:

$$\frac{\delta E_e}{kTC_\infty V n_t} = -\frac{1}{\frac{r_{max}}{r_{min}} - 1} \left\{ \frac{r_{max}}{r_{min}} e^{\frac{\alpha}{r_{max}}} - e^{\frac{\alpha}{r_{min}}} \right.$$

$$\left. + \frac{\alpha}{r_{min}} \left[\text{Ei}^*\left(\frac{\alpha}{r_{min}}\right) - \text{Ei}^*\left(\frac{\alpha}{r_{max}}\right)\right]\right\} + 1, \qquad (17.29)$$

where $\text{Ei}^*(z) = \text{Ei}(z) + \pi i$, $\text{Ei}(z)$ is the exponential integral defined for real argument x as $\text{Ei}(x) = \int_{-\infty}^{x} \frac{e^y}{y} dy$ for $x < 0$, and $\text{Ei}(x) = -\lim_{\varepsilon \to +0}\left[\int_{-x}^{-\varepsilon} \frac{e^{-t}}{t} dt + \int_{\varepsilon}^{+\infty} \frac{e^{-t}}{t} dt\right]$ for $x > 0$ [27]. In small particles, the vacancies concentration is always increased [21], i.e., $C_r > C_\infty$, then $\Delta v > 0$ and, consequently, $\alpha > 0$ what

Thermodynamics and stability of metallic nano-ensembles **Chapter | 17 371**

determines the choice of the exponential integral's branch in the calculations hereinafter.

Let us consider the asymptotic behavior[1] of δE_e with respect to α. With $\alpha \to 0$, a linear decrease in its absolute value takes place:

$$
\frac{\delta E_e}{kTC_\infty V n_t} \sim -\frac{1}{r_{max} - r_{min}} \left\{ r_{max}\left(1 + \frac{\alpha}{r_{max}}\right) - r_{min} - \frac{\alpha}{r_{min}} \right.
$$
$$
+ \alpha\left[\ln\frac{\alpha}{r_{min}} - \ln\frac{\alpha}{r_{max}} - \frac{\alpha}{r_{min}} + \frac{\alpha}{r_{max}}\right] - r_{max} + r_{min}\biggr\}
$$
$$
= -\frac{\alpha}{r_{max} - r_{min}}\ln\frac{r_{max}}{r_{min}}.
$$

Similar to this, with $\alpha \to +\infty$ and taking into account $\text{Ei}^*(x \to +\infty) \sim \frac{e^x}{x}\left(1 + \frac{1!}{x}\right)$, we obtain the exponential growth in its absolute value:

$$
\frac{\delta E_e}{kTC_\infty V n_t} \sim -\frac{r_{min}^2}{r_{max} - r_{min}}\frac{e^{\frac{\alpha}{r_{min}}}}{\alpha}
$$

which is typical for all distributions considered hereinafter.

17.2.3.2 Linear distributions

We calculate here the reduction of energy in the case of both linearly increasing and decreasing particles distributions on their radii. With the increasing distribution (17.15), the number of particles with greater radii prevails. Substitution of Eq. (17.15) into Eq. (17.12) gives rise to

$$
\frac{\delta E_{li}}{kTC_\infty V n_t} = -\int_{r_{min}}^{r_{max}} \left[n_{min} + (n_{max} - n_{min})\frac{r - r_{min}}{r_{max} - r_{min}}\right]\left(e^{\frac{\alpha}{r}} - 1\right)dr.
$$

The integral $\int_{r_{min}}^{r_{max}} r e^{\frac{\alpha}{r}}\,dr$ in the right part can be taken by using the same change of variable as before and integrating by parts (*see* [28] as well). Finally, we obtain:

$$
\frac{\delta E_{li}}{kTC_\infty V n_t} = -\frac{n_{min}r_{min}}{\left(\frac{r_{max}}{r_{min}} - 1\right)} \times \left\{ \left[\frac{r_{max}}{r_{min}} - \frac{n_{max}}{n_{min}}\right]\left[\frac{r_{max}}{r_{min}}e^{\frac{\alpha}{r_{max}}} - e^{\frac{\alpha}{r_{min}}}\right. \right.
$$
$$
+ \frac{\alpha}{r_{min}}\left(\text{Ei}^*\left(\frac{\alpha}{r_{min}}\right) - \text{Ei}^*\left(\frac{\alpha}{r_{max}}\right)\right)\right]
$$
$$
+ \frac{1}{2}\left[\frac{n_{max}}{n_{min}} - 1\right]\left[\frac{r_{max}^2}{r_{min}^2}e^{\frac{\alpha}{r_{max}}} - e^{\frac{\alpha}{r_{min}}}\right]
$$

---

[1] Symbol "∼" means hereinafter that the ratio of expressions on the left and on the right of the sign "=" is equal to 1 in the limit considered, i.e., the expression on the right is an asymptotic approximation for the expression on the left.

$$+\frac{\alpha}{r_{min}}\left(\frac{r_{max}}{r_{min}}e^{\frac{\alpha}{r_{max}}} - e^{\frac{\alpha}{r_{min}}}\right)$$

$$+\frac{\alpha^2}{r_{min}^2}\left(\text{Ei}^*\left(\frac{\alpha}{r_{min}}\right) - \text{Ei}^*\left(\frac{\alpha}{r_{max}}\right)\right)\Bigg]\Bigg\} + 1. \qquad (17.30)$$

By analogy, for the linearly decreasing distribution (17.16) when these are the smaller particles which prevail, its substitution into Eq. (17.12) results in:

$$\frac{\delta E_{ld}}{kTC_\infty Vn_t} = -\int_{r_{min}}^{r_{max}}\left[n_{max} - (n_{max} - n_{min})\frac{r - r_{min}}{r_{max} - r_{min}}\right]\left(e^{\frac{\alpha}{r}} - 1\right)dr$$

$$= -\frac{n_{min}r_{min}}{\left(\frac{r_{max}}{r_{min}} - 1\right)} \times \Bigg\{ \left[\frac{\frac{n_{max}}{n_{min}}\frac{r_{max}}{r_{min}} - 1}{\frac{r_{max}}{r_{min}} - 1}\right]\left[\frac{r_{max}}{r_{min}}e^{\frac{\alpha}{r_{max}}} - e^{\frac{\alpha}{r_{min}}}\right.$$

$$+\frac{\alpha}{r_{min}}\left(\text{Ei}^*\left(\frac{\alpha}{r_{min}}\right) - \text{Ei}^*\left(\frac{\alpha}{r_{max}}\right)\right)\Bigg]$$

$$-\frac{1}{2}\frac{\frac{n_{max}}{n_{min}} - 1}{\frac{r_{max}}{r_{min}} - 1}\left[\frac{r_{max}^2}{r_{min}^2}e^{\frac{\alpha}{r_{max}}} - e^{\frac{\alpha}{r_{min}}}\right.$$

$$+\frac{\alpha}{r_{min}}\left(\frac{r_{max}}{r_{min}}e^{\frac{\alpha}{r_{max}}} - e^{\frac{\alpha}{r_{min}}}\right)$$

$$+\frac{\alpha^2}{r_{min}^2}\left(\text{Ei}^*\left(\frac{\alpha}{r_{min}}\right) - \text{Ei}^*\left(\frac{\alpha}{r_{max}}\right)\right)\Bigg]\Bigg\} + 1. \qquad (17.31)$$

17.2.3.3 Exponential distribution

Substitution of Eq. (17.18) into Eq. (17.12) leads to the integral

$$\frac{\delta E_{ei}}{kTC_\infty Vn_t} = -\frac{1}{\gamma}\int_{r_{min}}^{r_{max}} e^{r/\gamma}\left(e^{\frac{\alpha}{r}} - 1\right)dr.$$

A closed expression for the integral $\int_{r_{min}}^{r_{max}} e^{\frac{r}{\gamma} + \frac{\alpha}{r}}\,dr$ is known neither in elementary nor in special functions. We consider first the integral properties. In case constants γ and α altogether with the normalization condition (17.13) resulted in that the integration limit r_{max} could be approximately considered as $+\infty$ for $y(r) = e^{\alpha/r}$ but limited for $e^{r/\gamma}$, the region of the integrand's sharp exponential growth would not be reached yet. Physically, such a distribution would lead to the same consequences as the linearly increasing distribution above. Provided the dynamic exponential growth of the integrand took place, i.e., $y(r) = e^{r/\gamma}$ had big magnitudes and the number of big particles exceeded the number of small particles by a factor of ten or more, the picture in Fig. 17.5 would take place.

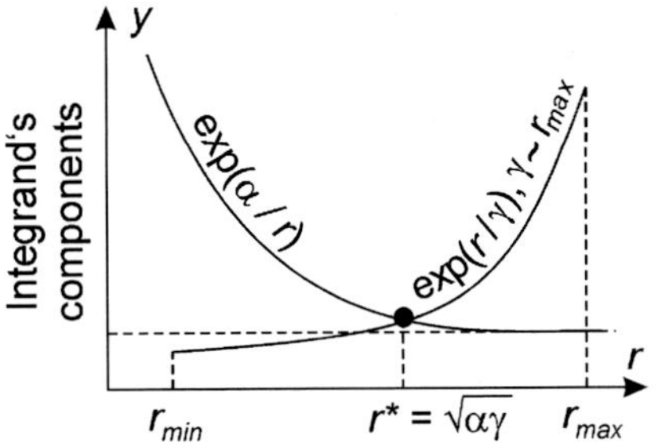

Thermodynamics and stability of metallic nano-ensembles **Chapter | 17 373**

FIGURE 17.5 Integrand's components in $\int_{r_{min}}^{r_{max}} e^{\frac{r}{\gamma}+\frac{\alpha}{r}}\,dr$ in the case of exponential distribution.

Approximately one can consider ($r^* = \sqrt{\alpha\gamma}$ is the intersection point of functions $y = e^{\alpha/r}$ and $y = e^{r/\gamma}$)

$$\int_{r_{min}}^{r_{max}} e^{\frac{r}{\gamma}+\frac{\alpha}{r}}\,dr \approx \int_{r_{min}}^{r^*} e^{\frac{\alpha}{r}}\,dr + \int_{r^*}^{r_{max}} e^{\frac{r}{\gamma}}\,dr \qquad (17.32)$$

since for $r_{min} < r < r^*$ the estimate $e^{r/\gamma} = O(1)$ is valid, whereas for $r^* < r < r_{max}$ we have $e^{\alpha/r} = O(1)$. The indefinite integrals in the right part of Eq. (17.32) can be easily taken by analogy with previous integrals, resulting in:

$$\frac{\delta E_{ei}}{kTC_\infty V n_t} \approx -\left\{\sqrt{\frac{\alpha}{\gamma}}e^{\sqrt{\alpha/\gamma}} - \frac{r_{min}}{\gamma}e^{\frac{\alpha}{r_{min}}} + \frac{\alpha}{\gamma}\left[\mathrm{Ei}^*\left(\frac{\alpha}{r_{min}}\right) - \mathrm{Ei}^*\left(\sqrt{\frac{\alpha}{\gamma}}\right)\right]\right.$$
$$\left. + e^{\frac{r_{max}}{\gamma}} - e^{\sqrt{\frac{\alpha}{\gamma}}}\right\} + 1. \qquad (17.33)$$

17.2.3.4 The normal (truncated) distribution

Substitution of Eq. (17.20) into Eq. (17.12) produces the equation

$$\frac{\delta E_G}{kTC_\infty V n_t} = -\frac{1}{\sigma}\int_{r_{min}}^{r_{max}} \exp\left[-\frac{(r-r_0)^2}{2\sigma^2}\right]\left(e^{\frac{\alpha}{r}} - 1\right)\,dr. \qquad (17.34)$$

No closed expression for integral $\int_{r_{min}}^{r_{max}} \exp\left[-\frac{(r-r_0)^2}{2\sigma^2}\right]e^{\frac{\alpha}{r}}\,dr$ is known in elementary or special functions either. To take the indefinite integral, we use the saddle point method [29]. The integrand is a product of two functions presented in Fig. 17.6.

We replace the Gaussian in Fig. 17.6 with a parabola in such a way that their vertices $(r_0, 1)$ coincide and the parabola's branches intersect the abscissa axis in points $r_0 \pm \sqrt{2}\sigma$. The parabola equation reads then:

$$y(r) = 1 - \frac{(r-r_0)^2}{2\sigma^2}.$$

374 Mechanics and Physics of Structured Media

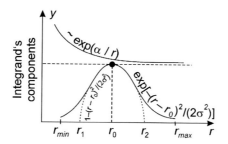

FIGURE 17.6 The integrand's components in Eq. (17.34) in the case of normal distribution.

Such a replacement under the integral sign gives the following expression:

$$\int_{r_{min}}^{r_{max}} \exp\left[-\frac{(r-r_0)^2}{2\sigma^2}\right] e^{\frac{\alpha}{r}} dr \approx \int_{r_1}^{r_2} \left[1 - \frac{(r-r_0)^2}{2\sigma^2}\right] e^{\frac{\alpha}{r}} dr \qquad (17.35)$$

where

$$r_1 = \begin{cases} r_0 - \sqrt{2}\sigma, & \text{if } r_0 - \sqrt{2}\sigma > r_{min}, \\ r_{min}, & \text{if } r_0 - \sqrt{2}\sigma \leq r_{min}, \end{cases}$$

$$r_2 = \begin{cases} r_0 + \sqrt{2}\sigma, & \text{if } r_0 + \sqrt{2}\sigma < r_{max}, \\ r_{max}, & \text{if } r_0 + \sqrt{2}\sigma \geq r_{max}. \end{cases}$$

By analogy with the consideration above, the integrals in the right part of Eq. (17.35) can be taken with the following result:

$$\frac{\delta E_G}{kTC_\infty V n_t} \approx -\frac{1}{\sigma} \left\{ \left[1 - \frac{r_0^2}{2\sigma^2}\right] \left[r_2 e^{\frac{\alpha}{r_2}} - r_1 e^{\frac{\alpha}{r_1}} + \alpha \left(\text{Ei}^*\left(\frac{\alpha}{r_1}\right) - \text{Ei}^*\left(\frac{\alpha}{r_2}\right)\right)\right] \right.$$
$$+ \frac{r_0}{2\sigma^2}\left[r_2^2 e^{\frac{\alpha}{r_2}} - r_1^2 e^{\frac{\alpha}{r_1}} + \alpha \left(r_2 e^{\frac{\alpha}{r_2}} - r_1 e^{\frac{\alpha}{r_1}}\right)\right.$$
$$+ \alpha^2 \left(\text{Ei}^*\left(\frac{\alpha}{r_1}\right) - \text{Ei}^*\left(\frac{\alpha}{r_2}\right)\right)\right]$$
$$- \frac{1}{6\sigma^2}\left[r_2^3 e^{\frac{\alpha}{r_2}} - r_1^3 e^{\frac{\alpha}{r_1}} + \frac{\alpha}{2}\left(r_2^2 e^{\frac{\alpha}{r_2}} - r_1^2 e^{\frac{\alpha}{r_1}}\right)\right.$$
$$\left.\left. + \frac{\alpha^2}{2}\left(r_2 e^{\frac{\alpha}{r_2}} - r_1 e^{\frac{\alpha}{r_1}}\right) + \frac{\alpha^3}{2}\left(\text{Ei}^*\left(\frac{\alpha}{r_1}\right) - \text{Ei}^*\left(\frac{\alpha}{r_2}\right)\right)\right]\right\} + 1.$$
(17.36)

The approximation we have used has a demerit that, because of an underestimate of the integral through replacement of function $\exp\left[-\frac{(r-r_0)^2}{2\sigma^2}\right] e^{\frac{\alpha}{r}}$ with the parabola (Fig. 17.7), the respective term contributes less at α big enough.

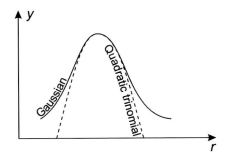

FIGURE 17.7 The saddle point method regarding calculation of the integral in Eq. (17.34).

A physically unreasonable dominance of the second integrand in Eq. (17.34) can lead to a wrong sign in the TP change, should the range of radii $[r_{min}, r_{max}]$ considered is not broad enough.

This effect has appeared to take place, in particular, in the calculations below for gold with the range $r_{max} - r_{min} < 15$ nm. To eliminate it, one could approximate the Gaussian with a parabola in the second integrand in Eq. (17.34) as well. However, this would cause a renormalization of the whole integral and solving some other equation (without the error function integrals) than Eq. (17.21) to determine dispersion. As a result, such steps would mean a study of a parabolic distribution rather than of the normal one. Therefore, an additional approximation was given up.

17.2.4 Reduction of TPs: results

The expressions obtained above for various particles distributions, were applied to nano-ensembles consisting of atoms of either gold or indium. For both metals, specific (per 1 "average" particle) reduction of TP $\frac{\delta E}{n_t V}$ was calculated at 300 K. To calculate the metal characteristics α, the data on pressure reduction depending on reverse radius of small metallic particles [21], were used. The following results were obtained: $\alpha_{In} = 19.2$ nm, $\alpha_{Au} = 67.0$ nm. Change in volume Δv due to the vacancy formation was considered equal to the volume of a sphere with radius of the respective atom ($r_{In} = 0.156$ nm, $r_{Au} = 0.174$ nm). Concentration of vacancies C_∞ defined as the ratio of the limiting number of equilibrium point defects to the total number of atoms in a parent phase was calculated by equation [30,31]:

$$C_\infty = \exp(S_\infty/k) \exp(E_a/(kT))$$

where S_∞ is vacancy formation entropy ($S_\infty^{In}/k = 3.95$, $S_\infty^{Au}/k = 3.15$ [31]), E_a stands for activation energy of a point defect formation ($E_a^{In} = 0.425$ eV, $E_a^{Au} = 1.0$ eV [31]). The following values for equilibrium concentration at 300 K were calculated: $C_\infty^{In} = 3.918 \cdot 10^{-6}$, $C_\infty^{Au} = 3.638 \cdot 10^{-16}$ (just for reference purposes, the melting temperature of indium 423 K and gold 1337 K).

376 Mechanics and Physics of Structured Media

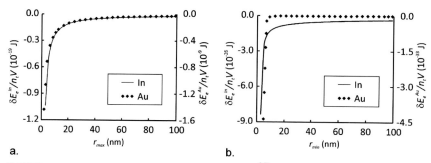

FIGURE 17.8 Specific (per 1 particle) reduction of TPs $\frac{\delta E_e}{V n_t}$ with even distribution (17.14) depending on: (a) uppermost radius r_{max} of particles in the ensemble, $r_{min} = 1$ nm; (b) lowermost radius r_{min}, $r_{max} = 100$ nm.

Fig. 17.8 presents reduction of TPs with even particles distribution on radii, calculated by Eq. (17.29).

For both metals, a considerable magnitude of reduction takes place for those ensembles whose particles do not exceed 20 nm in size (Fig. 17.8a). The greater the minimal radius in an ensemble is (Fig. 17.8b compared to Fig. 17.8a), the smaller in its absolute value the reduction occurs. This looks quite reasonable physically, since the greater the minimal radius is, the more the nano-particle resembles the parent phase. Elimination of particles with small radii appears especially sensitive for gold whose TP abruptly diminishes in its absolute value (Fig. 17.8b). Hereinafter, the reduction is shown to be more pronounced for gold than for indium at the same temperature. This is directly related to both an electronic subsystem determining the surface tension, and atomic volume. On the basis of Eq. (17.29), one can roughly imagine the dependence on them as $\sim e^{\alpha/r_{min}}$ (in absolute value), and α encompasses both B (related to surface tension and, therefore, to the electronic subsystem) and Δv. Each of the constants has a greater value for gold than for indium. As a result, the argument of exponential function appears for gold to be almost 3.5 times as much as for indium what determines a drastically greater reduction in its absolute value. What is a direct consequence of equilibrium vacancy concentration, is the abrupt reduction of TPs in the case of gold, seen in Fig. 17.8b. The melting temperature of gold is much more far from 300 K than that for indium, and while the particles in the ensemble still have a dimensional affinity to the massive crystal, the number of vacancies is not sufficient to let the vacancy-related effect reveal itself as it does in the case of indium. As ultra-small particles appear in the ensemble, the vacancy-related effect occurs abruptly unlike smooth reorganization process in the case of indium.

Fig. 17.9 shows the reduction of TPs with the linear distribution of particles, calculated by Eqs. (17.30) and (17.31).

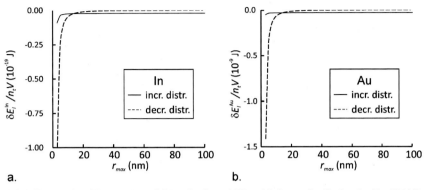

FIGURE 17.9 Specific (per 1 particle) reduction of TPs with linear distribution by Eq. (17.16). $r_{min} = 1$ nm.

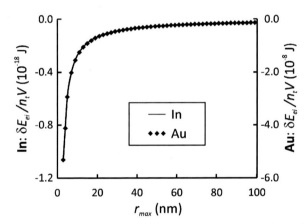

FIGURE 17.10 Specific (per 1 particle) reduction of TPs with exponentially increasing distribution (17.18). $r_{min} = 1$ nm.

The linearly increasing distribution providing more particles with greater radii appears physically reasonably to be closer to a massive crystal so that the reduction of TPs is smaller in its absolute value.

In Fig. 17.10, one can see the reduction of TPs with the exponential (increasing) distribution by Eq. (17.33).

In Fig. 17.11 we show the reduction of TPs with the normal distribution of particles due to Eq. (17.36).

Its data demonstrate the effect of shift of the most probable particle radius on the reduction. When shifted in direction of smaller radii, it makes the reduction more pronounced, and *vice versa*. With small r_{max}, the change constitutes from $-2.0 \cdot 10^{-20}$ J up to $-1.4 \cdot 10^{-19}$ J (Fig. 17.11a). The relative change of reduction is similar for indium and gold whereas the absolute change differs by 10 orders of magnitude.

Table 17.3 compares reduction of TPs for the distributions considered.

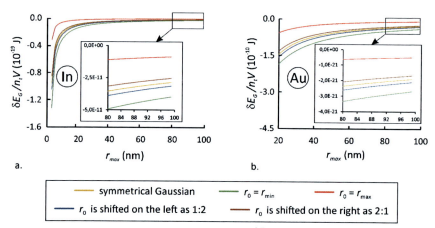

FIGURE 17.11 Specific (per 1 particle) reduction of TPs $\frac{\delta E_G}{V n_t}$ with the normal distribution (17.20). $r_{min} = 1$ nm. r_0 is the most probable particle's radius (the Gaussian's vertex). All distributions are asymmetrically truncated except for the symmetrical Gaussian.

TABLE 17.3 Orders of magnitude of TP reduction, J. $r_{min} = 1$ nm.

Distribution	In		Au	
	r_{max} close to 10 nm	r_{max} close to 100 nm	r_{max} close to 10 nm	r_{max} close to 100 nm
Even	10^{-20}	10^{-21}	10^{-10}	10^{-11}
Linearly increasing	10^{-21}	10^{-21}	10^{-11}	10^{-11}
Linearly decreasing	10^{-21}	10^{-23}	10^{-11}	10^{-13}
Exponentially increasing	10^{-18}	10^{-20}	10^{-8}	10^{-9}
Normal	10^{-20}	10^{-21}	10^{-10}	10^{-11}

Its data comply with the estimate of specific energy threshold 10^{-19}–10^{-18} J for chemical transformations [20].

Thus, out of all distributions considered, the most pronounced reduction of the TPs in its absolute value takes place with the exponentially increasing distribution (Fig. 17.12). The quasiholonomic constraint (17.19) results in rather big values of the distribution modulus γ. They restrain the exponential growth within the particle radius range considered.

The smallest in its absolute value reduction occurs with the linearly decreasing distribution (17.16).

17.3 Increase of the metallic nano-ensemble's TPs due to surface tension

On account of the sign of the relevant (second) term in Eq. (17.7), the increase of TPs (in particular, Gibbs energy) deteriorates the nano-system stability

Thermodynamics and stability of metallic nano-ensembles Chapter | 17 **379**

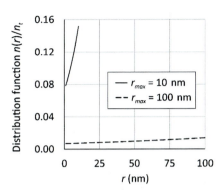

FIGURE 17.12 Distribution functions on particles radii for the exponentially increasing distribution.

and weakens its resistance with respect to external impacts, incl. fluctuational. Therefore, the surface tension effect is a factor competing with the vacancy-related effect. In this section, we determine its contribution to the whole stability balance of the nano-ensemble. Thereafter, particular conditions for configurations in which the nano-ensemble can benefit from this competition are to be commented.

17.3.1 Solution in quadrature of the problem of the TP increase due to surface tension

Vacancies in a dispersed particle affect its density. A massive crystal possessing at temperature T vacancies whose concentration is c_∞ is characterized by density ρ_∞ (a standard reference value). With a passage from the massive crystal to a nano-particle of the same nature, the vacancy concentration grows, the particle becomes fluffy and its density decreases. Let V_a be the atomic volume and V_v the volume of space (both per 1 atom) within the lattice (Fig. 17.13).

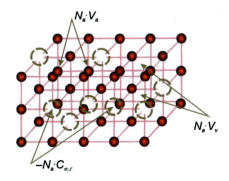

FIGURE 17.13 Vacancies and crystal density.

380 Mechanics and Physics of Structured Media

The density reference value ρ_∞ reads

$$\rho_\infty = \frac{N_A A - N_A A C_\infty}{N_A V_a + N_A V_v}. \tag{17.37}$$

In the numerator in Eq. (17.37), the weight of atoms of an ideal lattice (the first term) is corrected with respect to those atoms which left vacancies upon having been removed. In the denominator, the volume occupied by the lattice's atoms and a free space volume are summed up.

For a nano-particle, an equation similar to Eq. (17.37) reads:

$$\rho_r = \frac{N_A A - N_A A C_r}{N_A V_a + N_A V_v}. \tag{17.38}$$

Eq. (17.37) enables us to express V_v:

$$V_v = \frac{A (1 - C_\infty) - \rho_\infty V_a}{\rho_\infty}. \tag{17.39}$$

Substituting Eq. (17.39) into Eq. (17.38), we link the nano-particle's density with the density of a massive crystal of the same nature:

$$\rho_r = \rho_\infty \frac{1 - c_r}{1 - c_\infty}. \tag{17.40}$$

Taking into account Eqs. (17.6) and (17.7), we obtain the equation for a specific (per 1 atom) energy increase due to the surface tension effect:

$$\delta E_t = \frac{A B}{N_A \rho_\infty} \frac{1 - c_\infty}{1 - c_\infty \exp (\alpha / r)} \frac{1}{r}. \tag{17.41}$$

With a passage to the nano-ensemble of particles with various radii and while taking into account Eq. (17.10), Eq. (17.41) can be obviously transformed into the sum:

$$\delta E_T = \frac{A B V}{N_A \rho_\infty} \sum_{i=1}^{m} \frac{1 - c_\infty}{1 - c_\infty \exp (\alpha / r_i)} \frac{1}{r_i} \delta n_i (r_i). \tag{17.42}$$

We consider now the integrand's components. The functional properties of particles concentration have been already discussed in Section 17.2.1. Function $\frac{1 - C_\infty}{1 - C_\infty \exp(\alpha / r_i)} \frac{1}{r_i}$ asymptotically tends to zero with r_{max} big enough. In the issue, the sum in Eq. (17.42) is to be transformed into an integral like in Eq. (17.12):

$$\delta E_T \approx \frac{A B V (1 - C_\infty)}{N_A \rho_\infty} \int_{r_{min}}^{r_{max}} \frac{1}{1 - C_\infty \exp (\alpha / r)} \frac{n (r)}{r} dr. \tag{17.43}$$

Eq. (17.43) represents the solution in quadrature of the problem of TPs increase for the nano-substance, compared to a massive crystal. Although in the

Thermodynamics and stability of metallic nano-ensembles **Chapter | 17 381**

problem of vacancy-related reduction of TPs in Section 17.2 we were to perform integration up from 1 nm, should such small particles be present in an ensemble, the lower integration limit needs now to be discussed separately. At very small r, Eqs. (17.40)–(17.41) would lose applicability and lead to negative magnitudes of density as well as, technically, to an unlimited growth of the integrand (Fig. 17.14). The corresponding threshold value r_{th} can be calculated by using the equation

$$1 - c_\infty \exp\left(\frac{\alpha}{r_{th}}\right) = 0,$$

from where it instantly follows:

$$r_{th} = -\frac{\alpha}{\log c_\infty}.$$

For the parameters of indium and gold calculated in Section 17.2.4, we obtain $r_{th}^{In} = 1.54$ nm and $r_{th}^{Au} = 1.88$ nm. Since the singularity of $\frac{1}{1-C_\infty \exp(\alpha/r)}$ at $r \rightarrow r_{th}$ in Eq. (17.43) is not removable through integration, one should cut the integration off at $r = r_{th} + \delta r_{min}$, where $\delta r_{min} > 0$. The origin of this logarithmic divergence in Eq. (17.43) finds itself in Eq. (17.6) for an equilibrium concentration of vacancies in small particles which loses applicability in the case of ultra-small nano-particles. Roughly speaking, the particle with a zero radius possesses an infinite concentration of vacancies. However, particles of an atomic radius do not have collective features anymore, and their classic descriptive statistics needs to be replaced with something else.

Like Eq. (17.12), solution (17.43) withstands a passage to the limits $r_{min} \rightarrow r_{max} \rightarrow +\infty$.

17.3.2   Derivation of equations for TPs increase

We consider now consequences of Eq. (17.43) for particle distributions $n(r)$ introduced in Section 17.2.2.

17.3.2.1   Even distribution of particles on their radii

For the even distribution (17.14) and according to Eq. (17.43), the specific (per 1 particle) increase of TPs reads:

$$\frac{\delta E_T^e}{n_t V} \approx \frac{AB}{N_A \rho_\infty} \frac{1 - C_\infty}{r_{max} - r_{min}} \int_{r_{min}}^{r_{max}} \frac{1}{1 - C_\infty \exp(\alpha/r)} \frac{dr}{r}. \tag{17.44}$$

Let us analyze the integrand in Eq. (17.44) to determine correct cutting-off at small r_{min} at integration. Whereas the hyperbolic function $\sim 1/r$ has just a limited growth when the particle size decreases in considered area, function $f(r) = \frac{1}{1-C_\infty \exp(\alpha/r)}$ (Fig. 17.14) grows there abruptly for $r < 1.7$ nm for indium and $r < 2.0$ nm for gold, resulting in the divergence pointed out to above.

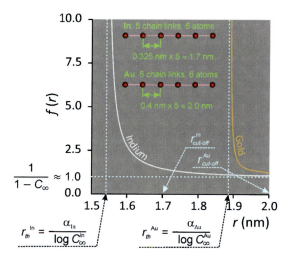

FIGURE 17.14 Removal of divergence in the integral in Eq. (17.44).

Therefore, cutting-off the divergent integral at these $r_{cut-off}$ will provide physically reasonable results. We make the following estimate as well. The cubic (Au) lattice constant is approximately equal to 0.4 nm and the smaller tetragonal (In) lattice constant is about 0.325 nm. Therefore, the abrupt functional growth of $f(r)$ would not be yet reached, were the gold nano-particles with a number of chains 2 nm/0.4 nm = 5, or with 6 atoms, brought into a line. The same result arises for In-nanoparticles: 1.7 nm/0.325 nm = 5. Finally, we note that this cutting-off condition remains valid for other particles distributions as well, since this is the integrand's component $f(r)$ only which contributes to the divergence independent of a distribution function.

To perform integration in Eq. (17.13), we change the variable $t \equiv \alpha/r$ and come to the expression

$$\frac{\delta E_T^e}{n_t V} \approx \frac{AB}{N_A \rho_\infty} \frac{1 - C_\infty}{r_{max} - r_{min}} \int_{\frac{\alpha}{r_{max}}}^{t_0} \frac{1}{1 - C_\infty e^t} \frac{dt}{t} = \cdots$$

where

$$t_0 = \begin{cases} \frac{\alpha}{r_{min}}, & \text{if } r_{min} \geq r_{cut-off}, \\ \frac{\alpha}{r_{cut-off}}, & \text{if } r_{min} < r_{cut-off}. \end{cases}$$

Since inequality $c_\infty e^t < 1$ is valid, taking into account the limitations introduced above, the fractional rational function with the exponent in the denominator can be expanded in a McLaurin series:

$$\cdots = \frac{AB}{N_A \rho_\infty} \frac{1 - c_\infty}{r_{max} - r_{min}} \int_{\frac{\alpha}{r_{max}}}^{t_0} \sum_{k=0}^{\infty} c_\infty^k \frac{e^{kt}}{t} dt.$$

Integration under the sum sign (the justification thereof is proposed further) leads to the final equation for TPs increase, when compared to a massive crystal, due to the surface tension effect, should the particles in the nano-ensemble be distributed evenly on their radii

$$\frac{\delta E_T^e}{n_t V} \approx \frac{AB}{N_A \rho_\infty} \frac{1 - C_\infty}{r_{max} - r_{min}} \left\{ \ln \frac{r_{max}}{\alpha / t_0} + \sum_{k=1}^{\infty} c_\infty^k \left[\mathrm{Ei}^* (k t_0) - \mathrm{Ei}^* (k \frac{\alpha}{r_{max}}) \right] \right\}.$$

$$(17.45)$$

The functional series items in Eq. (17.45) are essentially positive. To prove that the series converges, one can employ the d'Alembert criterion and asymptotic behavior of the exponential integral at big real arguments: $\mathrm{Ei}^*(x) \sim \exp(x)/x$ [27]:

$$\lim_{k \to \infty} \frac{c_\infty^{k+1} \left[\mathrm{Ei}^* ((k+1) t_0) - \mathrm{Ei}^* ((k+1) \frac{\alpha}{r_{max}}) \right]}{c_\infty^k \left[\mathrm{Ei}^* (k t_0) - \mathrm{Ei}^* (k \frac{\alpha}{r_{max}}) \right]}$$

$$= c_\infty \lim_{k \to \infty} \frac{\frac{\exp((k+1) t_0)}{(k+1) t_0}}{\frac{\exp(k t_0)}{k t_0}} = c_\infty \exp(t_0) \lim_{k \to \infty} \frac{k}{k+1} = c_\infty \exp(t_0) < 1.$$

Thus, according to Levy's theorem about monotonous convergence [32], the change of order of integration and summation is justified.

The study of properties of solution (17.45) shows that the series converges rather rapidly. At small r_{min}, should they be far enough from r_{max}, i.e., when the nano-particles size range is broad enough, the term $\mathrm{Ei}^* \left(k \frac{\alpha}{r_{max}} \right)$ does not practically influence the result at all. If the size range is narrow, so that $r_{min} \sim r_{max}$, and located in the neighborhood of r_{min}, one can restrict oneself in Eq. (17.45) with just one term with $\frac{\alpha}{r_{max}}$ at $k = 1$ and three first terms with $k t_0$ at $k = 1...3$. Further series terms, omitted thereby, appear to be negligible, compared to those taken into account. Should r_{min} increase, one has to consider more series terms with both $k \frac{\alpha}{r_{max}}$ and $k t_0$. Practically, however, five terms in each case do.

17.3.2.2 Linear distribution

For the linearly increasing distribution, substitution of Eq. (17.15) into Eq. (17.43) gives the expression

$$\frac{\delta E_T^{li}}{n_t V} \approx \frac{AB}{N_A \rho_\infty} \frac{1 - c_\infty}{r_{max} - r_{min}}$$

$$\times \left[(n_{min} r_{max} - n_{max} r_{min}) \int_{r_{min}}^{r_{max}} \frac{1}{1 - c_\infty \exp(\alpha/r)} \frac{dr}{r} \right.$$

$$\left. + (n_{max} - n_{min}) \alpha \int_{r_{min}}^{r_{max}} \frac{d(r/\alpha)}{1 - c_\infty \exp(\alpha/r)} \right].$$

384 Mechanics and Physics of Structured Media

The integral in the first term is identical to that in Eq. (17.44). The second term is to be considered separately with the change of variable $x \equiv 1/t$, introducing the cutting-off by analogy with that above:

$$I = \int_{r_{min}/\alpha}^{r_{max}/\alpha} \frac{dx}{1 - c_\infty \exp x} = \int_{\alpha/r_{max}}^{t_0} \frac{1}{1 - c_\infty \exp(t)} \frac{dt}{t^2} = \cdots.$$

Again, we represent the fractional rational function with the exponent by its McLaurin series, integrate [28] and keep the term with $k = 0$ apart:

$$\cdots = \frac{r_{max}}{\alpha} - \frac{1}{t_0} + \sum_{k=1}^{\infty} c_\infty^k \left\{ \frac{r_{max}}{\alpha} e^{k \frac{\alpha}{r_{max}}} - \frac{1}{t_0} e^{k t_0} + k \left[\text{Ei}^* (k t_0) - \text{Ei}^* (k \frac{\alpha}{r_{max}}) \right] \right\}.$$

Altogether, the results gained before enable us to obtain the final expression for the TPs increase in the case of linearly increasing distribution:

$$\frac{\delta E_T^{li}}{n_t V} \approx \frac{AB}{N_A \rho_\infty} \frac{1 - c_\infty}{r_{max} - r_{min}}$$

$$\times \left\{ (n_{min} r_{max} - n_{max} r_{min}) \sum_{k=1}^{\infty} c_\infty^k \left[\text{Ei}^* (k t_0) - \text{Ei}^* \left(k \frac{\alpha}{r_{max}} \right) \right] \right.$$

$$+ (n_{max} - n_{min}) \alpha \left[\frac{r_{max}}{\alpha} - \frac{1}{t_0} + \sum_{k=1}^{\infty} c_\infty^k \left(\frac{r_{max}}{\alpha} e^{k \frac{\alpha}{r_{max}}} - \frac{1}{t_0} e^{k t_0} \right. \right.$$

$$\left. \left. \left. + k \left[\text{Ei}^* (k t_0) - \text{Ei}^* \left(k \frac{\alpha}{r_{max}} \right) \right] \right) \right] \right\}. \tag{17.46}$$

By analogy, the substitution of the linearly decreasing distribution (17.16) into Eq. (17.43) leads to the final expression for the TPs increase:

$$\frac{\delta E_T^{ld}}{n_t V} \approx \frac{AB}{N_A \rho_\infty} \frac{1 - c_\infty}{r_{max} - r_{min}}$$

$$\times \left\{ (n_{max} r_{max} - n_{min} r_{min}) \sum_{k=1}^{\infty} c_\infty^k \left[\text{Ei}^* (k t_0) - \text{Ei}^* \left(k \frac{\alpha}{r_{max}} \right) \right] \right.$$

$$- (n_{max} - n_{min}) \alpha \left[\frac{r_{max}}{\alpha} - \frac{1}{t_0} + \sum_{k=1}^{\infty} c_\infty^k \left(\frac{r_{max}}{\alpha} e^{k \frac{\alpha}{r_{max}}} - \frac{1}{t_0} e^{k t_0} \right. \right.$$

$$\left. \left. \left. + k \left[\text{Ei}^* (k t_0) - \text{Ei}^* (k \frac{\alpha}{r_{max}}) \right] \right) \right] \right\}. \tag{17.47}$$

17.3.2.3 Exponential distribution

Substitution of the exponentially increasing distribution (17.18) into Eq. (17.43) gives the following TPs increase:

$$\frac{\delta E_T^{ei}}{n_t V} \approx \frac{AB}{N_A \rho_\infty} \frac{1 - c_\infty}{\gamma} \int_{r_{min}}^{r_{max}} \frac{\exp(r/\gamma)}{1 - c_\infty \exp(\alpha/r)} \frac{dr}{r} = \cdots .$$

Identically to the above, we cut off the integration and represent the exponent in the denominator through its McLaurin series:

$$\cdots = \frac{AB}{N_A \rho_\infty} \frac{1 - c_\infty}{\gamma} \left[\int_{r_{min}}^{r_{max}} e^{\frac{r}{\gamma}} \frac{dr}{r} + \sum_{k=1}^{\infty} c_\infty^k \int_{\alpha/t_0}^{r_{max}} e^{\frac{r}{\gamma} + k\frac{\alpha}{r}} \frac{dr}{r} \right] = \cdots .$$

To keep on further transformations, we employ estimates made in Section 17.2.3.3:

$$\cdots \approx \frac{AB}{N_A \rho_\infty} \frac{1 - c_\infty}{\gamma} \left[\int_{r_{min}}^{r_{max}} e^{\frac{r}{\gamma}} \frac{dr}{r} \right.$$
$$\left. + \sum_{k=1}^{\infty} c_\infty^k \left(\int_{\alpha/t_0}^{r*} e^{k\frac{\alpha}{r}} \frac{dr}{r} + \int_{r*}^{r_{max}} e^{\frac{r}{\gamma}} \frac{dr}{r} \right) \right] = \cdots .$$

The point of intersection of functions is determined now by the expression $r^* = \sqrt{\alpha\gamma k}$ (see Fig. 17.5 for reference purposes). A nonrealistic case $r^* \leq \alpha/t_0$ remains beyond consideration. Upon changing the variable $t \equiv \alpha/r$ in the integrals under the sum sign, we obtain:

$$\cdots \approx \frac{AB}{N_A \rho_\infty} \frac{1 - c_\infty}{\gamma} \left[\int_{r_{min}}^{r_{max}} e^{\frac{r}{\gamma}} \frac{dr}{r} + \sum_{k=1}^{\infty} c_\infty^k \left(\int_{\alpha/r*}^{t_0} \frac{e^{kt}}{t} dt + \int_{r*/\gamma}^{r_{max}/\gamma} \frac{e^t}{t} dt \right) \right].$$

Here we have standard integrals (see, e.g., [28]) and obtain the final expression for the specific increase of TPs:

$$\frac{\delta E_T^{ei}}{n_t V} \approx \frac{AB}{N_A \rho_\infty} \frac{1 - c_\infty}{\gamma}$$
$$\times \left\{ \text{Ei}^* \left(\frac{r_{max}}{\gamma} \right) - \text{Ei}^* \left(\frac{r_{min}}{\gamma} \right) \right.$$
$$\left. + \sum_{k=1}^{\infty} c_\infty^k \left[\text{Ei}^* (kt_0) + \text{Ei}^* \left(\frac{r_{max}}{\gamma} \right) - 2\text{Ei}^* \left(\sqrt{\frac{k\alpha}{\gamma}} \right) \right] \right\}. \qquad (17.48)$$

386 Mechanics and Physics of Structured Media

17.3.2.4 The normal distribution

Substitution of Eq. (17.20) into Eq. (17.43) leads to the following equation for the specific increase of TPs:

$$\frac{\delta E_T^G}{n_t V} \approx \frac{AB}{N_A \rho_\infty} \frac{1 - c_\infty}{\sigma} \int_{r_{min}}^{r_{max}} \frac{\exp\left[-\frac{(r-r_0)^2}{2\sigma^2}\right]}{1 - c_\infty \exp\left(\alpha/r\right)} \frac{dr}{r} = \cdots.$$

Like the transformations above, we cut off the integration and replace the exponent in the denominator with its McLaurin series. The term with $k = 0$ is being kept apart since the divergence at small r is not its concern and there is no need to cut off the integration in it:

$$\cdots = \frac{AB}{N_A \rho_\infty} \frac{1 - c_\infty}{\sigma} \left(\int_{r_{min}}^{r_{max}} e^{-\frac{(r-r_0)^2}{2\sigma^2}} \frac{dr}{r} \right.$$
$$\left. + \sum_{k=1}^{\infty} c_\infty^k \int_{\alpha/t_0}^{r_{max}} e^{-\frac{(r-r_0)^2}{2\sigma^2}} e^{k\frac{\alpha}{r}} \frac{dr}{r} \right) = \cdots. \tag{17.49}$$

The integrals in Eq. (17.49) cannot be represented through elementary functions, their possible solutions with the present integration limits are unknown to the author. Employment of the saddle point method by analogy with that in Section 17.2.3.4 enables us to obtain the following equation:

$$\cdots = \frac{AB}{N_A \rho_\infty} \frac{1 - c_\infty}{\sigma} \left\{ \int_{r_1^*}^{r_2} \left[1 - \frac{(r - r_0)^2}{2\sigma^2} \right] \frac{dr}{r} \right.$$
$$\left. + \sum_{k=1}^{\infty} c_\infty^k \int_{r_1}^{r_2} \left[1 - \frac{(r - r_0)^2}{2\sigma^2} \right] e^{k\frac{\alpha}{r}} \frac{dr}{r} \right\} \tag{17.50}$$

where

$$r_1 = \begin{cases} r_0 - \sqrt{2}\sigma, & \text{if } r_0 - \sqrt{2}\sigma > \alpha/t_0, \\ r_{min}, & \text{if } r_0 - \sqrt{2}\sigma \le \alpha/t_0, \end{cases}$$

$$r_2 = \begin{cases} r_0 + \sqrt{2}\sigma, & \text{if } r_0 + \sqrt{2}\sigma < r_{max}, \\ r_{max}, & \text{if } r_0 + \sqrt{2}\sigma \ge r_{max}, \end{cases}$$

$$r_1^* = \begin{cases} r_0 - \sqrt{2}\sigma, & \text{if } r_0 - \sqrt{2}\sigma > r_{min}, \\ r_{min}, & \text{if } r_0 - \sqrt{2}\sigma \le r_{min}, \end{cases}$$

and $(r_0 \pm \sqrt{2}\sigma)$ in these expressions are the intersection points of the parabola with the abscissa axis (see Fig. 17.6 for reference purposes). The integral in the first term of Eq. (17.50) can be expressed through the elementary functions. The

Thermodynamics and stability of metallic nano-ensembles **Chapter | 17 387**

integral under the sum sign decomposes into three standard integrals [28] with the change of variable $t \equiv \alpha/r$ as it was done earlier:

$$\int_{r_1}^{r_2} \left[1 - \frac{(r - r_0)^2}{2\sigma^2} \right] e^{k\frac{\alpha}{r}} \frac{dr}{r}$$

$$= \left(1 - \frac{1}{2} \frac{r_0^2}{2\sigma^2} \right) \int_{\alpha/r_2}^{\alpha/r_1} e^{kt} \frac{dt}{t} + \frac{\alpha r_0}{\sigma^2} \int_{\alpha/r_2}^{\alpha/r_1} e^{kt} \frac{dt}{t^2} - \frac{1}{2} \frac{\alpha^2}{\sigma^2} \int_{\alpha/r_2}^{\alpha/r_1} e^{kt} \frac{dt}{t^3}.$$

$$(17.51)$$

Integration in Eq. (17.50) with Eq. (17.51) enables us to obtain the final result for the specific increase of TPs in this case:

$$\frac{\delta E_T^G}{n_t V} \approx \frac{AB}{N_A \rho_\infty} \frac{1 - c_\infty}{\sigma} \left\{ \left(1 - \frac{1}{2} \frac{r_0^2}{2\sigma^2} \right) \ln \frac{r_2}{r_1^*} + \frac{r_0}{\sigma^2} (r_2 - r_1^*) - \frac{1}{4\sigma^2} \left(r_2^2 - r_1^{*2} \right) \right.$$

$$+ \sum_{k=1}^{\infty} c_\infty^k \left[\mathrm{Ei}^* \left(k \frac{\alpha}{r_1} \right) - \mathrm{Ei}^* \left(k \frac{\alpha}{r_2} \right) \right] \left[1 - \frac{1}{2} \frac{r_0^2}{2\sigma^2} + \frac{\alpha r_0}{\sigma^2} k - \frac{1}{4} \frac{\alpha^2}{\sigma^2} k^2 \right]$$

$$+ \frac{\alpha r_0}{\sigma^2} \sum_{k=1}^{\infty} c_\infty^k \left(\frac{r_2}{\alpha} e^{k\frac{\alpha}{r_2}} - \frac{r_1}{\alpha} e^{k\frac{\alpha}{r_1}} \right)$$

$$\left. - \frac{1}{4} \frac{\alpha^2}{\sigma^2} \sum_{k=1}^{\infty} c_\infty^k \left[\frac{r_2}{\alpha} e^{k\frac{\alpha}{r_2}} \left(\frac{r_2}{\alpha} + k \right) - \frac{r_1}{\alpha} e^{k\frac{\alpha}{r_1}} \left(\frac{r_1}{\alpha} + k \right) \right] \right\}. \qquad (17.52)$$

17.3.3 Increase of TPs: results

In addition to the metal parameters borrowed or calculated in Section 17.2.4 for the case of vacancy-related reduction of TPs, we additionally introduce here the following data: atomic weights 0.115 kg/mole (In) and 0.197 kg/mole (Au), massive crystal densities $7.31 \cdot 10^3$ kg/m^3 (In) and $19.31 \cdot 10^3$ kg/m^3 (Au). Constant B was calculated on the basis of the data [21] and constitutes 5.0 N/m for indium and 12.5 N/m for gold.

Fig. 17.15 presents the specific increase of TPs with the even distribution due to Eq. (17.45).

The increase of TPs appears to be more pronounced for gold than for indium. Varying r_{min} leads to $\frac{\delta E_T^e}{V n_t}$ decreasing faster than varying r_{max}. Indeed, greater r_{min} means that ultra-small nano-particles have been removed from the ensemble, affinity to a massive crystal has developed, and the effect itself has diminished. With r_{min} fixed and r_{max} varying, the affinity to a massive crystal develops on account of presence of bigger particles rather than of removal of ultra-small ones. The latter appears to have a smaller impact than the former. Thus, existence in an ensemble of ultra-small nano-particles with size 5–10 nm proves to be critical in order to reach a pronounced effect.

388 Mechanics and Physics of Structured Media

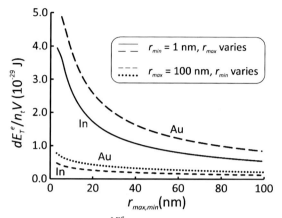

FIGURE 17.15 Specific increase of TPs $\frac{\delta E_T^e}{V n_t}$ due to Eq. (17.45) with even distribution (17.14).

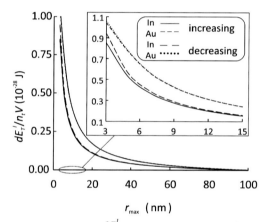

FIGURE 17.16 Specific increase of TPs $\frac{\delta E_T^l}{V n_t}$ due to Eqs. (17.46) and (17.47) with linear distributions of particles (17.15) and (17.16). $r_{min} = 1$ nm.

Fig. 17.16 shows the specific increase of TPs due to Eqs. (17.46) and (17.47) with linear distribution of particles. With the linearly decreasing distribution (lower concentrations correspond to bigger particles whereas higher concentrations to smaller particles), the increase of TPs temperately surpasses that with the linearly increasing distribution (lower concentrations correspond to smaller particles, and higher concentrations to bigger particles). This is reasonable as smaller particles prevail in the case of linearly decreasing distribution what produces less affinity to a massive crystal.

The increase of TPs is more pronounced for gold than for indium.

In Fig. 17.17 we present the specific increase of TPs with the exponentially increasing distribution function due to Eq. (17.48).

Although the effect is more pronounced for gold, the difference is not big.

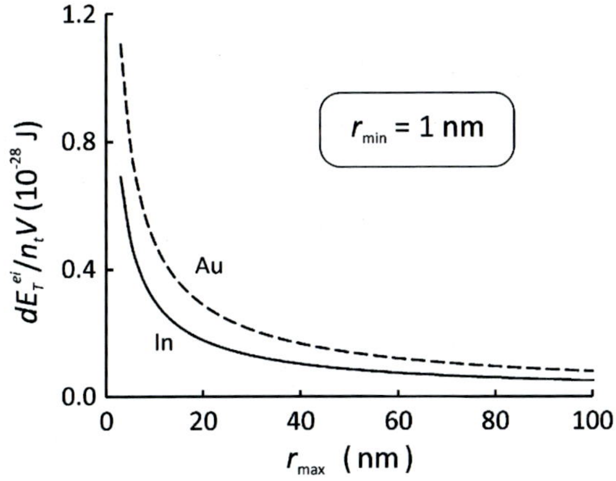

Thermodynamics and stability of metallic nano-ensembles Chapter | 17 **389**

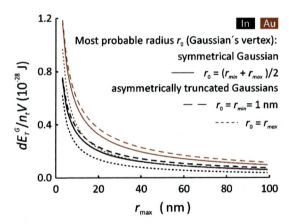

FIGURE 17.17 Specific increase of TPs due to Eq. (17.48) with exponentially increasing distribution (17.18).

FIGURE 17.18 Specific increase of TPs due to Eq. (17.52) with normal distribution (17.20).

Fig. 17.18 shows the specific increase of TPs due to Eq. (17.52) with the normal distribution of particles and, in particular, an influence of the most probable radius r_0 (the Gaussian's vertex) of particles on increase of TPs. Should r_0 move to the direction of smaller radii, $\frac{\delta E_T^G}{n_t V}$ increases, otherwise decreases.

For both metals, a relative change of TPs is pronounced here essentially weaker than in the case of vacancy-related effect, whereas the typical absolute values are close to each other.

Table 17.4 contains the data about the highest increase of TPs for all distributions considered in the area of small r_{max}.

In general, the most pronounced effect of TPs increase is reached with the linearly decreasing distribution for indium and with the asymmetrically truncated normal distribution with the Gaussian's vertex shifted towards smaller

390 Mechanics and Physics of Structured Media

TABLE 17.4 Specific increase of TPs for nano-ensembles on the basis of indium and gold with $r_{max} < 10$ nm for different distribution functions of particles on their radii. $r_{min} = 1$ nm.

Distribution		Distribution function	Specific increase of TPs, J	
			In	Au
Even		constant	$3.94 \cdot 10^{-29}$	$4.34 \cdot 10^{-29}$
Linear	increasing	linear	$8.51 \cdot 10^{-29}$	$1.04 \cdot 10^{-28}$
	decreasing		$9.44 \cdot 10^{-29}$	$1.06 \cdot 10^{-28}$
Exponentially increasing		exponent	$6.91 \cdot 10^{-29}$	$1.11 \cdot 10^{-28}$
Normal	symmetrical	Gaussian	$7.38 \cdot 10^{-29}$	$1.16 \cdot 10^{-28}$
	r_0 is shifted to r_{min}		$7.54 \cdot 10^{-29}$	$1.19 \cdot 10^{-28}$
	r_0 is shifted to r_{max}		$6.17 \cdot 10^{-29}$	$9.70 \cdot 10^{-29}$

particles, for gold. In each distribution, ultra-small nano-particles dominate in their number.

17.4 Balance of the vacancy-related and surface-tension effects

Comparison of both effects shows that the surface tension effect normally yields to the vacancy-related effect in absolute values. The difference can typically constitute several orders of magnitude. The situation changes, should ultra-small particles be removed from the ensemble (compare, for example, Figs. 17.8 and 17.15 for gold). In the absence of ultra-small particles, the vacancy-related effect diminishes by several orders of magnitude, and the surface tension effect begins to dominate. That the nano-effects arise mainly because of ultra-small nano-particles present in an ensemble, appears confirmed thereby and isn't a surprise. Difference in melting temperatures of indium and gold and, consequently, a sharply different number of vacancies have to be among the reasons. An essentially greater number of vacancies in the case of indium copes with maintenance of the vacancy-related effect as competitive with respect to its vis-à-vis. On gold, this does not appear to occur because of essentially smaller number of vacancies.

Finally, we consider a decomposition of both effects. Among affecting factors in Eqs. (17.12) and (17.43) are present atomic weight A, temperature factor kT, standard reference values for density ρ_∞ and for vacancy concentration C_∞ of respective massive crystals, and the dimensional factor itself represented through integrals having been calculated above for particular distributions and encompassing the energetic constant B as well (also in the form of a multiplica-

Thermodynamics and stability of metallic nano-ensembles **Chapter | 17** **391**

TABLE 17.5 Decomposition of specific change of TPs for the nano-ensemble with particles' radii from 1 nm (with a correction cutting-off the integration) up to 100 nm (ST – surface-tension effect, VR – vacancy-related effect)

Factor	Order of magnitude			
	In		Au	
	ST	VR	ST	VR
Overall order of magnitude	*-29*	*-20*	*-28*	*-10*
Atomic weight A (kg/mole)	-1		-1	
Temperature factor kT (J)		-21		-21
Density ρ_∞ (kg/m^3)	-4		-4	
Vacancy concentration C_∞ (dimensionless)		-6		-17
Dimensional factor (N/m^2 for ST and dimensionless for VR)	-24	+7	-23	+28

tive factor B/N_A for the surface tension effect). Table 17.5 presents this for the normal distribution with a symmetrical Gaussian.

Thus, the dimensional factor itself appears to be especially apparent at the vacancy-related effect and, indeed, to constitute its essence. One of the most destructive reasons for stability is likely to be related to nano-particles agglomeration [33,34] caused, in particular and when applicable, by their possibly too high concentrations.

17.5 Conclusions

In this chapter, the novel approach to determination of [in-]stability conditions for nano-particle systems treated as thermodynamic ensembles of multisized particles is proposed. The problem of nano-ensemble stability has been formulated and solved in the framework of a thermodynamic treatment extended, as compared to the Gibbs statistical mechanics, through the functional dependence of chemical potential on particle size.

The *in situ* factors determining properties of ultra-dispersed systems include those two acting oppositely, and namely a bulk- (or vacancy-) related factor that contributes to the system's stability and another one, the surface tension factor, deteriorating it. In terms of thermodynamics, the vacancy-related effect reveals itself through reduction of TPs (in particular, Gibbs energy) whereas the surface tension leads to their increase.

Calculations of change in TPs, compared to those for massive crystals, for ensembles of nano-particles have been performed (i) in quadrature and then (ii) analytically for particular nano-particle distributions on size (radii). Among those have been considered even, linear, exponential and normal distributions. The solution of the problem is presented in detail with relevant functional prop-

392 Mechanics and Physics of Structured Media

erties and approximate analytical methods what enables a reader to reproduce as well as to enhance the results obtained.

The equations derived have been applied to the nano-structures consisted of either indium or gold atoms. The biggest reduction of TPs due to the vacancy-related effect has been shown to arise for the exponential distribution. The specific (per 1 mole at 300 K) decrease in its absolute value can constitute up to 10^6 J for indium and 10^{16} J for gold. The most pronounced increase of TPs due to surface tension effect is reached with the linearly decreasing distribution on indium $(5.7 \cdot 10^{-5}$ J) and with normal distribution on gold $(7.2 \cdot 10^{-5}$ J) provided the Gaussian vertex (the most probable particle radius) were shifted to the area of smaller radii, producing an asymmetrically truncated Gaussian whose ultra-small nano-particles dominate in their number. The results obtained are related to ultra-small nano-particles prevailing in the ensembles and, therefore, physically reasonable. Both partial results as well as the overall effect of TP change are more pronounced on gold. This has been shown to be a consequence of the electronic subsystem (free electron "gas") properties determining the surface tension of metals, and atomic volume affecting both surface tension and bulk-(vacancy-) related effects.

The analysis has shown that existence of ultra-small particles in the ensemble has the vacancy-related effect to dominate over the surface tension effect by many orders of magnitude what contributes to the nano-system stability. Should concentration of ultra-small particles decrease, the vacancy-related effect appears then to diminish and can in certain system configurations yield to the surface tension effect (has been demonstrated with the example of gold). An overall Gibbs energy thereby increases so that the nano-system's stability deteriorates.

Generally, dispersing the substance in a form of nano-particles with a proper statistical distribution on radii in an ensemble is capable of promoting chemical transformations with the energy barrier calculated above. The results calculated find themselves in a good quantitative agreement with those obtained in the framework of substantially estimative calculations [20,26].

Prognostication of stability of artificially generated nano-ensembles seems to be one of the work's applied findings.

References

[1] IAAM, Advanced Materials Laureate 2019: Professor Herbert Gleiter – one of the highest awards in the field of Advanced Materials (press-release), Ulrika, Sweden, 28.11.2019, https://mb.cision.com/Main/17860/2975167/1152033.pdf.

[2] A. Nordmann, Invisible origins of nanotechnology: Herbert Gleiter, materials science, and questions of prestige, Perspectives on Science 17 (2) (2009) 123–143.

[3] H. Gleiter, P. Marquardt, Nanokristalline Strukturen – ein Weg zu neuen Materialien?, Zeitschrift für Metallkunde 75 (4) (1984) 263–267.

[4] R. Birringer, H. Gleiter, H.-P. Klein, P. Marquardt, Nanocrystalline materials an approach to a novel solid structure with gas-like disorder?, Physics Letters A 102 (8) (1984) 365–369.

Thermodynamics and stability of metallic nano-ensembles **Chapter | 17** **393**

[5] A.I. Rusanov, Termodinamika Poverkhnostnykh Yavleniy (Thermodynamics of Surface Phenomena), Khimiya, Leningrad, 1960.

[6] L.M. Scherbakov, O poverkhnostnom natyazhenii kapel malogo razmera (About the surface tension of small size droplets), Colloid Journal 14 (5) (1952) 379–382.

[7] L.M. Scherbakov, V.I. Rykov, O teplote ispareniya malykh kapel (About evaporation heat of small droplets), Colloid Journal 23 (2) (1961) 221–227.

[8] L.M. Scherbakov, O statisticheskoj otsenke izbytochnoy svobodnoy energii malykh obyektov v termodinamike mikrogeterogennykh sistem (About statistical estimate of redundant free energy of small objects in thermodynamics of micro-heterogeneous systems), Doklady Akademii Nauk SSSR 168 (2) (1966) 388–391.

[9] L.M. Scherbakov, O poverkhnostnom natyazhenii tverdogo tela po granitse razdela s sobstvennym rasplavom (About surface tension of a solid at its interface with own melt), Colloid Journal 23 (2) (1961) 215–220.

[10] P.E. Strebejko, O vliyanii izmelcheniya na temperaturu perekhoda (About the milling effect on transition temperature), Dissertation...doctor of sci., Institute of General and Inorganic Chemistry, Moscow, USSR, 1939, 257 p.

[11] P. Pawlow, Über die Abhängigkeit des Schmelzpunktes von der Oberflächenenergie eines festen Körpers, Zeitschrift für Physikalische Chemie 65 (1) (1909) 1–35, https://doi.org/10.1515/zpch-1909-6502.

[12] P. Pawlow, Über die Abhängigkeit des Schmelzpunktes von der Oberflächenenergie eines festen Körpers (Zusatz.), Zeitschrift für Physikalische Chemie 65 (1) (1909) 545–548, https://doi.org/10.1515/zpch-1909-6532.

[13] P. Pawlow, Über den Dampfdruck der Körner einer festen Substanz, Zeitschrift für Physikalische Chemie 68 (1) (1909) 316–322, https://doi.org/10.1515/zpch-1909-6824.

[14] R. Helmholtz, Untersuchungen über Dämpfe und Nebel, besonders über solche von Lösungen, Annalen der Physik 27 (5) (1886) 508–515.

[15] L. Houllevigue, Sur la chaleur de vaporisation et les dimensions moléculaires, Journal de Physique Théorique Et Appliquée 5 (1) (1896) 159–163.

[16] B. Sreznevsky, Ob isparenii chastits (About evaporation of liquids), Journal of Russian Physico-Chemical Society. Part Physics 15 (1883) 39.

[17] V.I. Vigdorovich, L.E. Tsygankova, A.Y. Osetrov, Nanostate of compound as basis for nanomaterial reactivity, Protection of Metals and Physical Chemistry of Surfaces 47 (3) (2011) 410–415, https://doi.org/10.1134/S207020511103018X.

[18] V.I. Vigdorovich, L.E. Tsygankova, Thermodynamics of nanostructured materials, Protection of Metals and Physical Chemistry of Surfaces 48 (5) (2012) 501–507, https://doi.org/10.1134/S2070205112050152.

[19] V.I. Vigdorovich, L.E. Tsygankova, N.V. Shel´, Dependence of binding energy of atoms in low-nuclear clusters on number of particles forming them. Cu_n, Ag_n, and Au_n clusters, Protection of Metals and Physical Chemistry of Surfaces 51 (4) (2015) 567–574, https://doi.org/10.1134/S2070205115040346.

[20] N.S. Lidorenko, S.P. Chizhik, N.T. Gladkikh, L.K. Grigorjeva, R.N. Kuklin, O roli razmernogo faktora v sdvige khimicheskogo ravnovesiya (About the role of dimensional factor in a chemical equilibrium shift), Doklady Akademii Nauk SSSR 257 (5) (1981) 1114–1118.

[21] I.D. Morokhov, S.P. Chizhik, N.T. Gladkikh, L.K. Grigorjeva, S.V. Stepanova, Razmernyy vakansionnyy effect (Dimensional vacancy-related effect), Doklady Akademii Nauk SSSR 248 (3) (1979) 603–604.

[22] V.I. Vigdorovich, K.O. Strel´nikova, Suppression of nanoscale effects in nanomaterials by gas- and liquid-phase adsorbates, Protection of Metals and Physical Chemistry of Surfaces 48 (3) (2012) 383–387, https://doi.org/10.1134/S2070205112030197.

[23] W. Stiller, Arrhenius Equation and Non-Equilibrium Kinetics, Teubner Texte zur Physik, B. S. Teubner Verlagsgesellschaft, 1989, p. 160.

[24] V.I. Vigdorovich, L.E. Tsygankova, N.V. Shel, P.N. Bernatsky, Teoreticheskie i Prikladnye Voprosy Nanotekhnologiy (Sovremennoe Sostoyanie i Problemy) (Theoretical and Applied

394 Mechanics and Physics of Structured Media

Issues of Nano-Technologies (Actual State and Problems)), R.V. Pershin's publishing house, Tambov, 2016.

[25] G. Guisbiers, M. Kazan, O. Van Overschelde, M. Wautelet, S. Pereira, Mechanical and thermal properties of metallic and semiconductive nanostructures, Journal of Physical Chemistry C 112 (2008) 4097–4103, https://doi.org/10.1021/jp077371n.

[26] M. Vigdorowitsch, L.E. Tsygankova, V.I. Vigdorovich, L.G. Knyazeva, To the thermodynamic properties of nano-ensembles, Materials Science & Engineering: B 263C (2021) 114897, https://doi.org/10.1016/j.mseb.2020.114897.

[27] E. Janke, F. Emde, F. Lösch, Tafeln höherer Funktionen, B.G. Teubner Verlagsgesellschaft, Stuttgart, 1960.

[28] I.S. Gradstein, I.M. Ryzhik, Tablitsy Integralov, Summ, Ryadov i Proizvedenij (Tables of Integrals, Sums, Series and Products), GIFML, Moscow, 1963, pp. 106–107.

[29] F.W.J. Olver, Asymptotics and Special Functions, Academic Press, New York, London, 1974.

[30] E.V. Goncharova, A.S. Makarov, R.A. Konchakov, N.P. Kobelev, V.A. Honik, Premelting generation of interstitial defects in polycrystalline indium, Journal of Experimental and Theoretical Physics Letters 106 (1) (2017) 35–39, https://doi.org/10.7868/S0370274X17130082.

[31] Ya. Kraftmakher, Equilibrium vacancies and thermophysical properties of metals, Physics Reports 299 (2–3) (1998) 79–188.

[32] A.N. Kolmogorov, S.V. Fomin, Elementy Teorii Funktsiy i Funktsionalnogo Analiza (Elements of Theory of Function and Functional Analysis), Nauka, Moscow, 1989, pp. 348–350.

[33] Y. Jazaa, Effect of nanoparticle additives on the tribological behavior of oil under boundary lubrication, PhD thesis in Mechanical Engineering, Iowa State University, 2018.

[34] V. Zin, S. Barison, F. Agresti, L. Colla, C. Paguraa, M. Fabrizioa, Improved tribological and thermal properties of lubricants by graphene based nano-additives, RSC Advances 6 (2016) 59477, https://www.doi.org/10.1039/c6ra12029f.

Chapter 18

Comparative analysis of local stresses in unidirectional and cross-reinforced composites

Alexander G. Kolpakov[a] and Sergei I. Rakin[a,b]

[a]*SysAn, Novosibirsk, Russia,* [b]*Siberian Transport University, Novosibirsk, Russia*

18.1 Introduction

The problem of computation of the effective characteristics and local stresses is the basic problem in the theory of composite materials, see, e.g. [1–4]. The rigor method for solving the problem was developed in the homogenization theory [5–7]. The homogenization theory reduces the original problems to the so-called periodicity cell problem [5–7], but does not supply a mechanical engineer with calculation formulas. For the mechanical engineer, the periodicity cell problem is not the final but starting point of investigation. Following [4], it would be correct to consider the periodicity cell problem as the point for "passing the torch" from the homogenization theory to the elasticity theory or structural mechanics. The solution to the periodicity cell problem with the elasticity theory methods provides the researcher with various and important information about the macro- and microscopic properties of composite material.

Before the homogenization theory and computers and software proper for the implementation of the homogenization theory procedures, numerous approaches to the computation of the composite materials were proposed (see [1,2,8–17] for the basic theories in the field). One of the most successful methods is based on the application of the complex variables theory to the analysis of planar and quasiplanar periodic elastic structures, see [18–28] (the list not completed, see also references in [25,26]). An important class of quasiplanar structures is composites reinforced with a unidirectional system of fibers [8,13,26]. The complex variables approach leads to computational formulas both for effective constants and local stress-strain state for such kind composites. The computational formulas usually have the form of series [20–22]. New methods and approaches based on the complex variables theory were developed recently [26–28].

In practice, the composites reinforced with unidirectional fibers are not widely used. The widely used are the cross-reinforced composites [29,30]

Mechanics and Physics of Structured Media. https://doi.org/10.1016/B978-0-32-390543-5.00023-2
Copyright © 2022 Elsevier Inc. All rights reserved.

396 Mechanics and Physics of Structured Media

formed by the stacking of the layers of unidirectional fibers. The following two-step procedure is often used for the computation of the cross-reinforced composites by engineers [29,30]:

- at the first step, every layer is changed by a proper homogeneous anisotropic layer;
- at the second step, the overall characteristics of the composite are computed in the accordance with the theory of laminated composites.

The first step may be carried out by using the methods based on the complex variables. As far as the second step, the problem of computation of the overall properties of the laminated composite has a well-known solution [4,31,32]. Each step above seems to be reasonable and correct separately. Do these two steps together lead to the right result? The qualitative analysis of microdeformations in fiber-reinforced composites presented in [4,33–35] does not confirm the two-step procedure. But it would be risky to use the analysis from [4,33–35] as an indisputable argument because it used some approximations and hypotheses.

Another often-used procedure is based on the Mori-Tanaka method [31] as applied to a fiber-reinforced composite.

We use numerical methods to answer the question above. Direct computation of composite material is impossible due to the large dimension of the problem (a typical composite structure can contain millions of constituent elements). The homogenization theory made it possible to reduce the dimension of the problem for composites of periodic structure and certain problems in the composites were solved numerically in 1980s. Mention the pioneering works [36–38]. Subsequent progress in the field is described in numerous publications, see, e.g., [39–41] and references in these books. Currently, computers have reached the power sufficient to implement the homogenization theory methods, and the ability to solve a specific problem is determined by the programming and computational skills of the researcher or engineer (or the supporting computational group).

The geometry of the cross-reinforced composites is rather complicated for constructing periodic FEM mesh. It is the only difficulty in the accurate numerical solution of the periodicity cell problem now. An approximate solution was given in [4,33–35] for the cross-reinforced composites with fibers of a square cross-section. The numerical calculations presented in the recent work [42] confirm [4,33–35] in general, and demonstrated that the local stress-strain state in the composite (especially in the matrix) is strongly inhomogeneous and depends on the geometry of the fibers and the reinforcement scheme.

There exists massive experimental data on the failure of fiber-reinforced composites. The experimental data demonstrate the great variety of the failure modes of the fiber-reinforced composites. For example, the paper [43] displays photos of the fiber-reinforced composites with the reinforcement scheme $0°$, $\pm45°$, and $90°$, which failure in very different modes under the same global load. Experiments usually demonstrate the final (the macroscopic) stage of failure and provide us with no information about the beginning (the microscopic) stage of failure.

This chapter presents the numerical solutions to periodicity cell problems for composites reinforced with orthogonal and with unidirectional systems of fiber. These solutions provide us with detailed information about the local stress-strain state in the components of composite (in the fibers and the matrix). We compare the solutions one with another, as well as with the predictions of "the two-step computational procedure" and the hypothesis of the Mori-Tanaka method.

18.2 Homogenization method as applied to composite reinforced with systems of fibers

We consider composite obtained by stacking layers of parallel fibers, Fig. 18.1. Denote $\mathbf{v}_s$ the directional vector of the fibers in the s-th layer. The fibers are embedded into the matrix. It is assumed the perfect connection of the fibers and matrix.

The circular fibers have diameter εR, the distance between the fibers in the layers is εh, the distance between the layers of fiber is $\varepsilon \delta$. The characteristic size ε of the elements of the composite is assumed to be small as compared to the size of the structure. In this chapter, the size of the structure is assumed to be 1, then $\varepsilon \ll 1$. To describe such two-scaled (1 and ε) material, "fast" (microscopic) variables $\mathbf{y} = \mathbf{x}/\varepsilon$ are introduced in addition to the "slow" (macroscopic) variables $\mathbf{x}$ [5], see Fig. 18.1.

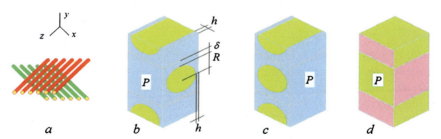

FIGURE 18.1 Cross-reinforced composite – a, the periodicity cells of the composites (in the "fast" variables $\mathbf{y}$): cross-reinforced – b, unidirectional – c, laminated – d.

The periodic stacking of the fibers in the layers and the layers in the composite lead to the periodicity of the composite structure as a whole. The periodicity cells of the composite can be selected in different ways, which does not affect the final results. One possible choice is shown in Fig. 18.1. In Fig. 18.1, the periodicity cell is rectangle $P = [0, h_1] \times [0, h_2] \times [0, h_3]$ in "fast" (microscopic) variables $\mathbf{y}$. The opposite faces of the periodicity cell P perpendicular to the Ox_i-axis (coordinates $x = x_1$, $y = x_2$, $z = x_3$) are rectangles $\Gamma_i = \{\mathbf{y} : y_i = 0\}$ and $\Gamma_i + h_i \mathbf{e}_i$.

We consider composite reinforced with unidirectional fibers and composite reinforced with the orthogonally oriented fibers. We confine ourselves to these

398 Mechanics and Physics of Structured Media

simple cases because it allows us to focus on the mechanical aspects of the problem. Note that the composites reinforced with the unidirectional and orthogonal fibers are still relevant for practice, see, e.g., [43–47].

We assume that the fibers in the unidirectionally reinforced composite are parallel to the Oz-axis; in the cross-reinforced composite, the fibers are parallel to the Ox-axis and the Oz-axis (correspondingly, $\mathbf{v}_1 = \mathbf{e}_1$ and $\mathbf{v}_1 = \mathbf{e}_3$; where $\mathbf{e}_1$, $\mathbf{e}_2$, $\mathbf{e}_3$ are basis vectors of the orthonormal $Oxyz$-coordinate system), Fig. 18.1. The layers are parallel to the Oxz-plane.

In the homogenization theory, solution to the elasticity theory problem for composite material of periodic structure is sought in the form [5]

$$\mathbf{u} = \mathbf{u}_0(\mathbf{x}) + \varepsilon \mathbf{u}_1(\mathbf{x}, \mathbf{x}/\varepsilon), \tag{18.1}$$

where $\mathbf{u}_0(\mathbf{x})$ is the homogenized (macroscopic) solution and $\varepsilon \mathbf{u}_1(\mathbf{x}/\varepsilon)$ is the corrector ($\mathbf{u}_1(\mathbf{x}, \mathbf{y})$ is a periodic function in $\mathbf{y}$ with the periodicity cell P) [5]. Since $\varepsilon \ll 1$, the corrector $\varepsilon \mathbf{u}_1(\mathbf{x}, \mathbf{x}/\varepsilon)$ makes a small contribution to the displacements (18.1), but a significant contribution to local strains and stresses because its derivatives $\frac{\partial \mathbf{u}_1}{\partial y_l}(\mathbf{x}, \mathbf{x}/\varepsilon)$ are not small. In the homogenization theory, it is proved [5] that $\mathbf{u}_1(\mathbf{x}, \mathbf{x}/\varepsilon)$ has the form $\frac{\partial u_{0k}}{\partial x_l}(\mathbf{x})\mathbf{N}^{kl}(\mathbf{x}/\varepsilon)$, where $\mathbf{N}^{kl}(\mathbf{y})$ is solution to the periodicity cell problem of the homogenization theory (see problem (18.3) below).

Introduce function

$$a_{ijkl}(\mathbf{y}) = \begin{cases} a_{ijkl}^F & \text{if fibers,} \\ a_{ijkl}^M & \text{in binder,} \end{cases} \tag{18.2}$$

where a_{ijkl}^F are the elastic constants of the fiber and a_{ijkl}^M are the elastic constants of the matrix. The function (18.2) describes the distribution of elastic constants over the periodicity cell P.

Periodicity cell problem of the homogenization theory has the form [5] (see also [6,7]): find the function $\mathbf{N}^{kl}(\mathbf{y})$ from the solution to the boundary value problem

$$\begin{cases} (a_{ijkl}(\mathbf{y})N_{k,ly}^{mn} - a_{ijmn}(\mathbf{y}))_{,iy} = 0 \text{ in } P, \\ \mathbf{N}^{kl}(\mathbf{y}) \text{ periodic in } \mathbf{y} \in P. \end{cases} \tag{18.3}$$

The boundary value problem (18.3) has many solutions that differ by rigid body displacements. In the homogenization theory, the zero average value condition for $\mathbf{N}^{kl}(\mathbf{y})$ is usually used to eliminate the rigid body displacements. This condition is not convenient in numerical computations and we eliminate the rigid body displacements by fixing some points of the periodicity cell. Both methods lead to the same results [42].

In the boundary value problem (18.3), indices $mn = 11, 22, 33$ correspond to macroscopic tension-compression along the corresponding axes, and indices $mn = 12, 13, 23$ correspond to macroscopic shifts in the corresponding planes.

Comparative analysis of local stresses Chapter | 18 **399**

Introduce the function $\mathbf{Z}^{mn}(\mathbf{y}) = \mathbf{N}^{mn}(\mathbf{y}) + y_m \mathbf{e}_n$ and denote $[f(\mathbf{y})]_i$, $\mathbf{y} \in \Gamma_i$ the "jump" of the function $f(\mathbf{y})$ on the opposite faces $\mathbf{y} \in \Gamma_i$ and $\mathbf{y} \in \Gamma_i + h_i \mathbf{e}_i$ of the periodicity cell P. In this notation, the boundary value problem (18.3) can be rewritten in the form (no sum in m, n here)

$$\begin{cases} (a_{ijkl}(\mathbf{y}) Z^{mn}_{k,ly})_{,iy} = 0 \text{ in } P, \\ [\mathbf{Z}^{mn}(\mathbf{y})]_i = \varepsilon_{mn}(\mathbf{x})[y_m \mathbf{e}_n]_i \end{cases} \tag{18.4}$$

The boundary value problem (18.4) is a problem in the "fast" variables $\mathbf{y}$. The macroscopic strains $\varepsilon_{mn}(\mathbf{x})$ in (18.4) are parameters. It leads to the so-called "separation of the fast and slow variables" in the homogenization method [5].

The homogenization method provides us information about both the homogenized (macroscopic) and local (microscopic) characteristics of composites. Namely, after the boundary value problem (18.4) be solved, the local stresses $\sigma^{mn}_{pq}(\mathbf{y})$ corresponding to $\mathbf{Z}^{mn}(\mathbf{y})$ are computed as [5]

$$\sigma^{mn}_{pq}(\mathbf{y}) = a_{pqkl}(\mathbf{y}) Z^{mn}_{k,l}(\mathbf{y}). \tag{18.5}$$

The local stresses $\sigma_{pq}(\mathbf{y})$ in the composite subjected to the arbitrary macroscopic strains $\varepsilon_{mn}(\mathbf{x})$ are computed as

$$\sigma_{pq}(\mathbf{y}) = \varepsilon_{mn}(\mathbf{x}) \sigma^{mn}_{pq}(\mathbf{y}) = \varepsilon_{mn}(\mathbf{x}) a_{pqkl}(\mathbf{y}) M^{mn}_{k,l}(\mathbf{y})$$
$$= \varepsilon_{mn}(\mathbf{x})[a_{pqmn}(\mathbf{y}) + a_{pqkl}(\mathbf{y}) N^{mn}_{k,l}(\mathbf{y})], \tag{18.6}$$

where $\mathbf{M}^{mn}(\mathbf{y})$ is solution to (18.4) corresponding to $\varepsilon_{mn}(\mathbf{x}) = 1$.

The effective (homogenized) constants A_{pqmn} of the composite are computed as [5]

$$A_{pqmn} = \frac{1}{mes\,P} \int_P (a_{pqmn}(\mathbf{y}) + a_{pqkl}(\mathbf{y}) N^{mn}_{k,l}(\mathbf{y})) d\mathbf{y}$$
$$= \frac{1}{mes\,P} \int_P a_{pqkl}(\mathbf{y}) M^{mn}_{k,l}(\mathbf{y}) d\mathbf{y}. \tag{18.7}$$

The homogenized stresses $\sigma_{kl}(\mathbf{x})$ and homogenized strains $\varepsilon_{mn}(\mathbf{x})$ are connected by the homogenized Hook's law: $\sigma_{kl}(\mathbf{x}) = A_{klmn} \varepsilon_{mn}(\mathbf{x})$ or vice versa $\varepsilon_{mn}(\mathbf{x}) = A^{-1}_{klmn} \sigma_{kl}(\mathbf{x})$, where A^{-1}_{klmn} is the inversion of the tensor A_{klmn} [48].

If the homogenized composite material is orthotropic, this is our case,

$$A_{mnmn} = \frac{1}{mes\,P} \int_P a_{pqkl}(\mathbf{y}) M^{mn}_{p,q}(\mathbf{y}) M^{mn}_{k,l}(\mathbf{y}) d\mathbf{y}$$
$$= \frac{1}{\varepsilon^2_{mn}} \frac{1}{mes\,P} \int_P a_{pqkl}(\mathbf{y}) Z^{mn}_{p,q}(\mathbf{y}) Z^{mn}_{k,l}(\mathbf{y}) d\mathbf{y}. \tag{18.8}$$

The integrands in (18.8) are the double local elastic energies.

400 Mechanics and Physics of Structured Media

18.3 Numerical analysis of the microscopic stress-strain state of the composite material

For a cross-reinforced composite, the local stress-strain state in the composite is of general form even if the macroscopic loads of the simplest form are applied to the composite. We consider the basic types of macroscopic strains and calculate the corresponding local stresses in the composite. In our computations, we use the following characteristics of the components. Fibers: Young's modulus $E_f = 170$ GPa; Poisson's ratio $\nu_f = 0.3$. Matrix: Young's modulus $E_b = 2$ GPa, Poisson's ratio $\nu_b = 0.36$. These material parameters correspond to carbon/epoxy composite [49].

The distance between the fibers in the layers is $h = 0.1$, the distance between the layers of fibers $\delta = 0.1$. The periodicity cell dimensions are $h_1 = 1.1$, $h_2 = 2$, $h_3 = 1.1$. The radius of the fiber is 0.45. These values are indicated in the "fast" (nondimensional) variables $\mathbf{y}$ and specify the relative dimensions of the elements of the periodicity cell. The corresponding actual (dimensional) values are computed by multiplying by the characteristic size ε.

In this chapter, we use programs written in APDL programming language of the ANSYS finite element complex [50]. The fibers are modeled by circular cylinders and the matrix occupies the remaining part of the periodicity cell. The ideal contact between the fibers and matrix is assumed. The finite elements SOLID186 are used both for the fibers and the matrix. The characteristic size of the finite elements is 0.02. The total number of finite elements is about 170000.

18.3.1 Macroscopic strain ε_{11} (tension-compression along the Ox-axis)

Fig. 18.2 displays the distribution of the numerically computed von Mises stress σ_M over the periodicity cell P. In the computations, the macroscopic strain $\varepsilon_{11} = 0.091$. Note that because of the linearity of the problem (18.4), the value of ε_{11} theoretically can be selected arbitrarily. When a specific computer program is used, the value of ε_{11} is selected from the conditions of the convenience of computations. Usually, it may be done in a fairly wide range. This remark is valid for other homogenized strains in the computations below. The choice of value $\varepsilon_{11} = 0.091$ (9.1%) here and other values of the homogenized strains below were determined only by the convenience of the programmer.

The numerically calculated distributions of the local von Mises stress over the periodicity cell of the cross-reinforced and the unidirectional reinforced composites are in Fig. 18.2. The minimal and the maximal values at end of the scale are zoomed. The similarities as well as differences of the distributions of the local von Mises stress are clearly visible.

The maximum von Mises stress occurs in the points indicated in Table 18.1. In the column "variation" the minimum and maximum values of von Mises stress in the fibers and matrix are presented, "No" means that the minimum

Comparative analysis of local stresses Chapter | 18 **401**

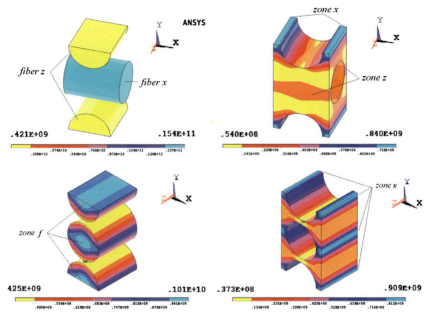

FIGURE 18.2 Local von Mises stress in the fibers and the matrix at macroscopic strain ε_{11}. Cross-reinforced composite (top) and unidirectional reinforced composite (bottom).

and maximum values approximately equal. The interval of variation characterizes the nonhomogeneity of the stress-strain states in the components of the composite. Recall that Eshelby solution shows that the stress-strain state in a single homogeneous ellipsoid embedded in a homogeneous matrix is constant [31] ("No" variation). This property forms the basis of the Mori-Tanaka method [31]. Our computations demonstrate that the stress-strain state in fibers cannot be accepted as a constant in some cases. To make this conclusion visual, we indicate in Table 18.1 the deviation of the maximum and minimum von Mises stress from the average value in percent (computed as $\pm\frac{1}{2}\frac{\max-\min}{(\max+\min)/2}\%$) for the fibers and the matrix.

The local stress-strain state in the matrix has a general form for all cases. Our calculations show that the stress-strain concentration takes place in several specific regions in the matrix. The positions of these regions depend on the macroscopic deformation of the composite. It is possible to introduce the "modes" of failure of the matrix corresponding to these regions.

The effective (homogenized) constants (18.7) may be computed by using the last formula in (18.8) as

$$A_{mnmn} = \frac{2}{\varepsilon_{mn}^2}\frac{1}{mes P}SENE, \qquad (18.9)$$

402 Mechanics and Physics of Structured Media

where $mes P = h_1 h_2 h_3$, *SENE* is ANSYS notation for the elastic energy, only $\varepsilon_{mn} \neq 0$.

In the case under consideration $mn = 11$ and $\varepsilon_{mn} = \varepsilon_{11}$. The values of the homogenized constant (18.9) are presented in Table 18.1. The maximum values of von Mises stress occur in the zone x where the distance between the fibers is minimal. Since the fibers are round in cross-section, the stresses quickly decrease with distance from these zones. Those, we meet the phenomenon of stress localization similar to that described in [51,52] and demonstrated in numerical computations [53,54]. The calculation of such stresses is possible only numerically.

TABLE 18.1 The maximum von Mises stress and A_{1111}.

Composite type		Zone	Value (GPa)	Variation (GPa)	Variation %	A_{1111} (GPa)
cross-reinforced	fiber	fiber x	15.4	No	0	55.9
	matrix	zone x	0.84	0.37-0.84	±39	
unidirectional	fiber	zone f	1.01	0.42-1.01	±35	10
	matrix	zone x	0.91	0.37-0.91	±42	

18.3.2 Macroscopic strain ε_{33} (tension-compression along the Oz-axis)

In the computations, the macroscopic strain $\varepsilon_{33} = 0.091$. The numerically calculated distributions of the local von Mises stress over the periodicity cell P are displayed in Fig. 18.2 for the cross-reinforced and the unidirectionally reinforced composites. The von Mises stress in the fibers for the unidirectional reinforced composite is the same on the left and the right of the scale. The multicolor picture results from the specific of ANSYS graphics output. The maximum von Mises stress occurs in the elements of the periodicity cell indicated in Fig. 18.2.

In the case under consideration $mn = 33$. The values of the homogenized constant (18.9) are displayed in Table 18.2.

TABLE 18.2 The maximum von Mises stress and A_{3333}.

Composite type		Zone	Value (GPa)	Variation (GPa)	Variation %	A_{3333} (GPa)
cross-reinforced	fiber	fibers z	15.4	No	0	55.9
	matrix	zone z	0.84	0.37-0.84	±39	
unidirectional	fiber	fibers f	1.53	No	0	102
	matrix	zone f	0.91	0.08-0.5	±72	

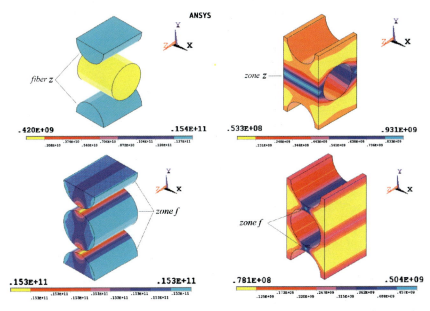

FIGURE 18.3 Local von Mises stress in the fibers and the matrix at macroscopic strain ε_{33}.

18.3.3 Macroscopic deformations ε_{22} (tension-compression along the Oy-axis)

In the computations, the homogenized strain $\varepsilon_{22} = 0.05$. Fig. 18.3 displays distribution of the numerically computed von Mises stress σ_M over the periodicity cell. The maximum von Mises stress occurs in the elements of the periodicity cell indicated in Fig. 18.3.

In the case under consideration $mn = 22$. The values of the homogenized constant (18.9) are displayed in Table 18.3.

TABLE 18.3 The maximum von Mises stress and A_{2222}.

Composite type		Zone	Value (GPa)	Variation (GPa)	Variation %	A_{2222} (GPa)
cross-reinforced	fiber	fiber x	0.89	0.19-0.89	±65	11.14
	matrix	zone t	0.64	0.37-0.84	±39	
unidirectional	fiber	fiber x	0.93	0.29-0.97	±54	12.95
	matrix	zone f	0.83	0.02-0.64	±94	

In the unidirectional composite, the fibers interaction is localized in the neighbor of line f, in the cross-reinforced composite – in the zone t, see Fig. 18.3. Despite this significant difference in the local stress-strain state, the

404 Mechanics and Physics of Structured Media

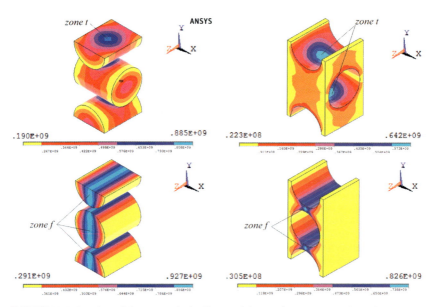

FIGURE 18.4 Local von Mises stress in the fibers and the matrix at macroscopic strain ε_{22}.

difference in homogenized stiffnesses A_{2222}, which are integral characteristics, is small.

Investigate the effective stiffnesses A_{2222} for the unidirectional and cross-reinforced composites in the dependence on Young's modulus of the matrix. The computations above have been done for $E_b = 2$ GPa. Table 18.4 displays the results of computations for $E_b = 20$ GPa, $E_b = 2$ GPa, and $E_b = 0.2$ GPa (other characteristics of the composites are not changed). It is seen that the values of A_{2222} for the cross-reinforced and for the unidirectional composites are close for all values of E_b.

TABLE 18.4 A_{2222} (GPa) in dependence on Young's modulus of matrix.

E_b	20 GPa	2 GPa	0.2 GPa
cross-reinforced	79.8	11.14	1.1
unidirectional	85	12.95	1.37

18.3.4 Macroscopic deformations ε_{13} (shift in the Oxz-plane)

In the numerical computations, the homogenized strain $\varepsilon_{13} = 0.091$. The numerical analysis, see Fig. 18.4, predicts the increase of the local stress in the fibers, and the matrix in each reinforcing layer. It is similar to that arising when a layer of unidirectional composite is tensile at an angle to the fibers' axis, see

Fig. 18.5. The presence of this type of microscopic strains was indicated in [5]. The maximum von Mises stress occurs in the elements of the periodicity cell indicated in Table 18.5.

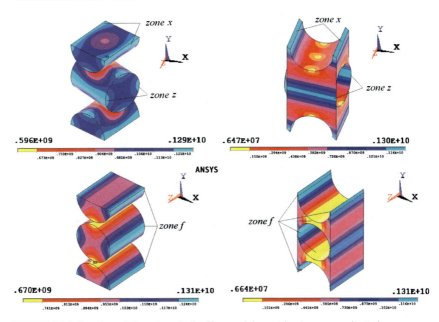

FIGURE 18.5 Local von Mises stress in the fibers and the matrix at macroscopic strain ε_{13}.

In the case under consideration $mn = 13$. The values of the homogenized constant (18.9) are displayed in Table 18.5. Hereafter "≈0" means a value less 0.01 GPa.

TABLE 18.5 The maximum von Mises stress and A_{1313}.

Composite type		Zone	Value (GPa)	Variation (GPa)	Variation %	A_{1313} (GPa)
cross-reinforced	fiber	fiber x	1.29	0.6-1.29	±37	10.47
	matrix	zone x	1.3	≈ 0-1.3	±100	
unidirectional	fiber	fiber x	1.31	0.67-1.31	±32	9.96
	matrix	zone x	1.31	≈ 0-1.29	±100	

Von Mises stress takes maximum values in the matrix between the fibers in the zones x and z, and in the parts, x and z of the fibers, see Fig. 18.5. In the matrix between the fiber layers, there is no significant increase in stresses. In particular, the stress in the matrix fragment, twisted by fibers lying one above the other, is small. In [5], the local stress in the matrix between the fiber layers can be significant. In this chapter, we consider fibers with a round cross-section.

406 Mechanics and Physics of Structured Media

We conclude that the geometry of the cross-section of the fibers can influence the local stresses.

18.3.5 Macroscopic strain ε_{12} (shift in the Oxy-plane)

In the computations, the macroscopic strain $\varepsilon_{12} = 0.05$. Fig. 18.6 displays distribution of the numerically computed von Mises stress σ_M over the periodicity cell P. The maximum von Mises stress occurs in the elements of the periodicity cell indicated in Table 18.6.

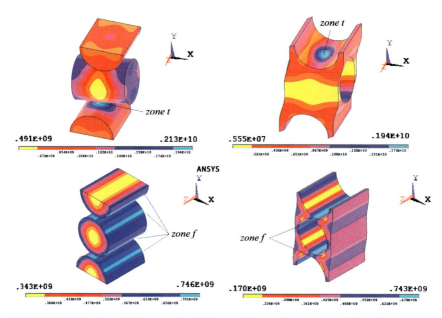

FIGURE 18.6 Local von Mises stress in the fibers and the matrix at macroscopic strain ε_{12}.

In the case under consideration $mn = 12$. The values of the homogenized constant (18.9) are displayed in Table 18.6.

TABLE 18.6 The maximum von Mises stress A_{1212}.

Composite type		Zone	Value (GPa)	Variation (GPa)	Variation %	A_{1212} (GPa)
cross-reinforced	fiber	fiber t	2.13	0.49-2.13	±63	40.6
	matrix	zone t	1.94	≈0-2.13	±100	
unidirectional	fiber	fiber f	0.74	0.34-0.74	±16	15.3
	matrix	zone f	0.74	≈0-0.74	±100	

18.3.6 Macroscopic strain ε_{23} (shift in the Oyz-plane)

In the computations, the macroscopic strain $\varepsilon_{23} = 0.05$. Fig. 18.7 displays distribution of the numerically computed von Mises stress σ_M over the periodicity cell P. The maximum von Mises stress occurs in the elements of the periodicity cell indicated in Table 18.7.

On our computations the interparticle distances δ/R and h/R is 0.22. The volume fraction of fibers is 0.53 (the maximum possible value for circular fibers is 0.79). This is the case of densely packed fibers. We meet the effects of the concentration and localization [55,56] in the necks between the closely placed fibers. Earlier, this effect in elastic systems was analyzed in [51–55] for particles reinforced composites.

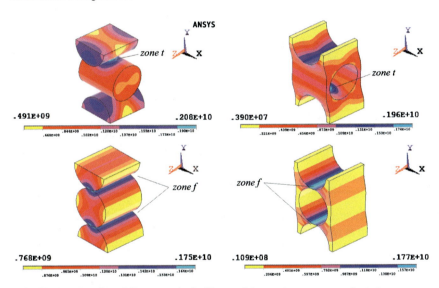

FIGURE 18.7 Local von Mises stress in the fibers and the matrix at macroscopic strain ε_{12}.

In the case under consideration $mn = 23$. The values of the homogenized constant (18.9) are displayed in Table 18.7.

TABLE 18.7 The maximum von Mises stress and A_{2323}.

Composite type		Zone	Value (GPa)	Variation (GPa)	Variation %	A_{2323} (GPa)
cross-reinforced	fiber	fiber t	2.08	0.49-2.05	±62	40.6
	matrix t	zone t	1.96	≈ 0	±100	
unidirectional	fiber	fiber f	1.75	0.77-1.75	±39	26.2
	matrix t	zone f	1.77	≈ 0-1.75	±100	

408 Mechanics and Physics of Structured Media

18.4 The "anisotropic layers" approach

In accordance with [4,31] the "anisotropic layers" approach, the cross-reinforced composite under consideration is modeled by the laminated composite forming by the periodically repeating orthotropic layers with the orientations 0 and 90°, see Fig. 18.1c. The elastic constants of the anisotropic layers are computed by using the complex variables method [18–28] or the homogenization theory [2–7]. The results of the computations with the complex variables method and the homogenization theory will be the same. Note that it the earlier papers on the fiber-reinforced composites the elastic constants of the anisotropic layers were computed by using the "mixture rule" or similar simplified models, see historical review in [32].

Thus, we use the homogenized elastic constants of the unidirectionally reinforced composite computed above as the elastic constants of the anisotropic layers. Let us compare the macroscopic and microscopic characteristics of the cross-reinforced and the laminated composites.

18.4.1 Axial overall elastic moduli A_{1111} and A_{3333}

In the cases described in Sections 18.3.1–18.3.6, the "energy per fiber" for the unidirectional composite is $SENE/2$. In accordance with (18.9)

$$SENE_{mnmn} = \frac{\varepsilon_{mn}^2}{2} A_{mnmn} mes\, P,$$

A_{mnmn} are indicated in the row "unidirectional" in the tables.

The energy at macroscopic strain ε_{11} per fiber in the unidirectional composite is $SENE_{1111}/2$. The energy at macroscopic strain ε_{22} per fiber in the unidirectional composite is $SENE_{2222}/2$. If we turn the central section of the periodicity cell in the unidirectional composite on 90°, we obtain the cross-reinforced composite, subjected to pure axial deformation ε_{11}. In the frameworks of the "anisotropic layers" model, the sum $E = SENE_{1111}/2 + SENE_{2222}/2$ is accepted as the energy of deformation of the periodicity cell of the cross-reinforced composite at the macroscopic strain ε_{11}. Thus, the "anisotropic layers" model predicts the effective stiffness $a_{1111} = \frac{2}{\varepsilon_{mn}^2} \frac{1}{mes\, P} E$. Compare with (18.9), we arrive at $a_{1111} = (A_{1111} + A_{3333})/2$. Similar formulas are valid for all other stiffness.

In the unidirectional composite $A_{1111} = 10$ GPa (Section 18.3.1) and $A_{3333} = 102$ GPa (Section 18.3.2). For the orthogonally reinforced composite, the "anisotropic layers" model predicts $a_{1111} = (10 + 102)/2 = 66$ GPa. The homogenization theory predicts (see Section 18.3.1) the stiffness of the cross-reinforced composite $A_{1111} = 55.9$ GPa. For both the models $A_{1111} = A_{3333}$ (this is true, see Sections 18.3.1 and 18.3.2) and $a_{1111} = a_{3333}$ due to the rotational symmetry of the problem. Numerical computations predict (see Section 18.3.5) $A_{1212} = 55.9$ GPa and (see Section 18.3.6) $A_{2323} = 55.9$ GPa.

18.4.2 Axial overall elastic modulus A_{2222}

In this case, the stiffness a_{2222} computed by using the "anisotropic layers" modes is equal to A_{2222} computed for the unilateral composite, $A_{2222} = 11.14$ GPa (see Table 18.3). The effective stiffness A_{2222} of the orthogonally-reinforced composite computed with the homogenization theory is 12.95 GPa (see Section 18.3.3).

18.4.3 Shift elastic moduli A_{1212} and A_{2323}

For the unidirectional composite, (see Section 18.3.5) $A_{1212} = 15.3$ GPa and (see Section 18.3.6) $A_{1313} = 26.2$ GPa. For the orthogonal-reinforced composite, the "anisotropic layers" model predicts the effective stiffness $a_{1212} = (15.3 + 26.2)/2 = 20.75$ GPa. The effective stiffness A_{1212} of the orthogonally-reinforced composite computed with the homogenization theory is 40.6 GPa, see Section 18.3.5. The equality $A_{1212} = A_{2323}$ follows from the symmetry of the problem. Numerical computations predict (see Section 18.3.5) $A_{1212} = 40.5$ GPa and (see Section 18.3.6) $A_{2323} = 40.6$ GPa.

18.4.4 Shift elastic modulus A_{1313}

In this case, the stiffness a_{1313} computed by using the "anisotropic layers" model is equal to A_{1313} computed for the unilateral composite, $A_{1313} = 9.96$ GPa (see Section 18.3.6). The effective stiffness A_{1313} of the orthogonally-reinforced composite computed with the homogenization theory is 10.47 GPa (see Section 18.3.6).

18.4.5 The local stresses

The components of the local stress-strain tensors In the laminated composites, separate into two groups. In one group, they follow Voigt rule, in another group – Reuss rule [4,29]. In both the groups, the local stress-strain tensors take constant values inside the layer [4,29]. The local stresses both in the unidirectionally reinforced composite and the cross-reinforced composite take values different from the constants, see Figs. 18.2–18.6.

18.5 The "multicomponent" approach by Panasenko

In our computations, we used the material parameters corresponding to carbon/epoxy composite. Such type composites are referred to as "stiff fibers in the soft matrix". Most models of the fiber-reinforced composite materials were developed for composites of this type (see [1,8–10,13,17,57–60] and references in these books and papers).

Based on our numerical computations, we conclude that the hypothesis of the uniform stress-strain state in the fibers (this is a hypothesis usually accepted

410 Mechanics and Physics of Structured Media

in the popular models of fiber-reinforced composites) can be accepted if the load is parallel to fibers. This hypothesis cannot be accepted if the load is not parallel to fibers. Fig. 18.5 displays the deformation of periodicity cells subjected to shift. The shift is equivalent to two orthogonal tension-compressions [48]. In the case under consideration, the tension-compressions are parallel to fibers. For these tension-compressions separately, the local stress-strain states in the fibers are uniform. But for the shift, the local stress-strain state in the fibers is not uniform. It is seen in Fig. 18.5 that von Mises stress in fibers changes in the interval from 0.6 GPa to 1.3 GPa (two times variation).

The local stress-strain state in the matrix is very not uniform for all types of macroscopic loading.

Comparing the local stress-strain state in the fibers and the matrix, we observe significant differences between these stress-strain states. The difference is so large that the fibers and the matrix may be qualified as two continua with different behaviors. The multicontinuum behavior for the high-contrast composite materials was predicted theoretically by Panasenko in [61,62].

Based on our numerical computations, we conclude that the "stiff fibers in the soft matrix" composites demonstrate the behavior similar to multicontinuum behavior predicted for the high-contrast composites in [61,62]. In our case, the contrast ratio $E_f/E_b = 85$.

In this chapter, we solve the periodicity cell problems directly, do not use the iterative schemes discussed in [61,62].

18.6 Solution to the periodicity cell problem for laminated composite

The homogenization procedure for the laminated composites is closely related to the problems analyzed above. To make the chapter self-content, we present solution to the periodicity cell problem (18.3) for laminated composite formed of anisotropic layers. The periodicity cell of the laminated composite is shown in Fig. 18.1c. The solution to (18.3) is a function of unique variable y. For function $\mathbf{N}^{kl}(y)$, the periodicity cell problem (18.3) takes the form

$$\begin{cases} (a_{i2k2}(y)N_k^{mn\prime} - a_{i2mn}(y))' = 0 \text{ in } P, \\ \mathbf{N}^{kl}(y) \text{ periodic in } y \in [-1, 1], \end{cases} \tag{18.10}$$

where a prime means the derivative d/dy. It is a boundary value problem for a system of ordinary differential equations.

The boundary value problem (18.10) may be solved in explicit form. From the differential equation (18.10), we have $a_{i2k2}(y)N_k^{mn\prime} + a_{i2mn}(y) = const_i$. The matrix a_{i2k2} is invertible, then

$$N_k^{mn\prime} = -\{a_{k2j2}\}^{-1}a_{j2mn} + \{a_{k2j2}\}^{-1}const_j, \tag{18.11}$$

where the superscript "−1" stands for the matrix inversion.

The periodicity condition from (18.10) gives $\langle \mathbf{N}^{kl\prime} \rangle = \frac{1}{2} \int_{-1}^{1} \mathbf{N}^{kl\prime}(y) dy = 0$. Substituting $N_k^{mn\prime}$ (18.11), we have $-\langle \{a_{k2j2}\}^{-1} a_{j2mn} \rangle + \langle \{a_{k2j2}\}^{-1} \rangle const_j = 0$. Thus $const_i = \langle \{a_{i2k2}\}^{-1} \rangle^{-1} \langle \{a_{k2j2}\}^{-1} a_{j2mn} \rangle$. Taking into account (18.11), we obtain

$$N_k^{mn\prime} = -\{a_{k2j2}\}^{-1} a_{j2mn} + \{a_{k2j2}\}^{-1} \langle \{a_{i2q2}\}^{-1} \rangle^{-1} \langle \{a_{q2p2}\}^{-1} a_{p2mn} \rangle.$$
(18.12)

In the case under consideration, the first formula form (18.7) becomes

$$A_{pqkl} = \langle a_{pqkl} + a_{pqm2} N_m^{kl\prime} \rangle.$$
(18.13)

Substituting (18.12) into (18.13), we obtain the following formula for calculating the homogenized elastic constants A_{pqkl} of laminated composite:

$$A_{pqkl} = \langle a_{pqkl} \rangle - \langle a_{pqm2} \{a_{m2j2}\}^{-1} a_{j2kl} \rangle$$
$$+ \langle a_{pqm2} \{a_{m2j2}\}^{-1} \rangle \langle \{a_{i2q2}\}^{-1} \rangle^{-1} \langle \{a_{q2p2}\}^{-1} a_{p2kl} \rangle.$$

The local stresses (18.5)

$$\sigma_{pq}(y) = \varepsilon_{mn}(\mathbf{x})[a_{pqmn}(y) + a_{pqk2}(y) N_k^{mn\prime}(y)],$$
(18.14)

where $N_k^{mn\prime}(y)$ are given by (18.12).

We assume that the layers are made of homogeneous materials. In this case, $a_{pqmn}(\mathbf{y})$ take constant values in the layers. It follows from (18.12) that the functions $N_k^{mn\prime}(y)$ also take constant values in the layers. Then, the local stresses $\sigma_{pq}(y)$ (18.14) take constant values in the layers. In particular, in the laminated composite, the variation of the local stresses in the layers is zero.

18.7 The homogenized strength criterion of composite laminae

By using the information about the local stresses in the composite's periodicity cell, we can derive the homogenized strength criterion of composite, i.e. the strength criterion for constitutive components of composite written in terms of the homogenized stresses or strains.

The strength of composite at the microlevel is determined by the local stresses $\sigma_{pq}(\mathbf{y})$ arising in the fibers and matrix of the composite when it is subjected to the homogenized strains $\varepsilon_{mn}(\mathbf{x})$. If macroscopic strains $\varepsilon_{mn}(\mathbf{x})$ are calculated, the local stresses $\sigma_{pq}(\mathbf{y})$ may be computed following the formula (18.6).

We assume that the strength criterion of the materials forming the periodicity cell P may be written in the form

$$f(\mathbf{y}, \sigma_{pq}(\mathbf{y})) < \sigma^*(\mathbf{y}),$$
(18.15)

412 Mechanics and Physics of Structured Media

where the function $\sigma^*(\mathbf{y})$ is defined as follows:

$$\sigma^*(\mathbf{y}) = \begin{cases} \sigma_F^* \text{ in fiber,} \\ \sigma_M^* \text{ in binder,} \end{cases} \qquad f(\mathbf{y}, \sigma_{pq}(\mathbf{y})) = \begin{cases} f_F(\sigma_{pq}(\mathbf{y})) \text{ in fiber,} \\ f_M(\sigma_{pq}(\mathbf{y})) \text{ in binder.} \end{cases}$$

(18.16)

In (18.16), is σ_F^* the strength limit of fibers, σ_M^* is the strength limit of matrix.

Then the condition (compare with (18.15))

$$\max_{\mathbf{y} \in P} \frac{f(\mathbf{y}, \sigma_{pq}(\mathbf{y}))}{\sigma^*(\mathbf{y})} < 1$$

(18.17)

ensures no damage at all points of the periodicity cell P. Substituting (18.6) into (18.17), we obtain

$$F(\varepsilon_{mn}(\mathbf{x})) = \max_{\mathbf{y} \in P} \frac{f(\mathbf{y}, \varepsilon_{mn}(\mathbf{x})[a_{pqmn}(\mathbf{y}) + a_{pqkl}(\mathbf{y})N_{k,l}^{mn}(\mathbf{y})])}{\sigma^*(\mathbf{y})} < 1.$$ (18.18)

The destruction of composite starts when the equality $F(\varepsilon_{mn}(\mathbf{x})) = 1$ is satisfied and occurs at point(s) $\mathbf{y}_0 \in P$ at which the maximum in (18.18) reaches values 1:

$$\frac{f(\mathbf{y}_0, \varepsilon_{mn}(\mathbf{x})[a_{pqmn}(\mathbf{y}_0) + a_{pqkl}(\mathbf{y})N_{k,l}^{mn}(\mathbf{y}_0)])}{\sigma^*(\mathbf{y}_0)} = 1.$$

(18.19)

It may be written in an alternative form $f(\mathbf{y}_0, \varepsilon_{mn}(\mathbf{x})[a_{pqmn}(\mathbf{y}_0) + a_{pqkl}(\mathbf{y}) \times N_{k,l}^{mn}(\mathbf{y}_0)]) = \sigma^*(\mathbf{y}_0)$.

Formula (18.18) is the strength criterion of constitutive elements of composite written in the terms of the homogenized strains $\varepsilon_{mn}(\mathbf{x})$. It proves the existence of the homogenized strength criterion of composites. The formulas (18.18) and (18.19) may be written in terms of the homogenized stresses $\sigma_{mn}(\mathbf{x})$. For this, it suffices to express the homogenized strains $\varepsilon_{mn}(\mathbf{x})$ in the form $\varepsilon_{mn}(\mathbf{x}) = A_{mnkl}^{-1}\sigma_{kl}(\mathbf{x})$ using the tensor of the homogenized elastic constants (18.7) and then substitute this expression to (18.18) and (18.19).

The equality $F(\varepsilon_{mn}(\mathbf{x})) = 1$ is the "first crack" condition, and $\mathbf{y}_0$ from (18.19) indicates the "weakest element" of the composite. Satisfying the equality $F(\varepsilon_{mn}(\mathbf{x})) = 1$ does not mean that the composite sample necessary divides into separate parts or even loses its carrying capacity. Since the "first cracks" occur in numerous periodicity cells simultaneously, the damage of composite will be massive.

The theoretical analysis presented is fulfilled for arbitrary composite materials. For the fiber-reinforced materials considered above, we have observed the specific regions of the stress-strain concentration and distinguish several modes of failure of the fibers and matrix. We denote the mentioned zones G (the zones are specified in Tables 18.1–18.7 in the column "zone"). This note allows to

Comparative analysis of local stresses **Chapter | 18 413**

change in (18.18) $\max_{\mathbf{y}\in P}$ for $\max_{\mathbf{y}\in G}$ and reduces the necessary computations. From the mechanic's point of view, it means that in the fiber-reinforced composite, one has to encounter only some typical modes of the damage of composite at the microlevel (in other words, the fiber-reinforced composites have some specific "weakest elements").

In some cases, see for example [42], stress-strain concentration has a specific form that allows one to construct the homogenized strength criterion in explicit form. In the general case (this is the case considered in this article), the strength criterion cannot be constructed in an explicit form. But all the calculations associated with (18.6), (18.18), (18.19) can be effectively implemented with a computer. Thus, the homogenized strength criterion can be constructed as a computer procedure. The main problem, which one meets, is the computation of the functions $\mathbf{N}^{mn}(\mathbf{y})$, which is also the main problem in the general homogenization theory [3,5,6].

18.8 Conclusions

We have computed the local stress-strain state for the two fiber-reinforced composites with the same characteristics of the fibers and the matrix (elastic moduli, radii of the fibers, and interfiber distances). The difference is the reinforcement schemes: unidirectional (0–0°) or orthogonal (0–90°) stacking of the fibers. In both the composites, the fibers are tightly packed.

The homogenized elastic moduli A_{iiii} ($i = 1, 2, 3$) and A_{1212}, A_{2323} computed with the homogenization theory and the "anisotropic layers" approach are close. The homogenized elastic moduli A_{1313} computed with the homogenization theory and the "anisotropic layers" approaches differ about twice.

In the unidirectionally reinforced composite and the cross-reinforced composite, the local stresses are inhomogeneous.

The local stresses in the unidirectionally reinforced composite and the cross-reinforced composite (see Figs. 18.2–18.7) do not show any similarity to the stresses in laminated composites. The values of the variation of von Mises stress in the fibers (see Tables 18.1–18.7) indicate that the basic hypothesis of the Mori-Tanaka method (the homogeneity of the stress-strain state in the fibers) cannot be accepted either for unidirectional and orthogonally reinforcement schemes with tightly packed fibers. The exceptions are the composites subjected to macroscopic tension-compressions strains along fibers. Note that the variation of the stress-strain state in the fibers decreases when the interparticle distance increase (the volume fraction of fibers decreases).

The numerical experiments show that for "stiff fibers in the soft matrix" the hypothesis of equivalent homogeneity fails. The behavior of the fiber-reinforced composites is similar to the multicontinuum behavior of high-contrast composites [60,62].

By using the results of the numerical solutions and the methodology from [42], one can develop the homogenized strength criterion for fiber-reinforced composites.

414 Mechanics and Physics of Structured Media

References

[1] J. Aboudi, Mechanics of Composite Material: A Unified Micromechanics Approach, Elsevier, Amsterdam, 1991.

[2] G.W. Milton, The Theory of Composites, Cambridge University Press, Cambridge, 2002.

[3] N.S. Bakhvalov, G.P. Panasenko, Homogenization: Averaging Processes in Periodic Media, Kluwer, Dordrecht, 1989.

[4] A.L. Kalamkarov, A.G. Kolpakov, Analysis, Design and Optimization of Composite Structures, Wiley, Chichester, 1997.

[5] E. Sanchez-Palencia, Non-Homogeneous Media and Vibration Theory, Springer, Berlin, 1980.

[6] A. Bensoussan, J-L. Lions, G. Papanicolaou, Asymptotic Analysis for Periodic Structures, North-Holland, Amsterdam, 1978.

[7] G.P. Panasenko, Multi-Scale Modeling for Structures and Composites, Springer, Berlin, 2005.

[8] R.M. Jones, Mechanics of Composite Materials, Scripta Book, Washington, DC, 1975.

[9] J.R. Vinson, R.L. Sierakowski, The Behavior of Structures Composed of Composite Materials, Kluwer, Dordrecht, 1987.

[10] A. Kelly, Yu.N. Rabotnov (Eds.), Handbook of Composites, North-Holland, Amsterdam, 1988.

[11] S. Nemat-Nasser, M. Hori, Micromechanics: Overall Properties of Heterogeneous Materials, North-Holland, Amsterdam, 1993.

[12] L.J. Gibson, M.F. Ashby, Cellular Solids. Structures and Properties, Cambridge University Press, Cambridge, 1997.

[13] C.T. Herakovich, Mechanics of Fibrous Composites, Wiley, New York, 1997.

[14] M. Sahimi, Heterogeneous Materials, Springer, New York, 2003.

[15] R.F. Gibson, Principles of Composite Material Mechanics, CRC Press, Boca Raton, FL, 2016.

[16] E.J. Barbero, Introduction to Composite Materials Design, CRC Press, Boca Raton, FL, 2017.

[17] B.D. Agarwal, L.J. Broutman, K. Chandrashekhara, Analysis and Performance of Fiber Composites, Wiley, Chichester, 2017.

[18] E.I. Grigolyuk, L.A. Fil'shtinskii, Elastic equilibrium of an isotropic plane with a doubly periodic system of inclusions, Sov. Appl. Mech. 2 (9) (1966) 1–5.

[19] E.I. Grigolyuk, L.A. Fil'shtinskii, Perforated Plates and Shells, Nauka, Moscow, 1970 (in Russian).

[20] E.I. Grigolyuk, L.A. Fil'shtinskij, Periodic Piecewise Homogeneous Elastic Structures, Nauka, Moscow, 1992 (in Russian).

[21] J.-K. Lu, Complex Variable Methods in Plane Elasticity, World Scientific, Singapore, 1995.

[22] V. Mityushev, S.V. Rogozin, Constructive Methods for Linear and Nonlinear Boundary Value Problems of Analytic Function Theory, Chapman & Hall/CRC, Boca Raton, FL, 2000.

[23] V. Mityushev, P.M. Adler, Longitudinal permeability of a doubly periodic rectangular array of cylinders, I, Z. Angew. Math. Mech. 82 (2002) 335–345.

[24] V. Mityushev, P.M. Adler, Longitudinal permeability of a doubly periodic rectangular array of cylinders. II. An arbitrary distribution of cylinders inside the unit cell, Z. Angew. Math. Phys. 53 (2002) 486–517.

[25] D.I. Bardzokas, M.L. Filshtinsky, L.A. Filshtinsky, Mathematical Models of Elastic and Piezoelectric Fields in Two-Dimensional Composites, Springer, Berlin, 2007.

[26] S. Gluzman, V. Mityushev, W. Nawalaniec, Computational Analysis of Structured Media, Academic Press, Amsterdam, 2018.

[27] V. Mityushev, P. Drygas, Effective properties of fibrous composites and cluster convergence, Multiscale Model. Simul. 17 (2) (2019) 696–715.

[28] P. Drygaś, S. Gluzman, V. Mityushev, W. Nawalaniec, Applied Analysis of Composite Media Analytical and Computational Results for Materials Scientists and Engineers, Elsevier, Amsterdam, 2020.

[29] G. Lubin (Ed.), Handbook of Composites, Van Nostrand, New York, 1982.

[30] R.M. Christensen, Mechanics of Composite Materials, Wiley, New York, 1979.

Comparative analysis of local stresses **Chapter | 18 415**

[31] G. Dvorak, Micromechanics of Composite Materials, Springer, Berlin, 2013.

[32] L.L. Vignolia, M.A. Savia, P.M.C.L. Pachecob, A.L. Kalamkarov, Comparative analysis of micromechanical models for the elastic composite laminae, Composites, Part B 174 (2019) 10696.

[33] B.D. Annin, A.L. Kalamkarov, A.G. Kolpakov, Analysis of local stresses in high modulus fiber composites, in: Localized Damage Computer-Aided Assessment and Control. V.2, Comput. Mechanics Publ., Southampton, 1990, pp. 131–144.

[34] A.L. Kalamkarov, A.G. Kolpakov, On the analysis and design of fiber-reinforced composite shells, J. Appl. Mech. 63 (4) (1996) 939–945.

[35] A.L. Kalamkarov, A.G. Kolpakov, Design problems for the fiber-reinforced composite materials, Composites, Part B 27B (1996) 485–492.

[36] J.-L. Lions, Notes on some computational aspects of the homogenization method in composite materials, in: J.-L. Lions, G.I. Marchuk (Eds.), Computational Methods in Mathematics, Geophysics and Optimal Control, Nauka, Novosibirsk, 1978, pp. 5–19 (in Russian).

[37] S.V. Sheshenin, Averaged modules of one composite, Vestn. Mosk. Univ., Ser. 1. Matem. Mekhan. 6 (1980) 78–83.

[38] G.P. Panasenko, Numerical solution of cell problems in averaging theory, USSR Comput. Math. Math. Phys. 28 (1) (1988) 183–186.

[39] L.L. Mishnaevsky Jr., Computational Mesomechanics of Composites: Numerical Analysis of the Effect of Microstructures of Composites of Strength and Damage Resistance, Wiley, Chichester, 2007.

[40] E.J. Barbero, Finite Element Analysis of Composite Materials Using ANSYS, CRC Press, Boca Raton, FL, 2013.

[41] F.L. Matthews, G.A.O. Davies, D. Hitchings, C. Soutis, Finite Element Modelling of Composite Materials and Structures, CRC Press, Boca Raton, FL, 2020.

[42] A.G. Kolpakov, S.I. Rakin, Homogenized strength criterion for composite reinforced with orthogonal systems of fibers, Mech. Mater. 148 (2020) 8.

[43] M. Cordin, Th. Bechtold, T. Pham, Effect of fibre orientation on the mechanical properties of polypropylene–lyocell composites, Cellulose 25 (2018) 7197–7210.

[44] G. de Botton, G. Shmuel, Mechanics of composites with two families of finitely extensible fibers undergoing large deformations, J. Mech. Phys. Solids 57 (2009) 1165–1181.

[45] Ch. El Hage, R. Younes, Z. Aboura, M.I. Benzeggagh, M. Zoaeter, Analytical and numerical modeling of mechanical properties of orthogonal 3D CFRP, Compos. Sci. Technol. 69 (2009) 111–116.

[46] L. Liu, Z.-M. Huang, Stress concentration factor in matrix of a composite reinforced with transversely isotropic fibers, J. Compos. Mater. 48 (1) (2014) 81–89.

[47] R.S. Lopes, C.S. Moreira, L.C.S. Nunes, Modeling of an elastic matrix reinforced with two families of fibers under simple shear: a mimic of annulus fibrosus, J. Braz. Soc. Mech. Sci. Eng. 41 (2019) 385–391.

[48] A.E.H. Love, A Treatise on the Mathematical Theory of Elasticity, Cambridge University Press, Cambridge, 2013.

[49] K.K. Chawla, Composite Materials, Science and Engineering, Springer, New York, 1998.

[50] M.K. Thompson, J.M. Thompson, ANSYS Mechanical APDL for Finite Element Analysis, Butterworth-Heinemann, Oxford, 2017.

[51] J.B. Keller, J.E. Flaherty, Elastic behavior of composite media, Commun. Pure Appl. Math. 25 (1973) 565–580.

[52] H. Kang, S. Yu, A proof of the Flaherty–Keller formula on the effective property of densely packed elastic composites, Calc. Var. 59 (2020) 22.

[53] A.A. Kolpakov, Numerical verification of the existence of the energy-concentration effect in a high-contrast heavy-charged composite material, J. Eng. Phys. Thermophys. 80 (4) (2007) 812–819.

[54] S.I. Rakin, Numerical verification of the existence of the elastic energy localization effect for closely spaced rigid disks, J. Eng. Phys. Thermophys. 87 (2014) 246–252.

416 Mechanics and Physics of Structured Media

[55] L. Berlyand, A. Kolpakov, Network approximation in the limit of small interparticle distance of the effective properties of a high-contrast random dispersed composite, Arch. Ration. Mech. Anal. 159 (2001) 179–227.

[56] A.A. Kolpakov, A.G. Kolpakov, Capacity and Transport in Contrast Composite Structures: Asymptotic Analysis and Applications, CRC Press, Boca Raton, FL, 2009.

[57] J.C. Halpin, J.L. Kardos, The Halpin-Tsai equations: a review, Polym. Eng. Sci. 16 (5) (1976) 344–352.

[58] C. Zweben, H.T. Hahn, T.-W. Chou, Delaware Composites Design Encyclopedia, CRC Press, Boca Raton, FL, 1989.

[59] J.C. Halpin, Primer on Composite Materials Analysis, CRC Press, Boca Raton, FL, 1992.

[60] J.N. Reddy (Ed.), Mechanics of Composite Materials: Selected Works of Nicholas J. Pagano, Kluwer, Dordrecht, 1994.

[61] G.P. Panasenko, Averaging of processes in strongly inhomogeneous structures, Dokl. Math. 33 (1) (1988) 20–22.

[62] G.P. Panasenko, Multicomponent homogenization for processes in essentially nonhomogeneous structures, Math. USSR Sb. 69 (1) (1991) 143–153.

Chapter 19

Statistical theory of structures with extended defects

Vyacheslav Yukalov[a,b] and Elizaveta Yukalova[c]

[a]*Bogolubov Laboratory of Theoretical Physics, Joint Institute for Nuclear Research, Dubna, Russia,* [b]*Instituto de Fisica de São Carlos, Universidade de São Paulo, São Carlos, São Paulo, Brazil,* [c]*Laboratory of Information Technologies, Joint Institute for Nuclear Research, Dubna, Russia*

19.1 Introduction

The structure of many solid-state materials is not represented by ideal crystalline lattices, but contains various defects [1–3]. By their size, defects can be classified into two types, point defects and extended defects. Examples of point defects are Schottky defects, Frenkel defects, vacancies, interstitials, and impurities. Examples of extended defects are dislocations, cracks, pores, heterophase embryos, and polymorphic inclusions. These defects are usually of nanosize scale, at least in some directions. For instance, a dislocation length can be of a macroscopic size, while a dislocation radius is of a nanoscale size [4,5].

In many cases, one is interested in the overall properties of a sample with nanosize defects, but not in the details of its internal structure. In that case, one needs a theory describing the sample as a whole. This implies that it is necessary to have an approach characterizing the average typical features of the sample. That is, there is a need in a kind of a statistical approach.

In this chapter, we present such a statistical approach for describing materials with nanosize defects. We keep in mind materials whose ideal crystalline structure is disturbed by the presence of extended defects with a disordered structure. These regions of disorder are randomly distributed across the sample. Usually, they are in local equilibrium or in quasiequilibrium with the surrounding crystalline matrix. In some cases they can move through the sample, as e.g. dislocations through a crystal. The fraction of particles forming the disordered regions, with respect to the total number of particles in the system, generally, is defined self-consistently from the conditions of the material stability.

Throughout the chapter, the system of units is used where the Planck and Boltzmann constants are set to one.

Mechanics and Physics of Structured Media. https://doi.org/10.1016/B978-0-32-390543-5.00024-4
Copyright © 2022 Elsevier Inc. All rights reserved.

19.2 Spatial separation of phases

The spatial regions filled by extended defects can be treated as the regions filled by a phase with disordered particle locations. We consider a sample containing a mixture of two different phases composed of the same kind of particles. Say, one phase, forms a solid with a crystalline lattice, and the other phase consists of regions of a disordered structure. Schematically, this is shown in Fig. 19.1. Using the Gibbs method of separating surfaces [6], the whole sample can be represented as consisting of the regions filled by the corresponding phases, so that the total sample volume is the sum of the volumes filled by each of the phases and the total number of particles is the sum of the particles in each phase,

$$V = V_1 + V_2, \qquad N = N_1 + N_2. \tag{19.1}$$

The regions of disorder are randomly distributed in space.

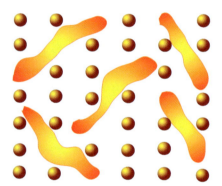

FIGURE 19.1 Crystal with randomly distributed extended nanosize defects.

Since we are interested in the average properties of the system, hence we need to describe the related averaging procedure. Below we present such a procedure based on the review articles [7,8]. For generality, we consider a quantum system. The same approach can also be realized for classical systems.

The spatial location of the phases can be characterized by the manifold indicator functions [9]

$$\xi_f(\mathbf{r}) = \begin{cases} 1, & \mathbf{r} \in V_f, \\ 0, & \mathbf{r} \notin V_f. \end{cases} \tag{19.2}$$

Here, for the simplicity of notation, we denote the spatial part, occupied by an f-th phase, and the volume of this part by the same letter V_f, with $f = 1, 2$.

The Hilbert space of microscopic states is the tensor product

$$\mathcal{H} = \mathcal{H}_1 \bigotimes \mathcal{H}_2 \tag{19.3}$$

of the Hilbert spaces associated with the phases enumerated by the index $f = 1, 2$. The spaces $\mathcal{H}_f$ can be defined as weighted Hilbert spaces [7,10]. The algebra of observables $\mathcal{A}(\xi)$ can be represented as a direct sum of the algebras of observables $\mathcal{A}_f(\xi_f)$ on the spaces $\mathcal{H}_f$ for the related phases,

$$\mathcal{A}(\xi) = \mathcal{A}_1(\xi_1) \bigoplus \mathcal{A}_2(\xi_2), \tag{19.4}$$

where $\mathcal{A}_f(\xi_f)$ is the representation of the algebra on the Hilbert space $\mathcal{H}_f$. Respectively, the energy operator (Hamiltonian) has the form

$$\hat{H}(\xi) = \hat{H}_1(\xi_1) \bigoplus \hat{H}_2(\xi_2). \tag{19.5}$$

And the number-of-particle operator is

$$\hat{N}(\xi) = \hat{N}_1(\xi_1) \bigoplus \hat{N}_2(\xi_2). \tag{19.6}$$

The general expression of the energy operator for an f-th phase reads as

$$\hat{H}_f(\xi_f) = \int \xi_f(\mathbf{r})\psi_f^{\dagger}(\mathbf{r}) \left[-\frac{\nabla^2}{2m} + U(\mathbf{r}) \right] \psi_f(\mathbf{r}) d\mathbf{r} +$$
$$+ \frac{1}{2} \int \xi_f(\mathbf{r})\xi_f(\mathbf{r}')\psi_f^{\dagger}(\mathbf{r})\psi_f^{\dagger}(\mathbf{r}')\Phi(\mathbf{r} - \mathbf{r}')\psi_f(\mathbf{r}')\psi_f(\mathbf{r}) d\mathbf{r} d\mathbf{r}' \tag{19.7}$$

and the number-of-particle operator for an f-th phase has the form

$$\hat{N}_f(\xi_f) = \int \xi_f(\mathbf{r})\psi_f^{\dagger}(\mathbf{r})\psi_f(\mathbf{r}) d\mathbf{r}. \tag{19.8}$$

Here $U(\mathbf{r})$ is an external potential and $\Phi(\mathbf{r})$ is an interaction potential.

19.3 Statistical operator of mixture

The statistical operator $\hat{\rho}(\xi)$ for a heterophase system depends on the configuration of the phases in the sample, which is denoted by ξ. This operator can be found from the principle of minimal information by minimizing the information functional under given additional constraints. These are: the normalization condition

$$\mathrm{Tr} \int \hat{\rho}(\xi)\mathcal{D}(\xi) = 1, \tag{19.9}$$

the definition of the system energy

$$\mathrm{Tr} \int \hat{\rho}(\xi)\hat{H}(\xi)\mathcal{D}(\xi) = E, \tag{19.10}$$

420 Mechanics and Physics of Structured Media

and the total number of particles

$$\text{Tr} \int \hat{\rho}(\xi)\hat{N}(\xi)\mathcal{D}(\xi) = N. \tag{19.11}$$

Here the trace is over the total Hilbert space (19.3). The notation $\mathcal{D}(\xi)$ implies the averaging over all admissible phase configurations that should be done in view of the random locations and shapes of the coexisting phases.

The corresponding information functional can be taken in the Kullback-Leibler [11,12] form

$$I[\hat{\rho}] = \text{Tr} \int \hat{\rho}(\xi) \ln \frac{\hat{\rho}(\xi)}{\hat{\rho}_0(\xi)} \mathcal{D}(\xi) + \alpha \left[\text{Tr} \int \hat{\rho}(\xi)\mathcal{D}(\xi) - 1 \right] +$$
$$+ \beta \left[\text{Tr} \int \hat{\rho}(\xi)\hat{H}(\xi)\mathcal{D}(\xi) - E \right] + \gamma \left[\text{Tr} \int \hat{\rho}(\xi)\hat{N}(\xi)\mathcal{D}(\xi) - N \right], \tag{19.12}$$

where α, $\beta \equiv 1/T$, and $\gamma \equiv -\beta\mu$ are Lagrange multipliers, T is temperature, and $\hat{\rho}_0(\xi)$ is a trial statistical operator prescribed by some a priori information, if any. In the case of no preliminary information, $\hat{\rho}_0(\xi)$ is a constant. Then the minimization of the information functional yields the statistical operator

$$\hat{\rho}(\xi) = \frac{1}{Z} \exp\{-\beta H(\xi)\}, \tag{19.13}$$

with the partition function

$$Z = \text{Tr} \int \exp\{-\beta H(\xi)\}\mathcal{D}(\xi), \tag{19.14}$$

where the grand Hamiltonain is

$$H(\xi) \equiv \hat{H}(\xi) - \mu\hat{N}(\xi). \tag{19.15}$$

Defining the effective renormalized Hamiltonian by the relation

$$\int \exp\{-\beta H(\xi)\}\mathcal{D}(\xi) = \exp\{-\beta\tilde{H}\} \tag{19.16}$$

results in the partition function

$$Z = \text{Tr} \exp\{-\beta\tilde{H}\}. \tag{19.17}$$

From here, the grand thermodynamic potential follows:

$$\Omega = -T \ln Z. \tag{19.18}$$

Recall that this thermodynamic potential characterizes a system where the regions of disorder are not frozen and their spatial locations are not fixed. This

Statistical theory of structures with extended defects **Chapter | 19** **421**

implies that the experiment with such a system does not provide information on spatial locations of the phase distribution, but the experimental data give a spatially averaged picture. This is why it is necessary to average over phase configurations, as is discussed above. In that sense, the overall properties of the averaged system correspond to an effectively equilibrium system, although at each moment of time the heterophase system is quasiequilibrium. A multiphase system equilibrium on average can be called a system that is in *heterophase equilibrium* [7].

19.4 Quasiequilibrium snapshot picture

As is explained above, we consider the case of a heterophase system equilibrium on average, where the regions of disorder are randomly distributed over the sample and, generally, can move in space. For example, dislocations, strictly speaking, are intrinsically nonequilibrium objects that are created, annihilated, and moving through the sample [13–18]. However a system with these dislocations, being averaged over their locations, represents an equilibrium crystal. The approach, we have started describing, is appropriate for the interpretation of those experiments that give the results averaged over random locations of phase configurations.

It is instructive to compare this approach with the description of a sample containing different phases whose locations are fixed in space. From the experimental point of view, the latter situation corresponds to the case where the observed data are taken in a snapshot way. This is called the case of spatially frozen phases. For the latter case, thermodynamics, generally, differs from the described above.

For a system with frozen phases, it is possible to define the statistical operator resorting to a quasiequilibrium picture [7,19]. The phase locations can again be described by a set of the manifold indicator functions (19.2). With the statistical operator of the whole system $\hat{\rho}(\xi)$, the statistical operators for each phase read as

$$\hat{\rho}_f(\xi_f) \equiv \mathrm{Tr}_{\mathcal{H}/\mathcal{H}_f}\hat{\rho}(\xi). \tag{19.19}$$

This operator has to be normalized,

$$\mathrm{Tr}_{\mathcal{H}_f}\hat{\rho}_f(\xi_f) = 1. \tag{19.20}$$

For each of the phases, the definitions of the energy

$$\mathrm{Tr}_{\mathcal{H}_f}\hat{\rho}_f(\xi_f)\hat{H}_f(\xi_f) = E_f(\xi_f) \tag{19.21}$$

and the number of particles

$$\mathrm{Tr}_{\mathcal{H}_f}\hat{\rho}_f(\xi_f)\hat{N}_f(\xi_f) = N_f(\xi_f) \tag{19.22}$$

are given.

422 Mechanics and Physics of Structured Media

The information functional takes the form

$$I[\hat{\rho}_f] = \mathrm{Tr}_{\mathcal{H}_f} \hat{\rho}_f(\xi_f) \ln \frac{\hat{\rho}_f(\xi_f)}{\hat{\rho}_f^0(\xi_f)} + \alpha_f(\xi_f) \left[\mathrm{Tr}_{\mathcal{H}_f} \hat{\rho}_f(\xi_f) - 1 \right] +$$
$$+ \beta_f(\xi_f) \left[\mathrm{Tr}_{\mathcal{H}_f} \hat{\rho}_f(\xi_f) \hat{H}_f(\xi_f) - E_f(\xi_f) \right]$$
$$+ \gamma_f(\xi_f) \left[\mathrm{Tr}_{\mathcal{H}_f} \hat{\rho}_f(\xi_f) \hat{N}_f(\xi_f) - N_f(\xi_f) \right], \tag{19.23}$$

where $\hat{\rho}_f^0(\xi_f)$ is a trial statistical operator. As earlier, we assume that no additional information is provided, so that the trial operator is constant. Now the Lagrange multipliers depend on the given phase distribution. For instance, temperature becomes

$$T_f(\xi_f) = \frac{1}{\beta_f(\xi_f)}. \tag{19.24}$$

Using the notation

$$\gamma_f(\xi_f) \equiv -\beta_f(\xi_f) \mu_f(\xi_f), \tag{19.25}$$

we minimize the information functional, which results in the statistical operator

$$\hat{\rho}_f(\xi_f) = \frac{1}{Z_f(\xi_f)} \exp \left\{ -\beta_f(\xi_f) H_f(\xi_f) \right\}, \tag{19.26}$$

with the partition function

$$Z_f(\xi_f) = \mathrm{Tr}_{\mathcal{H}_f} \exp \left\{ -\beta_f(\xi_f) H_f(\xi_f) \right\} \tag{19.27}$$

and the grand Hamiltonian

$$H_f(\xi_f) = \hat{H}_f(\xi_f) - \mu_f(\xi_f) \hat{N}_f(\xi_f). \tag{19.28}$$

Thus for the frozen phase distribution, for each phase, we get the grand thermodynamic potential

$$\Omega_f(\xi_f) = -T(\xi_f) \ln Z_f(\xi_f). \tag{19.29}$$

Averaging it over phase configurations, we obtain the thermodynamic potential for the whole system with the frozen phases,

$$\overline{\Omega} = \int \sum_f \Omega_f(\xi_f) \mathcal{D}(\xi). \tag{19.30}$$

As is clear, the grand thermodynamic potentials (19.18) and (19.30), in general, do not coincide. However, under some additional restrictions (to be mentioned below), the thermodynamic potential (19.18) can approximate potential (19.30).

19.5 Averaging over phase configurations

The averaging over phase configurations implies the functional integration over the manifold indicator functions (19.2). This integration has been explicitly formulated and accomplished in the series of papers [7,20–24]. Here we summarize the main results of this functional integration.

First it is necessary to consider the averaging of the functionals of the often met form

$$C_f(\xi_f) = \sum_{m=0}^{\infty} \int \xi_f(\mathbf{r}_1)\xi_f(\mathbf{r}_2)\ldots\xi_f(\mathbf{r}_m)C_f(\mathbf{r}_1, \mathbf{r}_2, \ldots, \mathbf{r}_m)d\mathbf{r}_1 d\mathbf{r}_2\ldots d\mathbf{r}_m.$$
(19.31)

Then we give the expression for the thermodynamic potential (19.18). And finally the averages of the operators corresponding to observable quantities are derived.

Theorem 19.1. *The averaging over phase configurations of functional (19.31) gives*

$$\int C_f(\xi_f)\mathcal{D}\xi = C_f(w_f),$$
(19.32)

where

$$C_f(w_f) = \sum_{m=0}^{\infty} w_f^m \int C_f(\mathbf{r}_1, \mathbf{r}_2, \ldots, \mathbf{r}_m)d\mathbf{r}_1 d\mathbf{r}_2\ldots d\mathbf{r}_m,$$
(19.33)

and the weight

$$w_f \equiv \frac{1}{V} \int \xi_f(\mathbf{r})d\mathbf{r}$$
(19.34)

defines the geometric probability of an f-th phase.

Theorem 19.2. *The thermodynamic potential*

$$\Omega = -T \ln \mathrm{Tr} \int \exp\{-\beta H(\xi)\}\mathcal{D}\xi,$$
(19.35)

after the averaging over phase configurations, becomes

$$\Omega = -T \ln \mathrm{Tr}\exp\{-\beta\tilde{H}\} = \sum_f \Omega_f,$$
(19.36)

where

$$\Omega_f = -T \ln \mathrm{Tr}_{\mathcal{H}_f} \exp\{-\beta H_f(w_f)\}.$$
(19.37)

424 Mechanics and Physics of Structured Media

The renormalized Hamiltonian is

$$\tilde{H} = \bigoplus_f H_f(w_f) \equiv \tilde{H}(w), \tag{19.38}$$

in which

$$H_f(w_f) = \lim_{\xi_f \to w_f} H_f(\xi_f). \tag{19.39}$$

The phase probabilities w_f are the minimizers of the thermodynamic potential

$$\Omega = \text{abs min } \Omega(w), \tag{19.40}$$

where

$$\Omega(w) = -T \ln \text{Tr} \left\{ -\beta \tilde{H}(w) \right\}. \tag{19.41}$$

The minimization is accomplished under the normalization conditions

$$\sum_f w_f = 1, \qquad 0 \le w_f \le 1. \tag{19.42}$$

Theorem 19.3. *The observable quantities, defined by the averages*

$$\langle \hat{A} \rangle = \text{Tr} \int \hat{\rho}(\xi) \hat{A}(\xi) \mathcal{D}\xi \tag{19.43}$$

of the operators

$$\hat{A}(\xi) = \bigoplus_f \hat{A}_f(\xi_f), \tag{19.44}$$

with

$$\hat{A}_f(\xi_f) = \sum_{m=0}^{\infty} \int \xi_f(\mathbf{r}_1)\xi_f(\mathbf{r}_2)\ldots\xi_f(\mathbf{r}_m)A_f(\mathbf{r}_1, \mathbf{r}_2, \ldots, \mathbf{r}_m)d\mathbf{r}_1 d\mathbf{r}_2\ldots d\mathbf{r}_m \tag{19.45}$$

after the averaging over phase configurations, reduce to the form

$$\langle \hat{A} \rangle = \text{Tr}\hat{\rho}(w)\hat{A}(w), \tag{19.46}$$

in which the renormalized operators are

$$\hat{A}(w) = \bigoplus_f \hat{A}_f(w_f), \tag{19.47}$$

with

$$\hat{A}_f(w_f) = \sum_{m=0}^{\infty} w_f^m \int A_f(\mathbf{r}_1, \mathbf{r}_2, \ldots, \mathbf{r}_m)d\mathbf{r}_1 d\mathbf{r}_2\ldots d\mathbf{r}_m, \tag{19.48}$$

and the renormalized statistical operator is

$$\hat{\rho}(w) = \frac{1}{Z} \exp\left\{-\beta\widetilde{H}(w)\right\}, \tag{19.49}$$

with the partition function

$$Z = \mathrm{Tr}\exp\left\{-\beta\widetilde{H}(w)\right\}. \tag{19.50}$$

Remark. If the thermodynamic potential (19.29), by expanding it in powers of the manifold indicator functions, can be represented in the form of the functional (19.31), then the thermodynamic potential (19.30) acquires the form of the thermodynamic potential (19.36), however with the weights w_f that are not the minimizers of the thermodynamic potential, but are given by the values prescribed by the corresponding volumes occupied by the frozen phases.

19.6 Geometric phase probabilities

According to definition (19.34), the weights w_f are the geometric phase probabilities

$$w_f = \frac{V_f}{V}. \tag{19.51}$$

Since

$$\sum_f N_f = N, \qquad \sum_f V_f = V, \tag{19.52}$$

probabilities (19.51) meet the normalization conditions (19.42).

For illustration, let us consider the case of two coexisting phases, $f = 1, 2$. Then, to satisfy conditions (19.42), we may set

$$w \equiv w_1, \qquad w_2 = 1 - w. \tag{19.53}$$

When w_f are not identically 0 or 1, they are defined by minimizing the thermodynamic potential, so that

$$\frac{\partial\Omega}{\partial w} = 0, \qquad \frac{\partial^2\Omega}{\partial w^2} > 0. \tag{19.54}$$

The first of these conditions gives

$$\left\langle\frac{\partial\widetilde{H}}{\partial w}\right\rangle = 0, \tag{19.55}$$

with the Hamiltonian (19.38).

Keeping in view Hamiltonians generated by form (19.7), we have the replica Hamiltonians

$$H_f(w_f) = \hat{H}_f(w_f) - \mu\hat{N}_f(w_f), \tag{19.56}$$

426 Mechanics and Physics of Structured Media

with the energy Hamiltonians

$$\hat{H}_f(w_f) = w_f \int \psi_f^\dagger(\mathbf{r}) \left[-\frac{\nabla^2}{2m} + U(\mathbf{r}) \right] \psi_f(\mathbf{r})d\mathbf{r} +$$
$$+ \frac{w_f^2}{2} \int \psi_f^\dagger(\mathbf{r})\psi_f^\dagger(\mathbf{r}')\Phi(\mathbf{r} - \mathbf{r}')\psi_f(\mathbf{r}')\psi_f(\mathbf{r})d\mathbf{r}d\mathbf{r}' \quad (19.57)$$

and the number-of-particle operators

$$\hat{N}_f(w_f) = w_f \int \psi_f^\dagger(\mathbf{r})\psi_f(\mathbf{r})d\mathbf{r}. \quad (19.58)$$

Introducing the notations

$$\hat{K}_f \equiv \int \psi_f^\dagger(\mathbf{r}) \left[-\frac{\nabla^2}{2m} + U(\mathbf{r}) \right] \psi_f(\mathbf{r})d\mathbf{r},$$
$$\hat{\Phi}_f \equiv \int \psi_f^\dagger(\mathbf{r})\psi_f^\dagger(\mathbf{r}')\Phi(\mathbf{r} - \mathbf{r}')\psi_f(\mathbf{r}')\psi_f(\mathbf{r})d\mathbf{r}d\mathbf{r}',$$
$$\hat{R}_f \equiv \int \psi_f^\dagger(\mathbf{r})\psi_f(\mathbf{r})d\mathbf{r}, \quad (19.59)$$

makes it straightforward to represent Hamiltonians (19.56) in the simple way

$$H_f(w_f) = w_f \hat{K}_f + \frac{w_f^2}{2}\hat{\Phi}_f - \mu w_f \hat{R}_f. \quad (19.60)$$

As is seen, the replica Hamiltonians (19.60) depend on w_f explicitly. At the same time, they also depend on w_f implicitly through the dependence of the field operators on $H_f(w_f)$, as far as the field operators satisfy the Heisenberg equations

$$i\frac{\partial}{\partial t}\psi_f(\mathbf{r}, t) = \left[\psi_f(\mathbf{r}, t), H_f(w_f)\right]. \quad (19.61)$$

The other equivalent form of these equations [25,26] reads as

$$i\frac{\partial}{\partial t}\psi_f(\mathbf{r}, t) = \frac{\delta H_f(w_f)}{\delta \psi_f^\dagger(\mathbf{r}, t)}. \quad (19.62)$$

In order to grasp the feeling of the structure of the equations for the phase probabilities, let us for a while neglect the implicit dependence on w_f, assuming that the explicit dependence prevails. Then Eq. (19.55) yields

$$w = \frac{\Phi_2 + K_2 - K_1 + \mu(R_1 - R_2)}{\Phi_1 + \Phi_2}, \quad (19.63)$$

where

$$K_f = \langle \hat{K}_f \rangle, \qquad \Phi_f = \langle \hat{\Phi}_f \rangle, \qquad R_f = \langle \hat{R}_f \rangle. \tag{19.64}$$

The second of Eqs. (19.54) yields the stability condition

$$\left\langle \frac{\partial^2 \widetilde{H}}{\partial w^2} \right\rangle = \beta \left\langle \left(\frac{\partial \widetilde{H}}{\partial w} \right)^2 \right\rangle, \tag{19.65}$$

for which the necessary condition is

$$\Phi_1 + \Phi_2 > 0. \tag{19.66}$$

It is useful to connect the geometric phase probabilities with the typical densities in the system. The particle density of an f-th phase is given by the ratio

$$\rho_f \equiv \frac{N_f}{V_f} = \frac{R_f}{V}. \tag{19.67}$$

The average density in the system is

$$\rho \equiv \frac{N}{V} = \sum_f w_f \rho_f. \tag{19.68}$$

The fraction of particles in an f-th phase reads as

$$n_f \equiv \frac{N_f}{N}, \tag{19.69}$$

with the evident conditions

$$\sum_f n_f = 1, \qquad 0 \le n_f \le 1. \tag{19.70}$$

From the relation

$$w_f \rho_f = n_f \rho, \tag{19.71}$$

it follows

$$\sum_f \rho_f = \rho \sum_f \frac{n_f}{w_f}. \tag{19.72}$$

The situation simplifies, when the phases are distinguished not by their densities, but by some other properties, like magnetic, electric, or orientational features, while the phase densities are equal. In that case, the phase probabilities coincide with the phase fractions,

$$w_f = n_f \qquad (\rho_f = \rho). \tag{19.73}$$

428 Mechanics and Physics of Structured Media

And the relations

$$R_f = \frac{N_f}{w_f} = \frac{N_f}{n_f} = N \qquad (19.74)$$

hold. Then in the equation for the phase probability (19.63), the term with the chemical potential vanishes.

19.7 Classical heterophase systems

For generality, we have considered above quantum systems. Of course, heterophase systems do not need to be necessarily quantum, but classical systems can also be heterophase. In the present section, we show how the theory is applied to classical systems.

Classical N particle systems are characterized by the position coordinates and momentum variables

$$q = \{\mathbf{q}_1, \mathbf{q}_2, \ldots, \mathbf{q}_N\}, \qquad p = \{\mathbf{p}_1, \mathbf{p}_2, \ldots, \mathbf{p}_N\}. \qquad (19.75)$$

For concreteness, three-dimensional spaces of position and momentum coordinates are kept in mind, so that there are in total $6N$ variables in the *space of microstates* $\{q, p\}$. In that space, a measure $\mu(q, p)$ is given, making it a *measurable phase space*, or simply a *phase space*

$$\mathbb{G} = \{q, p, \mu(q, p)\}. \qquad (19.76)$$

Classical mechanics usually is equipped with the differential measure

$$d\mu(q, p) = \frac{dq\,dp}{N!(2\pi\hbar)^{3N}}. \qquad (19.77)$$

Recall that the Planck constant $\hbar$ is included here in order to make the measure dimensionless, which is necessary for making dimensionless the effective number of states

$$W(E) = \int_{H(q,p)<E} d\mu(q, p)$$

in the phase volume bounded by the energy surface corresponding to the energy E. This, in turn, is required for making dimensionless the expression under the logarithm in the Boltzmann formula for entropy

$$S = k_B \ln W(E).$$

The use of $\hbar$ for the purpose of making dimensionless the phase-space measure and the effective number of states is dictated by the necessity of guaranteeing smooth transition between quantum and classical statistics. This inclusion of $\hbar$ in classical statistical mechanics is the standard commonly employed method

Statistical theory of structures with extended defects Chapter | 19 **429**

that does not influence the expressions of either observable quantities or thermodynamic characteristics (see textbooks, e.g. [27–29]), but correctly defines the Boltzmann entropy.

Suppose that the system volume V is filled by a mesoscopic mixture of several thermodynamic phases enumerated by $f = 1, 2, \ldots$ and that are distinguished by different order parameters or order indices [30,31]. The spatial location of these phases is characterized by the manifold indicator functions $\xi_f(\mathbf{r})$, defined in Eq. (19.2), for which

$$\int \xi_f(\mathbf{r})d\mathbf{r} = V_f, \qquad \sum_f V_f = V. \tag{19.78}$$

For each phase, there exists a probability density $\rho_f(q, p, \xi_f)$ normalized as

$$\int \rho_f(q, p, \xi_f)d\mu(q, p) = 1. \tag{19.79}$$

The phase space, complemented by the probability density, composes a *statistical ensemble*

$$\mathbb{E}_f = \left\{ \mathbb{G}, \rho_f(q, p, \xi_f) \right\}. \tag{19.80}$$

On a statistical ensemble (19.80), the representatives of observables $A_f(q, p, \xi_f)$ are defined, whose averages yield the observable quantities

$$\langle A_f \rangle = \int A_f(q, p, \xi_f)\rho_f(q, p, \xi_f)\mathcal{D}\xi d\mu(q, p). \tag{19.81}$$

The statistical ensemble of the whole system is the Cartesian product

$$\mathbb{E} = \mathsf{x}_f \mathbb{E}_f = \{\mathbb{G}, \rho(q, p, \xi)\}, \tag{19.82}$$

with the density distribution

$$\rho(q, p, \xi) = \mathsf{x}_f \rho_f(q, p, \xi_f). \tag{19.83}$$

The representatives of observables for the system are

$$A(q, p, \xi) = \bigoplus_f A_f(q, p, \xi), \tag{19.84}$$

so that the system observable quantities read as

$$\langle A \rangle = \int A(q, p, \xi)\rho(q, p, \xi)\mathcal{D}\xi d\mu(q, p). \tag{19.85}$$

For example, the system Hamiltonian, similarly to Eq. (19.5), is

$$H(q, p, \xi) = \bigoplus_f H_f(q, p, \xi_f), \tag{19.86}$$

430 Mechanics and Physics of Structured Media

in which the phase-replica Hamiltonians are

$$H_f(q, p, \xi_f) = \sum_{i=1}^{N} \xi_f(\mathbf{r}_i) \left[\frac{\mathbf{p}_i^2}{2m} + U(\mathbf{r}_i) \right] + \frac{1}{2} \sum_{i \neq j}^{N} \xi_f(\mathbf{r}_i) \xi_f(\mathbf{r}_j) \Phi(\mathbf{r}_i - \mathbf{r}_j).$$

(19.87)

The density distribution of the system is normalized,

$$\int \rho(q, p, \xi) \mathcal{D}\xi \, d\mu(q, p) = 1.$$

(19.88)

The average system energy is

$$\int \rho(q, p, \xi) H(q, p, \xi) \mathcal{D}\xi \, d\mu(q, p) = E.$$

(19.89)

The number of particles in the system, N, is fixed.

The information functional acquires the form

$$I[\rho] = \int \rho(q, p, \xi) \ln \frac{\rho(q, p, \xi)}{\rho_0(q, p, \xi)} \mathcal{D}\xi \, d\mu(q, p) +$$
$$+ \alpha \left[\int \rho(q, p, \xi) \mathcal{D}\xi \, d\mu(q, p) - 1 \right]$$
$$+ \beta \left[\int \rho(q, p, \xi) H(q, p, \xi) \mathcal{D}\xi \, d\mu(q, p) - E \right].$$

(19.90)

The minimization of this functional, in the case of no preliminary information, gives

$$\rho(q, p, \xi) = \frac{1}{Z} \exp\{-\beta H(q, p, \xi)\},$$

(19.91)

with the partition function

$$Z = \int \exp\{-\beta H(q, p, \xi)\} \mathcal{D}\xi \, d\mu(q, p).$$

(19.92)

After averaging over phase configurations, as is described in the previous sections, we come to the renormalized Hamiltonian

$$H(q, p, w) = \bigoplus_f H_f(q, p, w_f),$$

(19.93)

with the phase-replica terms

$$H_f(q, p, w_f) = w_f \sum_{i=1}^{N} \left[\frac{\mathbf{p}_i^2}{2m} + U(\mathbf{r}_i) \right] + \frac{w_f^2}{2} \sum_{i \neq j}^{N} \Phi(\mathbf{r}_i - \mathbf{r}_j).$$

(19.94)

Respectively, the functions representing observable quantities become

$$A(q, p, w) = \bigoplus_f A_f(q, p, w_f).$$ (19.95)

The distribution function reads as

$$\rho(q, p, w) = x_f \rho_f(q, p, w_f),$$ (19.96)

whose factors are

$$\rho_f(q, p, w_f) = \frac{1}{Z_f} \exp\left\{-\beta H_f(q, p, w_f)\right\},$$ (19.97)

with the normalization

$$\int \rho_f(q, p, w_f) d\mu(q, p) = 1$$ (19.98)

and the partition functions

$$Z_f = \int \exp\left\{-\beta H_f(q, p, w_f)\right\} d\mu(q, p).$$ (19.99)

Integrating over momenta gives

$$Z_f = \left(\frac{mk_B T}{2\pi \hbar w_f}\right)^{3N/2} \int \exp\left\{-\frac{w_f^2}{2k_B T} \sum_{i \neq j}^N \Phi(\mathbf{r}_i - \mathbf{r}_j) - \frac{w_f}{k_B T} \sum_{i=1}^N U(\mathbf{r}_i)\right\} \frac{dq}{N!}.$$ (19.100)

The system free energy takes the form

$$F = \sum_f F_f = F(w),$$ (19.101)

with the terms

$$F_f = -T \ln Z_f.$$ (19.102)

The phase probabilities are the minimizers of the free energy,

$$F = \text{abs min } F(w).$$ (19.103)

The observable quantities are the averages

$$\langle A \rangle = \sum_f \langle A_f \rangle,$$ (19.104)

where

$$\langle A_f \rangle = \int \rho_f(q, p, w_f) A_f(q, p, w_f) d\mu(q, p).$$ (19.105)

432 Mechanics and Physics of Structured Media

19.8 Quasiaverages in classical statistics

In order to distinguish different thermodynamic phases of quantum systems, there are several methods, such as the Frenkel [32], method of restricted phase space that is a classical analog of the Brout [33] method of restricted trace for quantum systems. The idea of these methods is to integrate not over all phase space, that is, over the whole range of spatial coordinates and momentum variables, but to limit the integration over a restricted region of the phase space, such that would provide the description for the required thermodynamic phase. Details of this approach can be found in the review article [7].

In a more general picture, it is possible to define *weighted phase spaces*, following the idea of introducing *weighted Hilbert spaces* [7,10]. For this purpose, the differential measure of a phase space is weighted with an auxiliary distribution $\varphi_f(q, p)$, so that this measure becomes

$$d\mu_f(q, p) = \frac{dq\,dp}{N!(2\pi\hbar)^{3N}} \varphi_f(q, p). \qquad (19.106)$$

The auxiliary distribution weights the points of the phase space so that to obtain the description characterizing the f-th thermodynamic phase. The weighted phase space for an f-th phase is

$$\mathbb{G}_f = \{q, p, \mu_f(q, p)\}. \qquad (19.107)$$

The phase space of a heterophase classical system takes the form

$$\mathbb{G} = \mathsf{x}_f \mathbb{G}_f. \qquad (19.108)$$

In the case where the auxiliary distribution is either absent or a constant, the method of weighted phase space reduces to the method of restricted phase space.

Technically, the selection of a phase space, required for a correct description of a needed thermodynamic phase, can be done by imposing constraints, such as symmetry breaking, on the corresponding distribution (19.97). A convenient way of symmetry breaking is by introducing infinitesimal sources, as was mentioned by Kirkwood [34] and developed by Bogolubov [35–37] into the method of quasiaverages.

Let us illustrate this for the case of distinguishing a periodic crystalline phase from a uniform disordered phase, being based on the Hamiltonians (19.93) and (19.94). For concreteness, let us label the periodic crystalline phase by $f = 1$, while the uniform phase by $f = 2$. As an order characteristic, one can accept the particle density equipped with the related symmetry properties. For the mixture of two phases, the representative of the observable density is

$$\hat{\rho}(\mathbf{r}, \xi) = \hat{\rho}_1(\mathbf{r}, \xi_1) \bigoplus \hat{\rho}_2(\mathbf{r}, \xi_2), \qquad (19.109)$$

where

$$\hat{\rho}_f(\mathbf{r}, \xi_f) = \sum_{i=1}^{N} \xi_f(\mathbf{r}) \delta(\mathbf{r} - \mathbf{r}_i). \tag{19.110}$$

After averaging over phase configurations, we have

$$\hat{\rho}(\mathbf{r}, w) = \hat{\rho}_1(\mathbf{r}, w_1) \bigoplus \hat{\rho}_2(\mathbf{r}, w_2), \tag{19.111}$$

with

$$\hat{\rho}_f(\mathbf{r}, w_f) = w_f \sum_{i=1}^{N} \delta(\mathbf{r} - \mathbf{r}_j). \tag{19.112}$$

The observable particle density reads as

$$\rho(\mathbf{r}, w) = \langle \hat{\rho}(\mathbf{r}, w) \rangle = \rho_1(\mathbf{r}, w_1) + \rho_2(\mathbf{r}, w_2), \tag{19.113}$$

where

$$\rho_f(\mathbf{r}, w_f) = \int \rho_f(q, p, w_f) \hat{\rho}_f(\mathbf{r}, w_f) d\mu(q, p). \tag{19.114}$$

Integrating out the momentum variables results in the expression

$$\rho_f(\mathbf{r}, w_f) = w_f \int \sum_{i=1}^{N} \delta(\mathbf{r} - \mathbf{r}_i) g_f(q, w_f) \frac{dq}{N!}, \tag{19.115}$$

in which

$$g_f(q, w_f) = \frac{1}{Q_f} \exp \left\{ -\frac{w_f^2}{2k_B T} \sum_{i \neq j}^{N} \Phi(\mathbf{r}_i - \mathbf{r}_j) - \frac{w_f}{k_B T} \sum_{i=1}^{N} U(\mathbf{r}_i) \right\}. \tag{19.116}$$

In the absence of any external potential, when $U(\mathbf{r}) \equiv 0$, the integral in the right-hand side of Eq. (19.115) gives a constant value, which is okay for the uniform phase but is not suitable for the crystalline phase. To overcome this problem in the case of a crystalline phase, it is possible to set

$$U(\mathbf{r}) = \varepsilon U_L(\mathbf{r}), \tag{19.117}$$

assuming a lattice potential periodic over the appropriate crystalline lattice,

$$U_L(\mathbf{r} + \mathbf{a}) = U_L(\mathbf{r}). \tag{19.118}$$

Here ε is a small parameter. To stress that this parameter enters the expression of the related density, we shall denote the latter as $\rho_1(\mathbf{r}, w_1, \varepsilon)$. Also, let us recall

434 Mechanics and Physics of Structured Media

the definition of the thermodynamic limit

$$N \to \infty, \qquad V \to \infty, \qquad \frac{N}{V} \to const. \qquad (19.119)$$

The correct density, periodic over the crystalline lattice, is given by the limiting procedure producing the quantity

$$\rho_1(\mathbf{r}, w_1) = \lim_{\varepsilon \to 0} \lim_{N \to \infty} \rho_1(\mathbf{r}, w_1, \varepsilon) = \rho_1(\mathbf{r} + \mathbf{a}, w_1) \qquad (19.120)$$

called a quasiaverage [35–37]. At the same time, by setting $U_L(\mathbf{r}) \equiv 0$, one obtains a uniform density

$$\rho_2(\mathbf{r}, w_2) = \rho_2 = const \qquad (U(\mathbf{r}) \equiv 0) \qquad (19.121)$$

corresponding to a disordered phase.

Instead of the double limiting procedure (19.120), it is possible to define a single limiting procedure by employing the source

$$U(\mathbf{r}) = \frac{1}{N^\gamma} U_L(\mathbf{r}) \qquad (0 < \gamma < 1) \qquad (19.122)$$

called thermodynamic quasiaverage [38,39].

In any case, the introduction of an auxiliary potential, is equivalent to the definition of a weighted phase space with the auxiliary distribution

$$\varphi_1(q, \varepsilon) = \exp\left\{ -\frac{w_f}{k_B T} \sum_{i=1}^{N} \varepsilon U_L(\mathbf{r}_i) \right\}. \qquad (19.123)$$

19.9 Surface free energy

When on microscopic level, there is a spatial phase separation, then on macroscopic level, there appears the concept of the surface free energy and the related thermodynamic quantities. This also happens in the case of the mesoscopic phase separation.

The surface free energy is defined [40–44] as the difference between the actual free energy of the system and the sum of the free energies of macroscopically separated pure Gibbs phases,

$$F_{sur} = F(w) - F_G. \qquad (19.124)$$

Here the free energy of a heterophase system with mesoscopic nanosize phase separation in two intermixed phases, according to the above theory, has the form

$$F(w) = F_1(w_1) + F_2(w_2). \qquad (19.125)$$

Statistical theory of structures with extended defects Chapter | 19 **435**

But the separation into two pure Gibbs phases, occupying the volumes V_1 and V_2, leads to the free energy

$$F_G = F_1^G + F_2^G = w_1 F_1(1) + w_2 F_2(1). \tag{19.126}$$

In this way, the surface free energy for a heterophase mixture of two phases is

$$F_{sur} = F_1(w_1) + F_2(w_2) - w_1 F_1(1) - w_2 F_2(1). \tag{19.127}$$

The Gibbs macroscopic mixture of pure phases (19.126) corresponds to a linear combination of the related free energies of pure phases. However, the effective free energy of a mesoscopic mixture (19.125) is not a linear combination of the pure-phase free energies. As has been shown in the above sections, the free energy of a heterophase mixture is only by the form looks as a sum of two terms. However these terms correspond not to pure phases, but to effective renormalized expressions nonlinearly depending on phase probabilities.

19.10 Crystal with regions of disorder

To illustrate the above theory, let us consider the model of a crystalline solid with nanosize regions of disorder. As examples, we can keep in mind solids with pores and cracks [45–47], crystals with dislocations [16–18], optical lattices with regions of broken periodicity [48], crystals with amorphous inclusions [49], and quantum crystals with vacancy clusters [50–53].

The ordered and disordered phases are distinguished by their densities

$$\rho_f \equiv \frac{N_f}{V_f} = \frac{1}{V} \int \langle \psi_f^\dagger(\mathbf{r}) \psi_f(\mathbf{r}) \rangle d\mathbf{r}, \tag{19.128}$$

so that the density of the ordered phase ρ_1 is larger than the density ρ_2 of the disordered phase:

$$\rho_1 > \rho_2. \tag{19.129}$$

Following the theory expounded above, the renormalized grand Hamiltonian of a two phase mixture has the form

$$\tilde{H} = H_1 \bigoplus H_2, \tag{19.130}$$

with the replica Hamiltonians

$$H_f = w_f \int \psi_f^\dagger(\mathbf{r}) \left[\hat{H}_L(\mathbf{r}) - \mu \right] \psi_f(\mathbf{r}) d\mathbf{r} +$$

$$+ \frac{w_f^2}{2} \int \psi_f^\dagger(\mathbf{r}) \psi_f^\dagger(\mathbf{r}') \Phi(\mathbf{r} - \mathbf{r}') \psi_f(\mathbf{r}') \psi_f(\mathbf{r}) d\mathbf{r} d\mathbf{r}', \tag{19.131}$$

436 Mechanics and Physics of Structured Media

where

$$\hat{H}_L(\mathbf{r}) = -\frac{\nabla^2}{2m} + U(\mathbf{r}).$$

We keep in mind a solid state with well localized particles, both in the ordered crystalline phase as well as in the disordered phase. Therefore the field operators can be expanded over localized orbitals φ_{nj} as

$$\psi_f(\mathbf{r}) = \sum_{nj} e_{jf} c_{nj} \varphi_{nj}(\mathbf{r}), \tag{19.132}$$

where the index $j = 1, 2, \ldots, N_L$ enumerates lattice sites, n is the index of quantum states, c_{nj} is an annihilation operator of a particle in the state n at the lattice site j, and e_{jf} equals one or zero depending on whether the site j is occupied or free.

We assume that particles are well localized in their lattice sites, so that their hopping between different sites can be neglected, and that each lattice site can be occupied not more than by one particle. These conditions imply the unipolarity properties

$$\sum_n c_{nj}^{\dagger} c_{nj} = 1, \qquad c_{mj} c_{nj} = 0. \tag{19.133}$$

The absence of hopping between the lattice sites means that only diagonal matrix elements survive,

$$E_0 \equiv \langle nj|\hat{H}_L|jn\rangle, \qquad \Phi_{ij} \equiv \langle ni, nj|\Phi|jn, in\rangle. \tag{19.134}$$

The constant term E_0 can be included in the chemical potential. As a result, substituting the field-operator expansion (19.132) into Hamiltonian (19.131) yields

$$H_f = \frac{w_f^2}{2} \sum_{i \neq j}^{N_L} \Phi_{ij} e_{if} e_{jf} - \mu w_f \sum_{j=1}^{N_L} e_{jf}. \tag{19.135}$$

Resorting to the canonical transformation

$$e_{jf} = \frac{1}{2} + S_{jf}^z \qquad (e_{jf} = 0, 1),$$

$$S_{jf}^z = e_{jf} - \frac{1}{2} \qquad \left(S_{jf}^z = \pm \frac{1}{2}\right), \tag{19.136}$$

the above Hamiltonian can be rewritten in the pseudospin representation

$$H_f = \frac{N_L}{8}\left(w_f^2\Phi - 4w_f\mu\right) + \frac{1}{2}\left(w_f^2\Phi - 2w_f\mu\right)\sum_{j=1}^{N_L} S_{jf}^z +$$

$$+ \frac{w_f^2}{2}\sum_{i\neq j}^{N_L}\Phi_{ij}S_{if}^z S_{jf}^z, \tag{19.137}$$

in which the parameter

$$\Phi \equiv \frac{1}{N_L}\sum_{i\neq j}^{N_L}\Phi_{ij} > 0 \tag{19.138}$$

has to be positive in view of the stability condition (19.66).

We use the mean-field approximation

$$S_{if}^z S_{jf}^z = S_{if}^z\langle S_{jf}^z\rangle + \langle S_{if}^z\rangle S_{jf}^z - \langle S_{if}^z\rangle\langle S_{jf}^z\rangle \tag{19.139}$$

and introduce the notation

$$s_f \equiv 2\langle S_{jf}^z\rangle = \frac{2}{N_L}\sum_{j=1}^{N_L}\langle S_{jf}^z\rangle. \tag{19.140}$$

Then we have the relations

$$s_f = 2\langle e_{jf}\rangle - 1, \qquad \langle e_{jf}\rangle = \frac{1+s_f}{2}. \tag{19.141}$$

The density of the f-th phase (19.128) has the form

$$\rho_f = \frac{1}{V}\sum_{j=1}^{N_L}\langle e_{jf}\rangle. \tag{19.142}$$

It is convenient to define the dimensionless density fractions

$$x_f \equiv \frac{1}{N_L}\sum_{j=1}^{N_L}\langle e_{jf}\rangle. \tag{19.143}$$

Introducing the lattice filling factor

$$\nu \equiv \frac{N}{N_L} \tag{19.144}$$

transforms the density fractions (19.143) into

$$x_f = \frac{\rho_f}{\rho}\nu = \frac{1+s_f}{2}. \tag{19.145}$$

438 Mechanics and Physics of Structured Media

For the grand thermodynamic potential we obtain

$$\frac{\Omega}{N_L} = \frac{1}{8} \sum_f w_f^2 (1 - s_f^2) \Phi - 2T \ln 2 - \frac{\mu}{2} - $$
$$- T \sum_f \ln \cosh \left[\frac{w_f^2 (1 + s_f) \Phi - 2 w_f \mu}{4T} \right]. \tag{19.146}$$

Minimizing the thermodynamic potential with respect to s_f gives

$$s_f = \tanh \left[\frac{2 w_f \mu - w_f^2 (1 + s_f) \Phi}{4T} \right]. \tag{19.147}$$

And minimizing with respect to w_f, under the normalization condition

$$w_1 + w_2 = 1, \tag{19.148}$$

results in the equation

$$\frac{\mu}{\Phi} = \frac{w_1 (1 + s_1)^2 - w_2 (1 + s_2^2)}{2(s_1 - s_2)}. \tag{19.149}$$

Using the relation

$$\frac{N_f}{N_L} = w_f x_f = \frac{w_f}{2}(1 + s_f), \tag{19.150}$$

for the filling factor we have

$$\nu = \frac{N}{N_L} = \frac{N_1 + N_2}{N_L} = \frac{1}{2} \sum_f w_f (1 + s_f). \tag{19.151}$$

From the latter it follows

$$w_1 = \frac{2\nu - 1 - s_2}{s_1 - s_2}, \qquad w_2 = \frac{2\nu - 1 - s_1}{s_2 - s_1}. \tag{19.152}$$

The role of the order parameters is played by the densities (19.129) or by the dimensionless density fractions (19.143). According to condition (19.129), distinguishing the ordered and disordered phases,

$$x_1 > x_2. \tag{19.153}$$

Taking into account that the filling factor (19.151) can be written in the form

$$\nu = w_1 x_1 + w_2 x_2 \tag{19.154}$$

and that $0 \le w_f \le 1$, it follows that

$$0 \le x_2 \le \nu \le x_1 \le 1. \tag{19.155}$$

To analyze the system stability, it is necessary to minimize the thermodynamic potential. For convenience, we define the dimensionless free energy

$$F \equiv \frac{\Omega + \mu N}{N_L \Phi} \tag{19.156}$$

and the related dimensionless specific heat and isothermal compressibility

$$C_V = -T \left(\frac{\partial^2 F}{\partial T^2} \right)_V, \qquad \kappa_T = \frac{1}{v^2} \left(\frac{\partial^2 F}{\partial v^2} \right)_T^{-1}. \tag{19.157}$$

Stability conditions require that the latter quantities satisfy the inequalities

$$0 \le C_V < \infty, \qquad 0 \le \kappa_T < \infty. \tag{19.158}$$

19.11 System existence and stability

To prove that the described above solid system presenting a crystal with nanoscopic regions of disorder really can exist and can be stable, we need to accomplish numerical investigations. For the following, it is useful to simplify the notation by setting for the ordered phase

$$x_1 \equiv x, \qquad w_1 \equiv w \tag{19.159}$$

and for the disordered phase

$$x_2 \equiv y, \qquad w_2 \equiv 1 - w. \tag{19.160}$$

Then the inequalities (19.155) become

$$0 \le y \le v \le x \le 1. \tag{19.161}$$

The phase probabilities reduce to

$$w = \frac{v - y}{x - y}, \qquad 1 - w = \frac{x - v}{x - y}. \tag{19.162}$$

For the chemical potential (19.149), we get

$$\frac{\mu}{\Phi} = \frac{w(x^2 + y^2) - y^2}{x - y}. \tag{19.163}$$

Also, let us measure temperature in units of Φ. Then the free energy (19.156) takes the form

$$F = \frac{1}{2} \left[w^2 x (1 - x) + (1 - w)^2 y (1 - y) \right] + \\ + \frac{T}{2} \ln \left[x (1 - x) y (1 - y) \right] + \left(v - \frac{1}{2} \right) \frac{w x^2 - (1 - w)^2 y^2}{x - y}. \tag{19.164}$$

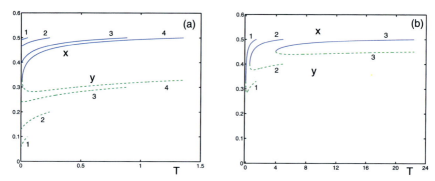

FIGURE 19.2 Dimensionless densities of the ordered phase $x(t)$ (solid line) and of the disordered phase $y(t)$ (dashed line) as functions of dimensionless temperature for different lattice filling factors: (a) $\nu = 0.1$ (line 1), $\nu = 0.2$ (line 2), $\nu = 0.3$ (line 3), and $\nu = 0.329$ (line 4); (b) $\nu = 0.33$ (line 1), $\nu = 0.4$ (line 2), and $\nu = 0.45$ (line 3).

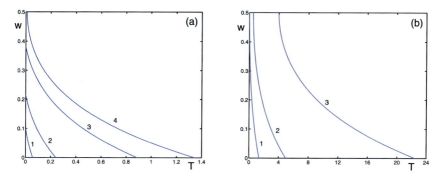

FIGURE 19.3 Geometric weight (geometric probability) of the ordered phase $w(T)$ as a function of dimensionless temperature for different lattice filling factors: (a) $\nu = 0.1$ (line 1), $\nu = 0.2$ (line 2), $\nu = 0.3$ (line 3), and $\nu = 0.329$ (line 4); (b) $\nu = 0.33$ (line 1), $\nu = 0.4$ (line 2), and $\nu = 0.45$ (line 3).

And for the order parameters (19.147), we obtain

$$2x = 1 + \tanh\left\{\frac{wy[wx - (1-w)y]}{2T(x-y)}\right\},$$
$$2y = 1 + \tanh\left\{\frac{(1-w)x[wx - (1-w)y]}{2T(x-y)}\right\}. \quad (19.165)$$

The system of Eqs. (19.162) and (19.165) is solved numerically (see Figs. 19.2 and 19.3) under restriction (19.161) and stability conditions (19.158). The solution exists for the filling factor in the interval

$$0 < \nu < \frac{1}{2} \quad (19.166)$$

TABLE 19.1 Lower and upper nucleation temperatures (in units of Φ) for different filling factors.

ν	T_n	T_n^*
0.1	0	0.0157
0.2	0	0.240
0.3	0	0.885
0.329	0.01	1.344
0.330	0.0125	1.371
0.40	0.521	4.933
0.45	0.958	22.425

and in the temperature range

$$T_n < T < T_n^*. \tag{19.167}$$

The temperature $T_n \geq 0$ shows where the heterophase system appears, because of which this is called the lower nucleation temperature. When temperature rises, the heterophase state disappears at the upper nucleation temperature T_n^* where the probability $w(T_n^*)$ becomes zero, which gives

$$T_n^* = \frac{\nu}{(1 - 2\nu)[\ln(1 - \nu) - \ln \nu]}. \tag{19.168}$$

Table 19.1 shows the nucleation temperatures in-between which the crystal state with regions of disorder can exist.

In the region of existence of the heterophase state, the free energy (19.164) is lower than the free energy of the pure ordered state, where $w \equiv 1$. Therefore such a system with the regions of disorder, describing extended defects, in the intervals of the filling factor (19.166) and temperature (19.167), corresponds to a stable heterophase system.

19.12 Conclusion

The description of structures with randomly distributed extended defects is notoriously difficult, since such materials are heterophase and strongly nonuniform. However, the description can be simplified when one is interested in the averaged properties of a sample as a whole. In that case, it is possible to develop an approach based on the averaging over phase configurations. As a result, it is possible to reduce the consideration to the study of the replicas of the separate phases, described by effective renormalized Hamiltonians. As a price for this simplification, one comes to the necessity of dealing with these more complicated effective Hamiltonians and to the need of calculating the geometric weights, or geometric probabilities of the phases. Nevertheless, it is worth

442 Mechanics and Physics of Structured Media

paying the price because the overall problem becomes treatable, while the calculational complications are rather minor. The approach is illustrated by a lattice model with regions of disorder. It is shown that such nonideal structures can exist as stable statistical systems.

It is important to emphasize that the developed approach can describe metastable as well as unstable states of matter. The system of equations, defining the order parameters and phase probabilities, is usually strongly nonlinear and displays several solutions. The solutions, for which the stability conditions, such as the positivity of specific heat, compressibility, and of other available susceptibilities, are not valid, correspond to unstable states. The solutions for which these stability conditions are satisfied, but the related free energy does not define an absolute minimum, describe metastable states. The solutions, leading to the lowest free energy and satisfying all stability conditions, correspond to stable states. Such a situation happens for the example treated in Sections 19.10 and 19.11. This is why, it has been stressed that we need to choose those solutions that satisfy the stability conditions (19.158) and minimize the free energy (19.156). Only these solutions describe stable states, while other solutions, for which some of the stability conditions are broken, or free energy is not minimal, correspond to unstable or metastable states.

References

[1] G.E.R. Schulze, Metallophysics, Academic, Berlin, 1967.
[2] J.M. Ziman, Models of Disorder, Cambridge University, Cambridge, 1979.
[3] E.I. Grigolyuk, L.A. Filshtinsky, Regular Piecewise Homogeneous Structures with Defects, Fizmatgiz, Moscow, 1994.
[4] J.P. Hirth, J. Lothe, Theory of Dislocations, Wiley, New York, 1982.
[5] D. Hull, D. Bacon, Introduction to Dislocations, Elsevier, Oxford, 2001.
[6] J.W. Gibbs, Collected Works, Longmans, New York, 1928.
[7] V.I. Yukalov, Phase transitions and heterophase fluctuations, Phys. Rep. 208 (1991) 395–489.
[8] V.I. Yukalov, Mesoscopic phase fluctuations: general phenomenon in condensed matter, Int. J. Mod. Phys. B 17 (2003) 2333–2358.
[9] N. Bourbaki, Théorie des Ensembles, Hermann, Paris, 1958.
[10] V.I. Yukalov, Systems with symmetry breaking and restoration, Symmetry 2 (2010) 40–68.
[11] S. Kullback, R.A. Leibler, On information and sufficiency, Ann. Math. Stat. 22 (1951) 79–86.
[12] S. Kullback, Information Theory and Statistics, Wiley, New York, 1959.
[13] A.H. Cottrell, Dislocations and Plastic Flow in Crystals, Oxford University, London, 1953.
[14] J. Friedel, Dislocations, Pergamon, Oxford, 1967.
[15] J. Hirth, J. Lothe, Theory of Dislocations, McGraw Hill, New York, 1968.
[16] J.S. Langer, Thermal effects in dislocation theory, Phys. Rev. E 94 (2016) 063004.
[17] J.S. Langer, Thermodynamic theory of dislocation-enabled plasticity, Phys. Rev. E 96 (2017) 053005.
[18] J.S. Langer, Statistical thermodynamics of crystal plasticity, J. Stat. Phys. 175 (2019) 531–541.
[19] V.I. Yukalov, Theory of cold atoms: basics of quantum statistics, Laser Phys. 23 (2013) 062001.
[20] V.I. Yukalov, Theory of melting and crystallization, Phys. Rev. B 32 (1985) 436–446.
[21] V.I. Yukalov, Effective Hamiltonians for systems with mixed symmetry, Physica A 136 (1986) 575–587.

[22] V.I. Yukalov, Renormalization of quasi-Hamiltonians under heterophase averaging, Phys. Lett. A 125 (1987) 95–100.

[23] V.I. Yukalov, Procedure of quasiaveraging for heterophase mixtures, Physica A 141 (1987) 352–374.

[24] V.I. Yukalov, Lattice mixtures of fluctuating phases, Physica A 144 (1987) 369–389.

[25] V.I. Yukalov, Nonequilibrium representative ensembles for isolated quantum systems, Phys. Lett. A 375 (2011) 2797–2801.

[26] V.I. Yukalov, Basics of Bose-Einstein condensation, Phys. Part. Nucl. 42 (2011) 460–513.

[27] K. Huang, Statistical Mechanics, Wiley, New York, 1963.

[28] R. Kubo, Statistical Mechanics, North-Holland, Amsterdam, 1965.

[29] A. Isihara, Statistical Physics, Academic, New York, 1971.

[30] V.I. Yukalov, Matrix order indices in statistical mechanics, Physica A 310 (2002) 413–434.

[31] V.I. Yukalov, Order indices and entanglement production in quantum systems, Entropy 22 (2020) 565.

[32] J.I. Frenkel, Kinetic Theory of Liquids, Clarendon, Oxford, 1946.

[33] R. Brout, Phase Transitions, Benjamin, New York, 1965.

[34] J.G. Kirkwood, Quantum Statistics and Cooperative Phenomena, Gordon and Breach, New York, 1965.

[35] N.N. Bogolubov, Lectures on Quantum Statistics, vol. 1, Gordon and Breach, New York, 1967.

[36] N.N. Bogolubov, Lectures on Quantum Statistics, vol. 2, Gordon and Breach, New York, 1970.

[37] N.N. Bogolubov, Quantum Statistical Mechanics, World Scientific, Singapore, 2015.

[38] V.I. Yukalov, Statistical theory of heterophase fluctuations, Physica A 108 (1981) 402–416.

[39] V.I. Yukalov, Method of thermodynamic quasiaverages, Int. J. Mod. Phys. B 5 (1991) 3235–3253.

[40] S. Ono, S. Kondo, Molecular Theory of Surface Tension in Liquids, Springer, Berlin, 1960.

[41] A.I. Rusanov, Thermodynamics of solid surfaces, Surf. Sci. Rep. 23 (1996) 173–247.

[42] A.I. Rusanov, Surface thermodynamics revisited, Surf. Sci. Rep. 37 (2005) 111–239.

[43] S. Kjelstrup, D. Bedeaux, Non-Equilibrium Thermodynamics of Heterogeneous Systems, World Scientific, Singapore, 2008.

[44] D. Bedeaux, S. Kjelstrup, Fluid-fluid interfaces of multi-component mixtures in local equilibrium, Entropy 20 (2018) 250.

[45] V.I. Yukalov, Properties of solids with pores and cracks, Int. J. Mod. Phys. B 3 (1989) 311–326.

[46] V.I. Yukalov, Properties of crystals with local symmetry breaking, in: W. Lulek, B. Lulek, M. Mucha (Eds.), Symmetry and Structural Properties of Condensed Matter, World Scientific, Singapore, 1991.

[47] V.I. Yukalov, Chaotic lattice-gas model, Physica A 213 (1995) 482–499.

[48] V.I. Yukalov, E.P. Yukalova, Optical lattice with heterogeneous atomic density, Laser Phys. 25 (2015) 035501.

[49] A.S. Bakai, Polycluster Amorphous Solids, Sinteks, Kharkov, 2013.

[50] M. Boninsegni, A.B. Kuklov, L. Pollet, N.V. Prokof'ev, B.V. Svistunov, M. Troyer, Fate of vacancy-induced supersolidity in ^{4}He, Phys. Rev. Lett. 97 (2006) 080401.

[51] P.N. Ma, L. Pollet, M. Troyer, F.C. Zhang, A classical picture of the role of vacancies and interstitials in helium-4, J. Low Temp. Phys. 152 (2008) 156–163.

[52] A.K. Singh, E.S. Penev, B.I. Yakobson, Vacancy clusters in graphane as quantum dots, ACS Nano 4 (2010) 3510–3514.

[53] V.I. Yukalov, Saga of superfluid solids, Physics 2 (2020) 49–66.

Chapter 20

Effective conductivity of 2D composites and circle packing approximations

Roman Czapla and Wojciech Nawalaniec

Institute of Computer Science, Pedagogical University of Cracow, Kraków, Poland

20.1 Introduction

We consider fibrous composites with unidirectional fibers, i.e., two-dimensional (2D) composites represented by a section perpendicular to the axis of fibers displayed in Fig. 20.1. Conductivity and elasticity of fibrous composites was theoretically studied sufficiently well in the case of circular inclusions, see the recent books [9,10] and references therein. Analytical formulas from [9,10] for the effective properties of such composites were derived by means of advanced symbolic computations. These computations yield a set of general formulas for regular and random composites with circular nonoverlapping inclusions where the number of inclusions N per representative cell, the centers of circles a_k and their radii r_k $(k = 1, 2, \ldots, N)$ are given in symbolic form. The computational effectiveness of the analytical formulas was confirmed by applications to $AlSi7/SiCp$ and to other similar composites [13,14] where the structural and macroscopic properties of composites with the huge number $N \sim 10^5$ were analyzed.

The exact and approximate analytical formulas were obtained for the circular [9,10] shapes of inclusions by the method of functional equations. Investigation of an arbitrary shape can be carried out by the generalized alternating method of Schwarz [10, page 59] which coincides with the method of contrast expansion [3]. However, implementation of the method of Schwarz in symbolic form requires huge computer memory for complex shapes of inclusions in high contrast and densely packed composites, whereas symbolic computations for circular inclusions are instantly performed in practice. Numerical computations according to an analytical formula (algorithm) were performed in the above works during hours for $N \sim 10^5$ disks on a standard laptop.

In the present chapter, we develop an alternative method for various shapes based on the circle packing. One can find the theory of optimal packing problems in [16] and works cited therein. We focus our attention on the general

446 Mechanics and Physics of Structured Media

possibility to approximate a domain by a set of nonoverlapping disks and the relation of such an approximation to computation of the effective constants of composites. A similar question was discussed in [15] for classification of random composites in the following way. First, a cluster consisting of about 20 rigidly linked equal disks was introduced. Next, Monte Carlo experiments to locate clusters according to a prescribed probabilistic distribution were performed and the corresponding structural sums were computed. This scheme was applied to composites with various shapes of clusters in order to classify random composites by different clusters. This method was used in [12] to study the collective behavior of bacteria by disks approximation of stadiums.

The present chapter concerns accurate cluster approximations when the number of disks per cluster reaches 241. In order to improve the precision of geometrical approximations we extend the algorithm to disks of different radii. We discuss 2D conductivity (transport) properties equivalent to the antiplane elasticity following [10]. A similar approach was developed in [9] and can be applied to plane elastic problems.

The present chapter is organized as follows. The general method for polydispersed structures of nonoverlapping disks is outlined in Section 20.2. Section 20.3 is devoted to the test example when a singular disk is approximated by a cluster consisting of smaller disks. The known exact formula for the effective conductivity of the regular hexagonal array of disks is compared to approximations obtained for packing of disks. Two methods of approximations by identical and by different disks are considered. Moreover, two variants with the fixed concentration of approximating and approximated domains, and with the inscribed disks are discussed. Sections 20.4 and 20.5 are devoted to application of the method to the checkerboard and to the hexagonal regular array of triangular inclusions. Some concluding remarks are selected in Section 20.6.

20.2 General polydispersed structure of disks

We begin from the most general structures of nonoverlapping disks. Following the homogenization theory we consider a representative cell generated by two fundamental translation vectors ω_1 and ω_2. Let the area of the parallelogram $\mathcal{Q}$ spanned by vectors ω_1 and ω_2 hold unity. These vectors form the doubly periodic lattice $\{m_1\omega_1 + m_2\omega_2\}$ where m_1 and m_2 run over integers. Let centers a_k of nonoverlapping circular inclusions of radii r_k ($k = 1, 2, \ldots, N$) be located in $\mathcal{Q}$. It is convenient to express these centers in terms of complex numbers. Introduce the concentration of inclusions $f = \sum_{k=1}^{N} \pi r_k^2$.

Let the conductivity of inclusions and of matrix be λ_1 and $\lambda = 1$, respectively. The contact between different materials is supposed to be perfect. Introduce the contrast parameter

$$\varrho = \frac{\lambda_1 - 1}{\lambda_1 + 1}. \tag{20.1}$$

We now proceed to discuss the effective conductivity tensor

$$\Lambda = \begin{pmatrix} \lambda_{11} & \lambda_{12} \\ \lambda_{21} & \lambda_{22} \end{pmatrix}. \tag{20.2}$$

The tensor Λ is symmetric, i.e., $\lambda_{12} = \lambda_{21}$. Analytical formulas for Λ are summarized in the book [10]. Its components can be calculated by the series in f

$$\lambda_{11} - i\lambda_{12} = 1 + 2\varrho f \sum_{q=0}^{\infty} B_q \, f^q, \tag{20.3}$$

where i denotes the imaginary unit. An analogous formula for $\lambda_{22} + i\lambda_{21}$ was derived in [10].

A symbolic-numerical algorithm for the coefficients B_q is summarized in [9, Chapter 2]. Its first goal consists in exact expressions of B_q in terms of the structural sums. These sums depend on the differences $a_k - a_j$ for $k, j = 1, 2, \ldots, N$ ($k \neq j$) and on the polydispersity constants

$$f_j = \left(\frac{r_j}{r}\right)^2, \quad j = 1, 2, \ldots, N, \tag{20.4}$$

where for definiteness r is taken as the maximal radius. Let $\eta = \sum_{j=1}^{N} f_j$. Consider the Eisenstein functions $E_k(z)$ of order $k = 1, 2, \ldots$, the special functions related to the classical elliptic functions by Weierstress [10]. Let $\mathcal{C}$ denote the operator of complex conjugation. The structural sum is defined by the following expression

$$
\begin{aligned}
e_{p_1, p_2, p_3, \ldots, p_n} &= \frac{1}{\eta^{1 + \frac{1}{2}(p_1 + \cdots + p_n)}} \\
&\times \sum_{k_0, k_1, \ldots, k_n} f_{k_0}^{t_0} f_{k_1}^{t_1} f_{k_2}^{t_2} \cdots f_{k_n}^{t_n} \\
&\times E_{p_1}(a_{k_0} - a_{k_1}) \mathcal{C} E_{p_2}(a_{k_1} - a_{k_2}) \\
&\times E_{p_3}(a_{k_2} - a_{k_3}) \ldots \mathcal{C}^{n+1} E_{p_n}(a_{k_{n-1}} - a_{k_n}),
\end{aligned}
\tag{20.5}
$$

where $t_0 = 1$, $t_j = p_j - t_{j-1}$ ($j = 1, 2, \ldots, n$), and $t_n = 1$. For instance, $f_{k_0}^{t_0}$ in (20.5) means f_{k_0} raised to power t_0.

The coefficients B_q are found by the symbolic computational scheme

$$
\begin{aligned}
B_0 &= 1, \quad B_1 = \pi^{-1} \varrho e_2, \quad B_2 = \pi^{-2} \varrho^2 e_{2,2}, \\
B_q &= \pi^{-1} \beta B_{q-1}, \quad q = 3, 4, 5 \ldots,
\end{aligned}
\tag{20.6}
$$

448 Mechanics and Physics of Structured Media

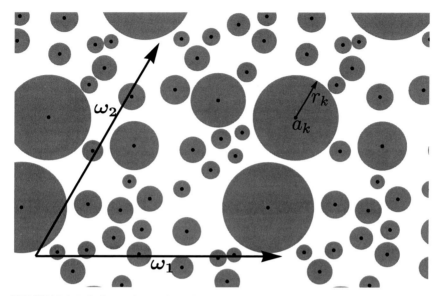

FIGURE 20.1 Polydispersed two-dimensional composite model structured in the two-periodic hexagonal lattice.

where β is the *substitution operator* modifying every structural sum in B_{q-1} according the transformation rule:

$$e_{p_1,p_2,\ldots,p_n} \longmapsto \varrho e_{2,p_1,p_2,\ldots,p_n} - \frac{p_2}{p_1-1} e_{p_1+1,p_2+1,p_3,\ldots,p_n}. \qquad (20.7)$$

The algorithm (20.6)–(20.7) yields explicit formulas for B_q as a finite linear combination of structural sums for any fixed q. Reasonable numerical values of B_q were obtained for $q \sim 20$ and $N \sim 10^3$ with a standard laptop. The details of computations are available in [8] for monodispersed and in [9] for polydispersed composites.

The effective conductivity of macroscopically isotropic 2D composites is determined by the scalar λ_e from equation $\Lambda = \lambda_e I$ where I denotes the unit matrix. It is calculated by the same formula (20.3)

$$\lambda_e = 1 + 2\varrho f \sum_{q=0}^{\infty} B_q f^q. \qquad (20.8)$$

Below, for definiteness we discuss only macroscopically isotropic composites using the notation λ_e for the effective conductivity for shortness.

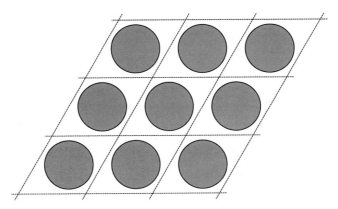

FIGURE 20.2 Hexagonal array of large disks.

20.3 Approximation of hexagonal array of disks

Consider the cell formed by the fundamental translation vectors $\omega_1 = (1, 0)$ and $\omega_2 = \left(\frac{1}{2}\sqrt[4]{\frac{4}{3}}, \frac{\sqrt{3}}{2}\sqrt[4]{\frac{4}{3}}\right)$. In the case $N = 1$ we arrive at the regular hexagonal (triangular) array of circular inclusion displayed in Fig. 20.2. The effective conductivity is given by the exact formula (4.2.29) and the numerical formulas (7.5.44), (7.9.77) from [10]. These known formulas are used in study of the test approximation problem when a large disk of the regular hexagonal array is approximated by various clusters of small disks as shown in Figs. 20.4, 20.5, and 20.8.

We take the contrast parameter $\varrho = 1$ corresponding to the perfectly conducting disks, see (20.1). The concentration $f = 0.5$ is numerically fixed in order to demonstrate numerical dependencies. It can be taken arbitrarily $0 < f < 1$, perhaps with application of the renormalization technique developed in [9,10] for some percolating structures. In the considered case, the effective conductivity of the regular hexagonal array of disks holds $\lambda_e \approx 3.004726$. This value is calculated by the polynomial approximation of the series (20.8) which is written in the numerical form (7.2.6) in the book [10]. The convergence of the partial sums of the series (20.8) to the exact value calculated with (4.2.29) and (7.2.6) from [10] is demonstrated in Fig. 20.3.

We now proceed to construct clusters of small disks approximating the large disk. Two types of approximations by identical and by different disks are considered. In order to study the effectiveness of the method we compare the exact value $\lambda_e \approx 3.004726$ with the estimated values for approximating clusters. We consider two methods of packing approximations:

Monodispersed Method (MM) concerns approximation of the large disk by a cluster being a subset of the hexagonal array of small identical disks;

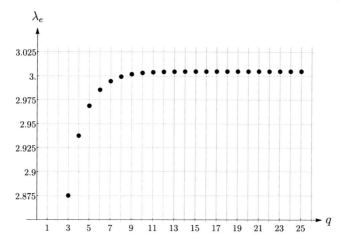

FIGURE 20.3 Approximation of λ_e by polynomials of order q truncating the series (20.8). Data are for the hexagonal array, $f = 0.5$ and $\varrho = 1$.

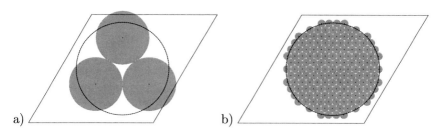

FIGURE 20.4 Examples of the approximation of large disk by MM-A with the fixed concentration $f = 0.5$. The number of disks holds 3 in (a) and 128 in (b).

Polydispersed Method (PM) concerns approximation of the large disk by a cluster of polydispersed small disks.

Every method is applied in two different variants:

Variant A. A cluster has the concentration $f = 0.5$ exactly, see Fig. 20.4 for MM-A.

Variant B. A cluster belongs to the large domain, hence $f < 0.5$, see Fig. 20.5 for MM-B and Fig. 20.8 for PM-B.

Hence, we consider four cases denoted below for shortness as MM-A, PM-A, MM-B, PM-B.

It is worth noting that the maximal concentration of the hexagonal array of small disks used in MM is equal to $\frac{1}{2}\frac{\pi\sqrt{3}}{6} \approx 0.45345$. Disks of different radii are used in PM. Such clusters of small disks achieve larger densities.

The approximated values of λ_e calculated with the accuracy $O(f^{21})$ by MM are compared to the exact value $\lambda_e \approx 3.004726$ in Fig. 20.6. The corresponding

Effective conductivity of 2D composites Chapter | 20 **451**

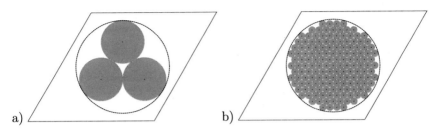

FIGURE 20.5 Approximations of large disk by MM-B. The number of disks holds 3 in (a) and 128 in (b).

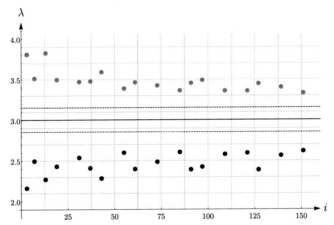

FIGURE 20.6 λ_e calculated with fixed accuracy $O(f^{21})$ by MM and i disks in the cluster. Data are for: the hexagonal array of disks (solid black line), approximation MM-A (gray points), approximation MM-B (black points). Dashed black lines denote 5% error.

results for PM are presented in Fig. 20.7. One can see that the application of PM yields much better approximation than of MM. For that reason PM will be applied in the next sections in order to estimate λ_e of other structures.

Moreover, we observe in Fig. 20.9 the strong dependence of the relative error $e_\lambda = \frac{\Delta \lambda_e}{\lambda_e}$ calculated for approximating clusters on the relative deviation of concentration $e_f = \frac{\Delta f}{f}$. Here, the relative values are computed as the ratio of absolute errors to the fixed values $\lambda_e = 3.004726$ and $f = 0.5$, respectively.

20.4 Checkerboard

The present section is devoted to the famous checkerboard structure displayed in Fig. 20.10. The cell is formed by the fundamental translation vectors $\omega_1 = (1, 0)$ and $\omega_2 = (0, 1)$. The famous Dykhne's formula [4] for the effective conductivity

452 Mechanics and Physics of Structured Media

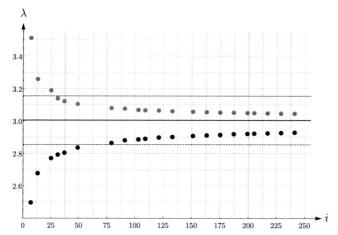

FIGURE 20.7 Approximation of λ_e calculated with fixed accuracy $O(f^{21})$ by PM and i disks in the cluster. Data are for: the hexagonal array of inclusions (solid black line), approximation PM-A (gray points), approximation PM-B (black points). Dashed black lines denote 5% error.

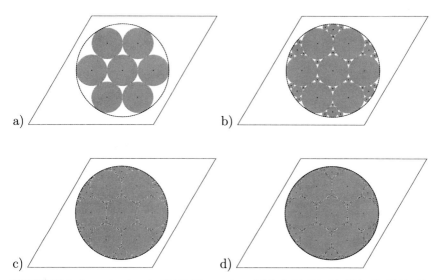

FIGURE 20.8 Approximations by PM-B. The number of disks holds 7 in (a), 37 in (b), 151 in (c), and 241 in (d).

reads in terms of (20.1) as

$$\lambda_e = \sqrt{\lambda_1 \lambda} = \sqrt{\frac{1+\varrho}{1-\varrho}}. \tag{20.9}$$

Effective conductivity of 2D composites Chapter | 20 **453**

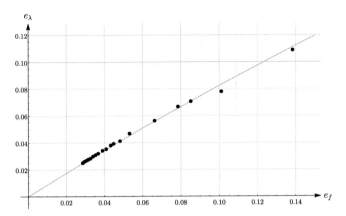

FIGURE 20.9 Dependence of the error e_λ on e_f for $\lambda_e = 3.004726$ and $f = 0.5$.

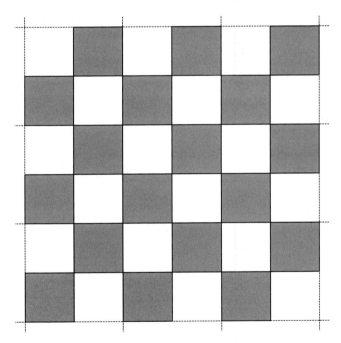

FIGURE 20.10 Infinite checkerboard structure

Its various extensions to four-phase square checkerboard and other similar structures can be found in [1], [2].

Following the above scheme we approximate the checkerboard by various numbers of disks displayed in Fig. 20.11. The corresponding approximations are shown in Figs. 20.12 and 20.13.

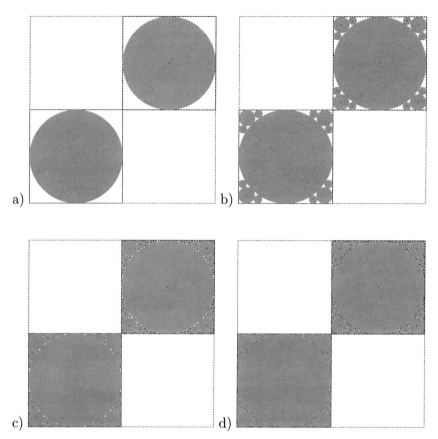

FIGURE 20.11 Approximations of the checkerboard structure by nonoverlapping disks. The number of disks holds 2 in (a), 26 in (b), 194 in (c), and 274 in (d).

20.5 Regular array of triangles

The present section concerns application of the method to the regular array of triangles (Fig. 20.14 and Fig. 20.15) with the fixed concentration $f = 0.5$. The results of calculations for λ_e with accuracy $O(f^{21})$ are presented in Fig. 20.16. We suppose that the interval [3.044, 3.154] includes the real value of λ_e for the regular array of triangle.

20.6 Discussion and conclusions

Analytical exact and approximate formulas for the effective properties of 2D composites with circular nonoverlapping inclusions were derived in [9,10]. Using the observation that any domain can be approximated by circle packing we apply the formulas from [9,10] in order to develop the circle packing method for

Effective conductivity of 2D composites Chapter | 20 **455**

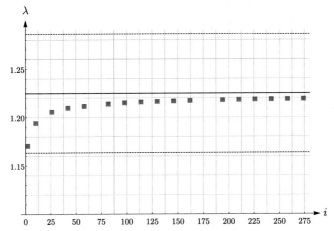

FIGURE 20.12 Approximation of λ_e calculated with fixed accuracy $O(f^{21})$ by PM-B and i disks in the cluster for $\rho = 0.2$. Solid line is for the exact result. Dashed black lines denote 5% error.

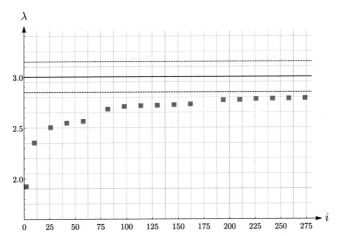

FIGURE 20.13 Approximation of λ_e calculated with fixed accuracy $O(f^{21})$ by PM-B and i disks in the cluster for $\rho = 0.8$. Solid line is for the exact result. Dashed black lines denote 5% error.

composites with inclusions of various shapes. The packing methods by equal disks (MM) and by different disks (PM) are compared and the better results for the PM are observed. Moreover, the packing variants, A and B, with the fixed and increasing concentrations are considered. The expected results that λ_e of the regular array of triangles is greater than λ_e of the hexagonal array of circles for the same concentration and others are established.

The precision of approximations may depend on the value of the contrast parameter (20.1) as demonstrated in Figs. 20.12 and 20.13. This result has to be treated in the following way. In Section 20.5, we consider triangles with acute

456 Mechanics and Physics of Structured Media

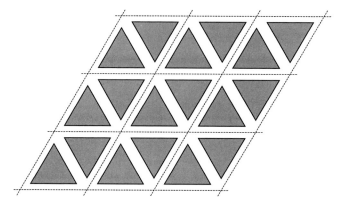

FIGURE 20.14 Regular array of triangles with the concentration $f = 0.5$.

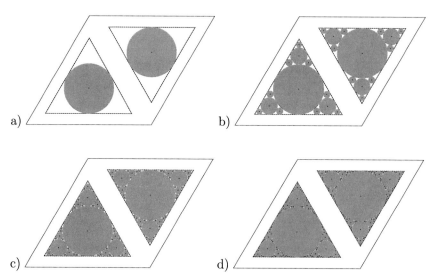

FIGURE 20.15 B-type approximations by disks lying in triangles. The number of disks holds 2 in (a), 26 in (b), 86 in (c), and 188 in (d).

corners and their approximations with smoothed corners. Hence, actually, two different structures are considered with different λ_e for large ϱ. The additional effect of percolation complicates approximation of domains with acute corners by smooth ones. It is known that singular flux occurs in small gaps between the high contrast phases. One can find asymptotically exact formulas in book [11]. The question of numerical investigations of such structures was systematically studied in [5], [6], [7] by a method of integral equations.

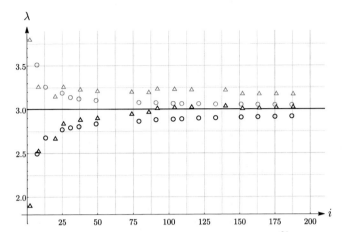

FIGURE 20.16 Approximation of λ_e calculated with fixed accuracy $O(f^{21})$ by PM-B and i disks in the cluster. Data are for: the hexagonal array of inclusions (solid black line), approximation variant A (gray); approximations: variant B (black), regular array of triangles (triangles), hexagonal array (circles).

References

[1] R.V. Craster, Y.V. Obnosov, Four phase periodic composites, SIAM J. Appl. Math. 61 (2001) 1839–1856.
[2] R.V. Craster, Y.V. Obnosov, A three phase tessellation: solution and effective properties, Proc. R. Soc. Lond. A 460 (2004) 1017–1037.
[3] P. Drygaś, V. Mityushev, Contrast expansion method for elastic incompressible fibrous composites, Adv. Math. Phys. 2017 (2017) 4780928, https://doi.org/10.1155/2017/4780928.
[4] A.M. Dykhne, Conductivity of a two-dimensional two-phase system, Sov. Phys. JETP 32 (1971) 63–65.
[5] J. Helsing, A fast and stable solver for singular integral equations on piecewise smooth curves, SIAM J. Sci. Comput. 33 (2011) 153–174.
[6] J. Helsing, The effective conductivity of random checkerboards, J. Comput. Phys. 230 (2011) 1171–1181.
[7] J. Helsing, The effective conductivity of arrays of squares: large random unit cells and extreme contrast ratios, J. Comput. Phys. 230 (2011) 7533–7547.
[8] R. Czapla, W. Nawalaniec, V. Mityushev, Effective conductivity of random two-dimensional composites with circular non-overlapping inclusions, Comput. Mater. Sci. 63 (2012) 118–126.
[9] P. Drygaś, S. Gluzman, V. Mityushev, W. Nawalaniec, Applied Analysis of Composite Media Analytical and Computational Results for Materials Scientists and Engineers, Elsevier, 2020.
[10] S. Gluzman, V. Mityushev, W. Nawalaniec, Computational Analysis of Structured Media, Academic Press, Elsevier, Amsterdam, 2018.
[11] A.A. Kolpakov, A.G. Kolpakov, Capacity and Transport in Contrast Composite Structures: Asymptotic Analysis and Applications, CRC Press / Taylor & Francis, Boca Raton, 2009.
[12] R. Czapla, Random sets of stadiums in square and collective behavior of bacteria, IEEE/ACM Trans. Comput. Biol. Bioinform. (2018).
[13] P. Kurtyka, N. Rylko, Structure analysis of the modified cast metal matrix composites by use of the RVE theory, Arch. Metall. Mater. 58 (2013) 357–360.
[14] P. Kurtyka, N. Rylko, T. Tokarski, A. Wójcicka, A. Pietras, Cast aluminium matrix composites modified with using FSP process – changing of the structure and mechanical properties, Compos. Struct. 133 (2015) 959–967.

458 Mechanics and Physics of Structured Media

[15] W. Nawalaniec, Classifying and analysis of random composites using structural sums feature vector, Proc. R. Soc. Lond. A 1978 (2019) 211–232.

[16] P. Brass, W.O.J. Moser, J. Pach, Research Problems in Discrete Geometry, Springer, New York, 2005.

Chapter 21

Asymptotic homogenization approach applied to Cosserat heterogeneous media

Victor Yanes[a], Federico J. Sabina[b], Yoanh Espinosa-Almeyda[b], José A. Otero[c], and Reinaldo Rodríguez-Ramos[d]

[a]*Faculty of Physics, University of Havana, Havana, Cuba,* [b]*Applied Mathematics and Systems Research Institute, National Autonomous University of Mexico (UNAM), Mexico City, Mexico,* [c]*Tecnologico de Monterrey, School of Engineering and Sciences, Mexico State, Nuevo León, Mexico,* [d]*Faculty of Mathematics and Computer Science, University of Havana, Havana, Cuba*

21.1 Introduction

The first steps in the micropolar theory of elasticity were taken in 1887 by Voigt [1], who incorporated the couple-stresses in addition to the classical stresses and generalized the symmetric classical theory of elasticity to an asymmetric theory. The asymmetric theory was developed by the Cosserat brothers in 1909 [2]. They included three additional degrees of freedom (independent from the displacement field) to describe the microrotations (local reorientation of the microstructure); therefore Cosserat's theory involved six degrees of freedom for every element of the body: three displacements and three microrotations. In 1966, Eringen [3] defined the micropolar elasticity theory after extending the Cosserat linear theory to include the microinertia effects. Since then, the interest to Cosserat's media is growing and many works have been published. An extensive bibliography related to micropolar elasticity theories can be found in Refs. [4–6].

One of the major challenges in the micropolar theory is to define the constitutive parameters that describe the deformation response of composites. Some homogenization works using different methods have been performed to determine the elastic moduli for specific Cosserat structures, for example: In the Ganghoffer's works [7–11], the nonclassical constitutive equations are established and the elastic moduli corresponding to repetitive lattices and trabecular structures are determined. Forest and Sab [12] developed an alternative methodology to the classical homogenization method in order to find the effective properties of 2D Cauchy medium by minimizing the elastic strain energy with

460 Mechanics and Physics of Structured Media

respect to displacement fields using the finite element method and generalizes the scheme to Cosserat materials. Forest et al. [13] applied the multiscale homogenization method for periodic heterogeneous linear elastic Cosserat media. Gorbachev and Emel'yanov [14,15] used the integral formulas for the displacements and rotations as the key of the homogenization process. Li et al. [16] applied a generalized Hill's lemma for micro-macro homogenization modeling of heterogeneous gradient-enhanced Cosserat continuum; and Bigoni and Drugan [17] derived closed-form formulas for Cosserat moduli via homogenization of a dilute suspension of elastic spherical inclusions in 3D and 2D embedded in an isotropic elastic matrix. Anisotropic cases with additional micropolar parameters can be found in [18] and [19].

On the other hand, the Eringen's microcontinuum field theories (known as micromorphic continuum) [20,21] constitute extensions of the classical field ones to microscopic space and time scales. From this, two fundamental statements are defined: (i) the microstretch continuum which is constrained to undergo microrotations and microstretches without microshearing, and (ii) the micropolar continuum (the model we are dealing with) which is supposed to have rigid directors, i.e., the micromotion is only a rigid-body rotation (microrotation). It can be proved (see Chen et al. [22]) that the micromorphic theory can be reduced to Mindlin's microstructure theory [23] which prompts the assumption that micromorphic theory and, its special case, the micropolar theory are derived from more elaborate considerations of microstructures and micromotion. The above-mentioned theories are closed but their main difference is that the micromotion in Eringen's theories is not treated as the gradient of the macrostrain. For example, Malyi [24] computes closed-form effective properties for polycrystalline materials. This work is focused on the analytical determination of the elastic constants of the Toupin-Mindlin gradient elasticity for polycrystalline materials based on micropolar elasticity theory, which is not the aim of the present work. Alexandrov et al. [25] proposed a gradient theory of plasticity involving space derivatives of a measure of strain rate (strain-rate gradient theory of plasticity). There are several approaches of Cosserat or generalized continua models which consider solid materials with complex microstructure of real properties like porous media and foams, granular and powder like materials, polycrystalline and composite materials [26–33], among others.

In this work, in comparison with the aforementioned works [12–15] and [24], whose procedures are different of the present one, the heterogeneous periodic Cosserat media are studied by the two-scale asymptotic homogenization method (AHM). The AHM general scheme is performed to determine the local problems, the homogenized problem and the effective properties of micropolar media. The local problems are solved and the analytical expressions of the local functions and the effective properties are reported for Cosserat multilaminated composites whose laminate distribution is perpendicular to the x_3 axis. Particular case of composite materials is considered, i.e., centro-symmetric materials for which the expressions of the local functions to be well defined and

Asymptotic homogenization approach Chapter | 21 **461**

are reported as well as the corresponding effective coefficients. An example of Cosserat bilaminated composite with cubic-symmetric constituents is analyzed and the values of their dimensionless effective stiffness and torque moduli are computed as a function of the volume fraction. The effective behavior of Cosserat bilaminate composites with cubic-symmetric constituents is characterized by 9 stiffness and 9 torque effective constants respectively. It is related to an orthotropic symmetry group referred as rotations by 90° [19], which implies invariance of C_{ijpq} and D_{ijpq} under rotations of 90°.

The novelty of this work is to determine analytical expressions for the effective coefficients of the micropolar theory for a 3D periodically heterogeneous Cosserat media as functions of the solutions of the local problems. These effective properties are determined by the geometry of the heterogeneities through double scale asymptotic expansions for the displacements and the microrotations by AHM. Also, as an example, Cosserat laminate composites are studied by AHM and focusing on the work for centro-symmetry materials. Analytical expressions of the effective properties are reported for composites made of N-layers. This work generalizes to previous ones in which AHM has been used to 2D or 3D Cauchy medium, see, for instance, Bakhvalov and Panasenko [34], Pobedrya [35], Sanchez-Palencia [36], Forest and Sab [12], among others. In addition, the herein proposed methodology complements the developed study by Forest and Sab [12] and Gorbachev and Emel'yanov [14] in Cosserat media. Former determine the effective Cauchy medium by minimizing the elastic strain energy with respect to displacement fields using the finite element method and extended the scheme to Cosserat material, in particular for a multilayered material, as a simple application and capability of the model. The latter finds the effective properties by an integral formulation and a representation in Taylor series for displacements and microrotations. To the best of our knowledge, very few of works have been found tackled the AHM procedure applied to elastic micropolar theory. The present study constitutes a valuable result and a methodology for further investigations.

21.2 Basic equations for micropolar media. Statement of the problem

In this section, the basic equations of the linear micropolar (Cosserat) elasticity theory for a three-dimensional Cosserat continuum are presented. Fundamentals of this theory can be found in Refs. [37–40].

Let us consider a micropolar continuum defined in a periodic domain Ω bounded by an infinitely smooth surface $\partial\Omega$ at the Cartesian coordinate system $\{x_1, x_2, x_3\}$. The basic equations for heterogeneous media Ω are described through a system of partial differential equations given by the linear equilibrium and balance equations,

$$\sigma_{ji,j} + f_i = 0, \qquad \mu_{ji,j} + \epsilon_{ijk}\sigma_{jk} + g_i = 0, \text{ in } \Omega \qquad (21.2.1)$$

462 Mechanics and Physics of Structured Media

under the action of the finite body force f_i and moment g_i with $i, j, k = 1, 2, 3$. Here, σ_{ji} and μ_{ji} are the components of the force stress tensor and couple stress tensor, respectively. ϵ_{ijk} is the Levi-Civita tensor and the comma notation represents the partial derivate relative to the x_j component.

The linear anisotropic constitutive equations are given by

$$\sigma_{ji} = C_{ijmn} e_{nm} + B_{ijmn} \psi_{nm}, \qquad \mu_{ji} = B_{ijmn} e_{nm} + D_{ijmn} \psi_{nm}, \qquad (21.2.2)$$

where C_{ijmn}, D_{ijmn}, and B_{ijmn} ($m, n = 1, 2, 3$) satisfy the symmetry properties

$$C_{ijmn} = C_{mnij}, \qquad D_{ijmn} = D_{mnij},$$

and

$$C_{ijmn} \neq C_{jimn} \neq C_{ijnm}, \qquad D_{ijmn} \neq D_{jimn} \neq D_{ijnm},$$
$$B_{ijmn} \neq B_{mnij} \neq B_{jimn} \neq B_{ijnm}.$$

In Eq. (21.2.2), C_{ijmn} and D_{ijmn} are the elastic and torque moduli, respectively, which 45 components each, and B_{ijmn} is the coupling moduli with 81 components. The tensors e_{mn} and ψ_{mn} of second orders are the asymmetric strain and the couple strain, respectively. In case of isotropic materials, the moduli C_{ijmn} and D_{ijmn} are described by three independents components each and B_{ijmn} vanish. The existence of B_{ijmn} contradicts the postulate of centro-symmetry. In general, for centro-symmetric material, the coupling moduli B_{ijkl} are zeros, and the properties C_{ijmn} and D_{ijmn} have the same structure, see for instance [19,20,41].

For a linear micropolar continuum, the displacement field vector $\boldsymbol{u} = u_m$ is complemented by a microrotation field vector $\boldsymbol{\omega} = \omega_m$ independent of the displacement field; therefore, a micropolar deformation is fully described by the asymmetric strain e_{mn} and the couple strain tensor ψ_{mn}, as follows

$$e_{nm} = u_{m,n} + \epsilon_{mns} \omega_s, \qquad \psi_{nm} = \omega_{m,n}, \qquad (21.2.3)$$

then, replacing (21.2.3) into (21.2.1) and (21.2.2), It yields to the coupled system of partial differential equations on Ω, such as

$$\left[C_{ijmn} \left(u_{m,n} + \epsilon_{mns} \omega_s \right) + B_{ijmn} \omega_{m,n} \right]_{,j} + f_i = 0, \qquad (21.2.4a)$$

$$\left[B_{ijmn} \left(u_{m,n} + \epsilon_{mns} \omega_s \right) + D_{ijmn} \omega_{m,n} \right]_{,j}$$
$$+ \epsilon_{ijk} \left[C_{kjmn} \left(u_{m,n} + \epsilon_{mns} \omega_s \right) + B_{kjmn} \omega_{m,n} \right] + g_i = 0, \qquad (21.2.4b)$$

that together with the boundary conditions

$$u_i \mid_{\partial \Omega_1} = 0, \qquad \sigma_{ji} n_j \mid_{\partial \Omega_2} = F_i, \qquad \omega_i \mid_{\partial \Omega_3} = 0, \qquad \mu_{ji} n_j \mid_{\partial \Omega_4} = G_i,$$
$$(21.2.5)$$

Asymptotic homogenization approach Chapter | 21 **463**

represents the heterogeneous classical boundary value problem with rapidly oscillating coefficients associate to the linear theory of micropolar elasticity. Here, $\partial\Omega_1$, $\partial\Omega_2$, $\partial\Omega_3$, and $\partial\Omega_4$ are disjoint sets of the $\partial\Omega$ partition, such as, $\partial\Omega = \partial\Omega_1 \bigcup \partial\Omega_2 \bigcup \partial\Omega_3 \bigcup \partial\Omega_4$, n_j is the unit outer normal vector to $\partial\Omega$, and the functions F_i and G_i are the components of the surface forces and torques, respectively.

21.2.1 Two-scale asymptotic expansions

In this section, the AHM is applied to the family of problems ((21.2.4a)–(21.2.5)) for heterogeneous Cosserat media. Fundamentals and rigorous mathematical background of AHM can be found in Refs. [34–36], and [42], among others. The AHM versatility allows one to formally homogenize a great variety of models or equations posed in a periodic domain [43]. Then, in order to obtain the corresponding local problems, homogenized problem and the effective properties of a Cosserat media, the solution of the boundary value problem ((21.2.4a)–(21.2.5)) is sought through two-scale asymptotic expansions, as follows:

The starting point is to consider the following *two-scale asymptotic expansion* (also called an *ansatz*), for the solutions u_i and ω_i of the system (21.2.4a)–(21.2.5)

$$u_m^\varepsilon = \sum_{i=0}^{+\infty} \varepsilon^i u_m^{(i)}(\boldsymbol{x}, \boldsymbol{y}), \qquad \omega_m^\varepsilon = \sum_{i=0}^{+\infty} \varepsilon^i \omega_m^{(i)}(\boldsymbol{x}, \boldsymbol{y}), \qquad (21.2.6)$$

where the terms $u_i(\boldsymbol{x}, \boldsymbol{y})$ and $\omega_i(\boldsymbol{x}, \boldsymbol{y})$ are infinitely differentiable functions of both variables $\boldsymbol{x}$ and $\boldsymbol{y}$, and Y-periodic functions with respect to $\boldsymbol{y}$. Herein, the $\boldsymbol{x} = \{x_1, x_2, x_3\}$ and $\boldsymbol{y} = \{y_1, y_2, y_3\}$ denote the macro (slow variable) and micro (fast variable) scales, which are related by $\boldsymbol{y} = \boldsymbol{x}/\varepsilon$ through a very small geometric parameter $\varepsilon = l/L$, see Fig. 21.1. The parameter ε represents the ratio between the characteristic dimension of the representative volume element (l), generally at the microscopic level, and the representative length of the entire composite (L), generally at the macroscopic level, which characterizes the effective dimension used to measure the composite's property in question. The series (21.2.6) are plugged into the system (21.2.4a)–(21.2.4b) and the following derivation rule is used for both solutions:

$$\frac{\partial(f^\varepsilon(\boldsymbol{x}, \boldsymbol{y}))}{\partial x_j} = \frac{\partial(f(\boldsymbol{x}, \boldsymbol{y}))}{\partial x_j} + \varepsilon^{-1} \frac{\partial(f(\boldsymbol{x}, \boldsymbol{y}))}{\partial y_j} = f(\boldsymbol{x}, \boldsymbol{y})_{,j} + \varepsilon^{-1} f(\boldsymbol{x}, \boldsymbol{y})_{|j},$$

$$(21.2.7)$$

where $f(\boldsymbol{x}, \boldsymbol{y})_{,j}$ and $f(\boldsymbol{x}, \boldsymbol{y})_{|j}$ are the derivatives of $f(\boldsymbol{x}, \boldsymbol{y})$ with respect to x_j and y_j, respectively.

For example, $u_{m,n}^\varepsilon(\boldsymbol{x}, \boldsymbol{y})$ is defined by:

$$u_{m,n}^\varepsilon(\boldsymbol{x}, \boldsymbol{y}) = \varepsilon^{-1} u_{m|n}^{(0)} + \sum_{i=0}^{+\infty} \varepsilon^i \left(u_{m,n}^{(i)} + u_{m|n}^{(i+1)} \right), \qquad (21.2.8)$$

464 Mechanics and Physics of Structured Media

and the derivation rule for the coefficients $C_{ijmn}(\mathbf{y})$, $B_{ijmn}(\mathbf{y})$, and $D_{ijmn}(\mathbf{y})$ is

$$(C, B, D)_{ijmn,j} = (C, B, D)_{ijmn|j}\frac{\partial y_j}{\partial x_j} = \varepsilon^{-1}(C, B, D)_{ijmn|j}. \qquad (21.2.9)$$

From now on, the dependency related to $\mathbf{x}$ and $\mathbf{y}$ is omitted, exceptionally, it is written. Thus, replacing the series (21.2.6) into (21.2.4a) and (21.2.4b) and neglecting the second or higher order terms, the explicit form of the system turns in

$$
\begin{aligned}
\varepsilon^{-1}C_{ijmn|j}&\left[\varepsilon^{-1}u_{m|n}^{(0)} + \left(u_{m,n}^{(0)} + u_{m|n}^{(1)}\right) + \varepsilon\left(u_{m,n}^{(1)} + u_{m|n}^{(2)}\right)\right.\\
&\left.+ \epsilon_{mns}\left(\omega_s^{(0)} + \varepsilon\omega_s^{(1)} + \varepsilon^2\omega_s^{(2)}\right)\right]\\
&+ C_{ijmn}\left[\varepsilon^{-2}u_{m,n|j}^{(0)} + \varepsilon^{-1}\left(u_{m|n,j}^{(0)} + u_{m,n|j}^{(0)} + u_{m|nj}^{(1)}\right)\right.\\
&\left.+ \left(u_{m,nj}^{(0)} + u_{m|n,j}^{(1)} + u_{m,n|j}^{(1)} + u_{m|nj}^{(2)}\right)\right]\\
&+ C_{ijmn}\,\epsilon_{mns}\left[\varepsilon^{-1}\omega_{s|j}^{(0)} + \left(\omega_{s,j}^{(0)} + \omega_{s|j}^{(1)}\right) + \varepsilon\left(\omega_{s,j}^{(1)} + \omega_{s|j}^{(2)}\right)\right]\\
&+ \varepsilon^{-1}B_{ijmn|j}\left[\varepsilon^{-1}\omega_{m|n}^{(0)} + \left(\omega_{m,n}^{(0)} + \omega_{m|n}^{(1)}\right) + \varepsilon\left(\omega_{m,n}^{(1)} + \omega_{m|n}^{(2)}\right)\right]\\
&+ B_{ijmn}\left[\varepsilon^{-2}\omega_{m|nj}^{(0)} + \varepsilon^{-1}\left(\omega_{m|n,j}^{(0)} + \omega_{m,n|j}^{(0)} + \omega_{m|nj}^{(1)}\right)\right.\\
&\left.+ \left(\omega_{m,nj}^{(0)} + \omega_{m|n,j}^{(1)} + \omega_{m,n|j}^{(1)} + \omega_{m|nj}^{(2)}\right)\right] + f_i = 0, \qquad (21.2.10a)
\end{aligned}
$$

and

$$
\begin{aligned}
\varepsilon^{-1}B_{ijmn|j}&\left[\varepsilon^{-1}u_{m|n}^{(0)} + \left(u_{m,n}^{(0)} + u_{m|n}^{(1)}\right) + \varepsilon\left(u_{m,n}^{(1)} + u_{m|n}^{(2)}\right)\right.\\
&\left.+ \epsilon_{mns}\left(\omega_s^{(0)} + \varepsilon\omega_s^{(1)} + \varepsilon^2\omega_s^{(2)}\right)\right]\\
&+ B_{ijmn}\left[\varepsilon^{-2}u_{m,n|j}^{(0)} + \varepsilon^{-1}\left(u_{m|n,j}^{(0)} + u_{m,n|j}^{(0)} + u_{m|nj}^{(1)}\right)\right.\\
&\left.+ \left(u_{m,nj}^{(0)} + u_{m|n,j}^{(1)} + u_{m,n|j}^{(1)} + u_{m|nj}^{(2)}\right)\right]\\
&+ B_{ijmn}\,\epsilon_{mns}\left[\varepsilon^{-1}\omega_{s|j}^{(0)} + \left(\omega_{s,j}^{(0)} + \omega_{s|j}^{(1)}\right) + \varepsilon\left(\omega_{s,j}^{(1)} + \omega_{s|j}^{(2)}\right)\right]\\
&+ \varepsilon^{-1}D_{ijmn|j}\left[\varepsilon^{-1}\omega_{m|n}^{(0)} + \left(\omega_{m,n}^{(0)} + \omega_{m|n}^{(1)}\right) + \varepsilon\left(\omega_{m,n}^{(1)} + \omega_{m|n}^{(2)}\right)\right]\\
&+ D_{ijmn}\left[\varepsilon^{-2}\omega_{m|nj}^{(0)} + \varepsilon^{-1}\left(\omega_{m|n,j}^{(0)} + \omega_{m,n|j}^{(0)} + \omega_{m|nj}^{(1)}\right)\right.\\
&\left.+ \left(\omega_{m,nj}^{(0)} + \omega_{m|n,j}^{(1)} + \omega_{m,n|j}^{(1)} + \omega_{m|nj}^{(2)}\right)\right]\\
&+ \epsilon_{ijk}\left\{C_{kjmn}\left[\varepsilon^{-1}u_{m|n}^{(0)} + \left(u_{m,n}^{(0)} + u_{m|n}^{(1)}\right) + \varepsilon\left(u_{m,n}^{(1)} + u_{m|n}^{(2)}\right)\right.\right.\\
&\left.\left.+ \epsilon_{mns}\left(\omega_s^{(0)} + \varepsilon\omega_s^{(1)} + \varepsilon^2\omega_s^{(2)}\right)\right]\right.
\end{aligned}
$$

$$+ B_{kjmn} \left[\varepsilon^{-1} \omega_{m|n}^{(0)} + \left(\omega_{m,n}^{(0)} + \omega_{m|n}^{(1)} \right) + \varepsilon \left(\omega_{m,n}^{(1)} + \omega_{m|n}^{(2)} \right) \right] \Bigg\} + g_i = 0,$$

$$(21.2.10b)$$

where $f_{m,nj} = \frac{\partial^2 f_m}{\partial x_j \partial x_n}$, $f_{m|n,j} = \frac{\partial^2 f_m}{\partial x_j \partial y_n}$, $f_{m,n|j} = \frac{\partial^2 f_m}{\partial y_j \partial x_n}$, and $f_{m|nj} = \frac{\partial^2 f_m}{\partial y_j \partial y_n}$ are the second order derivatives of f_m.

Now, collecting terms with respect to power of ε (i.e., ε^i), in this case is enough to consider the cases $i = -2, -1, 0$, we have

$$\varepsilon^{-2} \left[C_{ijmn|j}\, u_{m|n}^{(0)} + C_{ijmn}\, u_{m|nj}^{(0)} + B_{ijmn|j}\, \omega_{m|n}^{(0)} + B_{ijmn}\, \omega_{m|nj}^{(0)} \right]$$

$$+ \varepsilon^{-1} \Big[C_{ijmn|j} \left(u_{m,n}^{(0)} + u_{m|n}^{(1)} + \epsilon_{mns}\, \omega_s^{(0)} \right)$$

$$+ C_{ijmn} \left(u_{m|n,j}^{(0)} + u_{m,n|j}^{(0)} + u_{m|nj}^{(1)} + \epsilon_{mns}\, \omega_{s|j}^{(0)} \right)$$

$$+ B_{ijmn|j} \left(\omega_{m,n}^{(0)} + \omega_{m|n}^{(1)} \right) + B_{ijmn} \left(\omega_{m|n,j}^{(0)} + \omega_{m,n|j}^{(0)} + \omega_{m|nj}^{(1)} \right) \Big]$$

$$+ \varepsilon^0 \Big\{ C_{ijmn|j} \left(u_{m,n}^{(1)} + u_{m|n}^{(2)} + \epsilon_{mns}\, \omega_s^{(1)} \right)$$

$$+ C_{ijmn} \Big[u_{m,nj}^{(0)} + u_{m|n,j}^{(1)} + u_{m,n|j}^{(1)} + u_{m|nj}^{(2)} + \epsilon_{mns} \left(\omega_{s,j}^{(0)} + \omega_{s|j}^{(1)} \right) \Big]$$

$$+ B_{ijmn|j} \left(\omega_{m,n}^{(1)} + \omega_{m|n}^{(2)} \right)$$

$$+ B_{ijmn} \left(\omega_{m,nj}^{(0)} + \omega_{m|n,j}^{(1)} + \omega_{m,n|j}^{(1)} + \omega_{m|nj}^{(2)} \right) + f_i \Big\} = 0, \qquad (21.2.11a)$$

and

$$\varepsilon^{-2} \left[B_{ijmn|j}\, u_{m|n}^{(0)} + B_{ijmn}\, u_{m|nj}^{(0)} + D_{ijmn|j}\, \omega_{m|n}^{(0)} + D_{ijmn}\, \omega_{m|nj}^{(0)} \right]$$

$$+ \varepsilon^{-1} \Big[B_{ijmn|j} \left(u_{m,n}^{(0)} + u_{m|n}^{(1)} + \epsilon_{mns}\, \omega_s^{(0)} \right)$$

$$+ B_{ijmn} \left(u_{m|n,j}^{(0)} + u_{m,n|j}^{(0)} + u_{m|nj}^{(1)} + \epsilon_{mns}\, \omega_{s|j}^{(0)} \right)$$

$$+ D_{ijmn|j} \left(\omega_{m,n}^{(0)} + \omega_{m|n}^{(1)} \right) + D_{ijmn} \left(\omega_{m|n,j}^{(0)} + \omega_{m,n|j}^{(0)} + \omega_{m|nj}^{(1)} \right)$$

$$+ \epsilon_{ijk}\, C_{kjmn}\, u_{m|n}^{(0)} + \epsilon_{ijk}\, B_{kjmn}\, \omega_{m|n}^{(0)} \Big]$$

$$+ \varepsilon^0 \Big\{ B_{ijmn|j} \left(u_{m,n}^{(1)} + u_{m|n}^{(2)} + \epsilon_{mns}\, \omega_s^{(1)} \right)$$

$$+ B_{ijmn} \Big[u_{m,nj}^{(0)} + u_{m|n,j}^{(1)} + u_{m,n|j}^{(1)} + u_{m|nj}^{(2)} + \epsilon_{mns} \left(\omega_{s,j}^{(0)} + \omega_{s|j}^{(1)} \right) \Big]$$

$$+ D_{ijmn|j} \left(\omega_{m,n}^{(1)} + \omega_{m|n}^{(2)} \right) + D_{ijmn} \left(\omega_{m,nj}^{(0)} + \omega_{m|n,j}^{(1)} + \omega_{m,n|j}^{(1)} + \omega_{m|nj}^{(2)} \right)$$

$$+ \epsilon_{ijk}\, C_{kjmn} \left(u_{m,n}^{(0)} + u_{m|n}^{(1)} + \epsilon_{mns}\, \omega_s^{(0)} \right)$$

$$+ \epsilon_{ijk}\, B_{kjmn} \left(\omega_{m,n}^{(0)} + \omega_{m|n}^{(1)} \right) + g_i \Big\} = 0. \qquad (21.2.11b)$$

466 Mechanics and Physics of Structured Media

Identifying each coefficient of ε in (21.2.11a) and (21.2.11b) as an individual equation, a sequence of problems given by partial differential equations is obtained. Each contribution is assumed zero for all values of ε. The three first equations are enough for our purpose.

From (21.2.11a) and (21.2.11b), the terms corresponding to ε^{-2} can be written as follows

$$\left(C_{ijmn}\, u_{m|n}^{(0)} + B_{ijmn}\, \omega_{m|n}^{(0)}\right)_{|j} = 0, \tag{21.2.12a}$$

$$\left(B_{ijmn}\, u_{m|n}^{(0)} + D_{ijmn}\, \omega_{m|n}^{(0)}\right)_{|j} = 0. \tag{21.2.12b}$$

Thus, (21.2.12a) and (21.2.12b) is a system of partial differential equations with unknowns $u_{m|n}^{(0)}$ and $\omega_{m|n}^{(0)}$, which are defined as a function of $\boldsymbol{x}$ and $\boldsymbol{y}$. It can be checked that there exists a unique solution of these equations up to a constant (i.e., a function of $\boldsymbol{x}$ independent of $\boldsymbol{y}$). This implies that the nonperturbed terms $u_m^{(0)}$ and $\omega_m^{(0)}$ are independent functions of the local variable $\boldsymbol{y}$, i.e.,

$$u_m^{(0)}(\boldsymbol{x}, \boldsymbol{y}) \equiv u_m(\boldsymbol{x}), \qquad \omega_m^{(0)}(\boldsymbol{x}, \boldsymbol{y}) \equiv \omega_m(\boldsymbol{x}). \tag{21.2.13}$$

From (21.2.11a) and (21.2.11b), the system corresponding to ε^{-1} can be expressed by

$$C_{ijmn|j}\left(u_{m,n}^{(0)} + u_{m|n}^{(1)} + \epsilon_{mns}\,\omega_s^{(0)}\right)$$
$$+ C_{ijmn}\left(u_{m|n,j}^{(0)} + u_{m,n|j}^{(0)} + u_{m|nj}^{(1)} + \epsilon_{mns}\,\omega_{s|j}^{(0)}\right)$$
$$+ B_{ijmn|j}\left(\omega_{m,n}^{(0)} + \omega_{m|n}^{(1)}\right) + B_{ijmn}\left(\omega_{m|n,j}^{(0)} + \omega_{m,n|j}^{(0)} + \omega_{m|nj}^{(1)}\right) = 0, \tag{21.2.14a}$$

$$B_{ijmn|j}\left(u_{m,n}^{(0)} + u_{m|n}^{(1)} + \epsilon_{mns}\,\omega_s^{(0)}\right)$$
$$+ B_{ijmn}\left(u_{m|n,j}^{(0)} + u_{m,n|j}^{(0)} + u_{m|nj}^{(1)} + \epsilon_{mns}\,\omega_{s|j}^{(0)}\right)$$
$$+ D_{ijmn|j}\left(\omega_{m,n}^{(0)} + \omega_{m|n}^{(1)}\right) + D_{ijmn}\left(\omega_{m|n,j}^{(0)} + \omega_{m,n|j}^{(0)} + \omega_{m|nj}^{(1)}\right)$$
$$+ \epsilon_{ijk}\, C_{kjmn}\, u_{m|n}^{(0)} + \epsilon_{ijk}\, B_{kjmn}\, \omega_{m|n}^{(0)} = 0, \tag{21.2.14b}$$

then, taking into account (21.2.13), the derivatives of $u_m^{(0)}(\boldsymbol{x})$ and $\omega_m^{(0)}(\boldsymbol{x})$ are nulls with respect to the fast variable $\boldsymbol{y}$, hence this turns (21.2.14a)–(21.2.14b) into

$$\left(C_{ijmn}\, u_{m|n}^{(1)}\right)_{|j} + C_{ijmn|j}\left(u_{m,n}^{(0)} + \epsilon_{mns}\,\omega_s^{(0)}\right)$$
$$+ \left(B_{ijmn}\, \omega_{m|n}^{(1)}\right)_{|j} + B_{ijmn|j}\, \omega_{m,n}^{(0)} = 0, \tag{21.2.15a}$$

$$\left(B_{ijmn}\, u_{m|n}^{(1)}\right)_{|j} + B_{ijmn|j}\left(u_{m,n}^{(0)} + \epsilon_{mns}\,\omega_s^{(0)}\right)$$
$$+ \left(D_{ijmn}\,\omega_{m|n}^{(1)}\right)_{|j} + D_{ijmn|j}\,\omega_{m,n}^{(0)} = 0. \tag{21.2.15b}$$

Let us write the expressions for the strains and couple strains in the Cosserat theory of elasticity (21.2.3) once again to perform the series expansion of themselves

$$e_{nm} = u_{m,n} + \epsilon_{mns}\,\omega_s, \qquad \psi_{nm} = \omega_{m,n}. \tag{21.2.16}$$

Then, substituting the ansatz (21.2.6) into (21.2.16), the two-scale series expansion of the strains is

$$e_{nm} = \varepsilon^{-1} u_{m|n}^{(0)} + \varepsilon^0 u_{m,n}^{(0)} + \varepsilon^0 \epsilon_{mns}\,\omega_s^{(0)} + \varepsilon^0 u_{m|n}^{(1)} + \varepsilon^1 u_{m,n}^{(1)}$$
$$+ \varepsilon^1 \epsilon_{mns}\,\omega_s^{(1)} + \varepsilon^1 u_{m|n}^{(2)} + \dots. \tag{21.2.17}$$

Notice that the first term in (21.2.17) is zero because (21.2.13), therefore, (21.2.17) can be written as follows

$$e_{nm} = e_{nm}^{(0)} + \varepsilon e_{nm}^{(1)} + \dots, \tag{21.2.18}$$

where

$$e_{nm}^{(k)} = u_{m|n}^{(k+1)} + u_{m,n}^{(k)} + \epsilon_{mns}\,\omega_s^{(k)}. \tag{21.2.19}$$

The first term in (21.2.18), i.e., $e_{nm}^{(0)}$ is

$$e_{nm}^{(0)} = u_{m|n}^{(1)} + u_{m,n}^{(0)} + \epsilon_{mns}\,\omega_s^{(0)}, \tag{21.2.20}$$

then, similar to Sanchez-Palencia [36] (see Eq. (2.10) in page 89), Eq. (21.2.20) is rewritten as

$$e_{nm}^{(0)} = \overset{y}{e}_{nm}(u_m^{(1)}) + \overset{x}{e}_{nm}(u_m^{(0)}), \tag{21.2.21}$$

where

$$\overset{y}{e}_{nm}(u_m^{(1)}) \equiv u_{m|n}^{(1)}, \qquad \overset{x}{e}_{nm}(u_m^{(0)}) \equiv u_{m,n}^{(0)} + \epsilon_{mns}\,\omega_s^{(0)}. \tag{21.2.22}$$

Analogously, the two-scale series expansion of the couple strains can be found and here is omitted. Therefore, the first term is shown directly, i.e.,

$$\psi_{nm}^{(0)} = \omega_{m|n}^{(1)} + \omega_{m,n}^{(0)}, \tag{21.2.23}$$

then, rewrite (21.2.23) as in (21.2.20), we have

$$\psi_{nm}^{(0)} = \overset{y}{\psi}_{nm}(\omega_m^{(1)}) + \overset{x}{\psi}_{nm}(\omega_m^{(0)}), \tag{21.2.24}$$

468 Mechanics and Physics of Structured Media

where

$$\overset{y}{\psi}_{nm}(\omega_m^{(1)}) \equiv \omega_{m|n}^{(1)}, \quad \overset{x}{\psi}_{nm}(\omega_m^{(0)}) \equiv \omega_{m,n}^{(0)}. \tag{21.2.25}$$

Then, using (21.2.22) and (21.2.25) into (21.2.15a) and (21.2.15b) we have

$$\left(C_{ijmn}\,\overset{y}{e}_{nm}(u_m^{(1)})\right)_{|j} + C_{ijmn|j}\,\overset{x}{e}_{nm}(u_m^{(0)})$$

$$+ \left(B_{ijmn}\,\overset{y}{\psi}_{nm}(\omega_m^{(1)})\right)_{|j} + B_{ijmn|j}\,\overset{x}{\psi}_{nm}(\omega_m^{(0)}) = 0, \tag{21.2.26a}$$

$$\left(B_{ijmn}\,\overset{y}{e}_{nm}(u_m^{(1)})\right)_{|j} + B_{ijmn|j}\,\overset{x}{e}_{nm}(u_m^{(0)})$$

$$+ \left(D_{ijmn}\,\overset{y}{\psi}_{nm}(\omega_m^{(1)})\right)_{|j} + D_{ijmn|j}\,\overset{x}{\psi}_{nm}(\omega_m^{(0)}) = 0. \tag{21.2.26b}$$

The problem ((21.2.26a) and (21.2.26b)) is the same as the piezoelectric medium but with the Cosserat strains instead of the Cauchy strains and with the couple strains instead of the electric field vector. As in the above system associated to ε^{-2}, it is possible to transform ((21.2.26a) and (21.2.26b)) into an equivalent system which is elliptic and admits a solution in the class of Y-periodic functions with respect to $\mathbf{y}$. Therefore, one expects to obtain solutions of the same form.

Due to the linearity of the problem, the solutions for $u_m^{(1)}(\mathbf{x}, \mathbf{y})$ and $\omega_m^{(1)}(\mathbf{x}, \mathbf{y})$ exist and they are unique, see for instance [44], thus

$$u_m^{(1)}(\mathbf{x}, \mathbf{y}) = {}_{pq}\hat{N}_m(\mathbf{y})\,\overset{x}{e}_{pq}(u_m^{(0)}) + {}_{pq}\hat{U}_m(\mathbf{y})\,\overset{x}{\psi}_{pq}(\omega_m^{(0)}) + \tilde{u}_m^{(1)}, \tag{21.2.27}$$

$$\omega_m^{(1)}(\mathbf{x}, \mathbf{y}) = {}_{pq}V_m(\mathbf{y})\,\overset{x}{e}_{pq}(u_m^{(0)}) + {}_{pq}M_m(\mathbf{y})\,\overset{x}{\psi}_{pq}(\omega_m^{(0)}) + \tilde{\psi}_m^{(1)}, \tag{21.2.28}$$

where ${}_{pq}\hat{N}_m(\mathbf{y})$, ${}_{pq}\hat{U}_m(\mathbf{y})$, ${}_{pq}V_m(\mathbf{y})$, and ${}_{pq}M_m(\mathbf{y})$ are Y-periodic functions on the variable $\mathbf{y}$, which are known as local functions; $\tilde{u}_m^{(1)}$ and $\tilde{\psi}_m^{(1)}$ are constant vectors.

Then, replacing (21.2.27) and (21.2.28) into (21.2.26a) and (21.2.26b), and collecting by the terms $u_{p,q}^{(0)} + \epsilon_{pqk}\,\omega_k^{(0)}$ and $\omega_{p,q}^{(0)}$, we obtain:

$$\left[C_{ijpq} + C_{ijmn}\,{}_{pq}\hat{N}_{m|n} + B_{ijmn}\,{}_{pq}V_{m|n}\right]_{|j} \left(u_{p,q}^{(0)} + \epsilon_{pqk}\,\omega_k^{(0)}\right)$$

$$+ \left[B_{ijpq} + C_{ijmn}\,{}_{pq}\hat{U}_{m|n} + B_{ijmn}\,{}_{pq}M_{m|n}\right]_{|j}\,\omega_{p,q}^{(0)} = 0, \tag{21.2.29a}$$

$$\left[B_{ijpq} + B_{ijmn}\,{}_{pq}\hat{N}_{m|n} + D_{ijmn}\,{}_{pq}V_{m|n}\right]_{|j} \left(u_{p,q}^{(0)} + \epsilon_{pqk}\,\omega_k^{(0)}\right)$$

$$+ \left[D_{ijpq} + B_{ijmn}\,{}_{pq}\hat{U}_{m|n} + D_{ijmn}\,{}_{pq}M_{m|n}\right]_{|j}\,\omega_{p,q}^{(0)} = 0. \tag{21.2.29b}$$

In (21.2.29a) and (21.2.29b), the strains are related to microrotations however the couple strains does not with the displacements, therefore, they are assumed in similar form to (21.2.16). Then, $_{pq}\hat{N}_{m|n}$ and $_{pq}\hat{U}_{m|n}$ are defined, as follows

$$_{pq}\hat{N}_{m|n} = {}_{pq}N_{m|n} + \epsilon_{mnk}\,{}_{pq}V_k, \tag{21.2.30}$$

$$_{pq}\hat{U}_{m|n} = {}_{pq}U_{m|n} + \epsilon_{mnk}\,{}_{pq}M_k, \tag{21.2.31}$$

where $_{pq}N_m$ y $_{pq}U_m$ are Y-periodic functions.

Now, replacing (21.2.30) and (21.2.31) into (21.2.29a) and (21.2.29b), results

$$\left[C_{ijpq} + C_{ijmn}\left(_{pq}N_{m|n} + \epsilon_{mnk}\,{}_{pq}V_k\right) + B_{ijmn}\,{}_{pq}V_{m|n}\right]_{|j}\left(u^{(0)}_{p,q} + \epsilon_{pqk}\,\omega^{(0)}_k\right)$$
$$+ \left[B_{ijpq} + C_{ijmn}\left(_{pq}U_{m|n} + \epsilon_{mnk}\,{}_{pq}M_k\right) + B_{ijmn}\,{}_{pq}M_{m|n}\right]_{|j}\omega^{(0)}_{p,q} = 0, \tag{21.2.32a}$$

$$\left[B_{ijpq} + B_{ijmn}\left(_{pq}N_{m|n} + \epsilon_{mnk}\,{}_{pq}V_k\right) + D_{ijmn}\,{}_{pq}V_{m|n}\right]_{|j}\left(u^{(0)}_{p,q} + \epsilon_{pqk}\,\omega^{(0)}_k\right)$$
$$+ \left[D_{ijpq} + B_{ijmn}\left(_{pq}U_{m|n} + \epsilon_{mnk}\,{}_{pq}M_k\right) + D_{ijmn}\,{}_{pq}M_{m|n}\right]_{|j}\omega^{(0)}_{p,q} = 0, \tag{21.2.32b}$$

then, as $u^{(0)}_{p,q} + \epsilon_{pqk}\,\omega^{(0)}_k$ and $\omega^{(0)}_{p,q}$ are not null in (21.2.32a) and (21.2.32b) due to the strains and couple strains are different from zero, thus, their coefficients (the magnitudes between square brackets) satisfy that

$$\left(C_{ijpq} + C_{ijmn}\left(_{pq}N_{m|n} + \epsilon_{mnk}\,{}_{pq}V_k\right) + B_{ijmn}\,{}_{pq}V_{m|n}\right)_{|j} = 0,$$
$$\left(B_{ijpq} + B_{ijmn}\left(_{pq}N_{m|n} + \epsilon_{mnk}\,{}_{pq}V_k\right) + D_{ijmn}\,{}_{pq}V_{m|n}\right)_{|j} = 0, \tag{21.2.33}$$

and for the couple strains

$$\left(B_{ijpq} + C_{ijmn}\left(_{pq}U_{m|n} + \epsilon_{mnk}\,{}_{pq}M_k\right) + B_{ijmn}\,{}_{pq}M_{m|n}\right)_{|j} = 0,$$
$$\left(D_{ijpq} + B_{ijmn}\left(_{pq}U_{m|n} + \epsilon_{mnk}\,{}_{pq}M_k\right) + D_{ijmn}\,{}_{pq}M_{m|n}\right)_{|j} = 0, \tag{21.2.34}$$

where $_{pq}N_m$, $_{pq}V_m$, $_{pq}U_m$, and $_{pq}M_m$ are Y-periodic local functions which depend on **y**. The expressions (21.2.33) and (21.2.34) represent the so-called local problems on the periodic cell Y relate to the micropolar theory of elasticity, defined as $_{pq}\mathcal{L}^1$ and $_{pq}\mathcal{L}^2$, respectively.

Following a similar procedure as above, from (21.2.11a) and (21.2.11b), the system corresponding to ε^0 can be defined by

$$C_{ijmn|j}\left(u^{(1)}_{m,n} + u^{(2)}_{m|n} + \epsilon_{mns}\,\omega^{(1)}_s\right)$$
$$+ C_{ijmn}\left[u^{(0)}_{m,nj} + u^{(1)}_{m|n,j} + u^{(1)}_{m,n|j} + u^{(2)}_{m|nj} + \epsilon_{mns}\left(\omega^{(0)}_{s,j} + \omega^{(1)}_{s|j}\right)\right]$$

$$+ B_{ijmn|j}\left(\omega_{m,n}^{(1)} + \omega_{m|n}^{(2)}\right)$$

$$+ B_{ijmn}\left(\omega_{m,nj}^{(0)} + \omega_{m|n,j}^{(1)} + \omega_{m,n|j}^{(1)} + \omega_{m|nj}^{(2)}\right) + f_i = 0, \qquad (21.2.35a)$$

$$B_{ijmn|j}\left(u_{m,n}^{(1)} + u_{m|n}^{(2)} + \epsilon_{mns}\,\omega_s^{(1)}\right)$$

$$+ B_{ijmn}\left[u_{m,nj}^{(0)} + u_{m|n,j}^{(1)} + u_{m,n|j}^{(1)} + u_{m|nj}^{(2)} + \epsilon_{mns}\left(\omega_{s,j}^{(0)} + \omega_{s|j}^{(1)}\right)\right]$$

$$+ D_{ijmn|j}\left(\omega_{m,n}^{(1)} + \omega_{m|n}^{(2)}\right) + D_{ijmn}\left(\omega_{m,nj}^{(0)} + \omega_{m|n,j}^{(1)} + \omega_{m,n|j}^{(1)} + \omega_{m|nj}^{(2)}\right)$$

$$+ \epsilon_{ijk}\, C_{kjmn}\left(u_{m,n}^{(0)} + u_{m|n}^{(1)} + \epsilon_{mns}\,\omega_s^{(0)}\right)$$

$$+ \epsilon_{ijk}\, B_{kjmn}\left(\omega_{m,n}^{(0)} + \omega_{m|n}^{(1)}\right) + g_i = 0, \qquad (21.2.35b)$$

rearranging the system (21.2.35a)–(21.2.35b), we have

$$\left(C_{ijmn}\, u_{m|n}^{(2)}\right)_{|j} + \left(C_{ijmn}\, u_{m,n}^{(1)}\right)_{|j} + \left(C_{ijmn}\,\epsilon_{mns}\,\omega_s^{(1)}\right)_{|j} + \left(B_{ijmn}\,\omega_{m|n}^{(2)}\right)_{|j}$$

$$+ \left(B_{ijmn}\,\omega_{m,n}^{(1)}\right)_{|j} + C_{ijmn}\, u_{m|n,j}^{(1)} + C_{ijmn}\left(u_{m,nj}^{(0)} + \epsilon_{mns}\,\omega_{s,j}^{(0)}\right)$$

$$+ B_{ijmn}\,\omega_{m|n,j}^{(1)} + B_{ijmn}\,\omega_{m,nj}^{(0)} + f_i = 0, \qquad (21.2.36a)$$

$$\left(B_{ijmn}\, u_{m|n}^{(2)}\right)_{|j} + \left(B_{ijmn}\, u_{m,n}^{(1)}\right)_{|j} + \left(B_{ijmn}\,\epsilon_{mns}\,\omega_s^{(1)}\right)_{|j}$$

$$+ \left(D_{ijmn}\,\omega_{m|n}^{(2)}\right)_{|j} + \left(D_{ijmn}\,\omega_{m,n}^{(1)}\right)_{|j} + B_{ijmn}\, u_{m|n,j}^{(1)}$$

$$+ B_{ijmn}\left(u_{m,nj}^{(0)} + \epsilon_{mns}\,\omega_{s,j}^{(0)}\right) + D_{ijmn}\,\omega_{m|n,j}^{(1)} + D_{ijmn}\,\omega_{m,nj}^{(0)}$$

$$+ \epsilon_{ijk}\left[C_{kjmn}\, u_{m|n}^{(1)} + C_{kjmn}\left(u_{m,n}^{(0)} + \epsilon_{mns}\,\omega_s^{(0)}\right)\right.$$

$$\left. + B_{kjmn}\,\omega_{m|n}^{(1)} + B_{kjmn}\,\omega_{m,n}^{(0)}\right] + g_i = 0, \qquad (21.2.36b)$$

then, applying the average operator into (21.2.36a)–(21.2.36b) and taking into account the periodicity conditions of $C_{ijmn}(\mathbf{y})$, $B_{ijmn}(\mathbf{y})$, $D_{ijmn}(\mathbf{y})$, $u_i^{(1)}(\mathbf{x}, \mathbf{y})$, $u_i^{(2)}(\mathbf{x}, \mathbf{y})$, $\omega_i^{(1)}(\mathbf{x}, \mathbf{y})$, and $\omega_i^{(2)}(\mathbf{x}, \mathbf{y})$ with respect to $\mathbf{y}$ in the unit cell $Y = (0, 1) \times (0, 1) \times (0, 1)$, turns out

$$\left\langle\left(C_{ijmn}\, u_{m|n}^{(2)}\right)_{|j}\right\rangle_Y = 0, \quad \left\langle\left(C_{ijmn}\, u_{m,n}^{(1)}\right)_{|j}\right\rangle_Y = 0,$$

$$\left\langle\left(C_{ijmn}\,\epsilon_{mns}\,\omega_s^{(1)}\right)_{|j}\right\rangle_Y = 0, \quad \left\langle\left(B_{ijmn}\, u_{m|n}^{(2)}\right)_{|j}\right\rangle_Y = 0,$$

$$\left\langle\left(B_{ijmn}\, u_{m,n}^{(1)}\right)_{|j}\right\rangle_Y = 0, \quad \left\langle\left(B_{ijmn}\,\epsilon_{mns}\,\omega_s^{(1)}\right)_{|j}\right\rangle_Y = 0,$$

Asymptotic homogenization approach **Chapter | 21 471**

$$\left\langle \left(B_{ijmn}\, \omega^{(2)}_{m|n} \right)_{|j} \right\rangle_Y = 0, \quad \left\langle \left(B_{ijmn}\, \omega^{(1)}_{m,n} \right)_{|j} \right\rangle_Y = 0,$$

$$\left\langle \left(D_{ijmn}\, \omega^{(2)}_{m|n} \right)_{|j} \right\rangle_Y = 0, \quad \left\langle \left(D_{ijmn}\, \omega^{(1)}_{m,n} \right)_{|j} \right\rangle_Y = 0, \qquad (21.2.37)$$

then, it is obtained

$$\left\langle C_{ijmn}\, u^{(1)}_{m|n,j} + C_{ijmn} \left(u^{(0)}_{m,nj} + \epsilon_{mns}\, \omega^{(0)}_{s,j} \right) \right.$$
$$\left. + B_{ijmn}\, \omega^{(1)}_{m|n,j} + B_{ijmn}\, \omega^{(0)}_{m,nj} \right\rangle_Y + f_i = 0, \qquad (21.2.38a)$$

$$\left\langle B_{ijmn}\, u^{(1)}_{m|nj} + B_{ijmn} \left(u^{(0)}_{m,nj} + \epsilon_{mns}\, \omega^{(0)}_{s,j} \right) + D_{ijmn}\, \omega^{(1)}_{m|nj} + D_{ijmn}\, \omega^{(0)}_{m,nj} \right\rangle_Y$$
$$+ \epsilon_{ijk} \left\langle C_{kjmn}\, u^{(1)}_{m|n} + C_{kjmn} \left(u^{(0)}_{m,n} + \epsilon_{mns}\, \omega^{(0)}_s \right) \right.$$
$$\left. + B_{kjmn}\, \omega^{(1)}_{m|n} + B_{kjmn}\, \omega^{(0)}_{m,n} \right\rangle_Y + g_i = 0. \qquad (21.2.38b)$$

Replacing (21.2.27)–(21.2.28) into (21.2.38a)–(21.2.38b), and grouping terms conveniently, it is obtained

$$\left\langle C_{ijpq} + C_{ijmn} \left({}_{pq}N_{m|n} + \epsilon_{mnk}\, {}_{pq}V_k \right) + B_{ijmn}\, {}_{pq}V_{m|n} \right\rangle_Y \left(u^{(0)}_{p,q} + \epsilon_{pqk}\, \omega^{(0)}_k \right)_{,j}$$
$$+ \left\langle B_{ijpq} + C_{ijmn} \left({}_{pq}U_{m|n} + \epsilon_{mnk}\, {}_{pq}M_k \right) + B_{ijmn}\, {}_{pq}M_{m|n} \right\rangle_Y \omega^{(0)}_{p,qj} + f_i = 0, \qquad (21.2.39a)$$

$$\left\langle B_{ijpq} + B_{ijmn} \left({}_{pq}N_{m|n} + \epsilon_{mnk}\, {}_{pq}V_k \right) + D_{ijmn}\, {}_{pq}V_{m|n} \right\rangle_Y \left(u^{(0)}_{p,q} + \epsilon_{pqk}\, \omega^{(0)}_k \right)_{,j}$$
$$+ \left\langle D_{ijpq} + B_{ijmn} \left({}_{pq}U_{m|n} + \epsilon_{mnk}\, {}_{pq}M_k \right) + D_{ijmn}\, {}_{pq}M_{m|n} \right\rangle_Y \omega^{(0)}_{p,qj}$$
$$+ \epsilon_{ijs} \left\langle C_{sjpq} + C_{sjmn} \left({}_{pq}N_{m|n} + \epsilon_{mnk}\, {}_{pq}V_k \right) + B_{sjmn}\, {}_{pq}V_{m|n} \right\rangle_Y$$
$$\times \left(u^{(0)}_{p,q} + \epsilon_{pqk}\, \omega^{(0)}_k \right)$$
$$+ \epsilon_{ijs} \left\langle B_{sjpq} + C_{sjmn} \left({}_{pq}U_{m|n} + \epsilon_{mnk}\, {}_{pq}M_k \right) + B_{sjmn}\, {}_{pq}M_{m|n} \right\rangle_Y \omega^{(0)}_{p,q}$$
$$+ g_i = 0. \qquad (21.2.39b)$$

Now, rewritten (21.2.39a)–(21.2.39b), we have the **homogenized problem** in a simpler form,

$$C^{\text{eff}}_{ijpq} \left(u^{(0)}_{p,q} + \epsilon_{pqk}\, \omega^{(0)}_k \right)_{,j} + B^{\text{eff}}_{ijpq}\, \omega^{(0)}_{p,qj} + f_i = 0, \qquad (21.2.40a)$$

$$B^{\text{eff}}_{ijpq} \left(u^{(0)}_{p,q} + \epsilon_{pqk}\, \omega^{(0)}_k \right)_{,j} + D^{\text{eff}}_{ijpq}\, \omega^{(0)}_{p,qj}$$
$$+ \epsilon_{ijl} \left[C^{\text{eff}}_{ljpq} \left(u^{(0)}_{p,q} + \epsilon_{pqk}\, \omega^{(0)}_k \right) + B^{\text{eff}}_{ljpq}\, \omega^{(0)}_{p,q} \right] + g_i = 0, \qquad (21.2.40b)$$

472 Mechanics and Physics of Structured Media

where the corresponding **effective coefficients** are defined by

$$C_{ijpq}^{\text{eff}} = \left\langle C_{ijpq} + C_{ijmn} \left({}_{pq}N_{m|n} + \epsilon_{mnk}\, {}_{pq}V_k \right) + B_{ijmn}\, {}_{pq}V_{m|n} \right\rangle_Y, \quad (21.2.41)$$

$$B_{ijpq}^{\text{eff}} = \left\langle B_{ijpq} + C_{ijmn} \left({}_{pq}U_{m|n} + \epsilon_{mnk}\, {}_{pq}M_k \right) + B_{ijmn}\, {}_{pq}M_{m|n} \right\rangle_Y,$$
$$(21.2.42)$$

$$B_{ijpq}^{\text{eff}} = \left\langle B_{ijpq} + B_{ijmn} \left({}_{pq}N_{m|n} + \epsilon_{mnk}\, {}_{pq}V_k \right) + D_{ijmn}\, {}_{pq}V_{m|n} \right\rangle_Y, \quad (21.2.43)$$

$$D_{ijpq}^{\text{eff}} = \left\langle D_{ijpq} + B_{ijmn} \left({}_{pq}U_{m|n} + \epsilon_{mnk}\, {}_{pq}M_k \right) + D_{ijmn}\, {}_{pq}M_{m|n} \right\rangle_Y.$$
$$(21.2.44)$$

It is important to notice that the resulting effective properties ((21.2.41)–(21.2.44)) are the same as the reported by Gorbachev (see [14], eqs. 3.27–3.29, page 77).

In summary, we just arrived to the effective properties ((21.2.41)–(21.2.44)) of periodic micropolar heterogeneous media Ω with periodic unit cell Y of edge ε. In this procedure, we proposed the solutions for the displacements and the rotations (21.2.6) in power series form in terms of the small parameter ε, which relates both macro-micro scales. After performed the substitution of those series into the Cosserat elasticity equations (21.2.4a) and (21.2.4b), the terms with equal ε were collected and important conclusions as the local problems (21.2.33)–(21.2.34), the homogenized problem (21.2.40a)–(21.2.40b), and the effective coefficients (21.2.41)–(21.2.44) are obtained from them.

As can be seen in (21.2.41)–(21.2.44), the local pq-displacements ${}_{pq}N_m$ and ${}_{pq}U_m$ and the local pq-microrotations ${}_{pq}V_m$ and ${}_{pq}M_m$ are required to find the effective properties. This way, the solutions of the local problems ${}_{pq}\mathscr{L}^1$ and ${}_{pq}\mathscr{L}^2$ are needed. The solution of the local problem ${}_{pq}\mathscr{L}^1$ (${}_{pq}\mathscr{L}^2$) consists on find the local functions ${}_{pq}N_m$ and ${}_{pq}V_m$ (${}_{pq}U_m$ and ${}_{pq}M_m$) on the periodic cell Y, that satisfied the system (21.2.33) ((21.2.34)) subject to homogenized contact conditions and null average conditions of the local functions, as listed below:

Local problems ${}_{pq}\mathscr{L}^1$

$$\left(C_{ijpq} + C_{ijmn} \left({}_{pq}N_{m|n} + \epsilon_{mnk}\, {}_{pq}V_k \right) + B_{ijmn}\, {}_{pq}V_{m|n} \right)_{|j} = 0, \quad \text{in Y,}$$

$$\left(B_{ijpq} + B_{ijmn} \left({}_{pq}N_{m|n} + \epsilon_{mnk}\, {}_{pq}V_k \right) + D_{ijmn}\, {}_{pq}V_{m|n} \right)_{|j} = 0, \quad \text{in Y,}$$

$$\| {}_{pq}N_m \| = 0, \quad \| {}_{pq}V_m \| = 0 \quad \text{over } \Gamma,$$

$$\| {}_{pq}\sigma_{ji}^1\, n_i \| = - \| C_{ijpq} \| n_i, \quad \| {}_{pq}\mu_{ji}^1\, n_i \| = - \| B_{ijpq} \| n_i \quad \text{over } \Gamma,$$

$$\left\langle {}_{pq}N_m \right\rangle_Y = 0, \quad \left\langle {}_{pq}V_m \right\rangle_Y = 0,$$

$$_{pq}N_m(y_i)|_{y_i=0} = {}_{pq}N_m(y_i)|_{y_i=1}, \quad {}_{pq}V_m(y_i)|_{y_i=0} = {}_{pq}V_m(y_i)|_{y_i=1},$$
$$(21.2.45)$$

where $_{pq}\sigma_{ji}^1 = C_{ijmn}\left(_{pq}N_{m|n} + \epsilon_{mnk}\,_{pq}V_k\right) + B_{ijmn}\,_{pq}V_{m|n}$ and $_{pq}\mu_{ji}^1 = B_{ijmn}\left(_{pq}N_{m|n} + \epsilon_{mnk}\,_{pq}V_k\right) + D_{ijmn}\,_{pq}V_{m|n}$. Here, the double bar $\|f\|$ means the jump of the function f across the interface surface Γ, n_i is the i-th component of the unit outer normal vector to Γ, and $\langle\bullet\rangle_Y$ is the average operator over the periodic cell, such as: $\langle\bullet\rangle_Y = \int_Y(\bullet)\,dy$.

Local problems $_{pq}\mathcal{L}^2$

$$\left(B_{ijpq} + C_{ijmn}\left(_{pq}U_{m|n} + \epsilon_{mnk}\,_{pq}M_k\right) + B_{ijmn}\,_{pq}M_{m|n}\right)_{|j} = 0, \text{ in Y},$$

$$\left(D_{ijpq} + B_{ijmn}\left(_{pq}U_{m|n} + \epsilon_{mnk}\,_{pq}M_k\right) + D_{ijmn}\,_{pq}M_{m|n}\right)_{|j} = 0, \text{ in Y},$$

$$\|_{pq}U_m\| = 0, \quad \|_{pq}M_m\| = 0 \quad \text{over } \Gamma,$$

$$\|_{pq}\sigma_{ji}^2\,n_i\| = -\|B_{ijpq}\|n_i, \quad \|_{pq}\mu_{ji}^2\,n_i\| = -\|D_{ijpq}\|n_i \quad \text{over } \Gamma,$$

$$\langle_{pq}U_m\rangle_Y = 0, \qquad \langle_{pq}M_m\rangle_Y = 0,$$

$$_{pq}U_m(y_i)|_{y_i=0} = _{pq}U_m(y_i)|_{y_i=1}, \quad _{pq}M_m(y_i)|_{y_i=0} = _{pq}M_m(y_i)|_{y_i=1},$$
$$\tag{21.2.46}$$

where $_{pq}\sigma_{ji}^2 = C_{ijmn}\left(_{pq}U_{m|n} + \epsilon_{mnk}\,_{pq}M_k\right) + B_{ijmn}\,_{pq}M_{m|n}$ and $_{pq}\mu_{ji}^2 = B_{ijmn}\left(_{pq}U_{m|n} + \epsilon_{mnk}\,_{pq}M_k\right) + D_{ijmn}\,_{pq}M_{m|n}$.

21.3 Example. Effective properties of heterogeneous periodic Cosserat laminate media

Let us now consider a bounded periodically Cosserat laminated composite Ω with boundary $\partial\Omega$ in the Cartesian system $\{x_1, x_2, x_3\}$, see Fig. 21.1a. The laminate Ω is defined by repetitions of the periodic cell Y, in which the layered direction is along the x_3-axis, i.e., the x_3-axis is perpendicular to the face surfaces of the layer and the lower face surface corresponds to the value $x_3 = 0$. The periodic cell Y ($Y = \{(y_1, y_2, y_3) \in R^3 : 0 \le y_i \le l_i\}$) with $i = 1, 2, 3$ is defined by a bilaminated composite in the Cartesian system of coordinates $\{y_1, y_2, y_3\}$, at the microscale level, where l_i is the length of the periodic cell in the x_i direction and L denotes the plate thickness (see Fig. 21.1). Also, a perfect interface region is assumed between the layers, denoted by Γ. In this case, the Cosserat coefficients C_{ijkl}, B_{ijkl}, and D_{ijkl} are functions of the coordinate x_3. Therefore, the desired local functions $_{pq}N_m$, $_{pq}U_m$, $_{pq}V_m$, and $_{pq}M_m$ depend only on y_3 as well.

Under these assumptions the local problems $_{pq}\mathcal{L}^1$ (21.2.45) and $_{pq}\mathcal{L}^2$ (21.2.46) become into an ordinary integro-differential equation systems subject to homogenized perfect conditions, as follows:

474 Mechanics and Physics of Structured Media

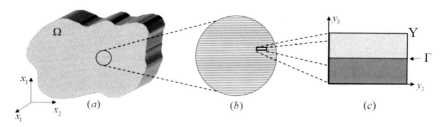

FIGURE 21.1 a) Heterogeneous Cosserat laminated composite; b) blow-up of periodic structure; c) the cross section of the periodic cell Y at the plane Oy_2y_3.

Local problems $_{pq}\mathcal{L}^1$

$$\left(C_{i3pq} + C_{i3m3\ pq}N'_m + C_{i3mn}\,\epsilon_{mnk\ pq}V_k + B_{i3m3\ pq}V'_m\right)' = 0, \quad \text{in } Y,$$

$$\left(B_{i3pq} + B_{i3m3\ pq}N'_m + B_{i3mn}\epsilon_{mnk\ pq}V_k + D_{i3m3\ pq}V'_m\right)' = 0, \quad \text{in } Y,$$

$$\|_{pq}N_m\| = 0, \quad \|_{pq}V_m\| = 0 \quad \text{over } \Gamma,$$

$$\|_{pq}\sigma^1_{3i}\,n_i\| = -\|C_{i3pq}\|n_i, \quad \|_{pq}\mu^1_{3i}\,n_i\| = -\|B_{i3pq}\|n_i \quad \text{over } \Gamma,$$

$$\langle_{pq}N_m\rangle_Y = 0, \quad \langle_{pq}V_m\rangle_Y = 0,$$

(21.3.1)

where $_{pq}\sigma^1_{3i} = C_{i3m3\ pq}N'_m + C_{i3mn}\,\epsilon_{mnk\ pq}V_k + B_{i3m3\ pq}V'_m$ and $_{pq}\mu^1_{3i} = B_{i3m3\ pq}N'_m + B_{i3mn}\,\epsilon_{mnk\ pq}V_k + D_{i3m3\ pq}V'_m$. Here the prime denotes the ordinary derivative with respect to y_3.

Local problems $_{pq}\mathcal{L}^2$

$$\left(B_{i3pq} + C_{i3m3\ pq}U'_m + C_{i3mn}\,\epsilon_{mnk\ pq}M_k + B_{i3m3\ pq}M'_m\right)' = 0, \quad \text{in } Y,$$

$$\left(D_{i3pq} + B_{i3m3\ pq}U'_m + B_{i3mn}\,\epsilon_{mnk\ pq}M_k + D_{i3m3\ pq}M'_m\right)' = 0, \quad \text{in } Y,$$

$$\|_{pq}U_m\| = 0, \quad \|_{pq}M_m\| = 0 \quad \text{over } \Gamma,$$

$$\|_{pq}\sigma^2_{3i}\,n_i\| = -\|B_{i3pq}\|n_i, \quad \|_{pq}\mu^2_{3i}\,n_i\| = -\|D_{i3pq}\|n_i \quad \text{over } \Gamma,$$

$$\langle_{pq}U_m\rangle_Y = 0, \quad \langle_{pq}M_m\rangle_Y = 0,$$

(21.3.2)

where $_{pq}\sigma^2_{3i} = C_{i3mn\ pq}U'_m + C_{i3mn}\,\epsilon_{mnk\ pq}M_k + B_{i3m3\ pq}M'_m$ and $_{pq}\mu^2_{3i} = B_{i3m3\ pq}U'_m + B_{i3mn}\,\epsilon_{mnk\ pq}M_k + D_{i3m3\ pq}M'_m$.

Consequently, the equivalent homogenized equations of the problem (21.2.40a)–(21.2.40b) for a micropolar laminated composite is determined by

$$C^{\text{eff}}_{i3pq}\left(u^{(0)}_{p,q} + \epsilon_{pqk}\,\omega^{(0)}_k\right)' + B^{\text{eff}}_{i3pq}\left(\omega^{(0)}_{p,q}\right)' + f_i = 0, \quad (21.3.3)$$

$$B^{\text{eff}}_{i3pq}\left(u^{(0)}_{p,q} + \epsilon_{pqk}\,\omega^{(0)}_k\right)' + D^{\text{eff}}_{i3pq}\,\omega^{(0)}_{p,q}$$

$$+ \epsilon_{i3l}\left[C^{\text{eff}}_{l3pq}\left(u^{(0)}_{p,q} + \epsilon_{pqk}\,\omega^{(0)}_k\right) + B^{\text{eff}}_{l3pq}\,\omega^{(0)}_{p,q}\right] + g_i = 0, \quad (21.3.4)$$

Asymptotic homogenization approach **Chapter | 21 475**

where the corresponding effective properties are given as follows:

$$C^{\text{eff}}_{ijpq} = \langle C_{ijpq} + C_{ijm3\,pq} N'_m + C_{ijmn}\,\epsilon_{mnk\,pq} V_k + B_{ijm3\,pq} V'_m \rangle_Y, \quad (21.3.5)$$

$$B^{\text{eff}}_{ijpq} = \langle B_{ijpq} + C_{ijm3\,pq} U'_m + C_{ijmn}\,\epsilon_{mnk\,pq} M_k + B_{ijm3\,pq} M'_m \rangle_Y$$

$$= \langle B_{ijpq} + B_{ijm3\,pq} N'_m + B_{ijmn}\,\epsilon_{mnk\,pq} V_k + D_{ijm3\,pq} V'_m \rangle_Y, \quad (21.3.6)$$

$$D^{\text{eff}}_{ijpq} = \langle D_{ijpq} + B_{ijm3\,pq} U'_m + B_{ijmn}\,\epsilon_{mnk\,pq} M_k + D_{ijm3\,pq} M'_m \rangle_Y, \quad (21.3.7)$$

In addition, the boundary conditions of the laminate are given by

$$_{pq}N_m\big|_{y_3=0,l} = 0, \quad _{pq}U_m\big|_{y_3=0,l} = 0, \quad _{pq}V_m\big|_{y_3=0,l} = 0, \quad _{pq}M_m\big|_{y_3=0,l} = 0. \quad (21.3.8)$$

Now, we focus to find the solutions of the local problem $_{pq}\mathcal{L}^1$ in order to find the local functions $_{pq}N_m$ and $_{pq}V_m$. Then, integrating the first equation in (21.3.1), we have

$$C_{i3pq} + C_{i3m3\,pq}N'_m + C_{i3mn}\epsilon_{mnk\,pq}V_k + B_{i3m3\,pq}V'_m = {}_{pq}A_i = constant, \quad (21.3.9)$$

Note that, each term in Eq. (21.3.9), for fixed pq, is a vector of order 3.

Let us solve (21.3.9) for the vector $_{pq}N'_m$ by the multiplication of the inverse matrix of C_{m3i3} of order 3×3, which will be written as C^{-1}_{i3m3}. Thus

$$_{pq}N'_m = C^{-1}_{m3i3}\,{}_{pq}A_i - C^{-1}_{m3i3}\left(C_{i3pq} + C_{i3kl}\,\epsilon_{kls}\,{}_{pq}V_s + B_{i3c3}\,{}_{pq}V'_c\right). \quad (21.3.10)$$

Next, since, $\langle _{pq}N'_m \rangle = 0$ we take the average of (21.3.10) and find $_{pq}A_i$, so that,

$$_{pq}A_i = \left\langle C^{-1}_{i3k3}\right\rangle^{-1}\left\langle C^{-1}_{k3l3}\,C_{l3pq}\right\rangle + \left\langle C^{-1}_{i3k3}\right\rangle^{-1}\left\langle C^{-1}_{k3l3}\,C_{l3ab}\,\epsilon_{abs}\,{}_{pq}V_s\right\rangle$$

$$+ \left\langle C^{-1}_{i3k3}\right\rangle^{-1}\left\langle C^{-1}_{k3l3}\,B_{l3c3}\,{}_{pq}V'_c\right\rangle. \quad (21.3.11)$$

Then, we can write $_{pq}N'_m$ as a function of $_{pq}V_m$ and $_{pq}V'_m$ replacing (21.3.11) into (21.3.10) and collecting the appropriate terms, as follows

$$_{pq}N'_m = \left[C^{-1}_{m3l3}\left\langle C^{-1}_{l3k3}\right\rangle^{-1}\left\langle C^{-1}_{k3d3}\,C_{d3rs}\right\rangle - C^{-1}_{m3l3}\,C_{l3rs}\right]$$

$$+ \left[C^{-1}_{m3l3}\left\langle C^{-1}_{l3k3}\right\rangle^{-1}\left\langle C^{-1}_{k3d3}\,C_{d3ab}\,\epsilon_{abc}\,{}_{pq}V_c\right\rangle - C^{-1}_{m3l3}\,C_{l3ab}\,\epsilon_{abc}\,{}_{pq}V_c\right]$$

$$+ \left[C^{-1}_{m3l3}\left\langle C^{-1}_{l3k3}\right\rangle^{-1}\left\langle C^{-1}_{k3d3}\,B_{d3c3}\,{}_{pq}V'_c\right\rangle - C^{-1}_{m3l3}\,B_{l3c3}\,{}_{pq}V'_c\right]. \quad (21.3.12)$$

476 Mechanics and Physics of Structured Media

Note that the average values of the three expressions between brackets in (21.3.12) are zero. Therefore, the expression of $_{pq}N'_m$ can be written as

$$_{pq}N'_m = r_{mpq} + \hat{r}^{(1)}_{mpq} + \hat{r}^{(2)}_{mpq}, \qquad (21.3.13)$$

where

$$r_{mpq} = C^{-1}_{m3l3}\left\langle C^{-1}_{l3k3}\right\rangle^{-1}\left\langle C^{-1}_{k3d3}\,C_{d3pq}\right\rangle - C^{-1}_{m3l3}\,C_{l3pq},$$

$$\hat{r}^{(1)}_{mpq} = C^{-1}_{m3l3}\left\langle C^{-1}_{l3k3}\right\rangle^{-1}\left\langle C^{-1}_{k3d3}\,C_{d3ab}\,\epsilon_{abc}\,{}_{pq}V_c\right\rangle - C^{-1}_{m3l3}\,C_{l3ab}\,\epsilon_{abc}\,{}_{pq}V_c,$$

$$\qquad\qquad\qquad\qquad\qquad\qquad\qquad\qquad\qquad\qquad\qquad\qquad\qquad (21.3.14)$$

$$\hat{r}^{(2)}_{mpq} = C^{-1}_{m3l3}\left\langle C^{-1}_{l3k3}\right\rangle^{-1}\left\langle C^{-1}_{k3d3}\,B_{d3c3}\,{}_{pq}V'_c\right\rangle - C^{-1}_{m3l3}\,B_{l3c3}\,{}_{pq}V'_c.$$

Similarly, if the second equation of (21.3.1) is integrated in a similar form that (21.3.9) and the condition $\left\langle _{pq}V'_m\right\rangle = 0$ is applied, we have

$$_{pq}V'_m = D^{-1}_{m3l3}\Big[\left\langle D^{-1}_{l3k3}\right\rangle^{-1}\left\langle D^{-1}_{k3d3}\left(B_{d3rs} + B_{d3ab}\,\epsilon_{abc}\,{}_{pq}V_c + B_{d3c3}\,{}_{pq}N'_c\right)\right\rangle$$

$$- \left(B_{l3pq} + B_{l3ab}\,\epsilon_{abc}\,{}_{pq}V_c + B_{l3c3}\,{}_{pq}N'_c\right)\Big]. \qquad (21.3.15)$$

Here, likewise as in Eq. (21.3.9), we use the same notation for the inverse matrix D_{l3m3} of order 3×3, which is written as D^{-1}_{m3l3}.

So, replacing (21.3.13) into (21.3.15), it yields

$$_{pq}V'_m = -D^{-1}_{m3l3}\Big[B_{l3pq} + B_{l3ab}\,\epsilon_{abc}\,{}_{pq}V_c + B_{l3c3}\left(r_{cpq} + \hat{r}^{(1)}_{cpq} + \hat{r}^{(2)}_{cpq}\right)$$

$$- \left\langle D^{-1}_{l3k3}\right\rangle^{-1}\left\langle D^{-1}_{k3d3}\left(B_{d3rs} + B_{d3ab}\,\epsilon_{abc}\,{}_{pq}V_c\right.\right.$$

$$\left.\left. + B_{d3c3}\left(r_{cpq} + \hat{r}^{(1)}_{cpq} + \hat{r}^{(2)}_{cpq}\right)\right)\right\rangle\Big], \qquad (21.3.16)$$

and grouping conveniently in (21.3.16), we have

$$_{pq}V'_m = \Big[D^{-1}_{m3l3}\left\langle D^{-1}_{l3k3}\right\rangle^{-1}\left\langle D^{-1}_{k3d3}\left(B_{d3pq} + B_{d3c3}\,r_{cpq}\right)\right\rangle$$

$$- D^{-1}_{m3l3}\left(B_{l3pq} + B_{l3c3}\,r_{cpq}\right)\Big]$$

$$+ \Big[D^{-1}_{m3l3}\left\langle D^{-1}_{l3k3}\right\rangle^{-1}\left\langle D^{-1}_{k3d3}\left(B_{d3ab}\,\epsilon_{abk}\,{}_{pq}V_k + B_{d3c3}\,\hat{r}^{(1)}_{cpq}\right)\right\rangle$$

$$- D^{-1}_{m3l3}\left(B_{l3ab}\,\epsilon_{abk}\,{}_{pq}V_k + B_{l3c3}\,\hat{r}^{(1)}_{cpq}\right)\Big]$$

$$+ \Big[D^{-1}_{m3l3}\left\langle D^{-1}_{l3k3}\right\rangle^{-1}\left\langle D^{-1}_{k3d3}\,B_{d3c3}\,\hat{r}^{(2)}_{ck3}\right\rangle - D^{-1}_{m3l3}\,B_{l3c3}\,\hat{r}^{(2)}_{cpq}\Big].$$

$$\qquad\qquad\qquad\qquad\qquad\qquad\qquad\qquad\qquad\qquad\qquad\qquad\qquad (21.3.17)$$

Analogous to (21.3.12), the average value of the three expressions between brackets in (21.3.17) is zero, then, we can rewrite (21.3.17) in the same form as (21.3.13), i.e.,

$$_{pq}V'_m = e_{mpq} + \hat{e}^{(1)}_{mpq} + \hat{e}^{(2)}_{mpq}, \tag{21.3.18}$$

where

$$
\begin{aligned}
e_{mpq} &= D^{-1}_{m3l3}\left\langle D^{-1}_{l3k3}\right\rangle^{-1}\left\langle D^{-1}_{k3d3}\left(B_{d3pq} + B_{d3c3}\, r_{cpq}\right)\right\rangle \\
&\quad - D^{-1}_{m3l3}\left(B_{l3pq} + B_{l3c3}\, r_{cpq}\right),\\
\hat{e}^{(1)}_{mpq} &= D^{-1}_{m3l3}\left\langle D^{-1}_{l3k3}\right\rangle^{-1}\left\langle D^{-1}_{k3d3}\left(B_{d3ab}\,\epsilon_{abk\,pq}\, V_k + B_{d3c3}\,\hat{r}^{(1)}_{cpq}\right)\right\rangle \\
&\quad - D^{-1}_{m3l3}\left(B_{l3ab}\,\epsilon_{abk\,pq}\, V_k + B_{l3c3}\,\hat{r}^{(1)}_{cpq}\right),\\
\hat{e}^{(2)}_{mpq} &= D^{-1}_{m3l3}\left\langle D^{-1}_{l3k3}\right\rangle^{-1}\left\langle D^{-1}_{k3d3}\, B_{d3c3}\,\hat{r}^{(2)}_{ck3}\right\rangle - D^{-1}_{m3l3}\, B_{l3c3}\,\hat{r}^{(2)}_{cpq}. \tag{21.3.19}
\end{aligned}
$$

Let us analyze the structure of (21.3.12) and (21.3.17) and for simplicity the indexes will be omitted

$$
\begin{aligned}
\mathbf{N}'(y) &= \mathbf{a}^{(1)}(y) + \left(\mathbf{a}^{(2)}(y)\left\langle \mathbf{a}^{(3)}(y)\mathbf{V}(y)\right\rangle - \mathbf{a}^{(3)}(y)\mathbf{V}(y)\right) \\
&\quad + \left(\mathbf{a}^{(2)}(y)\left\langle \mathbf{a}^{(3)}(y)\mathbf{V}'(y)\right\rangle - \mathbf{a}^{(3)}(y)\mathbf{V}'(y)\right),\\
\mathbf{V}'(y) &= \mathbf{b}^{(1)}(y) + \left(\mathbf{b}^{(2)}(y)\left\langle \mathbf{b}^{(3)}(y)\mathbf{V}(y)\right\rangle - \mathbf{b}^{(3)}(y)\mathbf{V}(y)\right) \\
&\quad + \left(\mathbf{b}^{(2)}(y)\left\langle \mathbf{b}^{(3)}(y)\mathbf{V}'(y)\right\rangle - \mathbf{b}^{(3)}(y)\mathbf{V}'(y)\right),
\end{aligned}
$$

where $\mathbf{a}^{(n)}$ and $\mathbf{b}^{(n)}$ ($n = 1, 2, 3$) are functions on $y \equiv y_3 = x_3/\varepsilon$, and the symbols $\mathbf{N}$ and $\mathbf{V}$ stand for the local functions $_{pq}N_m$ and $_{pq}V_m$, respectively. Since the components of $\mathbf{a}^{(2)}(y)$ and $\mathbf{b}^{(2)}(y)$ are approximately 1, the second and third terms the expressions can be considered as deviations from the mean value of $\mathbf{V}$ and $\mathbf{V}'$ respectively, therefore; the main contribution to the expressions of the functions $_{pq}N'_m$ and $_{pq}V'_m$, i.e., (21.3.12) and (21.3.19) is given by the first term. In order to obtain approximate expressions for $_{pq}N'_m$ and $_{pq}V'_m$ we preserve only the first terms in expressions (21.3.12) and (21.3.19), i.e.,

$$_{pq}N'_m(y) \approx r_{mpq} = C^{-1}_{m3l3}\left\langle C^{-1}_{l3k3}\right\rangle^{-1}\left\langle C^{-1}_{k3d3}\, C_{d3pq}\right\rangle - C^{-1}_{m3l3}\, C_{l3pq}, \tag{21.3.20}$$

$$
\begin{aligned}
_{pq}N_m(y) &= \int_0^y {}_{pq}N'_m(z)\,\mathrm{d}z - \left\langle \int_0^y {}_{pq}N'_m(z)\,\mathrm{d}z\right\rangle \\
&\approx \int_0^y \left(C^{-1}_{m3l3}\left\langle C^{-1}_{l3k3}\right\rangle^{-1}\left\langle C^{-1}_{k3d3}\, C_{d3pq}\right\rangle - C^{-1}_{m3l3}\, C_{l3pq}\right)\mathrm{d}z,
\end{aligned}
$$

$$\tag{21.3.21}$$

$$_{pq}V'_m(y) \approx e_{mpq} = D^{-1}_{m3l3} \left\langle D^{-1}_{l3k3} \right\rangle^{-1} \left\langle D^{-1}_{k3d3} \left(B_{d3pq} + B_{d3c3}\, r_{cpq} \right) \right\rangle$$
$$- D^{-1}_{m3l3} \left(B_{l3pq} + B_{l3p3}\, r_{cpq} \right), \tag{21.3.22}$$

$$_{pq}V_m(y) = \int_0^y {}_{pq}V'_m(z)\,\mathrm{d}z - \left\langle \int_0^y {}_{pq}V'_m(z)\,\mathrm{d}z \right\rangle$$
$$\approx \int_0^y \left(\left(D^{-1}_{m3l3} \left\langle D^{-1}_{l3k3} \right\rangle^{-1} \left\langle D^{-1}_{k3d3} \left(B_{d3pq} + B_{d3c3}\, r_{cpq} \right) \right\rangle \right. \right.$$
$$\left. \left. - D^{-1}_{m3l3} \left(B_{l3pq} + B_{l3c3}\, r_{cpq} \right) \right) \right)\,\mathrm{d}z. \tag{21.3.23}$$

The expressions (21.3.20)–(21.3.23) are approximate solutions for the local problem $_{pq}\mathcal{L}^1$ (see (21.3.1)). In order to find the solution of the local problem $_{pq}\mathcal{L}^2$ (i.e., $_{pq}U_m$ and $_{pq}M_m$ in (21.3.2)), a procedure analogous to the one previously described is performed, therefore, only the solutions are pointed out here as follows

$$_{pq}U'_m(y) \approx C^{-1}_{m3l3} \left\langle C^{-1}_{l3k3} \right\rangle^{-1} \left\langle C^{-1}_{k3d3}\, B_{d3pq} \right\rangle - C^{-1}_{m3l3}\, B_{l3pq}, \tag{21.3.24}$$

$$_{pq}U_m(y) = \int_0^y {}_{pq}U'_m(z)\,\mathrm{d}z - \left\langle \int_0^y {}_{pq}U'_m(z)\,\mathrm{d}z \right\rangle$$
$$\approx \int_0^y \left(C^{-1}_{m3l3} \left\langle C^{-1}_{l3k3} \right\rangle^{-1} \left\langle C^{-1}_{k3d3}\, B_{d3pq} \right\rangle - C^{-1}_{m3l3}\, B_{l3pq} \right)\,\mathrm{d}z, \tag{21.3.25}$$

$$_{pq}M'_m(y) \approx D^{-1}_{m3l3} \left\langle D^{-1}_{l3k3} \right\rangle \left\langle D^{-1}_{k3d3} \left(D_{q3pq} + B_{d3c3}\, s_{cpq} \right) \right\rangle$$
$$- D^{-1}_{m3l3} \left(D_{l3pq} + B_{l3c3}\, s_{cpq} \right), \tag{21.3.26}$$

$$_{pq}M_m(y) = \int_0^y {}_{pq}M'_m(z)\,\mathrm{d}z - \left\langle \int_0^y {}_{pq}M'_m(z)\,\mathrm{d}z \right\rangle$$
$$\approx \int_0^y \left(D^{-1}_{m3l3} \left\langle D^{-1}_{l3k3} \right\rangle \left\langle D^{-1}_{k3d3} \left(D_{d3pq} + B_{d3c3}\, s_{cpq} \right) \right\rangle \right.$$
$$\left. - D^{-1}_{m3l3} \left(D_{l3pq} + B_{l3c3}\, s_{cpq} \right) \right)\,\mathrm{d}z, \tag{21.3.27}$$

where

$$s_{mpq} = C^{-1}_{m3l3} \left\langle C^{-1}_{l3k3} \right\rangle^{-1} \left\langle C^{-1}_{k3d3}\, B_{d3mn} \right\rangle - C^{-1}_{m3l3}\, B_{l3mn}. \tag{21.3.28}$$

Once the local functions $_{pq}N_m$, $_{pq}U_m$, $_{pq}V_m$, and $_{pq}M_m$ solutions of the local problems $_{pq}\mathcal{L}^1$ (21.3.1) and $_{pq}\mathcal{L}^2$ (21.3.2) are found, the effective coefficients of Cosserat homogeneous laminated media ((21.3.5)–(21.3.7)) can be computed by substituting the local functions and their derivatives into the expressions (21.3.5)–(21.3.7).

Asymptotic homogenization approach **Chapter | 21 479**

In the present work, we derive the effective properties for centro-symmetric laminated Cosserat materials. A material is considered to be centro-symmetric if $-\mathbf{I}$ ($\mathbf{I}$ is the second order identity tensor) is a symmetry transformation of their constitutive law. For this type of materials the coupling moduli B_{ijpq} are not considered because they do not satisfy the symmetry condition $B_{ijpq} = B_{pqij}$. This condition means that the stresses do not longer depend on the microcurvatures and that couple-stresses do not depend on the strains since B_{ijpq} is responsible of the coupling between stresses and microcurvatures and between couple stresses and strains. Let us write the condition explicitly

$$B_{ijpq} = 0. \tag{21.3.29}$$

If we apply (21.3.29) to (21.3.22) we obtain

$$_{pq}V'_m = 0. \tag{21.3.30}$$

Therefore, to determinate the local function $_{pq}V_m$ it is necessary to integrate (21.3.22) and taking into account that $\langle_{pq}V_m\rangle = 0$, thus

$$_{pq}V_m = \int_0^L {}_{pq}V'_m \, dy - \left\langle \int_0^L {}_{pq}V'_m \, dy \right\rangle \equiv 0. \tag{21.3.31}$$

Then, taking into account (21.3.29)–(21.3.31), the expressions (21.3.20) and (21.3.26) become

$$_{pq}N'_m = C_{m3l3}^{-1} \left\langle C_{l3k3}^{-1} \right\rangle^{-1} \left\langle C_{k3d3}^{-1} C_{d3pq} \right\rangle - C_{m3l3}^{-1} C_{l3pq}, \tag{21.3.32}$$

$$_{pq}M'_m = D_{m3l3}^{-1} \left\langle D_{l3k3}^{-1} \right\rangle \left\langle D_{k3d3}^{-1} D_{d3pq} \right\rangle - D_{m3l3}^{-1} D_{l3pq}. \tag{21.3.33}$$

Finally, replacing (21.3.29)–(21.3.33) into the statements (21.3.1)–(21.3.7), the effective properties of a micropolar centro-symmetric laminated composites can be found as follows

$$C_{ijpq}^{\text{eff}} = \left\langle C_{ijpq} + C_{ijm3} C_{m3l3}^{-1} \left(\left\langle C_{l3k3}^{-1} \right\rangle^{-1} \left\langle C_{k3d3}^{-1} C_{d3pq} \right\rangle - C_{l3pq} \right) \right\rangle, \tag{21.3.34}$$

$$D_{ijpq}^{\text{eff}} = \left\langle D_{ijpq} + D_{ijm3} D_{m3l3}^{-1} \left(\left\langle D_{l3k3}^{-1} \right\rangle \left\langle D_{k3d3}^{-1} D_{d3pq} \right\rangle - D_{l3pq} \right) \right\rangle, \tag{21.3.35}$$

$$B_{ijpq}^{\text{eff}} = 0. \tag{21.3.36}$$

Note that the expression (21.3.34) is reported for the Cauchy classical elasticity (see Ref. [35], eq. 1.12, page 145). The expression (21.3.36) is not obvious, it is important to notice that (21.3.2) does not depend only on B_{ijpq}, but also depend on $_{pq}V'_m$ and due to the fact that $_{pq}V'_m$ depends on B_{ijpq} in every single term then $_{pq}V'_m$ becomes zero what makes (21.3.3) zero.

480 Mechanics and Physics of Structured Media

There are 32 crystal groups and five transversal isotropy groups. If the constitutive law of the material is centro-symmetric only 13 groups are independent, denoting by C_n ($n = 1, 2, ..., 13$) the symmetry group of Cosserat centro-symmetric materials, see details in Refs. [20] and [45]. Including the full isotropy group C_0, in Ref. [45], tables for the material moduli for the 14 symmetry groups are found. In the next section, as an example, the numerical values of the effective properties for a Cosserat laminated composite with cubic symmetry are determined.

21.4 Numerical results

21.4.1 Cosserat laminated composite with cubic constituents

Now, the effective properties of a periodic Cosserat laminated composite with cubic constituents are computed. For that, the formulas of the effective Cosserat properties ((21.3.34)–(21.3.36)) are applied. Considering that each layer of the periodic Cosserat laminated composite posses cubic symmetry, we can write the constituents material elastic properties as follows:

$$
\begin{aligned}
C_{ijpq} &= \lambda\, \delta_{ij}\, \delta_{pq} + (\mu + \alpha)\, \delta_{ip}\, \delta_{jq} + (\mu - \alpha)\, \delta_{iq}\, \delta_{jp} + \eta\, \delta_{ijpq}, \\
D_{ijpq} &= \beta\, \delta_{ij}\, \delta_{pq} + (\gamma + \epsilon)\, \delta_{ip}\, \delta_{jq} + (\gamma - \epsilon)\, \delta_{iq}\, \delta_{jp} + \nu\, \delta_{ijpq}, \qquad (21.4.1) \\
B_{ijpq} &= 0,
\end{aligned}
$$

where $\delta_{ijpq} = 1$ if $i = j = p = q$, else $\delta_{ijpq} = 0$.

As can be seen in (21.4.1), a cubic-symmetric micropolar solid is characterized by eight independent constants in contrast to an elastic medium with cubic constituents where only three elastic constants are defined. Here, μ is the shear modulus, λ is the Lamé coefficient, and η is the classical cubic constant or anisotropy constant. Also, the five elastic moduli, i.e., α, β, γ, ϵ, and ν, are known as the Cosserat or micropolar elastic constants (α is also known as the micropolar couple modulus). Isotropic reduced Cosserat medium can be obtained considering η, β, γ, ϵ, and ν to be zero (a reduced Cosserat medium is one whose particles have independent translational and rotational degrees of freedom, but the medium does not react to the gradient of rotation of point bodies, no internal moments work on the gradient of angular velocity, see Ref. [46]); and if the aforementioned properties are null except λ and μ, i.e., $\eta = \beta = \gamma = \epsilon = \nu = \alpha = 0$, we have the classical isotropic medium. A more complete dissertation on the material elastic constants for the linear isotropic case can be found in Refs. [6] and [20]. The standard bulk, transversal, and axial shear moduli are given by $K = \lambda + \frac{2}{3}\mu + \frac{1}{3}\eta$, $M = \mu + \frac{1}{2}\eta$, and μ; and the micropolar torsional bulk, transversal, and axial stand as $\mathcal{K} = \beta + \frac{2}{3}\gamma + \frac{1}{3}\nu$, $\mathcal{M} = \gamma + \frac{1}{2}\nu$, and γ.

For Cosserat composite with cubic symmetry, the elastic moduli satisfy the following restrictions derived from the positiveness of the quadratic parts of the internal energy, as follows

$$\mu > 0, \quad \alpha > 0, \quad 2\mu + \eta > 0, \quad 3\lambda + 2\mu + \eta > 0,$$
$$\gamma > 0, \quad \epsilon > 0, \quad 2\gamma + \nu > 0, \quad 3\beta + 2\gamma + \nu > 0, \tag{21.4.2}$$

where the stiffness C_{ijpq} and the torque D_{ijpq} moduli in matrix form are given by

$C_{ijpq} =$

$\begin{array}{c}pq \rightarrow \\ \hline ij \downarrow\end{array}$	11	22	33	12	13	23	21	31	32
11	$\lambda + 2\mu + \eta$	λ	λ						
22	λ	$\lambda + 2\mu + \eta$	λ						
33	λ	λ	$\lambda + 2\mu + \eta$						
12				$\mu + \alpha$			$\mu - \alpha$		
13					$\mu + \alpha$			$\mu - \alpha$	
23						$\mu + \alpha$			$\mu - \alpha$
21				$\mu - \alpha$			$\mu + \alpha$		
31					$\mu - \alpha$			$\mu + \alpha$	
32						$\mu - \alpha$			$\mu + \alpha$

and

$D_{ijpq} =$

$\begin{array}{c}pq \rightarrow \\ \hline ij \downarrow\end{array}$	11	22	33	12	13	23	21	31	32
11	$\beta + 2\gamma + \nu$	β	β						
22	β	$\beta + 2\gamma + \nu$	β						
33	β	β	$\beta + 2\gamma + \nu$						
12				$\gamma + \epsilon$			$\gamma - \epsilon$		
13					$\gamma + \epsilon$			$\gamma - \epsilon$	
23						$\gamma + \epsilon$			$\gamma - \epsilon$
21				$\gamma - \epsilon$			$\gamma + \epsilon$		
31					$\gamma - \epsilon$			$\gamma + \epsilon$	
32						$\gamma - \epsilon$			$\gamma + \epsilon$

Particularly, for cubic-symmetric laminated composites, the nonnull effective properties C_{ijpq}^{eff} and D_{ijpq}^{eff} following (21.3.34), (21.3.35), and (21.4.1) are written as follows

$$C_{1111}^{\text{eff}} = C_{2222}^{\text{eff}}$$
$$= \langle \lambda + 2\mu + \eta \rangle + \left\langle \frac{\lambda}{\lambda + 2\mu + \eta} \right\rangle^2 \left\langle \frac{1}{\lambda + 2\mu + \eta} \right\rangle^{-1} - \left\langle \frac{\lambda^2}{\lambda + 2\mu + \eta} \right\rangle,$$
$$C_{3333}^{\text{eff}} = \left\langle \frac{1}{\lambda + 2\mu + \eta} \right\rangle^{-1},$$

482 Mechanics and Physics of Structured Media

$$C_{1122}^{\text{eff}} = \langle \lambda \rangle + \left\langle \frac{\lambda}{\lambda + 2\mu + \eta} \right\rangle^2 \left\langle \frac{1}{\lambda + 2\mu + \eta} \right\rangle^{-1} - \left\langle \frac{\lambda^2}{\lambda + 2\mu + \eta} \right\rangle,$$

$$C_{1133}^{\text{eff}} = C_{2233}^{\text{eff}} = \left\langle \frac{\lambda}{\lambda + 2\mu + \eta} \right\rangle \left\langle \frac{1}{\lambda + 2\mu + \eta} \right\rangle^{-1},$$

$$C_{1212}^{\text{eff}} = C_{2121}^{\text{eff}} = \langle \mu + \alpha \rangle, \quad C_{1313}^{\text{eff}} = C_{2323}^{\text{eff}} = \left\langle \frac{1}{\mu + \alpha} \right\rangle^{-1},$$

$$C_{3131}^{\text{eff}} = C_{3232}^{\text{eff}} = \langle \mu + \alpha \rangle + \left\langle \frac{\mu - \alpha}{\mu + \alpha} \right\rangle^2 \left\langle \frac{1}{\mu + \alpha} \right\rangle^{-1} - \left\langle \frac{(\mu - \alpha)^2}{\mu + \alpha} \right\rangle,$$

$$C_{1221}^{\text{eff}} = \langle \mu - \alpha \rangle, \quad C_{1331}^{\text{eff}} = C_{2332}^{\text{eff}} = \left\langle \frac{\mu - \alpha}{\mu + \alpha} \right\rangle \left\langle \frac{1}{\mu + \alpha} \right\rangle^{-1}, \tag{21.4.3}$$

and

$$D_{1111}^{\text{eff}} = D_{2222}^{\text{eff}}$$

$$= \langle \beta + 2\gamma + \nu \rangle + \left\langle \frac{\beta}{\beta + 2\gamma + \nu} \right\rangle^2 \left\langle \frac{1}{\beta + 2\gamma + \nu} \right\rangle^{-1} - \left\langle \frac{\beta^2}{\beta + 2\gamma + \nu} \right\rangle,$$

$$D_{3333}^{\text{eff}} = \left\langle \frac{1}{\beta + 2\gamma + \nu} \right\rangle^{-1},$$

$$D_{1122}^{\text{eff}} = \langle \beta \rangle + \left\langle \frac{\beta}{\beta + 2\gamma + \nu} \right\rangle^2 \left\langle \frac{1}{\beta + 2\gamma + \nu} \right\rangle^{-1} - \left\langle \frac{\beta^2}{\beta + 2\gamma + \nu} \right\rangle,$$

$$D_{1133}^{\text{eff}} = D_{2233}^{\text{eff}} = \left\langle \frac{\beta}{\beta + 2\gamma + \nu} \right\rangle \left\langle \frac{1}{\beta + 2\gamma + \nu} \right\rangle^{-1},$$

$$D_{1212}^{\text{eff}} = D_{2121}^{\text{eff}} = \langle \gamma + \epsilon \rangle, \quad D_{1313}^{\text{eff}} = D_{2323}^{\text{eff}} = \left\langle \frac{1}{\gamma + \epsilon} \right\rangle^{-1},$$

$$D_{3131}^{\text{eff}} = D_{3232}^{\text{eff}} = \langle \gamma + \epsilon \rangle + \left\langle \frac{\gamma - \epsilon}{\gamma + \epsilon} \right\rangle^2 \left\langle \frac{1}{\gamma + \epsilon} \right\rangle^{-1} - \left\langle \frac{(\gamma - \epsilon)^2}{\gamma + \epsilon} \right\rangle,$$

$$D_{1221}^{\text{eff}} = \langle \gamma - \epsilon \rangle, \quad D_{1331}^{\text{eff}} = D_{2332}^{\text{eff}} = \left\langle \frac{\gamma - \epsilon}{\gamma + \epsilon} \right\rangle \left\langle \frac{1}{\gamma + \epsilon} \right\rangle^{-1}. \tag{21.4.4}$$

The average operator is interpreted for a bilaminated composite as $\langle f \rangle = f^{(1)} V_1 + f^{(2)} V_2$, which is the Voigt's average (average operator) of the property f. Here, V_1 and V_2 are the volume fractions per unit length occupied by the layer 1 and 2, respectively, and $V_1 + V_2 = 1$. A graphic idea of the bilaminated composite is shown in Fig. 21.1c.

In order to compute the effective properties (21.4.3) and (21.4.4), equivalent dimensionless forms of $C_{ijpq}^{\text{eff}}/C_{1111}^{(1)}$ and $D_{ijpq}^{\text{eff}}/D_{1111}^{(1)}$ are defined. For that, let us declare the new constants in a proper form

$$\lambda = C_{1122}, \quad \mu = \frac{C_{1212} + C_{1221}}{2}, \quad \alpha = \frac{C_{1212} - C_{1221}}{2},$$

$$\eta = C_{1111} - C_{1122} - C_{1212} - C_{1221},$$

$$\beta = D_{1122}, \quad \gamma = \frac{D_{1212} + D_{1221}}{2}, \quad \epsilon = \frac{D_{1212} - D_{1221}}{2},$$
$$\nu = D_{1111} - D_{1122} - D_{1212} - D_{1221}. \tag{21.4.5}$$

Then, applying (21.4.5) into (21.4.2), the restrictions become

$$C_{1212} + C_{1221} > 0, \quad C_{1212} - C_{1221} > 0,$$
$$C_{1111} - C_{1122} > 0, \quad C_{1111} + 2C_{1122} > 0,$$
$$D_{1212} + D_{1221} > 0, \quad D_{1212} - D_{1221} > 0,$$
$$D_{1111} - D_{1122} > 0, \quad D_{1111} + 2D_{1122} > 0, \tag{21.4.6}$$

and therefore, the effective properties (21.4.3) are rewritten as (equal moduli will be omitted)

$$C_{1111}^{\text{eff}} = \langle C_{1111} \rangle + \left\langle \frac{C_{1122}}{C_{1111}} \right\rangle^2 \left\langle \frac{1}{C_{1111}} \right\rangle^{-1} - \left\langle \frac{C_{1122}^2}{C_{1111}} \right\rangle, \quad C_{3333}^{\text{eff}} = \left\langle \frac{1}{C_{1111}} \right\rangle^{-1},$$

$$C_{1122}^{\text{eff}} = \langle C_{1122} \rangle + \left\langle \frac{C_{1122}}{C_{1111}} \right\rangle^2 \left\langle \frac{1}{C_{1111}} \right\rangle^{-1} - \left\langle \frac{C_{1122}^2}{C_{1111}} \right\rangle,$$

$$C_{1133}^{\text{eff}} = \left\langle \frac{C_{1122}}{C_{1111}} \right\rangle \left\langle \frac{1}{C_{1111}} \right\rangle^{-1}, \quad C_{1212}^{\text{eff}} = \langle C_{1212} \rangle, \quad C_{1313}^{\text{eff}} = \left\langle \frac{1}{C_{1212}} \right\rangle^{-1},$$

$$C_{3131}^{\text{eff}} = \langle C_{1212} \rangle + \left\langle \frac{C_{1221}}{C_{1212}} \right\rangle^2 \left\langle \frac{1}{C_{1212}} \right\rangle^{-1} - \left\langle \frac{C_{1221}^2}{C_{1212}} \right\rangle$$

$$C_{1221}^{\text{eff}} = \langle C_{1221} \rangle, \quad C_{1331}^{\text{eff}} = \left\langle \frac{C_{1221}}{C_{1212}} \right\rangle \left\langle \frac{1}{C_{1212}} \right\rangle^{-1}. \tag{21.4.7}$$

Notice that, when a pure cubic elastic Cauchy composite (for $\alpha = 0$) is considered, the nine effective properties (21.4.7) become into six and match with equation (35) of Ref. [47]. Therefore, applying in (21.4.7) the average operator for a bilaminated composite, the effective properties can be expressed as

$$C_{1111}^{\text{eff}} = \left(C_{1111}^{(1)} V_1 + C_{1111}^{(2)} V_2 \right) + \left(\frac{C_{1122}^{(1)}}{C_{1111}^{(1)}} + \frac{C_{1122}^{(2)}}{C_{1111}^{(2)}} V_2 \right)^2 \left(\frac{V_1}{C_{1111}^{(1)}} + \frac{V_2}{C_{1111}^{(2)}} \right)^{-1}$$
$$- \left(\frac{C_{1122}^{(1)} C_{1122}^{(1)}}{C_{1111}^{(1)}} V_1 + \frac{C_{1122}^{(2)} C_{1122}^{(2)}}{C_{1111}^{(2)}} V_2 \right),$$

$$C_{3333}^{\text{eff}} = \left(\frac{V_1}{C_{1111}^{(1)}} + \frac{V_2}{C_{1111}^{(2)}} \right)^{-1},$$

484 Mechanics and Physics of Structured Media

$$C_{1122}^{\text{eff}} = \left(C_{1122}^{(1)}V_1 + C_{1122}^{(2)}V_2\right)$$
$$+ \left(\frac{C_{1122}^{(1)}}{C_{1111}^{(1)}}V_1 + \frac{C_{1122}^{(2)}}{C_{1111}^{(2)}}V_2\right)^2 \left(\frac{V_1}{C_{1111}^{(1)}} + \frac{V_2}{C_{1111}^{(2)}}\right)^{-1}$$
$$+ \left(\frac{C_{1122}^{(1)}C_{1122}^{(1)}}{C_{1111}^{(1)}}V_1 + \frac{C_{1122}^{(2)}C_{1122}^{(2)}}{C_{1111}^{(2)}}V_2\right),$$

$$C_{1133}^{\text{eff}} = \left(\frac{C_{1122}^{(1)}}{C_{1111}^{(1)}}V_1 + \frac{C_{1122}^{(2)}}{C_{1111}^{(2)}}V_2\right)\left(\frac{V_1}{C_{1111}^{(1)}} + \frac{V_2}{C_{1111}^{(2)}}\right)^{-1},$$

$$C_{1212}^{\text{eff}} = \left(C_{1212}^{(1)}V_1 + C_{1212}^{(2)}V_2\right), \quad C_{1313}^{\text{eff}} = \left(\frac{V_1}{C_{1212}^{(1)}} + \frac{V_2}{C_{1212}^{(2)}}\right)^{-1},$$

$$C_{3131}^{\text{eff}} = \left(C_{1212}^{(1)}V_1 + C_{1212}^{(2)}V_2\right)$$
$$+ \left(\frac{C_{1221}^{(1)}}{C_{1212}^{(1)}}V_1 + \frac{C_{1221}^{(2)}}{C_{1212}^{(2)}}V_2\right)^2 \left(\frac{V_1}{C_{1212}^{(1)}} + \frac{V_2}{C_{1212}^{(2)}}\right)^{-1}$$
$$- \left(\frac{C_{1221}^{(1)}C_{1221}^{(1)}}{C_{1212}^{(1)}}V_1 + \frac{C_{1221}^{(2)}C_{1221}^{(2)}}{C_{1212}^{(2)}}V_2\right),$$

$$C_{1221}^{\text{eff}} = \left(C_{1221}^{(1)}V_1 + C_{1221}^{(2)}V_1\right),$$

$$C_{1331}^{\text{eff}} = \left(\frac{C_{1221}^{(1)}}{C_{1212}^{(1)}}V_1 + \frac{C_{1221}^{(2)}}{C_{1212}^{(2)}}V_2\right)\left(\frac{V_1}{C_{1212}^{(1)}} + \frac{V_2}{C_{1212}^{(2)}}\right)^{-1}. \tag{21.4.8}$$

Then, rewriting the stiffness effective properties in the aforementioned dimensionless form, i.e., $\hat{C}_{ijpq}^{\text{eff}} \equiv C_{ijpq}^{\text{eff}}/C_{1111}^{(1)}$ so that $A_1 = C_{1111}^{(2)}/C_{1111}^{(1)}$, $A_2 = C_{1122}^{(1)}/C_{1111}^{(1)}$, $A_3 = C_{1122}^{(2)}/C_{1111}^{(1)}$, $A_4 = C_{1212}^{(1)}/C_{1111}^{(1)}$, $A_5 = C_{1212}^{(2)}/C_{1111}^{(1)}$, $A_6 = C_{1221}^{(1)}/C_{1111}^{(1)}$, and $A_7 = C_{1221}^{(2)}/C_{1111}^{(1)}$, the dimensionless effective properties are given by

$$\hat{C}_{1111}^{\text{eff}} = (V_1 + A_1 V_2) + \left(A_2 V_1 + A_1^{-1}A_3 V_2\right)^2 \left(V_1 + A_1^{-1}V_2\right)^{-1}$$
$$- \left(A_2^2 V_1 + A_1^{-1}A_3^2 V_2\right),$$

$$\hat{C}_{1122}^{\text{eff}} = (A_2 V_1 + A_3 V_2) + \left(A_2 V_1 + A_1^{-1}A_3 V_2\right)^2 \left(V_1 + A_1^{-1}V_2\right)^{-1}$$
$$- \left(A_2^2 V_1 + A_1^{-1}A_3^2 V_2\right),$$

$$\hat{C}_{1133}^{\text{eff}} = \left(A_2 V_1 + A_1^{-1}A_3 V_2\right)\left(V_1 + A_1^{-1}V_2\right)^{-1},$$

$$\hat{C}_{3333}^{\text{eff}} = \left(V_1 + A_1^{-1}V_2\right)^{-1},$$

$$\hat{C}^{\mathrm{eff}}_{1212} = (A_4 \mathrm{V}_1 + A_5 \mathrm{V}_2), \quad \hat{C}^{\mathrm{eff}}_{1313} = \left(A_4^{-1}\mathrm{V}_1 + A_5^{-1}\mathrm{V}_2\right)^{-1},$$

$$\hat{C}^{\mathrm{eff}}_{3131} = (A_4 \mathrm{V}_1 + A_5 \mathrm{V}_2) + \left(A_4^{-1} A_6 \mathrm{V}_1 + A_5^{-1} A_7 \mathrm{V}_2\right)^2 \left(A_4^{-1}\mathrm{V}_1 + A_5^{-1}\mathrm{V}_2\right)^{-1}$$
$$- \left(A_4^{-1} A_6^2 \mathrm{V}_1 + A_5^{-1} A_7^2 \mathrm{V}_2\right),$$

$$\hat{C}^{\mathrm{eff}}_{1221} = (A_6 \mathrm{V}_1 + A_7 \mathrm{V}_2),$$

$$\hat{C}^{\mathrm{eff}}_{1331} = \left(A_4^{-1} A_6 \mathrm{V}_1 + A_5^{-1} A_7 \mathrm{V}_2\right)\left(A_4^{-1}\mathrm{V}_1 + A_5^{-1}\mathrm{V}_2\right)^{-1}. \tag{21.4.9}$$

Analogously, the dimensionless effective torque moduli $\hat{D}^{\mathrm{eff}}_{ijpq} \equiv D^{\mathrm{eff}}_{ijpq}/D^{(1)}_{1111}$ are found as the stiffness effective properties above procedure, so the steps are omitted. Only remark a few brief outlined as follows $B_1 = D^{(2)}_{1111}/D^{(1)}_{1111}$, $B_2 = D^{(1)}_{1122}/D^{(1)}_{1111}$, $B_3 = D^{(2)}_{1122}/D^{(1)}_{1111}$, $B_4 = D^{(1)}_{1212}/D^{(1)}_{1111}$, $B_5 = D^{(2)}_{1212}/D^{(1)}_{1111}$, $B_6 = D^{(1)}_{1221}/D^{(1)}_{1111}$, $B_7 = D^{(2)}_{1221}/D^{(1)}_{1111}$, and the effective torque moduli are

$$\hat{D}^{\mathrm{eff}}_{1111} = (\mathrm{V}_1 + B_1 \mathrm{V}_2) + \left(B_2 \mathrm{V}_1 + B_1^{-1} B_3 \mathrm{V}_2\right)^2 \left(\mathrm{V}_1 + B_1^{-1} \mathrm{V}_2\right)^{-1}$$
$$- \left(B_2^2 \mathrm{V}_1 + B_1^{-1} B_3^2 \mathrm{V}_2\right),$$

$$\hat{D}^{\mathrm{eff}}_{3333} = \left(\mathrm{V}_1 + B_1^{-1} \mathrm{V}_2\right)^{-1},$$

$$\hat{D}^{\mathrm{eff}}_{1122} = (B_2 \mathrm{V}_1 + B_3 \mathrm{V}_2) + \left(B_2 \mathrm{V}_1 + B_1^{-1} B_3 \mathrm{V}_2\right)^2 \left(\mathrm{V}_1 + B_1^{-1} \mathrm{V}_2\right)^{-1}$$
$$- \left(B_2^2 \mathrm{V}_1 + B_1^{-1} B_3^2 \mathrm{V}_2\right),$$

$$\hat{D}^{\mathrm{eff}}_{1133} = \left(B_2 \mathrm{V}_1 + B_1^{-1} B_3 \mathrm{V}_2\right)\left(\mathrm{V}_1 + B_1^{-1} \mathrm{V}_2\right)^{-1},$$

$$\hat{D}^{\mathrm{eff}}_{1212} = (B_4 \mathrm{V}_1 + B_5 \mathrm{V}_2), \quad \hat{D}^{\mathrm{eff}}_{1313} = \left(B_4^{-1}\mathrm{V}_1 + B_5^{-1}\mathrm{V}_2\right)^{-1},$$

$$\hat{D}^{\mathrm{eff}}_{3131} = (B_4 \mathrm{V}_1 + B_5 \mathrm{V}_2) + \left(B_4^{-1} B_6 \mathrm{V}_1 + B_5^{-1} B_7 \mathrm{V}_2\right)^2 \left(B_4^{-1}\mathrm{V}_1 + B_5^{-1}\mathrm{V}_2\right)^{-1}$$
$$- \left(B_4^{-1} B_6^2 \mathrm{V}_1 + B_5^{-1} B_7^2 \mathrm{V}_2\right),$$

$$\hat{D}^{\mathrm{eff}}_{1221} = (B_6 \mathrm{V}_1 + B_7 \mathrm{V}_2),$$

$$\hat{D}^{\mathrm{eff}}_{1331} = \left(B_4^{-1} B_6 \mathrm{V}_1 + B_5^{-1} B_7 \mathrm{V}_2\right)\left(B_4^{-1}\mathrm{V}_1 + B_5^{-1}\mathrm{V}_2\right)^{-1}. \tag{21.4.10}$$

Restrictions in (21.4.6) imply

$$
\begin{array}{llll}
A_4 + A_6 > 0, & A_4 - A_6 > 0, & 1 - A_2 > 0, & 1 + 2A_2 > 0, \\
A_5 + A_7 > 0, & A_5 - A_7 > 0, & A_1 - A_3 > 0, & A_1 + 2A_3 > 0, \\
B_4 + B_6 > 0, & B_4 - B_6 > 0, & 1 - B_2 > 0, & 1 + 2B_2 > 0, \\
B_5 + B_7 > 0, & B_5 - B_7 > 0, & B_1 - B_3 > 0, & B_1 + 2B_3 > 0.
\end{array} \tag{21.4.11}
$$

486 Mechanics and Physics of Structured Media

The computation of the stiffness $\hat{C}^{\text{eff}}_{ijpq}$ and torque $\hat{D}^{\text{eff}}_{ijpq}$ effective properties are considered for an arbitrary fictitious bilaminated composite, where the values of the above ratios are taken following the restrictions (21.4.11) as

$$A_1 = 0.45, \quad A_2 = 0.65, \quad A_3 = 0.20, \quad A_4 = 0.35, \quad A_5 = 0.026,$$
$$A_6 = 0.30, \quad A_7 = 0.024. \tag{21.4.12}$$
$$B_1 = 10.0, \quad B_2 = -0.4, \quad B_3 = -3.7, \quad B_4 = 0.7, \quad B_5 = 7.5,$$
$$B_6 = 0.6, \quad B_7 = 6.3. \tag{21.4.13}$$

The values of the ratios (21.4.12) and (21.4.13) are not taken absolutely random. Herein, the restrictions (21.4.11) have been taken into account and we assume that the first medium has greater elastic properties, i.e., $|\lambda_1| > |\lambda_2|$, $|\mu_1| > |\mu_2|$, $|\alpha_1| > |\alpha_2|$, $|\eta_1| > |\eta_2|$. All ratios in (21.4.12) satisfy $0 < A_i < 1$ ($i = 1, \ldots, 7$) and, as expected from this condition, all the stiffness $\hat{C}_{ijrs}$ lays between zero and one ($0 < \hat{C}_{ijpq} < 1$); this is because all those constants are definite positive. In addition, it is assumed that the torque properties of the first medium are lower than the second medium, i.e., $|\beta_1| < |\beta_2|$, $|\gamma_1| < |\gamma_2|$, $|\epsilon_1| < |\epsilon_2|$, $|\nu_1| < |\nu_2|$, and the negative values in (21.4.13) are due to β is always defined negative.

In Tables 21.1 and 21.2, the variation of the dimensionless effective stiffness $\hat{C}^{\text{eff}}_{ijpq}$ and torque moduli $\hat{D}^{\text{eff}}_{ijpq}$ are shown for a bilaminated composite as a function of volume fraction V_1, running from 0.2 to 0.9. As can be seen, the dependence of the elastic and torque effective characteristics with respect to the volume fraction is nonlinear, except for $\hat{C}^{\text{eff}}_{1212}$ and $\hat{C}^{\text{eff}}_{1221}$. The values of the effective elastic coefficients increase with the volume fraction in all the cases. It's interesting to analyze the behavior of the effective torque. Some of their coefficients increase or decrease their values but in all the cases the absolute values tend to zero. We can conclude that the contribution of the elastic coefficients becomes more significant compared to the contribution of the torque coefficients as the volume fraction increases. An expected result considering that the medium V_1 has more significant elastic properties than torque properties.

The overall behavior of Cosserat bilaminate composites with cubic-symmetric constituents provides a homogenized material described by 18 independent moduli, 9 stiffness, and 9 torque effective constants, respectively, corresponding to the orthotropic symmetry group referred as rotations by 90° [19], which implies invariance of C_{ijpq} and D_{ijpq} under rotations of 90° about the unit vector $e_3 : O = \{e_3 \otimes e_3 \mp e_3 \times I\}$.

21.5 Conclusions

In the present work, a general Cosserat medium with homogeneous boundary conditions is considered. The heterogeneous micropolar material is assumed to be heterogeneities periodically distributed. In this framework the two-scale (micro and macro scales) asymptotic homogenization process is applied to obtain

TABLE 21.1 Dimensionless values of the effective stiffness properties as a function of the volume fraction.

Effective Coefficients ＼ Volume Fraction	0.2	0.3	0.4	0.5	0.6	0.7	0.8	0.9
$\hat{C}^{\text{eff}}_{1111}$	0.523596	0.564072	0.607692	0.655172	0.707463	0.765854	0.832143	0.908911
$\hat{C}^{\text{eff}}_{3333}$	0.505618	0.538922	0.576923	0.62069	0.671642	0.731707	0.803571	0.891089
$\hat{C}^{\text{eff}}_{1122}$	0.253596	0.284072	0.317692	0.355172	0.397463	0.445854	0.502143	0.568911
$\hat{C}^{\text{eff}}_{1133}$	0.245506	0.272754	0.303846	0.339655	0.381343	0.430488	0.489286	0.560891
$\hat{C}^{\text{eff}}_{1212}$	0.0908	0.1232	0.1556	0.188	0.2204	0.2528	0.2852	0.3176
$\hat{C}^{\text{eff}}_{1221}$	0.0792	0.1068	0.1344	0.162	0.1896	0.2172	0.2448	0.2724
$\hat{C}^{\text{eff}}_{3131}$	0.0480645	0.0599209	0.0726497	0.0867021	0.102905	0.122955	0.150969	0.200205
$\hat{C}^{\text{eff}}_{1313}$	0.0319074	0.0359968	0.0412886	0.0484043	0.0584833	0.0738636	0.10022	0.155822
$\hat{C}^{\text{eff}}_{1331}$	0.0290323	0.0325158	0.0370236	0.0430851	0.051671	0.0647727	0.0872247	0.134589

TABLE 21.2 Dimensionless values of the effective torque properties as a function of the volume fraction.

Volume Fraction Effective Coefficients	0.2	0.3	0.4	0.5	0.6	0.7	0.8	0.9
$\hat{D}_{1111}^{\text{eff}}$	7.57771	6.68192	5.83183	5.0050	4.19162	3.38673	2.58751	1.79230
$\hat{D}_{3333}^{\text{eff}}$	3.57143	2.7027	2.17391	1.81818	1.5625	1.36986	1.21951	1.0989
$\hat{D}_{1122}^{\text{eff}}$	-3.66229	-3.32808	-2.94817	-2.545	-2.12838	-1.70327	-1.27249	-0.837703
$\hat{D}_{1133}^{\text{eff}}$	-1.34286	-1.02432	-0.830435	-0.7	-0.60625	-0.535616	-0.480488	-0.436264
$\hat{D}_{1212}^{\text{eff}}$	6.14	5.46	4.78	4.1	3.42	2.74	2.06	1.38
$\hat{D}_{1221}^{\text{eff}}$	5.16	4.59	4.02	3.45	2.88	2.31	1.74	1.17
$\hat{D}_{1313}^{\text{eff}}$	2.54854	1.91606	1.53509	1.28049	1.09833	0.961538	0.855049	0.769795
$\hat{D}_{3131}^{\text{eff}}$	3.6165	2.96989	2.5	2.1189	1.7887	1.49038	1.21336	0.951246
$\hat{D}_{1331}^{\text{eff}}$	2.14951	1.61934	1.3	1.08659	0.933891	0.819231	0.729967	0.658504

an homogeneous body with equivalent characteristics as the original problem. The local problems for the micropolar elasticity are reported and analytical expressions for the effective properties of the new homogenized body are derived. The problem is reduced to study layered structures and the local problems are solved and approximated expressions for general (asymmetric) layered composite are reported. The problem is simplified assuming that the constituent of each laminate has cubic symmetry which implies that $B_{ijpq} = 0$, what uncouples the equations. Since experimental data for the characterization of Cosserat cubic-symmetric materials are not enough. The analysis of the problem under dimensionless perspective has been used, taking always into account the restrictions that the eight cubic constants satisfy the positiveness of the quadratic parts of the internal energy. We reported data from the effective elastic and torque moduli for such symmetry, and tables are analyzed and commented. As result from the homogenization process applied to a Cosserat bilaminated composite with cubic-symmetric constituents is obtained an homogenized Cosserat material described by 18 independent moduli, nine effective stiffness properties, and nine effective torque properties, respectively. The new homogenized Cosserat material belongs to an orthotropic symmetry group restricted with invariance under rotations of $90°$.

Acknowledgments

Y.E.A. gratefully acknowledges the Program of Postdoctoral Scholarships of DGAPA from UNAM, Mexico. F.J.S. thanks the funding of PAPIIT-DGAPA-UNAM IA100919. R.R.R. would like to thank the Department of Mathematics and Mechanics at IIMAS, UNAM.

References

[1] W. Voigt, Theoretische Studien über die Elastizitätsverhältnisse der Krystalle, Abhandlungen der Mathematischen Classe der Königlichen Gesellschaft der Wissenschaften zu Göttingen 34 (1887) 3–51.

[2] E. Cosserat, F. Cosserat, Théorie des corps déformables, Hermann et fils, Paris, 1909 (in French).

[3] A. Eringen, Linear theory of micropolar elasticity, Journal of Mathematics and Mechanics, JSTOR (1966).

[4] H. Altenbach, V.A. Eremeyev, On the linear theory of micropolar plates, Journal of Applied Mathematics and Mechanics (Zeitschrift für Angewandte Mathematik und Mechanik) 89 (4) (2009) 242–256.

[5] J. Altenbach, H. Altenbach, V. Eremeyev, On generalized Cosserat-type theories of plates and shells: a short review and bibliography, Archive of Applied Mechanics 80 (1) (2010) 73–92.

[6] S. Hassanpour, G. Heppler, Micropolar elasticity theory: a survey of linear isotropic equations, representative notations, and experimental investigations, Mathematics and Mechanics of Solids (2015).

[7] F. Dos Reis, J. Ganghoffer, Micropolar continua from the homogenization of repetitive planar lattices, Advanced Structured Materials (2011) 193–217.

[8] F. Dos Reis, J. Ganghoffer, Construction of micropolar continua from the asymptotic homogenization of beam lattices, Computers & Structures 112 (2012) 345–363.

490 Mechanics and Physics of Structured Media

[9] I. Goda, M. Assadi, S. Belouettar, J. Ganghoffer, A micropolar anisotropic constitutive model of cancellous bone from discrete homogenization, Journal of the Mechanical Behavior of Biomedical Materials 16 (2012) 87–108.

[10] I. Goda, M. Assadi, J. Ganghoffer, Cosserat anisotropic models of trabecular bone from the homogenization of the trabecular structure: 2D and 3D framework, Advanced Structured Materials (2013) 111–141.

[11] I. Goda, M. Assidi, J. Ganghoffer, A 3D elastic micropolar model of vertebral trabecular bone from lattice homogenization of the bone microstructure, Biomechanics and Modeling in Mechanobiology 13 (1) (2013) 53–83.

[12] S. Forrest, K. Sab, Cosserat overall modeling of heterogeneous media, Mechanics Research Communications 25 (4) (1998) 449–454.

[13] S. Forrest, F. Padel, K. Sab, Asymptotic analysis of heterogeneous Cosserat media, International Journal of Solids and Structures 38 (2001) 4585–4608.

[14] V.I. Gorbachev, A.N. Emel'yanov, Homogenization of the equations of the Cosserat theory of elasticity of inhomogeneous bodies, Mechanics of Solids 49 (1) (2014) 73–82.

[15] V.I. Gorbachev, A.N. Emel'yanov, Homogenization of Problems of Cosserat Theory of Elasticity of Composites, Additional Materials. Intern. Scientific Symposium in Problems of Mechanics of Deformable Solids Dedicated to A.A. Il'yushin on the Occasion of His 100th Birthday, January 2021, 2011 (Izd-vo MGU, Moscow, 2012), pp. 81–88 (in Russian).

[16] X. Li, J. Zhang, X. Zhang, Micro-macro homogenization of gradient-enhanced Cosserat media, European Journal of Mechanics. A, Solids 30 (2011) 362–372.

[17] D. Bigoni, W. Drugan, Analytical derivation of Cosserat moduli via homogenization of heterogeneous elastic materials, Journal of Applied Mechanics 74 (4) (2006) 741–753.

[18] J. Dyszlewicz, Micropolar Theory of Elasticity, Lecture Notes in Applied and Computational Mechanics, Springer-Verlag, Berlin, Germany, 2012.

[19] V. Eremeyev, W. Pietraszkiewicz, Material symmetry group of the non-linear polar-elastic continuum, International Journal of Solids and Structures, Elsevier 49 (14) (2012) 1993–2005.

[20] A.C. Eringen, Microcontinuum Field Theories I: Foundations and Solids, Springer, New York, 1999.

[21] A.C. Eringen, Microcontinuum Field Theories II: Fluent Media, Springer, New York, 2001.

[22] Y. Chen, J.D. Lee, A. Eskandarian, Micropolar theory and its applications to mesoscopic and microscopic problems, Computer Modeling in Engineering & Sciences 1 (2004) 35–43.

[23] R.D. Mindlin, Micro-structure in linear elasticity, Archive for Rational Mechanics and Analysis 15 (1964) 51–78.

[24] V.I. Malyi, Theoretical determination of the five physical constants of the Toupin-Mindlin gradient elasticity for polycrystalline materials, in: Nonlinear Dynamics of Discrete and Continuous Systems, Advanced Structured Materials, vol. 139, Springer Nature, 2021, pp. 145–154.

[25] S. Alexandrov, E. Lyamina, Y. Jeng, A strain-rate gradient theory of plasticity and its comparison with strain gradient theories, Applied Mechanics and Materials 284–287 (2013) 8–12.

[26] S. Diebels, A macroscopic description of the quasi-static behavior of granular materials based on the theory of porous media, Granular Matter 2 (2000) 143–152.

[27] B. Gulua, R. Janjgava, Some basic boundary value problems for plane theory of elasticity of porous Cosserat media with triple-porosity, Proceeding in Applied Mathematics and Mechanics 17 (2) (2017) 705–706.

[28] Z. Rueger, R.S. Lakes, Experimental Cosserat elasticity in open-cell polymer foam, Philosophical Magazine 96 (2016) 93–111.

[29] Z. Rueger, R.S. Lakes, Experimental study of elastic constants of a dense foam with weak Cosserat coupling, Journal of Elasticity 137 (2019) 101–115.

[30] N.A. Collins-Craft, I. Stefanou, J. Sulem, I. Einav, A Cosserat Breakage Mechanics model for brittle granular media, Journal of the Mechanics and Physics of Solids 141 (2020) 103975.

[31] I. Giorgio, F. Dell'isola, A. Misra, Chirality in 2D Cosserat media related to stretch-micro-rotation coupling with links to granular micromechanics, International Journal of Solids and Structures 202 (2020) 28–38.

[32] P. Neff, A finite-strain elastic-plastic Cosserat theory for polycrystals with grain rotations, International Journal of Engineering Science 44 (8–9) (2006) 574–594.

[33] M. Shirani, D.J. Steigmann, A Cosserat model of elastic solids reinforced by a family of curved and twisted fibers, Symmetry 12 (2020) 1133.

[34] N. Bakhvalov, G. Panasenko, Homogenization: Averaging Process in Periodic Media, Mathematics and Its Applications (Soviet Series), 1989.

[35] B.E. Pobedrya, Mechanics of Composite Materials, Izd-vo MGU, Moscow, 1984 (in Russian).

[36] E. Sanchez-Palencia, Non-Homogeneous Media and Vibration Theory, Springer-Verlag, 1980.

[37] W. Nowacki, The linear theory of micropolar elasticity, in: Micropolar Elasticity, 1974.

[38] H. Altenbach, V.A. Eremeyev, Generalized Continua from the Theory to Engineering Applications, CISM International Centre for Mechanical Sciences, 2013.

[39] V. Eremeyev, L. Lebedev, H. Altenbach, Foundations of Micropolar Mechanics, Springer-Verlag, 2013.

[40] R. Toupin, Elastic materials with couple-stresses, Archive for Rational Mechanics and Analysis 11 (1) (1962) 385–414.

[41] S. Kessel, Lineare Elastizitätstheorie des anisotropen Cosserat-Kontinuums, Abhandlungen der Braunschweigischen Wissenschaftlichen Gesellschaft 16 (1964) 1–22.

[42] E. Sanchez-Palencia, Homogenization Techniques for Composite Media, Springer-Verlag, 1985.

[43] A.A. Kolpakov, A.G. Kolpakov, Capacity and Transport in Contrast Composite Structures: Asymptotic Analysis and Applications, CRC Press, Boca Raton, 2010.

[44] J. Otero, J. Castillero, R. Ramos, Homogenization of heterogeneous piezoelectric medium, Mechanics Research Communications (1997).

[45] Q.S. Zheng, A.J.M. Spencer, On the canonical representation for Kronecker powers of orthogonal tensors with application to material symmetry problems, International Journal of Engineering Science 31 (4) (1993) 617–635.

[46] E. Grekova, R. Abreu, Isotropic linear viscoelastic reduced Cosserat medium: an acoustic metamaterial and a first steep to model geomedium, Advanced Structured Materials (2019).

[47] J. Castillero, J. Otero, R. Ramos, A. Bourgeat, Asymptotic homogenization of laminated piezocomposite materials, International Journal of Solids and Structures 35 (5–6) (1998) 527–541.

Appendix A

Finite clusters in composites

V. Mityushev

This Appendix contains remarks to the finite volume methodology discussed in Section 12.3.

The correct application of the self-consistency approach (SCA) [8–10] from Chapter 12, in particular, to the finite volume W from Fig. 12.1 of the discussed paper, yields the effective constants of the clusters W diluted in a composite. More precisely, the effective constants can be calculated with the precision at most $O(f)$, where f denotes the concentration of the clusters W in the whole composite. The precision may be increased to $O(f^2)$ for macroscopically isotropic composites.

This assertion is not an alternative approach or another model. It should be considered as a general theorem which demonstrates the misleading methodology of SCA applied to composites in general.

The main gap is the asymptotic disbalance in the precision in f because of the following reasons:

- The local concentration f_{loc} interior the cluster is improperly identified with the concentration f of the homogenized medium.
- The interactions between the elements of the cluster W interior the cluster are properly taken into account by solution to the corresponding boundary value problem. But the interactions between the elements of W and the exterior elements are replaced by interactions between the elements of W and the homogenized exterior medium. This an approximation neglects the terms $O(f^2)$, where f is the concentration of W in the composite.
- Let n denote the number of elements in a cluster W_n. "Improvement" of SCA by the limit $n \to \infty$ leads to a conditionally convergent (absolutely divergent) series. This is the source of "new models". One takes own method of summation and obtains "own new model" like Mori-Tanaka, etc.

To summarize, the local fields in the cluster W, considered as a part of composite, are properly computed far away from its boundary ∂W. The effective constants presented in the chapter are computed for clusters diluted in the composite. The change of the cluster W by addition of an element yields another cluster, hence, another dilute composite.

493

494 Finite clusters in composites

A formula for a dilute composite like Clausius-Mossotti approximation may sufficiently well approximate results derived for periodic composites when the percolation effect is absent.

Consider a simple example of the percolation effect in Fig. A.1 below with perfectly conducting inclusions. The periodic composite in Fig. A.1a has the infinite overall conductivity. The corresponding diluted composite in Fig. A.1c has a finite conductivity depending on the concentration f. Application of SCA in Fig. A.1b cannot lead to the infinite overall conductivity without additional "improvements". This example demonstrates inconsistency of self-consistency approach and its proper application.

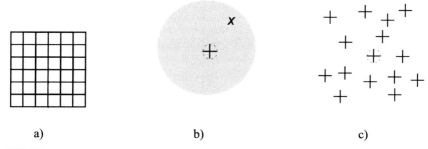

a)   b)   c)

FIGURE A.1 a) Double periodic composite with the perfectly conducting black rods has the infinite conductivity. b) The cell W bounded by the dashed circle represents the composite (a). According to SCA a single cell W is embedded in host of the unknown averaged effective conductivity X. c) This value X is actually equal to the effective conductivity of dilute clusters W.

The comments refer only to the finite volume concept represented in Fig. 12.1 and do not concern the considered advanced periodic problems.

Index

A

Abel integral equation, 56
Acoustic waves, 192
Airy function, 67, 73, 75, 76
Amplitudes, 147
Anisotropic layers, 408–410, 413
Ansatz, 463
Arbitrage, 337
  action, 338, 339
Assets, 329
Asymptotic analysis, 201
Asymptotic homogenization method
    (AHM), 460
Asymptotic solution, 50
Atiyah class, 312, 313, 320
Axial overall elastic modulus, 408, 409

B

Beltrami compatibility conditions, 123
Beltrami stress functions, 128
Bianalytic functions, 177, 186
Bilaminated composite, 473, 482, 483,
    486
  Cosserat, 461, 489
Birkhoff factorization, 303, 304, 306,
    308–310
Birkhoff strata, 310
Boundary value problem, 8, 42, 64, 70,
    74, 75, 77, 105, 111, 118, 126,
    398, 410, 463
  classic, 75
  Hilbert, 303
  inverse, 116
  Riemann, 21

C

Capital Asset Pricing Model (CAPM),
    327
Cash, 329
Cell
  coated sphere, 100
  periodic, 469, 472, 473
    double, 18
  periodicity, 11, 273, 275, 278, 397–400,
      402, 405–408, 410–412
    problem, 395, 396, 398
  radially inhomogeneous, 103, 107
  spherical, 100, 103, 106, 107
Characteristic class, 316, 321
Checkerboard, 451
Circular
  cylinders, 276, 277
  fibers, 397, 407
  hole, 77
  inclusions, 7, 8, 19, 93, 96, 445
    nonoverlapping, 445, 454
Classic
  action, 340
  boundary value problem, 75
  functions
    Eisenstein, 177
    elliptic, 21
  HSA, 100, 103
  method, 73
  theory, 71, 107
  thermodynamics, 359
  Weierstrass, 20, 177, 178
Classical
  heterophase systems, 428
  market model, 329
  statistics, 432

495

496 Index

Closed form solution, 65
Closed strategy, 336
Coated sphere, 99, 101, 107
   cell, 100
Coefficients
   Cosserat, 473
   effective, 472
   homogenized, 88
   sedimentation, 152
   torque, 486
Coherence, 340
Compact Lie group, 303, 306, 307
Compatibility conditions, 123–126, 128,
      132, 133
   Beltrami, 123
   for micropolar elasticity, 132
   for nonlinear problems, 132
Complex potentials, 18, 19, 23, 63, 64, 66,
      68, 70, 73, 74, 78, 80, 92, 95
   method, 14
Complexification, 306, 307, 313, 314, 317
Composites, 63
   fibrous, 7, 9, 65, 89, 96, 445
   laminae, 411
   laminated, 396, 408–410
   macroscopically isotropic, 23, 93, 96,
      99, 100, 157, 448
   magneto-electro-elastic, 63
   MEE, 63, 89, 95
   periodic, 233, 298
     two-phase, 272, 280
   piezoelectric, 63, 65
   random, 24, 25, 445, 446
   two-dimensional, 445
   unidirectional, 403, 404, 408, 409
     reinforced, 398, 402, 408, 409, 413
Compressibility factor, 151
Condition
   compatibility, 123–126, 128, 132, 133
     Beltrami, 123
     de Saint-Venant, 124, 125, 127
   linear conjugation, 306
   minimal derivative (sensitivity), 148
   minimal difference, 145, 148, 150–154
   periodicity, 470
   transmission, 306

Conductivity, 144, 146, 149, 160
   effective, 159, 160, 170, 271, 272,
     278–283, 291, 296
   macroscopic, 144
Connection, 305, 311–314, 316, 318–320,
     324
Constrained Lagrangian system, 338
Contrast parameter, 92, 93, 446, 455
   matrix, 91
Contribution
   by Eisenstein, 10
   by Natanzon, 14
   by Rayleigh, 12
   Filshtinsky's, 5, 27, 30, 32
Cosserat, 460
   coefficients, 473
   composites
     bilaminated, 461, 486, 489
     laminated, 461, 480
     multilaminated, 460
   continuum, 460
   elasticity equations, 472
   materials, 460
   media, 460, 461
     laminate, 473
   strains, 468
   structures, 459
   theory, 459
Counterexamples by Pobedrya and
     Georgievskii, 129
Couple
   strains, 467–469
   stress, 479
     tensor, 462
Cracks
   front, 267
   in MME plane, 55
   interface, 253, 254, 258, 259, 267
   opening, 53, 55, 236, 237, 240, 242,
     247, 265, 267, 268
   penny-shaped, 246
   problems, 29
   surfaces, 234, 236, 237, 240, 242, 244
   tips, 52, 253, 257, 258, 265, 268
Critical index, 144, 146
   superconductivity, 157
Crystal, 435

Cubic constituents, 480
Curvature, 337
  formula, 337

D
de Saint-Venant compatibility conditions,
    124, 125, 127
Decomposition
  Rylko, 91
  Straley-Milgrom, 90
Deflator, 331, 332, 336–338, 350
Dielectrics, 72
Discounted prices, 329
Dispersion, 201, 202
  relation, 191, 203
Displacements, 123, 130, 131, 133, 134,
    217, 219, 233, 257, 398,
    459–461, 469, 472
Distribution
  exponential, 372, 385
  linear, 371, 383
  normal, 373, 386
  particle, 363, 381
Domain
  fundamental, 178, 182
  unbounded, 42, 45
Double periodic, 7, 11, 15–18
  arrays, 19, 177
  cell, 18
  domain, 15
  function, 10, 16, 32
    bianalytic, 177, 186
    class, 6, 10
    polyanalytic, 177
    polyharmonic, 20, 21
    polymeromorphic, 188
  kernel, 6
  problems, 14, 19, 279
  stress tensor, 15
Double periodicity, 10, 21

E
Effective
  coefficients, 472
  compliance tensor, 241

conductivity, 159, 170, 271, 272,
    278–283, 291, 296
  external field, 238
  field method, 238
Ehrenfest, 344
Eisenstein
  approach, 10
  functions, 10, 11, 23, 177–179, 185,
    187
  method, 13
  series, 186, 188
    lattice, 23
  summation, 10, 13, 22, 24, 25,
    177–179, 188
Eisenstein-Natanzon-Filshtinsky
    functions, 24
Eisenstein-Rayleigh-Natanzon-
    Filshtinsky lattice sums,
    24
Elastic fields, 235
Elastic waves, 198
Elasticity, 1, 7, 9, 18, 25, 29, 33, 131–134,
    149, 150, 167, 214, 279, 459,
    469
  antiplane, 446
  Cosserat, 472
  plane, 6, 14, 15
  problems, 130, 132, 134, 150
    in stresses, 130
  theory, 25, 133, 461
Electrically impermeable crack, 265
Electrically permeable crack, 263
Electromagnetic waves, 197
Equal impedances, 194
Equation
  Cosserat elasticity, 472
  equilibrium, 42, 124, 125, 127, 128,
    130–133
  Euler-Lagrange, 340
  functional, 110
  integral
    Abel, 56
    Fredholm, 32, 33
    volume, 234, 235, 251
  linear algebraic, 12, 17, 21, 23, 30, 33,
    65, 78, 114, 115, 234, 243,
    261, 262
  local, 88

498 Index

MEE, 82
ordinary differential, 304
Schrödinger, 341, 348, 350
Equilibrium equations, 42, 124, 125, 127, 128, 130–133
Eshelby solution, 401
Euler-Lagrange equations, 340
Expansion factor, 165
Explicit asymptotic formulas, 114
Exponential distribution, 372, 385

F

Fast Fourier transform (FFT), 234
Feynman integrals, 348
Feynman's path integral, 350
Fibers
circular, 397, 407
interaction, 403
unidirectional, 395, 397, 445
Fibrous composites, 7, 9, 65, 89, 96, 445
Field
effective, 238
external, 238
elastic, 235
local, 8–10, 23, 24, 41, 43, 52, 66, 69, 73, 76, 79, 80, 89, 177, 233, 234
MEE, 28, 60, 64
values, 67
Filshtinsky's contribution, 5, 27, 30, 32
Formula
curvature, 337
explicit asymptotic, 114
Maxwell, 9
Rayleigh, 23
Fourier integral transformation, 131
Fredholm integral equations, 32, 33
Function
Airy, 67, 73, 75, 76
bianalytic, 177, 186
double periodic, 10, 16, 32
Eisenstein, 10, 11, 23, 177–179, 185, 187
Eisenstein-Natanzon-Filshtinsky, 24
Lagrange, 339
local, 84
Natanzon-Filshtinsky, 178, 185
generalized, 180

stress, 124, 125, 127, 130, 132
Airy, 133
Beltrami, 128
Maxwell's, 127
Morera's, 127
Weierstrass, 178
Functional equations, 110
Fundamental
domain, 178, 182
solution, 276, 284
periodical, 32

G

Gauge, 331
transform, 332
Geometric arbitrage theory, 327, 328, 350
background, 329
Geometric phase probabilities, 425
Gold, 375–377, 381, 382, 387, 388, 390, 392
Graphene-type composites, 159
with vacancies, 167, 170
GRHTP, 308, 310, 311, 314

H

Hamilton principle, 339
Hard-disks fluids, 151
Hashin-Shtrikman assemblage, 99, 100
Heisenberg representation, 344
Heisenberg's uncertainty relation, 346
Heterogeneous
classical boundary value problem, 463
inclusions, 235
media, 235, 241
periodic Cosserat media, 460
laminate, 473
Heterophase
equilibrium, 421
system, 419, 421, 428, 434, 441
Hexagonal array, 9, 24, 25, 449, 450, 455
of disks, 449
High-frequency viscosity, 150
Highly mismatched impedances, 196
Homogeneous
elastic media, 236
host media, 235
Homogenization, 2, 9, 10, 31, 82, 84, 89, 107, 233–235, 250, 460
approach, 31, 95

method, 397
process, 460, 489
theory, 30, 31, 283, 395, 396, 398, 408, 409, 413, 446
Homogenized
coefficients, 88
constant, 402, 403, 405–407
elastic, 411
problem, 471
strains, 399, 400, 411, 412
strength criterion, 411
stresses, 399, 412
HSA-type structure, 102

I

Inclusions
circular, 7, 8, 19, 93, 96, 445
heterogeneous, 235
periodic, 274
spherical, 248
rigid, 248
Indium, 375–377, 381, 387–390, 392
Instantaneous forward rate, 332
Integrable, 313, 318
Intensity factors (IF), 265
Interface crack, 253, 254, 258, 259, 267
problems, 253, 262
Inverse boundary value problem, 116
Inverse matrix, 273, 475, 476
Irreducible, 322
Ising model, 153, 155
Isolated singular point, 318
Iterated roots, 145

K

Kinetics, 360

L

Lagrange function, 339
Lagrangian, 339
Laminated composite, 396, 408–410
Lattice sums, 24
Eisenstein-Rayleigh-Natanzon-Filshtinsky, 24
Rayleigh, 17
Laurent series, 11, 20, 75

Layers
anisotropic, 408–410, 413
thickness, 224, 228
Lie algebra, 307, 309, 312, 313, 321
Lie group, 311
compact, 316
Linear
algebraic equations, 12, 17, 21, 23, 30, 33, 65, 78, 114, 115, 234, 243, 261, 262
conjugation condition, 306
distribution, 371, 383
elasticity problems, 132
Local
equations, 88
fields, 8–10, 23, 24, 41, 43, 52, 66, 69, 73, 76, 79, 80, 89, 177, 233, 234
functions, 84, 472, 473, 475, 477, 478
problems, 65, 460, 461, 463, 472–474, 478, 489
relations, 86
stresses, 395, 399, 400, 406, 409, 411, 413
Longitudinal flow, 271, 272, 275, 280, 286
Loop group, 303, 304, 307, 308

M

Macroscopic deformations, 403, 404
Macroscopic strain, 399, 400, 402, 406, 407, 411
Macroscopically isotropic, 25, 90, 92, 99
composites, 23, 93, 96, 99, 100, 157, 448
Magneto-electro-elasticity, 41, 65, 66
Market Fiber Bundle, 334
Market model, 331–333, 337, 338, 346
classical, 329
Nelson $\mathcal{D}$ weak differentiable, 336
Market portfolio, 328, 329, 338, 341–343, 345–347, 352
Massive crystal, 358, 363, 376, 377, 379, 380, 383, 387, 388
Maturity date, 331
Maxwell formula (MF), 9
McLaurin series, 382, 384–386
Measurable phase space, 428

500 Index

Media, 237
 heterogeneous, 235, 241
 homogeneous
  elastic, 236
  host, 235
 micropolar, 460, 461
 periodic, 191
MEE
 composites, 63, 89, 95
 equations, 82
 fields, 28, 60, 64
Membrane-fluid, 202
Method
 complex potentials, 14
 effective field, 238
 Eisenstein, 13
 homogenization, 397
 Nelson's, 348
 of solution, 219
 polydispersed, 449
 Rayleigh, 4, 17, 19, 21
 Ritz's, 219, 220
Micropolar elasticity, 463, 489
 theory, 459
Micropolar media, 460, 461
Microscopic stress-strain state, 400
Minimal derivative (sensitivity) condition,
  148
Minimal difference condition, 145, 148,
  150–154
Model
 Capital Asset Pricing, 327
 Ising, 153, 155
 market, 331–333, 337, 338, 346
  classical, 329
 novel, 25
 radially inhomogeneous, 102
Modified Dirichlet problem, 110
Monodispersed Method (MM), 449
Monodromy
 group, 318
 homomorphism, 317, 320, 323, 324
 matrix, 304
 representation, 324
Morera's stress functions, 127
Moving load, 201, 209, 210
 problem, 205
Multicomponent approach, 409

N

Nano-materials, 357
Nano-particle, 360, 368, 376, 379–383,
  387, 390–392
 radius, 358
Nano-substance, 357
Natanzon
 complex potentials, 18
 contribution, 14
 representations, 18
Natanzon-Filshtinsky functions, 178, 185
 generalized, 180
Nelson
 $\mathcal{D}$ weak differentiable market model,
  336
 method, 348
 stochastic derivatives, 353
  generalized, 353
No arbitrage (NA), 330, 337
No-free-lunch-with-vanishing-risk
  (NFLVR), 330
No-unbounded-profit-with-bounded-risk
  (NUPBR), 330
Nominal prices, 329
Normal distribution, 373, 386
Novel model, 25

O

Optimization, 143, 158, 159, 161, 164,
  166, 168
Ordinary differential equations (ODE),
  304
Oscillating singularity, 254, 258, 263, 268

P

Parametric vibrations, 214, 215, 229
Parent phase, 357
Partial indices, 308
Particle distributions, 363, 381
Penny-shaped cracks, 246
Periodic
 array, 275
 cell, 469, 472, 473
 composite, 233, 298
  Cosserat laminated, 480
  two-phase, 272, 278, 280
 inclusions, 274
 media, 191

Index **501**

structures, 178, 233, 272, 274, 276, 279
system, 246, 248
Periodically
  Cosserat laminated composite, 473
  heterogeneous Cosserat media, 461
Periodicity, 15, 177, 273, 274, 278, 280
  assumptions, 183
  cell, 11, 273, 275, 278, 397–400, 402,
      405–408, 410–412
    problem, 395–398, 410
  conditions, 470
  double, 10, 21
  structure, 271, 275, 283
Perturbation term, 109, 110, 116, 118, 119
Phase configurations, 420–424, 430, 433,
      441
Phase space, 428
  measurable, 428
Piezoelectric, 253–255, 267
  composites, 63, 65
  fibrous, 89
Piezoelectromagnetic, 253
Plane
  elastic, 64, 89, 96
    periodic, 6
    problems, 15, 21
  elasticity, 6, 14, 15
    problem, 14
  isotropy, 89
  piezoelectricity, 29
  problems, 29, 41, 63, 132
  strain, 89, 254
Poincaré type series, 22, 113
Polydispersed Method (PM), 449
Polydispersed structure, 446
Polynomial growth, 318
Potentials, 66
Principal bundles, 314
  fiber, 333
Problems
  boundary value, 8, 42, 64, 70, 74, 75,
      77, 105, 111, 118, 126, 398,
      410, 463
    classic, 75
    heterogeneous classical, 463
    Hilbert, 303
    inverse, 116
    Riemann, 21

cracks, 29
  interface, 253, 262
  double periodic, 19
  elasticity, 130, 132, 134, 150
    linear, 132
  for fibrous composites, 7
  homogenized, 471
  in stresses, 129
  local, 65, 460, 461, 463, 472–474, 478,
      489
  MEE, 64
  modified Dirichlet, 110
  moving load, 205
  periodicity cell, 397, 410
  plane, 29, 41, 63, 132
    elastic, 15, 21
    elasticity, 14
  Riemann-Hilbert
    classical, 304
    linear conjugation, 306
    monodromy, 303, 316, 320
    transmission, 303, 306–308
  stability, 213
  stationary, 110
  transmission, 278, 279, 284, 285

Q

Quasiaverages, 432
Quasiequilibrium snapshot picture, 421
Quasiholonomic constraint, 364–366, 368

R

R-functions theory, 213, 219, 220, 223
Radially inhomogeneous
  approximation, 100
  cell, 103, 107
  model, 102
  sphere, 100, 107
Random close packing (RCP), 157
Random composites, 24, 25, 445, 446
Random Sequential Adsorption (RSA),
      157
Rayleigh
  contribution, 12
  formula, 23
  lattice sums, 17
  method, 4, 17, 19, 21
Reduction of TPs, 375
Reiterated homogenization, 107

502 Index

Representative volume element (RVE), 233, 238
RHMP, 303–306
RHTP, 303, 306, 307
Riemann sphere, 303, 304, 306, 307, 311, 314, 315, 318, 320
Riemann surfaces, 321
Riemann-Hilbert linear conjugation problem (RHLCP), 306
Riemann-Hilbert problem
  classical, 304
  monodromy, 303, 316, 320
  transmission, 303, 307, 308
Riemann-Hilbert transmission problem (RHTP), 306
Rigid spheres, 152
Rigid spherical inclusions, 248
Ritz's method, 219, 220
Root approximants, 143, 144, 155, 158, 159, 161, 166
RVE, 241, 246
Rylko decomposition, 91

S

Schrödinger and Heisenberg representation, 344
Schrödinger equation, 341, 348, 350
Sedimentation coefficient, 152
Self-financing, 330
  constraint, 339
Shift elastic modulus, 409
Short rate, 333
Slipping zone, 258, 260, 262, 265, 268
  length, 262
Small scatterers, 194
Smeared stiffener theory, 30
Solution
  asymptotic, 50
  closed form, 65
  Eshelby, 401
  fundamental, 276, 284
    periodical, 32
Southwell's paradox, 126, 127
Space
  of microstates, 428
  weighted Hilbert, 432
  weighted phase, 432
Spatial separation, 418

Spherical
  cell, 100, 103, 106, 107
  crystalline particles, 357
  inclusions, 248
    nonoverlapping, 110
Stability, 213–215, 229
  problems, 213
Stable, 323
Stationary problems, 110
Statistical ensemble, 429
Statistical operator, 419
Stiff fibers in the soft matrix, 409, 410, 413
Stigler's law of eponymy, 124
Stochastic
  analogue, 340
  embedding, 340
  parallel transport, 335
Strains, 66, 70, 123, 132, 398, 411, 467, 469, 479
  Cauchy, 468
  Cosserat, 468
  couple, 467–469
  homogenized, 400, 411, 412
Straley-Milgrom decomposition, 90
Strategy, 330
  closed, 336
  weak $\mathcal{D}$-admissible, 336
Stratonovich integrals, 335, 352
Stress
  function, 124, 125, 127, 130, 132
    Airy, 133
    Beltrami, 128
    Maxwell's, 127
    Morera's, 127
  intensity factors, 6, 33, 41, 52–56, 60
  tensor, 14, 15, 17, 23, 27, 28, 83, 127, 133, 198, 237, 241, 242, 246, 462
Structurally-orthotropic theory, 30
Subsonic regime, 207
Substitution operator, 448
Superconductivity
  critical index, 157, 160, 163
  problem, 144
Supersonic regime, 209
Surface free energy, 434

Index **503**

Surface tension, 357, 360, 378, 379, 391, 392
  effect, 379, 380, 383, 390–392
  factor, 391
Symmetric matrix, 48, 90
Symmetry properties, 432, 462
System
  constrained Lagrangian, 338
  heterophase, 419, 421, 428, 434, 441
    classical, 428
  periodic, 246, 248

T

Taylor series, 12, 185
Tension-compression, 400, 402, 403
Term structure, 331
Theory
  classic, 71, 107
  Cosserat, 459
  elasticity, 25, 133, 461
    micropolar, 459
  geometric arbitrage, 327, 328, 350
  homogenization, 30, 31, 283, 395, 396, 398, 408, 409, 413, 446
  of integral equations, 5
  of magneto-electro-elasticity, 27
  of shells, 32
  R-functions, 213, 219, 220, 223
  smeared stiffener, 30
  structurally-orthotropic, 30
Thermodynamics, 358
  of small objects, 357
Torque coefficients, 486
TPs increase, 380, 381, 383–385, 389
TPs reduction, 370
Transmission condition, 306

Transmission problem, 278, 279, 284, 285
Two-phase periodic composite, 272, 278, 280
Two-scale asymptotic expansion, 463

U

Unbiased estimate, 155
Unbounded domain, 42, 45
Unidirectional
  composite, 403, 404, 408, 409
    reinforced, 398, 402, 408, 409, 413
  fibers, 395, 397, 445
    layers, 396
  infinite cylinders, 65, 89
Unitary, 322

V

Vacancy-related reduction, 362
Valuation date, 331
Variation, 338
Volume element, 238
Volume integral equations, 234, 235, 251

W

Waves
  acoustic, 192
  elastic, 198
  electromagnetic, 197
Weak $\mathcal{D}$-admissible strategy, 336
Weierstrass functions, 20, 24, 177–179
Weighted Hilbert spaces, 432
Weighted phase spaces, 432

Z

Zero curvature (ZC), 338

Printed in the United States
by Baker & Taylor Publisher Services